CAGE HYDROCARBONS

CAGE HYDROCARBONS

Edited by

George A. Olah

University of Southern California
Los Angeles, California

A WILEY-INTERSCIENCE PUBLICATION
John Wiley & Sons, Inc.
NEW YORK / CHICHESTER / BRISBANE / TORONTO / SINGAPORE

Library of Congress Cataloging in Publication Data

Cage hydrocarbons / edited by George A. Olah.

p. cm.

"Paul v. R. Schleyer, for thirty years of adamantane chemistry and his sixtieth birthday."

A Wiley-Interscience publication."

Includes bibliographical references.

ISBN 0-471-62292-3

1. Adamantane. 2. Hydrocarbons. 3. Schleyer, Paul von R., 1930- . I. Olah, George A. (George Andrew), 1927- . II. Schleyer, Paul von R., 1930- .

QD305.H6C24 1990

547.1'39--dc20 89-29217

Printed in the United States of America

10 9 8 7 6 5 4 3 2 1 CIP

Paul v. R. Schleyer for Thirty Years
of Adamantane Chemistry and His Sixtieth Birthday

Contributors

Dieter Born
Institute for Organic Chemistry
University of Giessen
Giessen 4-6300
Federal Republic of Germany

Wolf-Dieter Fessner
Department of Organic Chemistry and Biochemistry
University of Freiburg
7800 Freiberg
Federal Republic of Germany

M. Anthony McKervey
Department of Chemistry
University College
Cork, Ireland

Günther Maier
Institute for Organic Chemistry
University of Giessen
Giessen 4-6300
Federal Republic of Germany

Jiri Mareda
Department of Organic Chemistry
University of Geneva
Geneva 4, Switzerland

Armin de Meijere
Institute for Organic Chemistry
University of Göttingen
Göttingen
Federal Republic of Germany

Paul Müller
Department of Organic Chemistry
University of Geneva
Geneva 4, Switzerland

George A. Olah
Loker Hydrocarbon Research Institute
Department of Chemistry
University of Southern California
Los Angeles, California 90089

Leo A. Paquette
Evans Chemical Laboratories
The Ohio State University
Columbus, Ohio 43210

Horst Prinzbach
Department of Organic Chemistry and Biochemistry
University of Freiburg
7800 Freiburg
Federal Republic of Germany

Harald Rang
Institute for Organic Chemistry
University of Giessen
Giessen 4-6300
Federal Republic of Germany

John J. Rooney
Department of Chemistry
The Queen's University
Belfast, Northern Ireland

Paul von Ragué Schleyer
Institute for Organic Chemistry
University of Erlangen-Nurnberg
D-8520 Erlangen
Federal Republic of Germany

Ted S. Sorenson
Department of Chemistry
University of Calgary
Calgary, Alberta
Canada T2N 1N4

Steven M. Whitworth
Department of Chemistry
University of Calgary
Calgary, Alberta
Canada T2N 1N4

A. G. Yurchenko
Department of Chemical Technology
Kiev Polytechnic Institute
Kiev 252056, USSR

Preface

In March 1988 the Loker Hydrocarbon Research Institute of the University of Southern California, of which Paul is a Senior Fellow since its inception, held one of its semiannual research symposia. It was dedicated to Paul Schleyer on the thirtieth anniversary of his discovery of the fundamentally simple preparation of adamantane, which started the modern age of cage hydrocarbon chemistry. As not all of Paul's friends who are active in the field could attend, it was decided to publish a volume extended by other contributions to give an overview of cage hydrocarbons.

I would like to thank all of the contributors for making the project possible. Although other commitments prevented the materialization of some contributions, the ten chapters of the book give a representative and broad scope view of the field.

We dedicate this volume to Paul, our friend and colleague, as a token of our admiration, as well as to commemorate his sixtieth birthday.

Paul's work inspires all of us, and he continues to make significant contributions. The field of cage hydrocarbons is one of the most dynamic and fastest growing areas in chemistry and will continue to extend its significance in years to come.

On a personal note, my friendship and scientific interaction with Paul, spanning more than 30 years is most cherished. It is a real pleasure to have been able to contribute, together with other colleagues, to this volume. Many happy returns, Paul.

GEORGE A. OLAH

Los Angeles, California
March 1990

Mookie, the Olah's cocker spaniel, enlightens
an otherwise blank page.

(photo by Mark Sassaman)

Contents

Introduction

E. M. ARNETT
Department of Chemistry
Duke University
Durham, North Carolina

It is a pleasure to introduce this volume dedicated to Paul Schleyer with some personal remarks about Paul. It is, in fact, a special distinction in view of his numerous friends, collaborators, and scientific colleagues who are well qualified to introduce him.

Because of Paul's clearly teutonic name and his long-standing tenure of the Professorship at Erlangen-Nurnberg, the younger generation of audiences attending his frequent lectures in the United States are often amazed at Professor Schleyer's mastery of the English language, which he speaks without the slightest trace of an accent. This mystery is explained readily by the fact that he was born and raised in Cleveland and got his education in the Cleveland public schools. From there he went to Princeton, where he received his undergraduate education, and then to Harvard, where he obtained his doctoral degree with Professor P. D. Bartlett. He returned to Princeton University in 1956 and immediately plunged into his pioneering research on caged hydrocarbons, the topic of this volume of collected contributions by his friends.

When I arrived at Harvard, about a year after Paul's departure, the graduate student body was still reverberating from the stimulation of Paul's presence in the organic chemistry community. Not only had he played a commanding role within the Bartlett physical organic seminars shared by the Roberts, Swain, and Westheimer groups, he had also dominated the famous evening seminars of R. B. Woodward's research group, which was populated by some of the brightest and most ambitious young minds in synthetic chemistry, both from the United States and Europe.

In a very short time after his arrival at Princeton, Paul had put his stamp on the most active and competitive area of American physical organic chemistry—the elucidation of carbonium ion mechanisms and reactivities through the study of solvolysis kinetics. In addition to the relatively reserved and careful pronouncements of Bartlett and J. D. Roberts, and the more flamboyant and aggressive rhetoric of Saul Winstein at UCLA and Herbert C. Brown of Purdue, Paul brought to the fore a vigorous controversy concerning the structures of nonclassical carbonium ions as inferred primarily from the relative rates of solvolysis of carefully chosen test substrates.

In addition to Paul's valuable contribution of adamantyl systems to the clarification of carbonium ion stabilities, Paul and Herbert C. Brown committed themselves to a carefully reasoned discussion of the enormous mass of information that had become available, in the hope of reaching a rational solution to the problem, or at least clarifying it so that the average interested noncombatant could understand what all the fuss was about. Their joint book, *The Nonclassical Ion Problem*, is an enduring monument to their intense effort to provide a satisfactory resolution to this interesting problem, based primarily on solvolytic studies and their interpretation at that time. No resolution emerged from their debate in print, but the argument was presented in its clearest terms without the rancor or legalistic legerdemain that had characterized several previous public confrontations. Their book remains a unique contribution to the history of chemistry.

During the late 1960s and 70s Paul's relationship with George Olah at Case Western Reserve and John Pople's group at Carnegie–Mellon University led to a number of prolonged visits to both institutions and the enormously valuable collaboration with John Pople during which Paul taught organic chemistry to John, and John taught theoretical chemistry to Paul. I was fortunate to share in a number of intervisitations, in which Paul often participated, between the Olah group in Cleveland and Pople's group and mine in Pittsburgh.

With his appetite for computer time now thoroughly whetted, Paul became an increasingly expensive faculty member for Princeton, and after several encounters with the administration concerning research support, Paul took a typically creative solution to the problem. As a lone countercurrent migrant against the brain drain from Europe, he moved his operations completely to Germany and his present position at Erlangen-Nurnburg. From this base he has contributed and continues his constant flow of significant research publications in a wide variety of fields and his education of a large number of European, American, and Asian students and postdoctoral researchers.

Paul's unique contributions to chemistry stem from a number of important personal characteristics. The most obvious is his high intelligence coupled with an enormous capacity for hard work. However, there are many other bright, working scientists whose contributions have fallen far short of Paul Schleyer's example. Seen in retrospect, from our thirty-year friendship, to me, a major factor in Paul's leadership is his willingness to take big risks—to make an austere and often dangerous commitment where other less enterprising people would have considered the potential benefits too small and unlikely to justify the obvious hazards. Among these dangerous forays I can point to his move to Germany, his commitment to chemical calculations of complicated molecules, and his predictions of lithium chemistry that resulted from many of those calculations. I have every reason to believe that Paul's motivation for taking the risks, which have repeatedly established his leadership and success, lie in his deep and persistent love of chemistry. To follow his muse he has been ready to commit much, risk much, and often make sacrifices. However, when all is said and done, it is obvious that he has been greatly rewarded by a lifetime of intense involvement in what he most enjoys doing. What more could you ask than that?

1 My Thirty Years in Hydrocarbon Cages: From Adamantane to Dodecahedrane

PAUL VON RAGUÉ SCHLEYER

Institute for Organic Chemistry
University of Erlangen-Nürnberg
Erlangen, Federal Republic of Germany

The 1957 discovery of a simple way to prepare adamantane (**1**) by isomerization[1] marks the emergence of cage hydrocarbons and their derivatives as a recognizable class of chemical compounds. Before then, very few molecules with encapsulating carbon skeletons were known, and none were readily available. My own involvement did not begin deliberately, but resulted from a lucky accident. I have been exploring the chemistry of this class of molecules ever since. The accidental discovery involved the isomerization of tetrahydrodicyclopentadiene (**2**), a readily available and inexpensive starting material, to adamantane (**1**) in the presence of aluminum halide catalysts.[1] The yield in the initial experiments was only 10–15%,[1,2] but adamantane could be obtained easily from the product mixtures because of its exceptional crystallinity. The cage molecule bottleneck was broken. Effective work on this class of molecules could begin. Before I recount high points of my own cage hydrocarbon research, let us set the stage back to the mid 1950s.

The beautifully symmetrical structure represented by adamantane (**1**) has fascinated chemists for nearly a century, at least since the elucidation of the constitution of hexamethylenetetramine (**3**) in 1895[3] and the determination of the diamond structure in 1913.[4] The interlocking chair cyclohexane rings in **1** comprise the smallest fragment with the basic characteristics of the diamond and "adamantane" was named on this basis. Adamantane is the cage hydrocarbon prototype. While the "hole" in adamantane is too small to hold another atom, the tetrahedral structure continues to capture chemists' imaginations. Urotropine

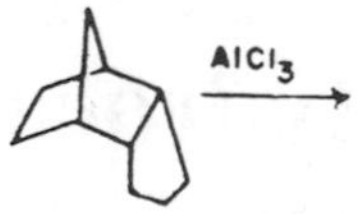

2
Schleyer (1957)

1

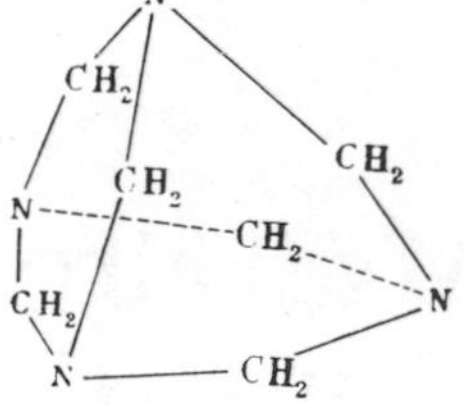

3
Duden and Scharff's formulation (1895)

(**3**, hexamethylenetetramane) is the oldest recognized example,[3] but there are numerous well-known heterocyclic and inorganic molecules possessing the adamantane skeleton.[5,6] While many of these were quite easy to prepare, the all-carbon analogue eluded early synthetic attempts. The key steps along the classical synthetic route to adamantane were taken by Meerwein, whose "ester" (**4**) provided a bicyclo[3.3.1]nonane derivative,[7] by Böttger, who succeeded where Meerwein and others had failed in constructing the adamantane skeleton, (**5**),[8] by Prelog, who synthesized adamantane in low yield by removing all the functional groups from **6**,[9] and by Stetter, who improved the preparative routes significantly.[6] But even before Böttger succeeded in closing the cage,[8] adamantane itself had been discovered in a most unexpected manner. It was isolated from a sample of Czechoslovakian petroleum taken from the village of Houdinin in Moravia.[10]

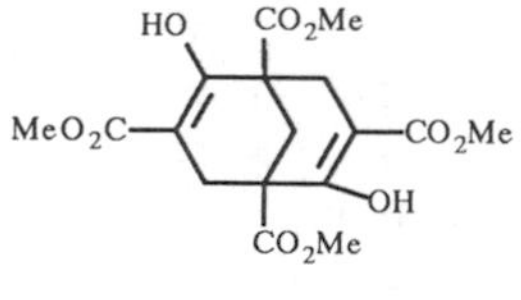

4
Meerwein (1922)

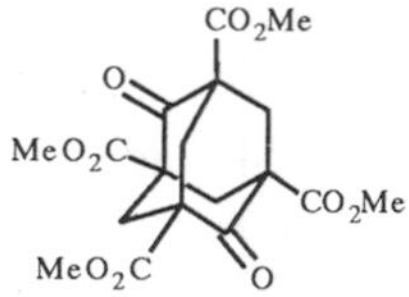

5
Bottger (1937)

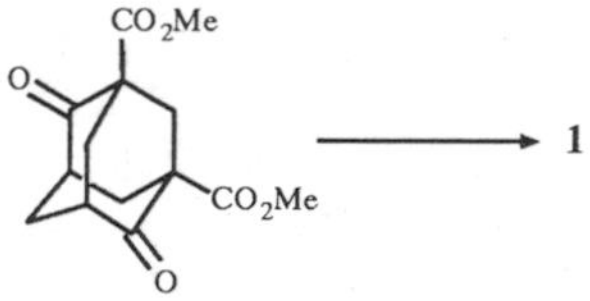

6
Prelog and Seiwerth (1941)

Akin to the far more extensive American Petroleum Institute project, a program had been undertaken in Prague to identify the compounds present in crude petroleum. Landa's predecessor in this project had bequeathed a room full of sample tubes containing the many distillation fractions. In one of these, crystals had formed. Their composition ($C_{10}H_{16}$) and physical properties (e.g., mp 270°C) led Lukes to deduce the diamond structure intuitively.[9] Prelog's synthesis[9] confirmed Landa's finding chemically, but did not provide a ready source of this hydrocarbon. Further investigations were not undertaken.

Landa did not follow up his discovery until well after World War II.[11] By processing large amounts of petroleum steam distillate, reasonable amounts of adamantane could be obtained. Although adamantane, as well as a number of its derivatives, were subsequently found to be present in crude petroleum worldwide, few chemists are willing to expend such efforts to obtain starting materials. In Aachen, Stetter had improved the synthetic routes to adamantane and had begun the investigation of its derivatives.[6] Normally, syntheses are designed to lead to functionally substituted molecules. The functional groups can then be transposed or further elaborated. But Landa made the key observation that adamantane reacts with elemental bromine to give 1-bromo-adamantane in high yield.[11] This strategy has become of great importance. Many cage hydrocarbons (like adamantane) are now easy to obtain by rearrangement; the functional groups can be introduced afterwards. But by 1954, when I began my academic career at Princeton, adamantane remained a "laboratory curiosity."

In 1957, *Chemical Abstracts* gave only four references to adamantane chemistry, which included the newly discovered rearrangement synthesis.[1] From 1957, the exponential growth in adamantane chemistry is documented by the number of citations listed in CAS for five-year periods. In the 1957–1962 half-decade, there were an average of 22 references per year to adamantane and its derivatives. This has increased to nearly 2000/year in the most current five-year period! Adamantane is now a common chemical. Every fine chemical catalog now offers dozens of adamantane derivatives. These have found widespread use and, for example, are becoming increasingly important in the pharmaceutical industry.

"Cage hydrocarbons" can be defined as having three or more rings arranged topologically so as to enclose space in the center of the molecular structure. Besides adamantane (**1**), other familiar cage hydrocarbons with a more or less spherical topology include cubane (**7**),[12] dodecahedrane (**8**),[13] and the currently much discussed C_{60} "buckminsterfullerane" or "footballane," believed to have structure (**9**).[14] Elongated topologies, exemplified by diamantane (congressane, **10**)[15] also are possible. Further variations are numerous.

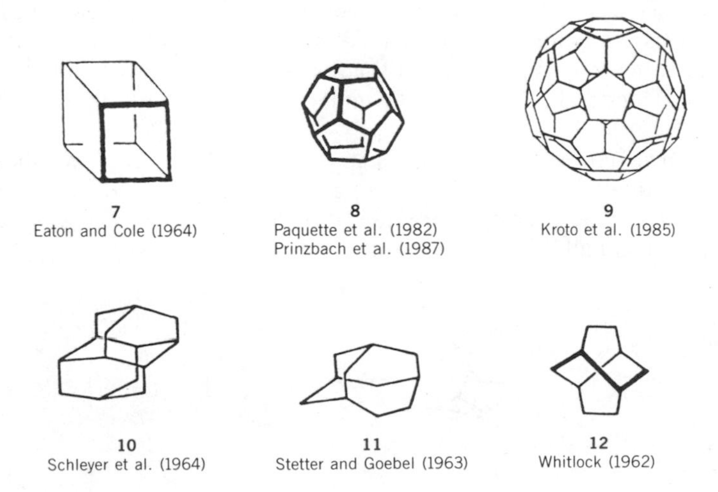

By 1964,[5a] when Ray Fort and I published the first of several reviews from Princeton,[5] cage hydrocarbon chemistry was still largely confined to adamantane, although homoadamantane (**11**)[16] and twistane (**12**)[17] (an adamantane isomer) had made their appearances. Thereafter, many more types of cage molecules were prepared and a recognizable field became established. Fort's 1976 book[5d] documents the progress in the next decade, but the field has now grown too large to be encompassed by a single review.[18] This volume can cover only some of the many pertinent topics. This contribution is a more personal, anecdotal account of my fascination with hydrocarbon cages.

The special characteristics of cage compounds have become so familiar that chemists tend to take them for granted. But cage molecules do have special structural characteristics that set them apart from acyclic molecules, from condensed ring systems, and even from their bridged ring counterparts. Cage molecules have rigid geometries, well-defined distances between functional groups, locked conformations, nonplanar bridgeheads, and globular shapes. Furthermore, graded sets of molecules are possible, for example, noradamantane, adamantane, and homoadamantane or adamantane, diamantane, and triadamantane, which facilitate the design of molecules with desired chemical, physical, or biological properties. Adamantane became available in an era when concerns over the "how and why" of chemistry predominated, even over interest in natural product syntheses. The structural characteristics of cage molecules proved to be ideal for physical organic and mechanistic studies. But these led in turn to synthetic advances and to the construction of new, often quite strained cage systems. Not only ring closures but also ring contractions and ring enlargement strategies were employed. Unexpectedly, medicinal applications of cage molecules were discovered. The globular structure of adamantyl amine contributes to the mode of its antiviral activity. Drugs containing adamantane groups are fat soluble, but are not readily degraded in the human body. Hence, they are more persistent and longer acting than other formulations based on long chain hydrocarbon groups.

As was the case with adamantane three decades ago, we are now witnessing a rapid development of the chemistry of cubane and of dodecahedrane. Both of these molecules are exhibiting unexpected and special characteristics attributed to their cage structures. The "ortho" lithiation of cubane carboxamide[19] and the *failure* of the 1-dodecahedryl cation to undergo rapid 1,2-hydrogen shifts[20] are examples.

After many unsuccessful "one-step" approaches, the synthesis of dodecahedrane finally was achieved by rearrangement of pagodane (**13**).[21] The first successful run on a sample provided by Prinzbach's group was achieved in Erlangen in the Fall of 1985, but the yield was very low.

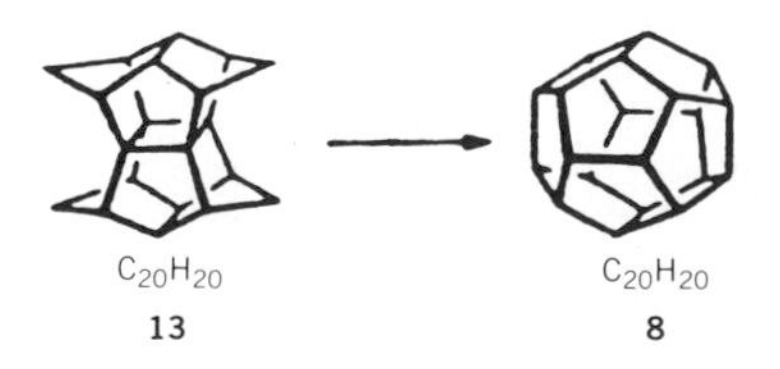

When dodecahedrane can be obtained more efficiently, the development of its chemistry will follow the historical pattern set by adamantane, and flourish. Groups led by Paquette at Ohio State[13,20] and by Prinzbach in Freiburg[20,22] already are demonstrating the richness of dodecahedrane chemistry with the limited amounts of material as yet available. Of course, a large number of cage molecule types have now been constructed, besides those that can be prepared by rearrangement. These often involve strained ring systems although more conventional but increasingly efficient synthetic routes are employed. Cubane (**7**) exemplifies this class of molecules. Spurred by interest in high energy density materials,[18i] the parent hydrocarbon is now available in kilogram quantities and cubane chemistry is developing rapidly.[12,19]

LUCKY ACCIDENTS AND THE PREPARED MIND[23]

In 1951 Paul D. Bartlett suggested the solvolysis of dicyclopentadiene derivatives for my doctoral project at Harvard. I read all the literature I could find, not only about this system but on bridged hydrocarbons in general. D.H.R Barton had left Harvard only the year before and regarding molecules in three dimensions (conformational analysis)[3] seemed obvious to my fellow students. R.B. Woodward stimulated mechanistic thinking. My contemporaries debated interpretations hotly. One of them even boasted that there was not a single organic reaction he could not rationalize. Indeed, we tried to find the wildest possible transformations to pose as problems in the Thursday night Woodward seminars. For me, terpene-based bicyclo[2.2.1.]heptane rearrangements were among the most challenging. Physical organic chemistry was providing exciting insights and dominated chemical thinking for two decades. Synthetic chemists began to approach their problems mechanistically as well.

The intellectual atmosphere of Harvard's chemistry department in the early 1950s was remarkable. The faculty included P. D. Bartlett, K. E. Bloch, L. F. Fieser, G. B. Kistiakowsky, J. Kochi, J. J. Lingane, W. N. Lipscomb, W. Moffitt, E. Rochow, G. Stork, F. H. Westheimer, G. Wilkinson, R. B. Woodward, E. B. Wilson, and P. Yates. A large percentage of my contemporaries—graduate students and postdoctorals—became distinguished, even quite famous scientists. We learned from one another. The stimulation and intellectual competition were equally important. I should have stayed at Harvard longer. Instead, I leaped at the opportunity to return to Princeton as an instructor, and arrived there, despite an unfinished Ph.D. thesis, in the fall of 1954. The climb up the academic ladder was leisurely then. After five years I might be promoted to assistant professor and possibly could achieve tenure after six more years. But the delay in finishing my Ph.D. thesis almost cost me my job. This ended up as a review of the entire bridged ring system literature: 600 pages, a compliment from Woodward (the second reader), but no publications. The benefit was personal. My mind had been well prepared.

An unexplained observation reported in 1903 was the basis of one of my first research projects at Princeton. Tetrahydrodicyclopentadiene was said to give an uncharacterized liquid product with concentrated sulfuric acid. In 1903 not even the structure of the starting material had been elucidated. But it seemed likely to me that the endo isomer (**2**) rearranged to the more stable exo form (**14**). These compounds were quite familiar to me from my thesis work. Indeed, this is just what happened, but the reaction was very slow. In the petroleum industry, aluminum halide catalysts were being used to induce chain branching by rearrangement. The carbocation mechanisms had been established. Indeed, when aluminum chloride was added to **2** the reaction warmed, and **14** was produced much more efficiently. But it was not an easy matter in those days to analyze hydrocarbon mixtures. Princeton's Chemistry Department possessed neither a gas chromatograph nor even a spinning band column. Separations were achieved by tedious, week-long fractional distillations through carefully packed and temperature controlled columns. The procedure required constant attention. When the ambient temperature changed, the necessary equilibrium conditions were disturbed. The apparatus was set up next to my desk in my Frick Laboratory–office, the best place to keep an eye on the distillation. After a week, all of the lower boiling rearrangement product (**14**) had distilled, and the pot was nearly empty. All at once white crystals formed in the take-off head capillary. I assumed this product to be the endo isomer (**2**, a low melting solid). I flamed the glass tube. Nothing happened. More heat. Still nothing. The crystals never did melt, but only popped!

$AlCl_3$ $AlCl_3$

2 **14** **1**

At first, I could hardly believe that adamantane (**1**) had been produced. The rearrangement would have to be even more deep seated than those with which I had posed as puzzles in the Woodward seminars. But the melting point left little doubt. Even its determination was remarkable. Sublimation could only be prevented by immersing the entire sealed part of the capillary under the heating bath oil. Infrared (IR) and mass spectral characterization sufficed for publication. It would have been tedious to repeat Prelog's synthesis[9] for a direct comparison! The original paper— Journal of the American Chemical Society (JACS) *Communications* were limited to 500 words in those days—is reproduced in Figure 1.[1] As McKervey put it, "Not only did this discovery make adamantane immediately available but it also established the pattern of much subsequent research in this area."[18b]

Quite a nice start for my scientific career, but I would have been even more pleased by a result I had anticipated beforehand.

A SIMPLE PREPARATION OF ADAMANTANE

Sir:

Because of the analogy with the structure of the diamond, the highly symmetrical molecule, adamantane (tricyclo[3.3.1.1^{3,7}]decane) (I), has occasioned interest for many years.[1] The total synthesis of this hydrocarbon, first isolated from petroleum naphtha[2] in minute yields,[3] has been accomplished several times, utilizing a number of modifications of the same general approach.[1,4] The over-all yields of even the best of these methods, involving a moderately large number of steps, did not exceed a few per cent.; therefore, investigations of the chemistry of this compound have been hindered to a large extent by its unavailability. We wish to report a facile two-step preparation of adamantane from the very readily available compound, dicyclopentadiene (II).

endo-Trimethylenenorbornane[5] (tetrahydrodicyclopentadiene, III), which can be prepared from II in essentially quantitative yield by hydrogenation,[6] was refluxed with 10 per cent. of its weight of $AlBr_3$ or $AlCl_3$ overnight.[7] At the end of this time the products were distilled directly from the reaction pot, with no attempt at fractionation. The precipitation of adamantane was completed by cooling the distillate to Dry-Ice temperature; a yield of about 10 per cent. of crude I could be obtained by filtration through a coarse filter. Additional I could be obtained by subjecting the filtrate to fractional distillation through an efficient column. The forecuts, b.p. to 185°, contained a large number of components. The main fraction, b.p. 185.0°, n^{20}_D 1.4871, consisted of *exo*-trimethylenenorbornane (IV),[8] the expected product of the reaction.[9] The yield of this material, about 50 per cent., could be improved considerably by conducting the isomerization at lower temperatures. From the higher boiling cuts, b.p. to 195°, approximately 5 per cent. additional crude I was recovered. After washing with ethanol, fractional sublimation of the combined samples of I gave a 12–13 per cent. yield of pure adamantane, m.p. 269.6–270.8° (sealed tube). Reported m.p. 268.5–270°.[1] *Anal.* Calcd. for $C_{10}H_{16}$: C, 88.16; H, 11.84. Found: C, 88.31; H, 11.99. The infrared spectrum[10] and mass spectral pattern[11] of the rearrangement product further established its identity as adamantane.

Since, as would be anticipated, IV, under the same conditions, gave a similar yield of I, it may be possible to improve the yield of I considerably The driving force for the rearrangement is undoubtedly the result of the fact that I, in contrast to III and IV, possesses an arrangement of atoms uniquely free from angular and conformational strain. Conceptually, it is possible to visualize several routes for the conversion of III or IV into I. The simplest of these necessitates only three steps involving carbon-to-carbon rearrangements. The possible mechanisms of these unusual transformations will be commented on in greater detail later.

(1) *Cf.* an excellent review, H. Stetter, *Angew. Chem.*, **66**, 217 (1954). More recent references will be found below.[3,4,10]

(2) S. Landa and V. Macháček, *Coll. Czech. Chem. Comm.*, **5**, 1 (1933).

(3) S. Landa, Š. Kriebel and E. Knobloch, *Chem. Listy*, **48**, 61 (1954) (*C. A.*, **49**, 1598 (1955)).

(4) V. Prelog and R. Seiwerth, *Ber.*, **74**, 1644, 1769 (1941); H. Stetter, O.-E. Bänder and W. Neumann, *ibid.*, **89**, 1922 (1956).

(5) The nomenclature used here will be that suggested previously (P. R. Schleyer and M. M. Donaldson, THIS JOURNAL, **78**, 5702 (1956)).

(6) For references, *cf.* E. Josephy and F. Radt, Eds., "Elsevier's Encyclopaedia of Organic Chemistry," Vol. 13, Elsevier Publishing Co., Inc., New York, N. Y., 1946, p. 1022.

(7) For a recent review of the action of Lewis acids upon alkanes, *cf.* H. Pines and J. Mavity in B. T. Brooks, *et al.*, Eds., "The Chemistry of Petroleum Hydrocarbons," Vol. III, Reinhold Publishing Corp., New York, N. Y., 1955, Chap. 39, pp. 9–58.

(8) H. Bruson and T. W. Riener, THIS JOURNAL, **67**, 723 (1945); *cf.* P. D. Bartlett and A. Schneider, *ibid.*, **68**, 6 (1946).

(9) M. M. Donaldson, unpublished results from this laboratory. *Cf.* J. F. Eykman, *Chem. Weekblad*, **1**, 7 (1903); **3**, 687 (1906).

(10) R. Mecke and H. Spiesecke, *Ber.*, **88**, 1997 (1955). The author is indebted to Dr. R. A. Dean, The British Petroleum Co., Ltd., for a copy of the spectrum of synthetic adamantane.

(11) Catalog of Mass Spectral Data, A.P.I. Research Project 44, Carnegie Institute of Technology, Pittsburgh, Pennsylvania, No. 939.

FRICK CHEMICAL LABORATORY
PRINCETON UNIVERSITY
PRINCETON, N. J.
PAUL VON R. SCHLEYER

RECEIVED MAY 17, 1957

Fig. 1. The 1957 adamantane synthesis communication (reproduced from Ref. 1a).

CAGE HYDROCARBONS BY REARRANGEMENT

The two major ways to exploit the discovery of the adamantane rearrangement—mechanistic and synthetic—provided my research group with research opportunities ever since. The rigid cage structure of adamantane affords splendid advantages for physical organic studies. The preparation and the solvolysis of both 1-[24] and 2-admantyl[25] derivatives (**15** and **16**) were among our first projects. These compounds not only provided structure–reactivity insights later to be expanded into quantitative methods for predicting carbocation

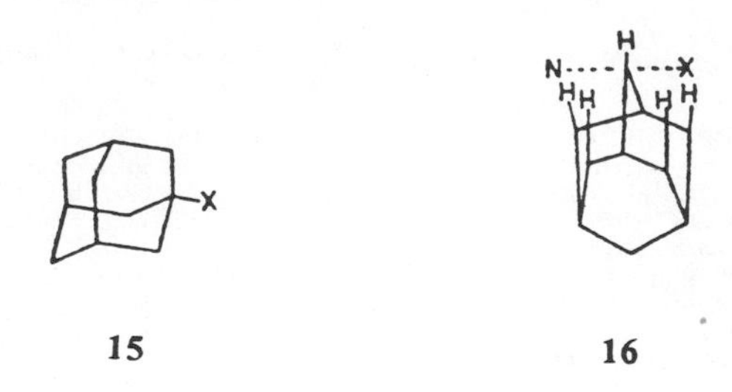

stabilities,[26–28] but also became key mechanistic standards for fully unassisted solvolysis.[29,30] Solvent participation in **15** and **16** does not take place. Attack at the rear is blocked. Participation of a skeletal C—C bond is also unfavorable due to the increase in strain.[31,32] We were to exploit these characteristics later and together with Olah to study many stable ions in cage systems (Fig. 2).[32] Degenerate carbocation rearrangements, also in the $(CH)_n^+$ series, were highly informative and often spectacular.[5g–i] Bridgehead cations, both as observable entities[32] and as reactive intermediates,[27,28,33] afforded us splendid opportunities to probe the effects of gradual structural changes. The increase in strain energy in going from the hydrocarbon to the corresponding bridgehead cation, as calculated by the molecular mechanics program we had developed,[27] correlated very well with the corresponding solvolysis rates. We predicted and later verified the reactivity of new systems.[27b,33] The rates of some of these were even enhanced rather than retarded because of the bridgehead location.[33f] Müller[28] refined our work impressively. By updating our molecular mechanics force field,[27] he showed (Fig. 3) that the entire set of bridgehead reactivities ranging over 10^{25} can be correlated by a *single* line. When we analyzed the thermochemical data,[27e] adamantane was found *not* to be "strain-free," as had long been taken for granted.

What other cage compounds can be prepared by rearrangement? What could we find out about the mechanism of the adamantane rearrangement or the reaction pathway? Our first approach was to use methyl groups as labels. Numerous $C_{11}H_{18}$ precursors including various methyltetrahydrodicyclopentadiene derivatives were treated with aluminum halides. But instead of giving 1- (**17**) or 2-methyladamantane (**18**) specifically, as we had hoped, mixtures were obtained.[34] Equilibration had taken place during the reactions, and we learned very little about the mechanism.[35] Instead, the reward was synthetic. The easily prepared tetramethylenenorbornane (**19**) gave spectacular results in the very first experiment. Aluminum bromide was added to the neat liquid hydrocarbon and stirred magnetically overnight. The next day, the entire contents of the flask had turned solid. The methyladamantane yield was more than quantitative—until we remembered to remove the stirring bar! The commercially available methylcyclopentadiene dimer, after hydrogenation, served as a good precursor for 1,3-dimethyladamantane (**20**).[35] The adamantane

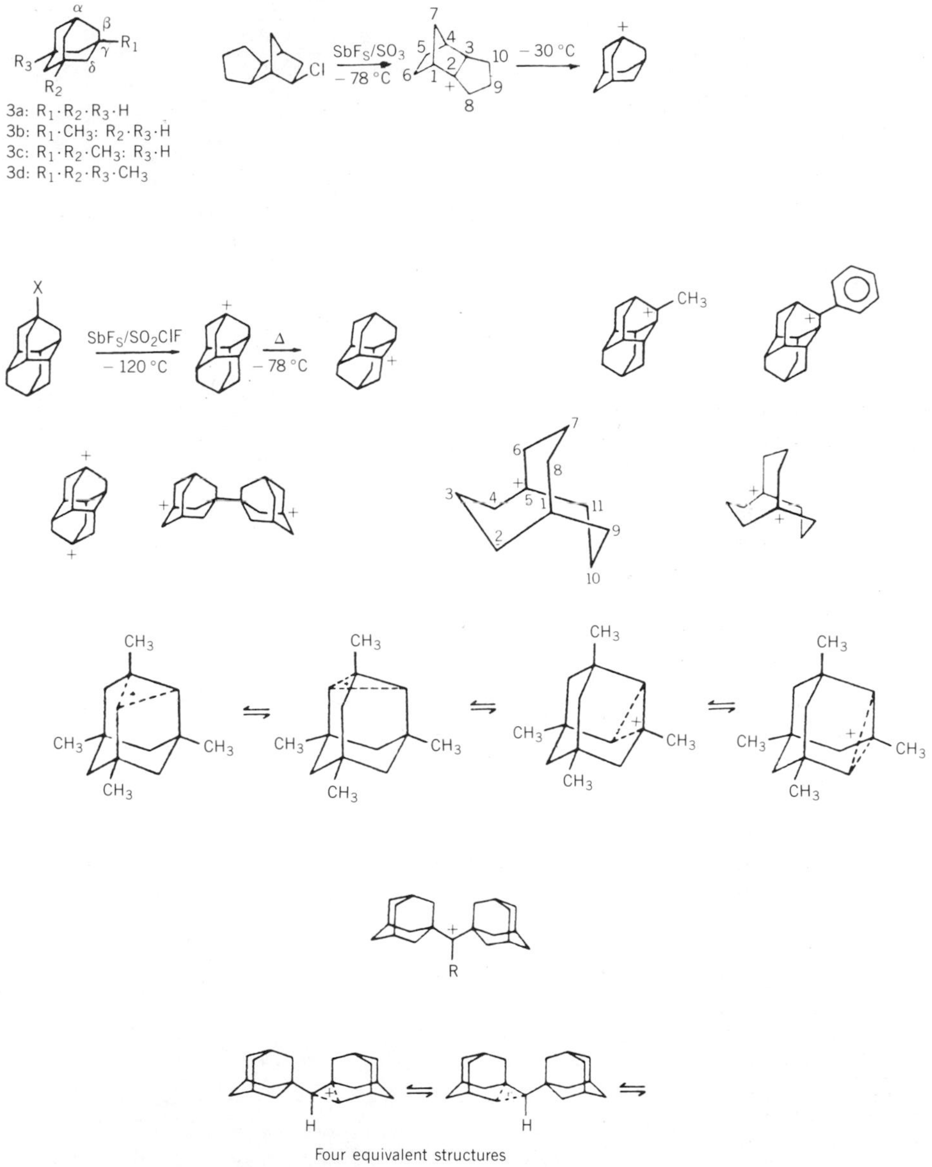

α β γ δ R_1 R_2 R_3

3a: $R_1 \cdot R_2 \cdot R_3 \cdot H$
3b: $R_1 \cdot CH_3$: $R_2 \cdot R_3 \cdot H$
3c: $R_1 \cdot R_2 \cdot CH_3$: $R_3 \cdot H$
3d: $R_1 \cdot R_2 \cdot R_3 \cdot CH_3$

Cl SbF_5/SO_3 − 78 °C 7 4 5 3 2 6 1 + 10 9 8 − 30 °C +

X SbF_5/SO_2ClF − 120 °C + Δ − 78 °C +

CH_3

6 7 8 3 4 + 5 1 11 9 2 10

CH_3 CH_3 CH_3 CH_3 ⇆ CH_3 CH_3 CH_3 CH_3 ⇆ CH_3 CH_3 CH_3 CH_3 ⇆ CH_3 CH_3 CH_3 CH_3

+ R

H ⇆ H ⇆

Four equivalent structures

Fig. 2. Stable carbocations in cage and related systems studied by Olah, Schleyer et al. (see Ref. 32).

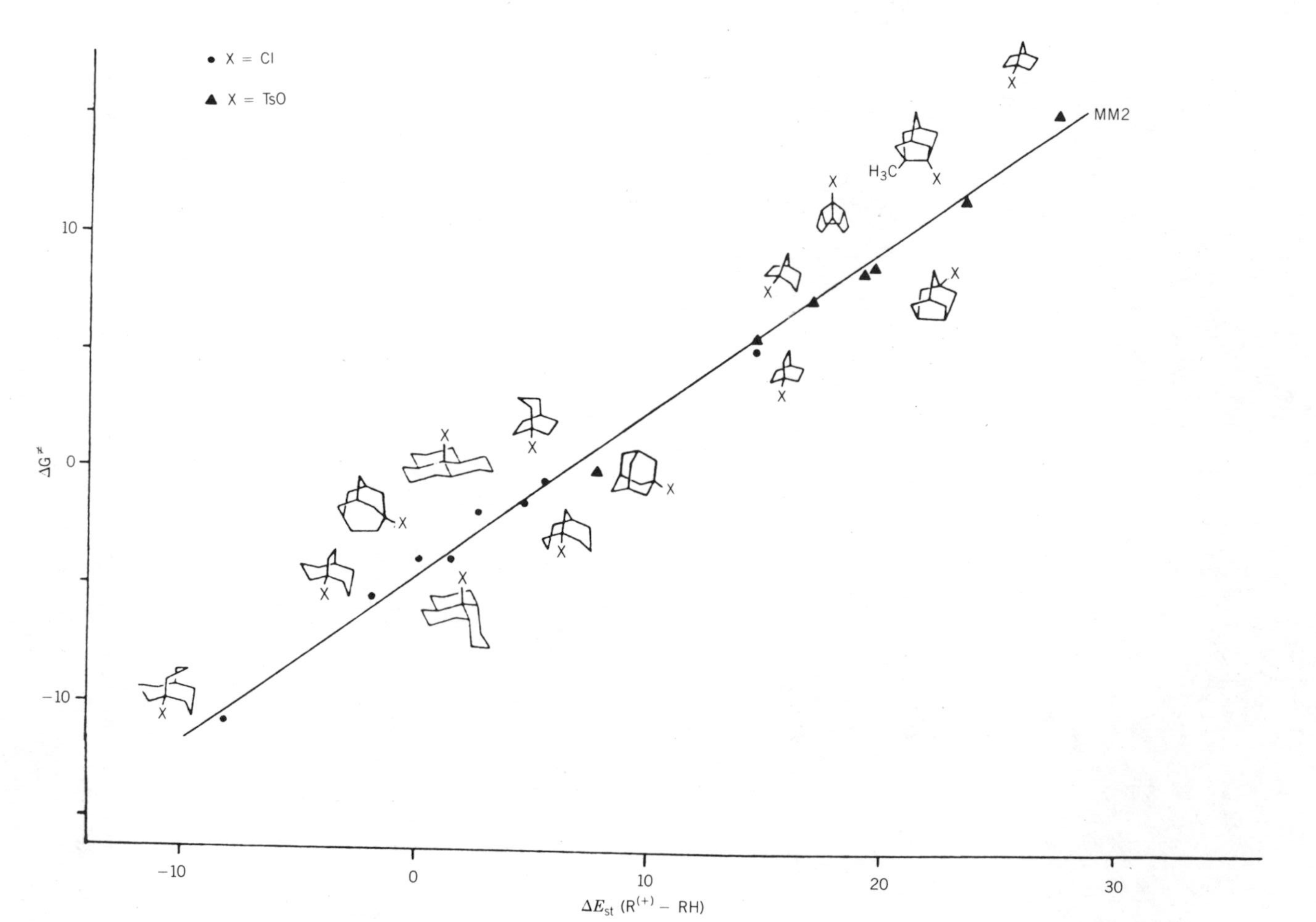

Fig. 3. Müller, Blanc, and Mareda's refined plot of solvolysis rates versus calculated steric energy differences between carbenium ions (R^+) and hydrocarbon (RH) (reproduced from Ref. 28b).

rearrangement was established to be general, and occurred even more readily with the higher homologues than with the parent. The intermediates involved are less strained when more carbon atoms are present.

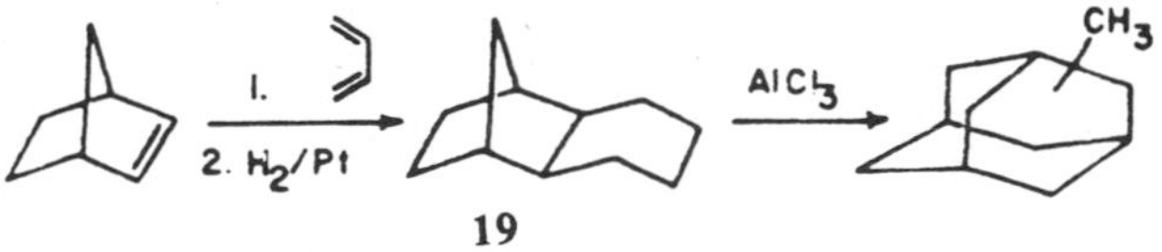

The first rearrangement syntheses of trimethyl- and of tetramethyladamantane (**21**),[36] accomplished by A. Schneider, Sun Oil Company, provided an important new understanding. We had assumed that these thermodynamically controlled rearrangements required strained starting materials to provide the driving force. Schneider (who also had worked for Bartlett, before my time at Harvard, on the mechanism of the aluminum chloride catalyzed hydride exchanges in hydrocarbons) showed that hydrogenated tricyclic aromatics rearranged quite efficiently to the isomeric polymethylated adamantanes. Since the trans–syn–trans isomer of perhydroanthacene (**22**) was strain-free, the driving force was due only to the greater degree of chain branching in the tetramethyladamantane product. By the late 1960s we were employing empirical force field (molecular mechanics) calculations routinely to provide quantitative thermochemical data on cage and other hydrocarbons. These predicted, for

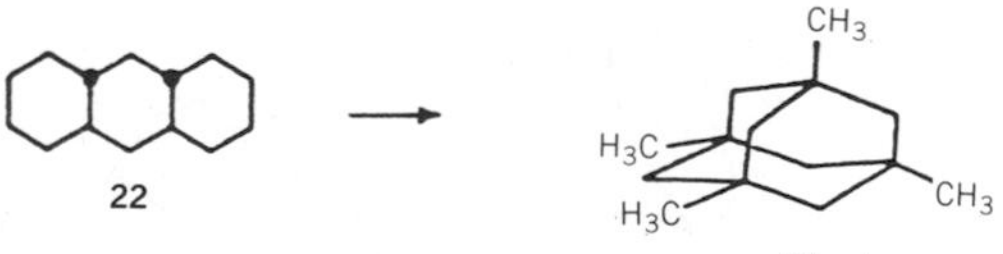

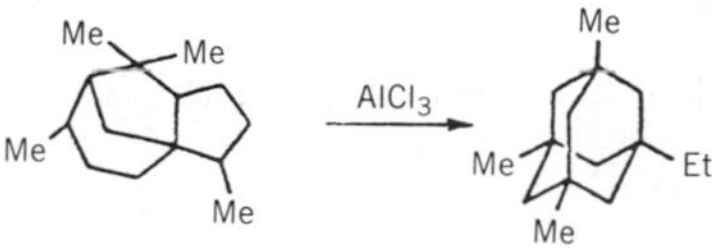

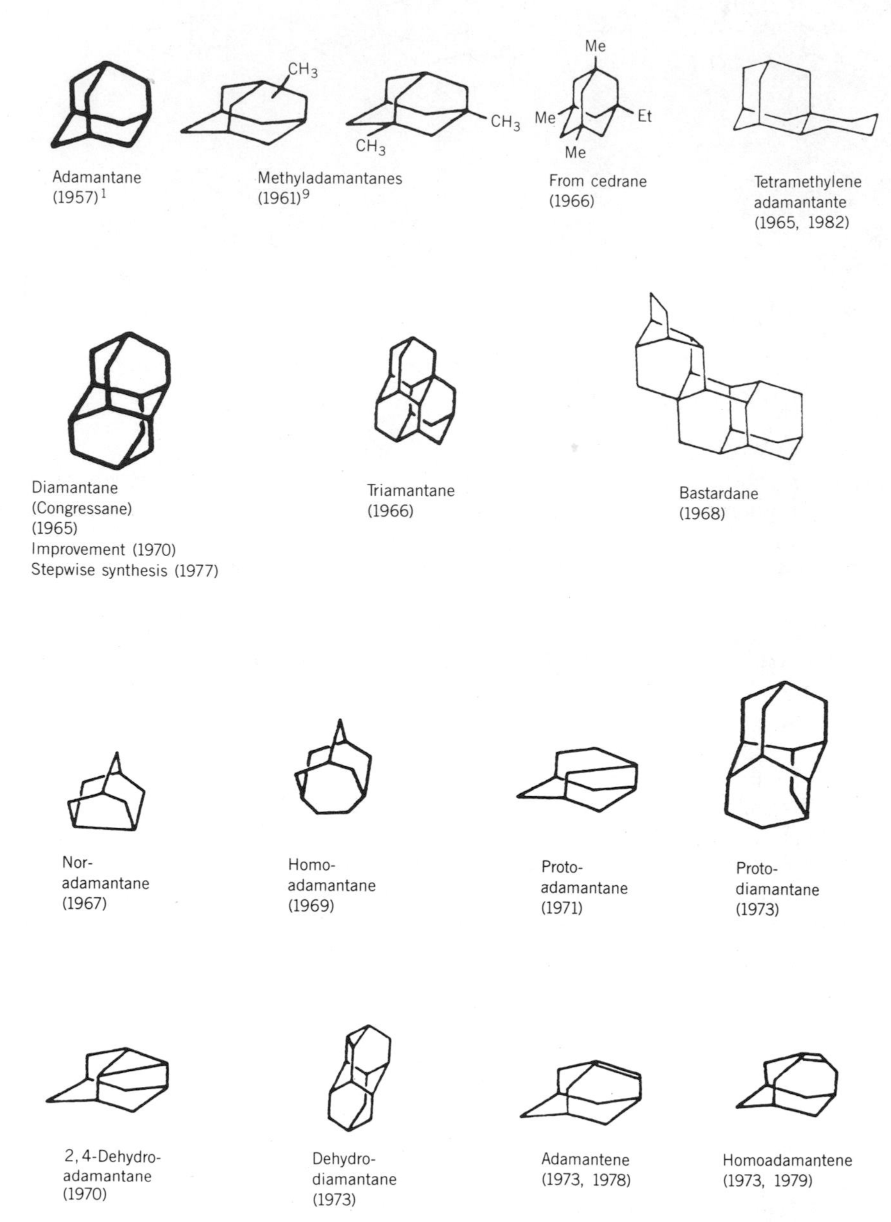

Fig. 4. Representative cage hydrocarbons prepared by the Schleyer group.

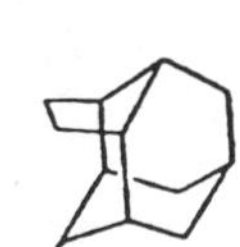

Ethanoadamantante
(1974)

Ethanodiamantanes
(1970)

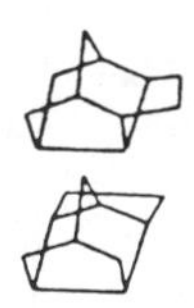

Ethanonoradamantanes
(1976)

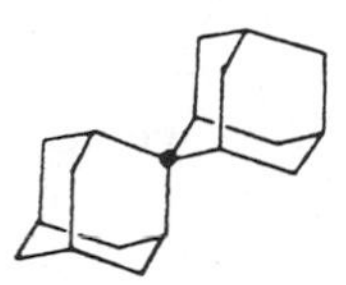

Spirodiamantane
(1972)

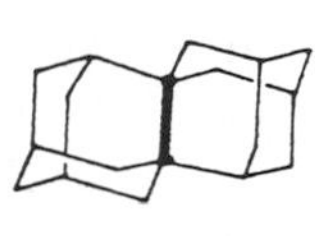

[2]Diamantane
(1973)

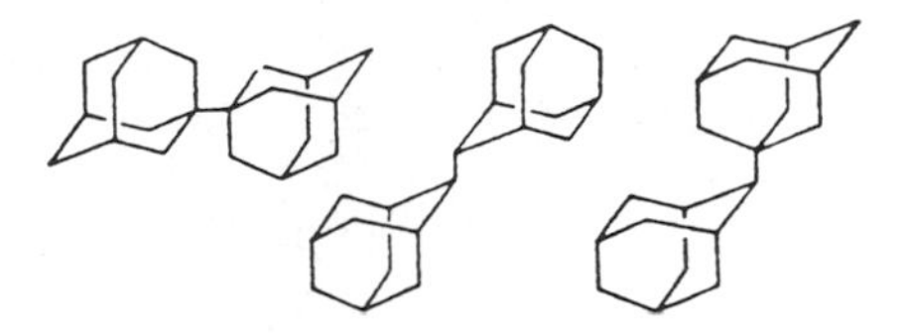

Biadamantane
by equilibration
(1973)

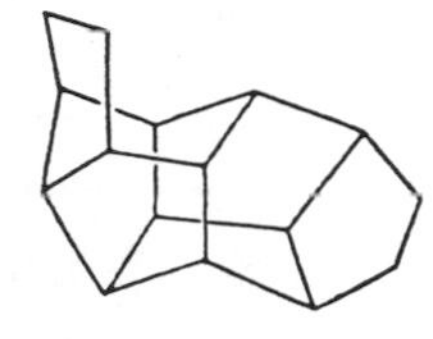

Bisethano
Noradamantane
(1976)

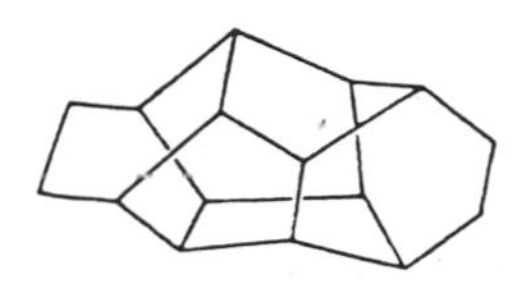

Bicyclo[2.2.2]octane
Dimer (D_{2d})
(1980)

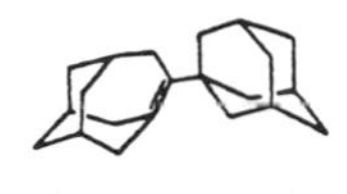

Adamantyl-homo-
adamantene
(1982)

(D_3)-Trishomo-
cubane
(1974)

Trimethylene-
adamantanes
(1982)

Dodecahedrane
(1987)
(PAQUETTE, ET AL 1982)

Fig. 4. *(Continued)*

example, that the most stable tricyclic $C_{15}H_{26}$ isomer would be 1,3,5-trimethyl-7-ethyladamantane (**23**) rather than 1,2,3,5,7-pentamethyladamantane.[37] The confirmation by rearrangement of an isomer (**24**) obtained by hydrogenating cedrene (a tricyclic terpene) marked one of the first verifications of a calculational prediction in my group.[37b] The bridgehead reactivities were another.[27] The true worth of a theory is not its ability to provide an after-the-fact rationalization, but to predict an otherwise unexpected result beforehand.[27a] We predicted many cage "stabilomers" as a challenge and guide for the experimentalist.[37a]

We then became bolder in our choice of starting materials. Further examination of crude petroleum samples in other laboratories had revealed the presence of other diamondoid hydrocarbons in addition to adamantane. These might have arisen in petroleum by similar isomerizations, or by even more deep-seated processes. Sure enough, not even isomers were needed if drastic conditions were employed. Straight chain hydrocarbons, and virtually every compound we treated with aluminum halides at higher temperatures gave mixtures of various polymethyladamantanes.[38] Butter would have worked as starting material! Whitlock and Siefken[17b] were right, "Adamantane may be conceived as a bottomless pit, into which rearranging molecules may irreversibly fall."

Rearrangement syntheses of many cage hydrocarbons have been achieved. Those contributed by my group are summarized in Figure 4. Also included are a few cage hydrocarbons we prepared by other methods.

CONGRESSANE AND FURTHER STEPS TOWARDS THE DIAMOND

The participants at the 1963 IUPAC meeting in London were issued a challenge: the synthesis of the official emblem, "Congressane" (**10**). I had not attended the meeting (travel money, especially for transatlantic flights, was not available to junior faculty members in those days), but soon set to work. During a consulting trip to Union Carbide in Charleston, West Virginia, Trecker gave me a large sample of a possible pentacyclic precursor (**25**), prepared by photodimerizing norbornene. Chris Cupas tried the rearrangement at Princeton. A large amount of tar was produced, but no indication of success. The second run looked equally unpromising, until Chris noticed that small crystals had sublimed on the cooler parts of the flask. The simple IR and nuclear magnetic resonance (NMR) spectra left little doubt concerning the constitution,[15,39] but we wanted to make doubly sure. I had heard that X-ray crystallographers at the Navy laboratory in Washington could solve crystal structures of molecules without any heavy atoms. Thus, the determination of the structure of Congressane[40] was an early application of the method that led to Karle's 1985 Nobel prize.

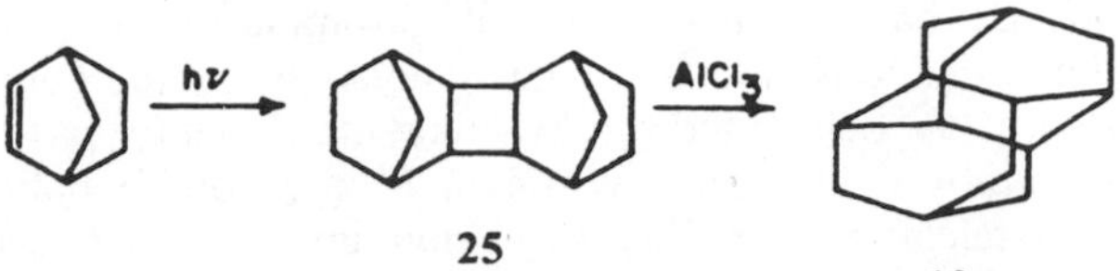

hν → 25 → $AlCl_3$ → 10

Trivial names like Congressane gain publicity,[18h] but more systematic names are preferable. The von Baeyer polycyclic nomenclature is unwieldy: Pentacyclo[7.3.1.1^{4,12}.0^{2,7}.0^{6,11}]tetradecane conveys little immediate information. Instead, we accepted a suggestion and renamed Congressane "diamantane." Accordingly, "triamantane" would be the name of the third number of the diamond series (**26**). This was our next synthetic goal. Success came rapidly.[41]

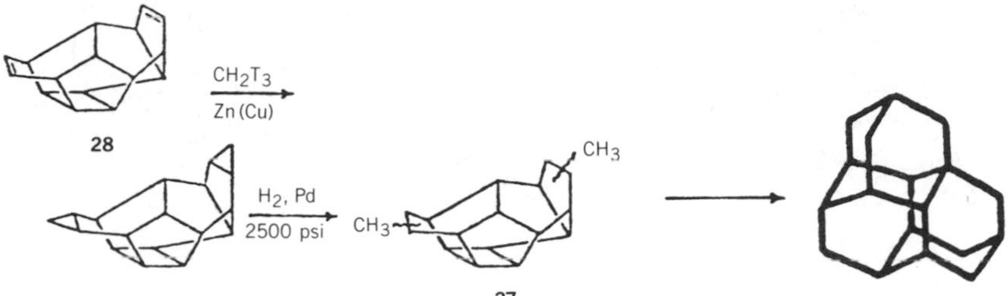

28 → CH_2T_3, Zn(Cu) → 29 → H_2, Pd, 2500 psi → 27 → 26

As always, a suitable precursor was needed. These should be readily available or easy to make. Our heptacyclooctadecane candidate (**27**, $C_{18}H_{24}$) was elaborated from cyclooctatetraene. This intriguing [**8**]annulene is now expensive and difficult to prepare, but generous samples once could be obtained from Bayer. Pilot plant quantities had been produced from acetylene, but no industrial use developed and production was abandoned. When heated, cyclooctatetraene dimerizes and then undergoes an intriguing series of thermal rearrangements. The end product, **28**, had the seven rings we needed, but possessed only 16 carbon atoms. Two more were added by cyclopropanation of the two double bonds (**29**) followed by hydrogenolysis of the strained cyclopropane ring C—C bonds (**27**). (Our later exploitations of this hydrogenolysis reaction are discussed below.)

With the triamantane starting material (**27**) in hand, we were ready to try the more effective "sludge" catalyst.[36] Aluminum chloride is not soluble in hydrocarbons, and aluminum bromide quickly "throws down" a precipitate and becomes less effective. The petroleum industry had overcome this problem by developing catalysts based on aluminum bromide-*t*-butyl bromide for isomerizations and alkylations. Such sludge catalysts improved contact and had been used effectively by Schneider for his alkyladamantane rearrangements;[36] our results were improved as well. While in Munich on sabbatical leave in 1965, I received a telegram from my Princeton group, "Congressane by the ton!" The yield of diamantane had been improved from 1 to over 10%. The sludge catalyst worked with **27** as well. The first synthesis of triamantane (**26**) was achieved in 2–5% yield shortly afterwards.[41]

Three tetramantanes (cf. **30**) are possible. These correspond in shape to isobutane, *anti*-butane, and *skew*-butane.[42] We started again with the cyclooctatetraene dimer (**28**), this time elaborating on it by a Diels–Alder reaction with cyclohexadiene to the C_{22} nonacyclo level (**31**). After hydrogenation, a nicely crystalline isomer formed upon treatment with the sludge catalyst. However, NMR quickly revealed that the symmetry of the product was too low to be any of the tetramantanes. The X-ray structure (**32**) showed that isomerization towards the diamond arrangement had only occurred

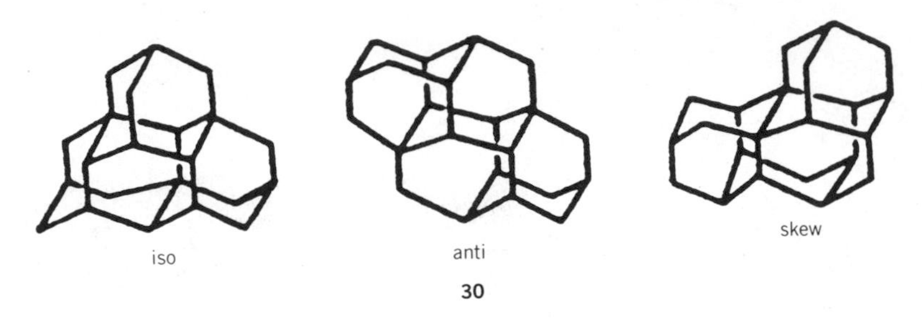

30

in part. A diamantane moiety was present, but also an ethano-bridged noradamantane. This "variation from standard" suggested the trivial name *JACS* was willing to publish, "bastardane."[43] I explained the more usual meaning of the English word to my Japanese co-worker, Eiji Osawa. "Ah," he exclaimed, "an unwanted child."[18h] We did not persevere. But the synthesis of "the wanted child," *anti*-tetramantane (**30**), was achieved by Tony McKervey,[44] our most effective competitor in the field of hydrocarbon rearrangements.[18a,b] Both of us had taken advantage of Schrauzer's transition metal complex catalyzed dimerization product of norbornadiene, "Binor-S" (**33**),[45] as a readily available polycyclic intermediate. Hydrogenolysis of the cyclopropane rings in **33** gave

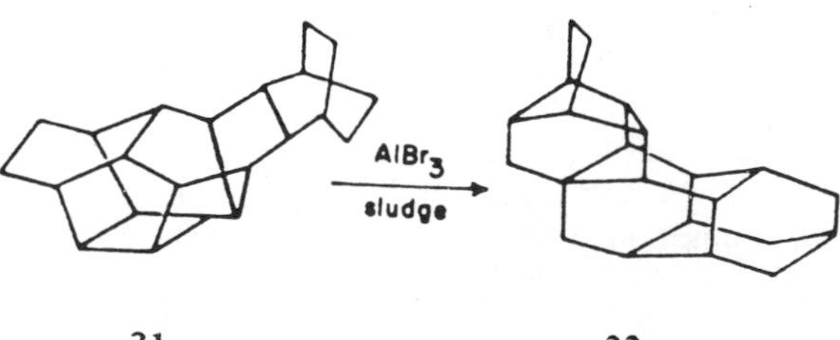

31 **32**

a $C_{14}H_{20}$ precursor (**34**), which could be isomerized to diamantane in high yield.[46,47] McKervey's metal catalyzed rearrangements of Binor-S led to precursors (e.g., **35**), which rearranged to triamantane in good yield.[18] This hydrocarbon is now available in quantity as well.

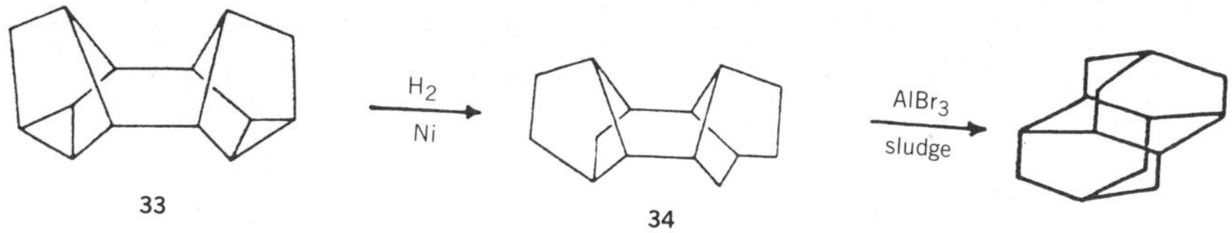

McKervey, with whom we were to collaborate productively, also developed even more effective isomerization catalysts and conditions.[18a,b] Following the lead of the petroleum industry, his group explored the use of a chlorinated platinum on alumina catalyst in the gas phase. Vapors of the precursor were simply passed down a hot tube filled with the catalyst. But there were other advantages as well, since these conditions effected ring closures that were not observed in solution. Furthermore, hydrogen, used as a carrier gas, exchanged with the hydrocarbon reversibly. Hence, olefinic starting materials that did not have the same number of rings as the desired products could be employed as precursors. Thus, McKervey's 1978 synthesis of *anti*-tetramantane (**30**) involved rearrangement of a $C_{22}H_{28}$ diene mixture (**36** and **37**).[44]

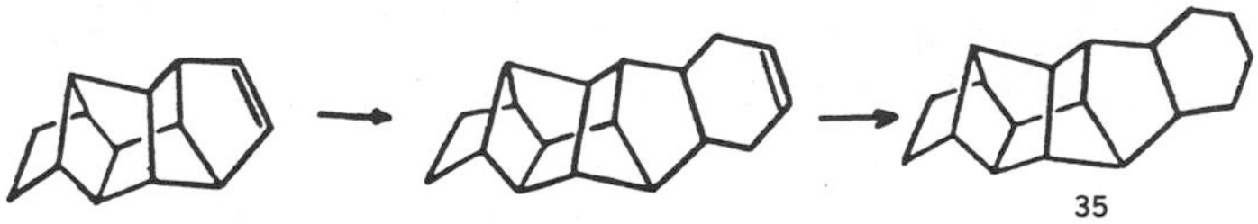

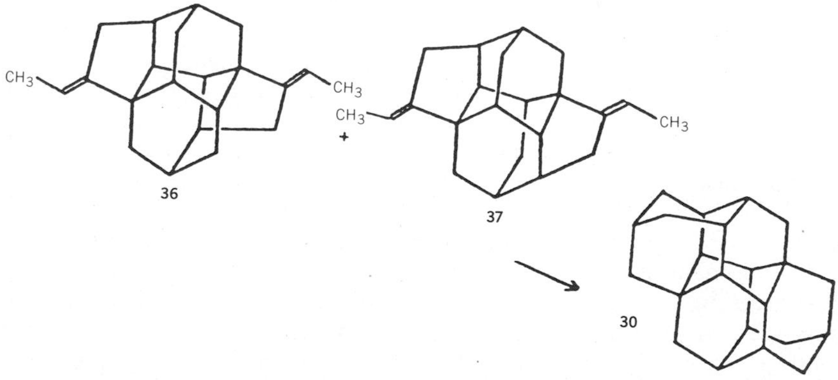

My group's approach to the elaboration of polymantanes was more mundane. Our "rational" (stepwise) synthesis of diamantane (**10**) from adamantane (**1**) involved two intramolecular carbon insertion processes.[48] The first gave ethanoadamantanone (**38**). (Ethanoadamantane has been identified as a component of petroleum; we also prepared it by rearrangement.)[49] The second ring closure gave a protodiamantanone (**39**). A similar route might be employed

N_2CH O 38 O N_2CH H O O 39 10

for the elaboration of triamantane (**26**) to give *iso* or *syn*-tetramantane and the conceptually attractive neopentamantane (**40**), cyclohexamantane (**41**), and even the "adamantane of adamantanes" (**42**).[42]

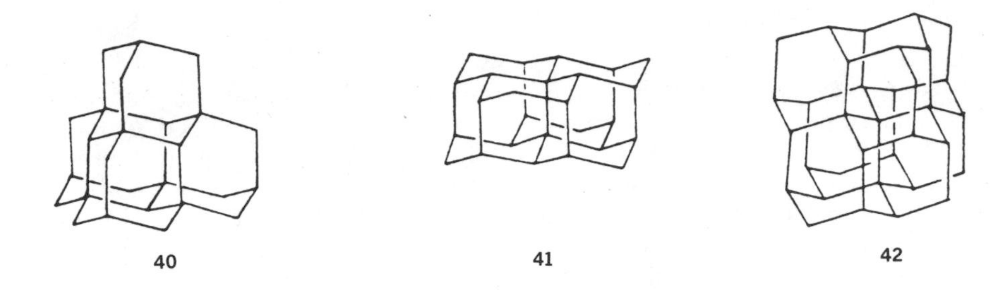

40 41 42

ADAMANTANE REARRANGEMENT MECHANISMS

Arrow-pushing mechanistic speculations for organic reactions had reached their heyday in the 1950s. Students can still be challenged to formulate a plausible mechanistic pathway from tetrahydrodicyclopentadiene to adamantane. Formally, only three C—C bonds needed to be broken and reconnected,[1] but this hypothetical process involves some unlikely migrations. More plausible pathways involving more steps can be formulated rather easily. But this is the problem. There are too many such "paper" possibilities. As described above, our first mechanistic investigation using methyl "tags" was discouraging.[34] Carbon labeling experiments were indicated, but the high symmetry of the adamantane end product (there are only two different kinds of carbon atoms) reduced the amount of information that could be expected from single isotopic substitutions. I wondered if the mechanism would ever be elucidated. My early pessimism has proven to be unfounded.

Whitlock, et al. made the first intellectual breakthrough.[17b] Examination of the structures of tricyclic $C_{10}H_{16}$ isomers revealed 16 that did not contain three- or four-membered rings and were not excessively strained. Assuming that only C—C 1,2- shifts were involved, Whitlock prepared a graph connecting tetrahydrodicyclopentadiene with adamantane. Despite the large number of possible mechanistic pathways (at least 2897!), Whitlock's well-conceived experiments also led him to deduce the most likely pathway. This is shown by the bold lines in our augmented version of the Whitlock graph (Fig. 5).

Our contribution was to consider the thermochemistry and to analyze the rearrangement steps in more detail. We employed our empirical force-field (molecular mechanics) programs, not only to calculate the energies of the intermediate hydrocarbons, but also of the carbenium ions.[27] The calculated heats of formation of the $C_{10}H_{16}$ isomers were added to Whitlock's graph. These values showed that 1,7-trimethylenenorbornane (**7** in Fig. 5) was the thermochemical bottleneck. It was the highest energy structure through which

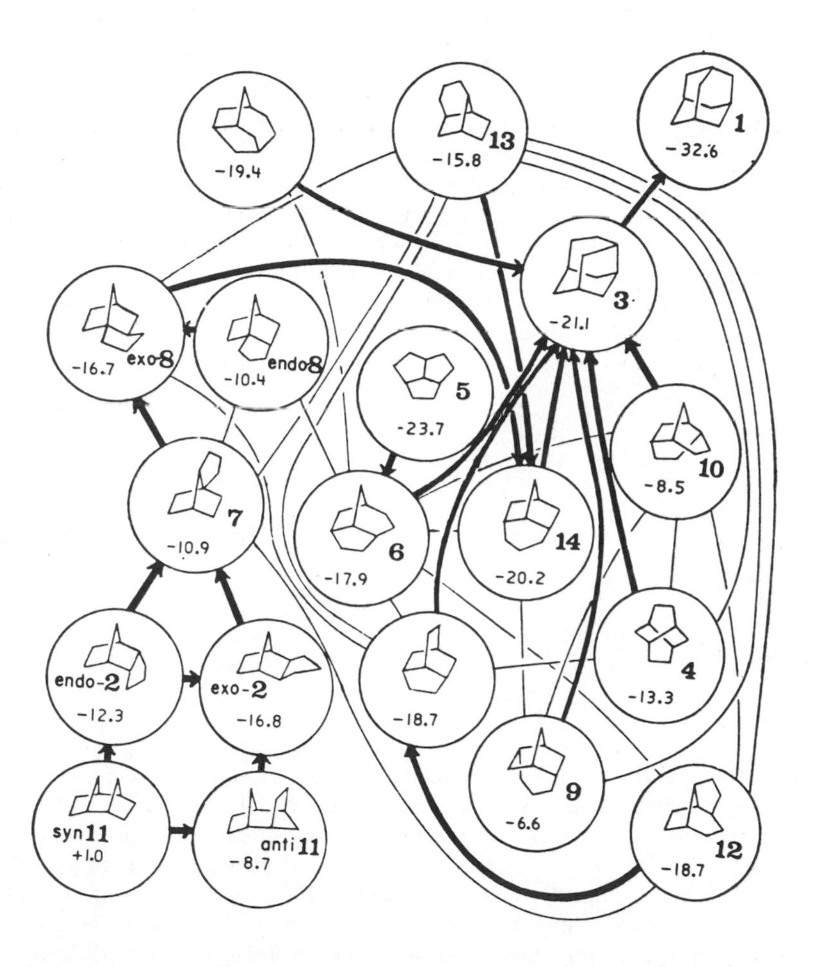

Fig. 5. The Princeton modification of the Whitlock–Siefken[17b] adamantane rearrangement graph. The heats of formation calculated by empirical force-field methods are given for each $C_{10}H_{16}$ isomer. The most likely pathways are indicated by bold lines (reproduced from Ref. 50).

the rearrangement had to proceed. However, the next intermediate, *exo*-**8** (Fig. 5), had several possibilities for further isomerization. The most likely candidate was identified by considering the energies of the carbocation intermediates, which would result from the skeletal transformation. This prediction was shown to be correct by synthesizing *exo*-**8**, and subjecting it to rearrangement.[51] Gas–liquid chromatography (GLC) could be used to follow the reaction; **14** was formed first, then **3**, and finally **1** (Fig. 5). These intermediates cannot be observed (except perhaps in trace quantities) when starting from **2**; the rate-limiting step is indeed the formation of the "bottleneck" **7**. In the ensuing years, my students and various research groups have prepared virtually all the other $C_{10}H_{16}$ isomers in "adamantaneland".[5d] All rearrange to adamantane. Judging from the intermediates that can be detected,[50–53] the pathways predicted in Figure 4 always appear to be followed. The bottleneck (**7**, Fig. 5), when it was finally prepared and rearranged, revealed two (not just one) rearrangement pathways leading to adamantane.[51]

A greater degree of refinement of the adamantane rearrangement mechanism has been achieved by following the reaction course of isotopically labeled adamantane isomers[52] and by studying the behavior of carbocation intermediates under short-lived conditions.[53] Stable ion investigations offer relatively limited information, as only the tertiary ions **43** and **44** (as well as the 1-adamantyl cation, **45**) can be observed directly.[32b] The additional tertiary candidate, (**46**), rearranges too rapidly to **44** to be detected. However, the rearrangement of **40** into **42** in superacid indicates that a sequential cascade of Wagner–Meerwein rearrangements[5f] as well as 1,2- and 1,3-hydride shifts take place intramolecularly.[5e,53]

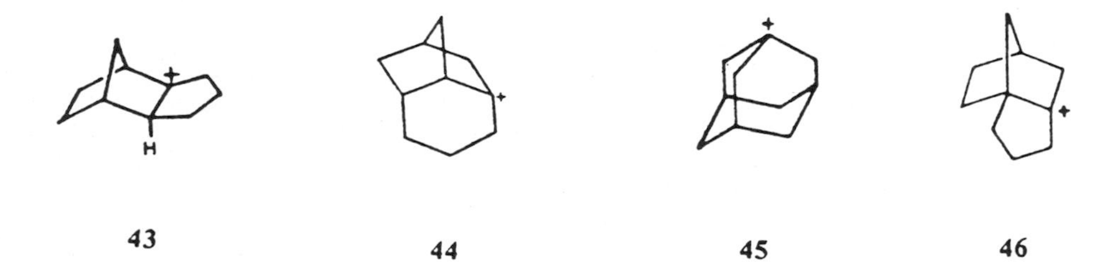

Our carbon labeling studies, although published more recently,[52] were actually carried out at Princeton. We had shown earlier that adamantane could be forced to undergo degenerate skeletal rearrangements via protoadamantane intermediates.[54] The same mechanism is involved in apparent 1,2-methyl shifts (Fig. 6).[55] Direct migration is precluded since the carbocation orbitals are misaligned. The extensive studies by Ganter and his Zürich group on the mechanism of the adamantane rearrangement refined our knowledge of this process considerably.[53] The behavior of carbocations that are generated solvolytically or trapped rapidly reveal the detailed behavior of individual

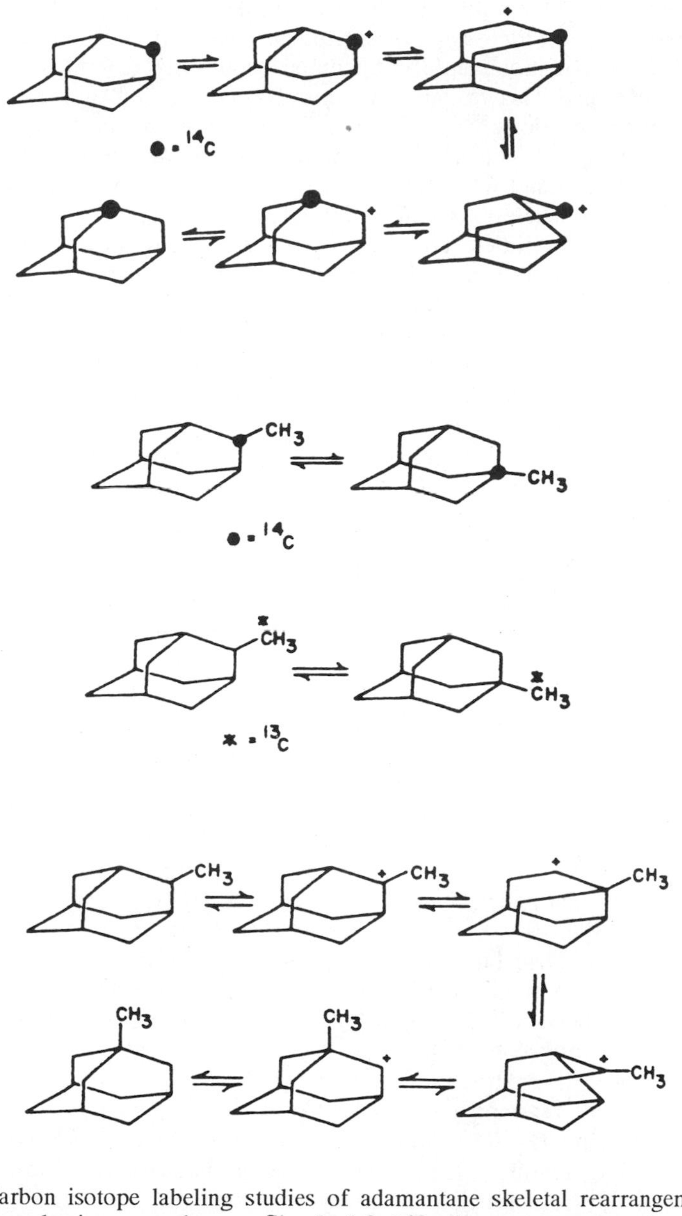

● = ^{14}C

● = ^{14}C

✱ = ^{13}C

Fig. 6. Carbon isotope labeling studies of adamantane skeletal rearrangements. The proposed mechanisms are shown. Simple 1,2-shifts do not occur over the adamantane framework because the orbitals are misaligned, that is, the highly deformed transition states would be too high in energy (see Ref. 31).

species. In a joint but still unpublished project Sorensen calculated all of the carbocation intermediates with semiempirical molecular orbital (MO) theory.

When more carbon atoms are present in the system, the graphical approach to mechanistic elucidation rapidly becomes frightfully complicated. With considerable patience Osawa and his co-workers worked out the tricycloundecane interconnections.[35a] Figure 7 is only a part of the full tricycloundecane graph! The methyladamantane products are not even shown. One can easily understand why our early methyl labeling rearrangement studies went astray.[34] Nevertheless, a new feature employed in the analysis reduces the number of possibilities considerably. Although these reactions are thermodynamically controlled, individual steps may be ruled out if they have activation energies too high to compete with those for more favorable processes. 1,2-Shifts are most likely when the "vacant" carbocation *p* orbital and the migrating C—C bond are aligned. The larger the dihedral angle in rigid systems, the higher the migration barrier is expected to be. We chose a 30° value (measured from models) as our "cut-off," and excluded steps that were unfavorable on this basis. This procedure still seems quite reasonable, although I would now carry out quantum mechanical calculations on the most probable transition states.

The elucidation of the diamantane rearrangement mechanism was even more complicated.[47] The location and identification of all viable $C_{14}H_{20}$ pentacyclic isomers was too enormous a task for paper and pencil. Structural representations of the same compound, when drawn from various perspectives, look completely different! Todd Wipke wrote a computer program based on the von Baeyer nomenclature scheme to identify all of the possible pentacyclotetradecanes. Highly strained and otherwise unlikely candidates were eliminated, for example, isomers with three- and four-membered rings, or with intertwined bonds. Even after this pruning, an unmanageably large number of isomers remained. Tamara Gund proposed a reasonable way to proceed. Instead of constructing the full graph, the most favorable step from each intermediate was identified by using the criteria we had developed. This is shown in Figure 8, taken from the original paper.[47] Thus all of the possible rearrangement products from tetrahydro-Binor-S (in Fig. 8) were considered, and the most likely pathway (leading to **24** in Fig. 8) was identified. Similarly, continuing from **24** gave **28**, then **29**, and finally diadamantane (**1**, all in Fig. 8).

Figure 8 also shows (at the top right) the consequences of a mechanistic "dead end." Following one of the other pathways from **28** leads to many new isomers, but none have structures suitable for rearrangement to diamantane. Although the actual rearrangement pathway of diamantane has received little attention experimentally, it is likely that such dead ends are common in polycyclic rearrangements. The detection of an isomeric hydrocarbon after partial reaction does not prove it to be an intermediate.

Fig. 7. Main portion of the tricycloundecane graph. As in Figure 5, the calculated heats of formation and most likely pathways are given. Identified intermediates are shown in darkened circles. Dashed circles represent compounds confirmed *not* to be intermediates (reproduced from Ref. 35a).

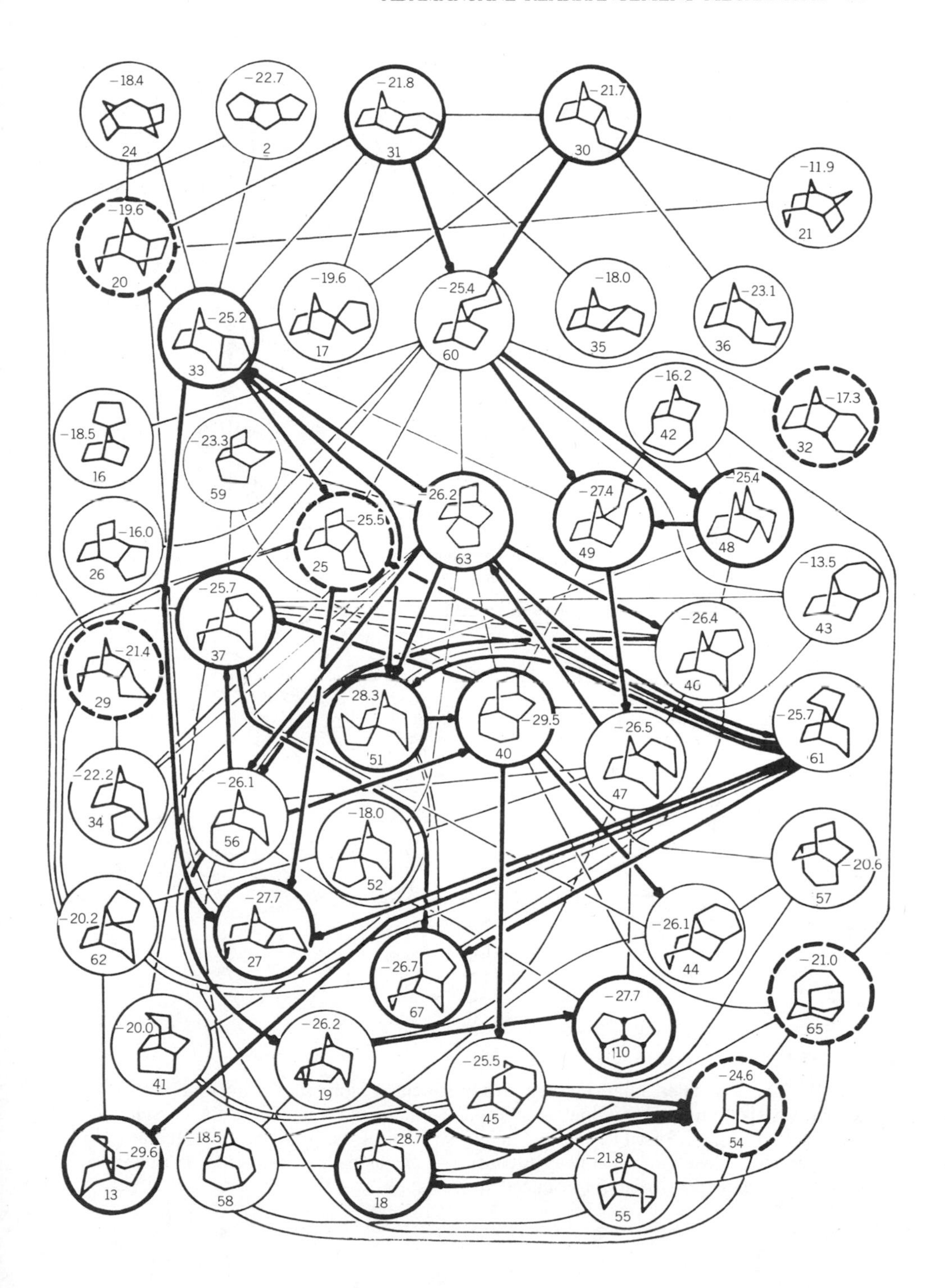

−18.4
24
−22.7
2
−21.8
31
−21.7
30
−11.9
21
−19.6
20
−19.6
17
−25.4
60
−18.0
35
−23.1
36
−25.2
33
−16.2
42
−17.3
32
−18.5
16
−23.3
59
−25.5
25
−26.2
63
−27.4
49
−25.4
48
−16.0
26
−13.5
43
−25.7
37
−26.4
46
−21.4
29
−28.3
51
−29.5
40
−26.5
47
−25.7
61
−22.2
34
−26.1
56
−18.0
52
−20.6
57
−20.2
62
−27.7
27
−26.1
44
−26.7
67
−21.0
65
−20.0
41
−26.2
19
−27.7
10
−25.5
45
−24.6
54
−29.6
13
−18.5
58
−28.7
18
−21.8
55

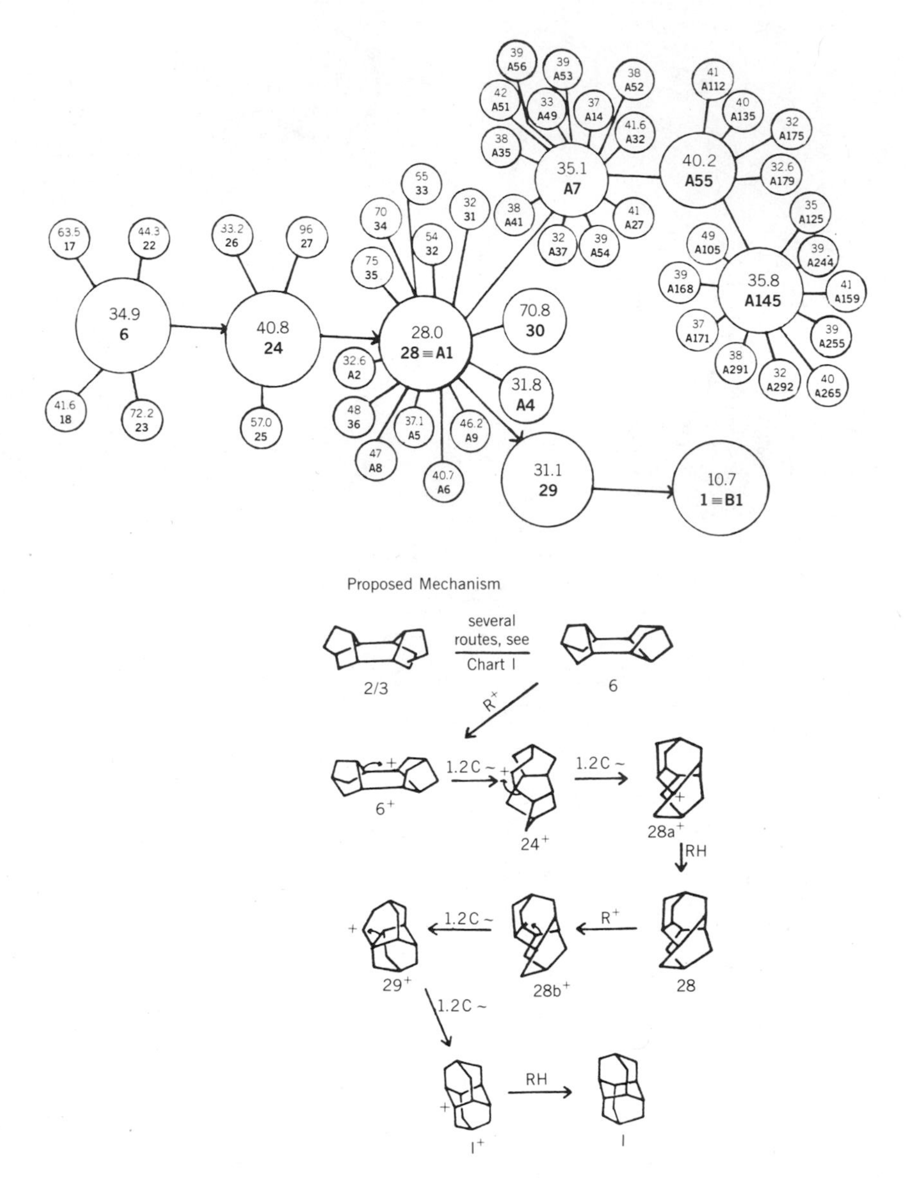

Fig. 8. The truncated diamantane rearrangement graph (top) and the proposed mechanism (bottom). Compound numbers are underlined and the strain energies are given in each circle. Possible rearrangements of each of the principal intermediates are shown and the most likely pathway identified a mechanistic "dead end" as illustrated in the top right-hand corner (reproduced from Ref. 47).

HYDROGENOLYSIS, SUPPORTED CATALYSTS, AND (FINALLY!) DODECAHEDRANE

While hydrogenolysis—the cleavage of chemical bonds by hydrogen—has never become a reaction of major synthetic importance, it has served cage molecule chemistry well. We had employed drastic conditions for the hydrogenation of the triamantane precursor (**29**),[41] but later found that such three membered ring cleavages could be realized under mild conditions. *gem*-Dimethyl (**47**) and *t*-butyl groups, even in strained, axial positions (**48**), could be introduced in this manner.[56] Hydrogenolysis of the three-membered rings in Binor-S (**33**)[45] gave the diamantane precursor (**34**).[46,47] Four-membered rings in cage molecules could also be opened: For example, basketane (**49**) gave twistane quite specifically.[57] Along with Osawa and Musso,[5e,57b] we considered the reasons for this selectivity. Although hydrogenolysis reactions are kinetically controlled, the thermodynamically most stable product is usually formed. The weakest bond tends to cleave preferentially. The "weakest bond" can be identified by force-field calculations or by the bond lengths in crystal structures.

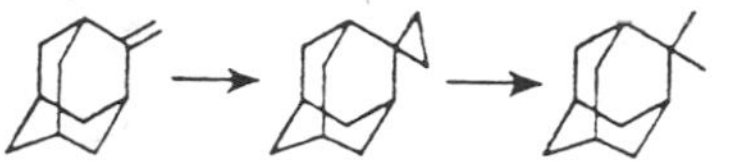

47

9 10 8 1 7 6 3 2 4 5

49

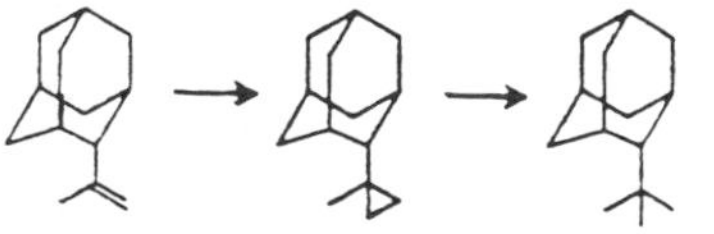

48

After my arrival in Erlangen, our research in this area expanded with a new goal. Efforts to synthesize dodecahedrane (**8**) by rearrangement were begun in earnest. We had already considered using the saturated $C_{20}H_{20}$ isomers as precursors while at Princeton, but highly strained molecules with this composition (e.g., the basketene dimer, **50**) could hardly be expected to survive strong Lewis acid conditions.[58]

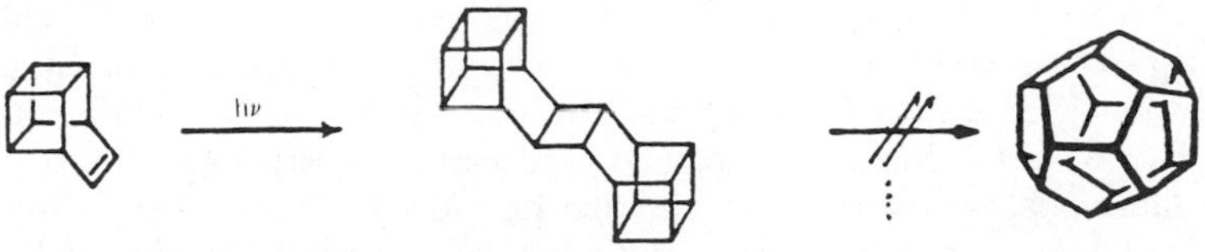

hν

8

50

Our Erlangen projects were stimulated by McKervey's success in using supported chlorinated platinum on alumina catalysts for cage molecule rearrangements.[18a,b] Our experience with the methyladamantanes and our knowledge of isomerization pathways suggested that C_{21} or C_{22} precursors should be more favorable. These might lead to methyl- and dimethyl-dodecahedranes much more readily since the intermediates on the rearrangement pathways should be less strained. Hence, another aspect of this project was to find ways to remove methyl groups from cage hydrocarbons. This proved to be much easier to achieve than making dodecahedrane!

Landa kindly sent me his reprints on a regular basis. As these often were published in Czech, in local and (for me) obscure journals, I would not have seen them otherwise. Landa reported that a commercially supported nickel catalyst in a hydrogen flow apparatus was capable of removing methyl groups from methyladamantane.[59] But the temperatures were high and the yields were low. Our "homemade" nickel/alumina catalyst was much more active. 2-Methyladamantane could be demethylated quantitatively; in addition, the yield of adamantane from the 1-methyl isomer was quite good.[60]

All kinds of novel reactions were discovered. The same catalyst was capable of removing (or in some cases reducing) many functional groups, not only from the adamantane skeleton (Fig. 9), but also from simple substrates.[61,62] Alkyl chains could be degraded stepwise with some selectivity. The direct conversion of triamantane (**26**) first into diamantane (**10**) and then into adamantane (**1**) was our most spectacular result![60] Would that diamantoid systems could be "built up" in the opposite direction with equal ease.

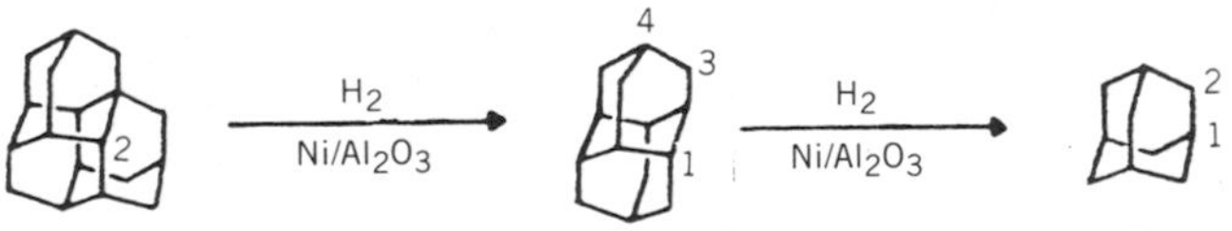

On a tonnage basis, most of chemistry is carried out not in the flasks of the academic or research laboratory, but in the hot tubes of industrial installations. The reasons are obvious. What could be simpler than passing an educt with a reactive gas over a supported catalyst and isolating practically pure product directly? Neither solvents nor byproducts need disposal. Small scale experiments can be carried out similarly with equal effectiveness. It seems strange to me, also in view of the employment prospects for most of our students, that practical laboratories do not provide experience with such reactions. Perhaps research chemists one day will overcome their prejudices and take advantage of the splendid potential of such vapor phase syntheses.

Led by Wilhelm Maier our catalytic research at Erlangen flourished.[60–62] But the main goal was still the preparation of dodecahedrane. Peter Grubmüller synthesized a number of precursors, but none gave any indication of polyquinoid

X = OH, Y = H
X, Y = O

(1) (2)

X = CN, CH_2Br

X = CH_3, Y = OH

(4)

CH_3 CH_3

(3) (5)

No.	Reactant X	Y	T [°C]	Product	Yield [%]
(1a)	Br		200	*(4)*	97
(1b)	NH_2		240	*(4)*	99
(1c)	OH		180	*(4)*	97
(1d)	$OSiMe_3$		150	*(4)*	99
(1e)	CH_2OH		200	*(4)*	99
(1f)	COOH		240	*(4)*	90
(1g)	$COOCH_3$		240	*(4)*	71
(1h)	COCl		260	*(4)*	86
(1i)	CN		180	*(3)*	99
(1k)	CH_2Br		180	*(3)*	99
(1l)	$CO_2Si(CH_3)_3$		220	*(4)*	55
(3)			280	*(4)*	80
(2a)	=O		220	*(4)*	93
(2b)	OH	H	170	*(4)*	99
(2c)	CH_3	OH	150	*(5)*	99
(5)			235	*(4)*	99

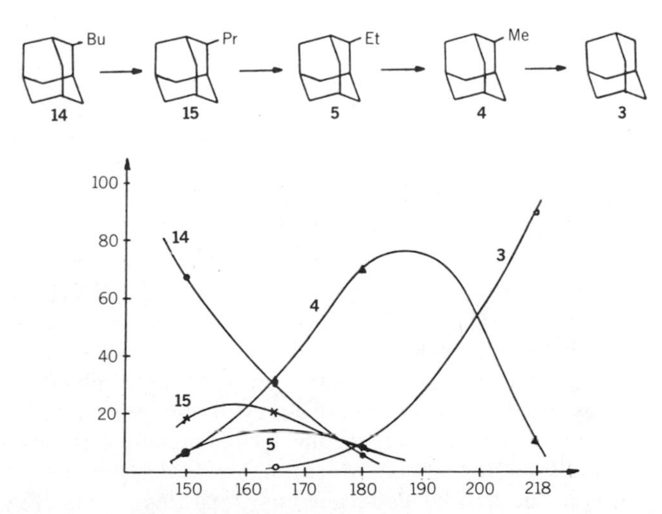

Fig. 9. Representative Erlangen hydrogenolysis results in cage systems. These were generally obtained by passing educt vapors in a hydrogen atmosphere over a nickel–alumina catalyst ~ 200–250°C (see Refs. 60–62).

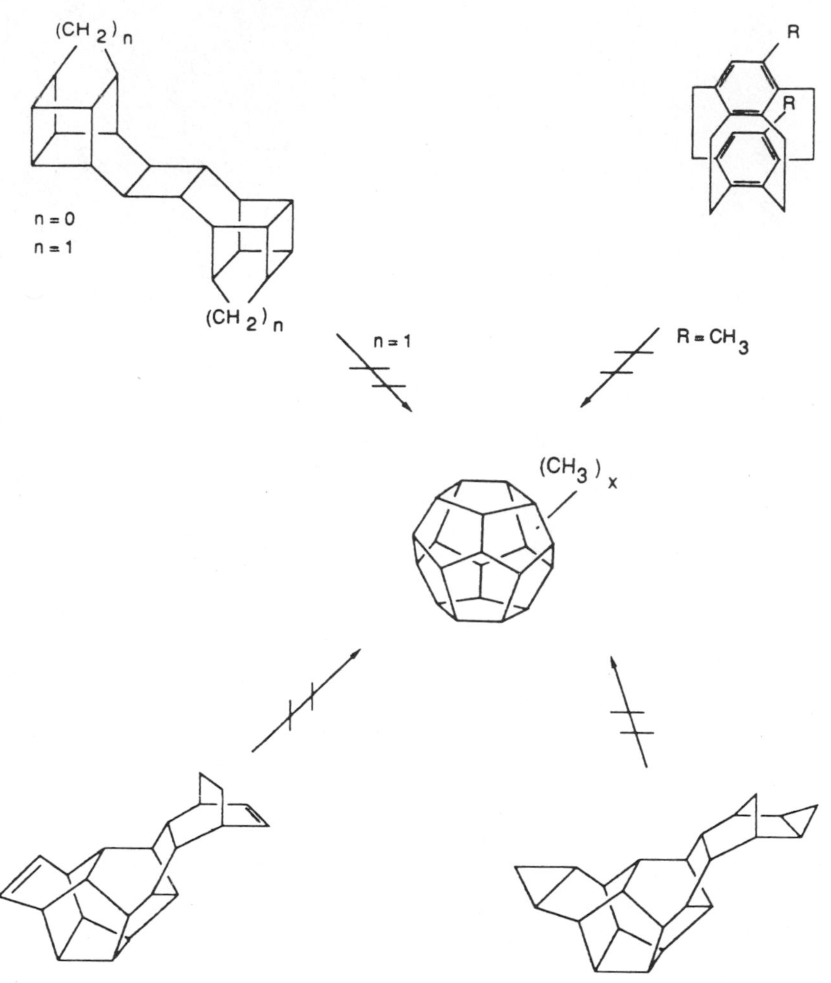

Fig. 10. Dodecahedrane failures. None of these compounds gave any NMR evidence for the formation of polyquinoid part structures when rearranged either with Lewis acids or with supported catalysts in the gas phase.[63,64]

part structures upon rearrangement either with the old Lewis acid or with the new supported catalysts (Fig. 10).[63]

We then explored polybridged cyclophanes, cage molecules in their own right, as rearrangement precursors. Tetra- and pentacyclophane have the same globular topology as dodecahedrane. Tetracyclophane is at least 50 kcal/mol less stable than its isomer, dodecahedrane. Unfortunately, these molecules not only were difficult to prepare on a reasonable scale, but also proved to be intractable towards the desired dodecahedrane rearrangements (Fig. 10).[64]

Along with other research groups, we had been competing with Leo Paquette. His progress towards dodecahedrane was slow but steady. Ours was hit or miss. Like my students at Christmas parties, Leo teased me by showing crystalline samples labeled "dodecahedrane" during a visit to Ohio State. After Leo really did succeed[13] and my research grant had run out, I was ready to give up. But then Horst Prinzbach visited Erlangen and offered us a sample of pagodane.[65] He had attempted the isomerization to dodecahedrane without success, but was aware that the gas-phase methodology we had developed was likely to be more effective.

We had only a few milligrams of pagodane to work with, but Wolfgang Roth's very first experiment over a palladium on carbon catalyst was a partial success.[66] The product was a mixture containing only a few percent of dodecahedrane, but the sharp ^{13}C and proton signals at the characteristic chemical shifts[13] were unmistakable. Subsequent runs, also using different catalysts, gave similar but not better results. Dodecahedrane had finally been synthesized "in one step," but in very low yield.

The next somewhat larger sample of pagodane from Freiburg required quite a long time to prepare. In the meantime, Maier had established his catalytic research group in Berkeley. His doctoral students included Alan McEwen, who had already spent a sabbatical with me in Erlangen exploring cage molecule chemistry experimentally (polyhalogenation of adamantane)[67a] and computationally (hyperstable olefins[67b] and unconventional aromaticity, e.g., in dodecahedrapentaene).[67c] The Berkeley group had developed effective methods and had more sophisticated and better equipment for exploring various catalysts and conditions. Prinzbach's sample was more than adequate for an extensive survey. Tantalizingly good yields of dodecahedrane were obtained erratically, but only ~ 8% could be achieved reproducibly. This was the result we decided to publish.[21]

Paquette[13,20] and Prinzbach[22] are developing the chemistry of dodecahedrane beautifully. Reading their papers makes me wistful. Pagodane derivatives do convert more readily to substituted dodecahedranes.[22] Cage molecules continue to provide insights, unexpected behavior, and synthetic challenges to be overcome.

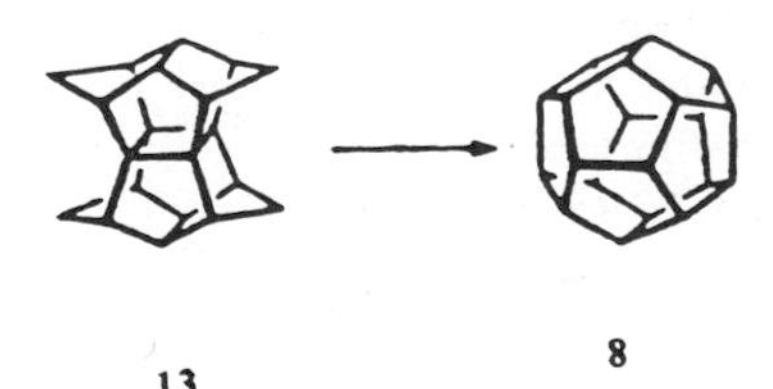

FILLING HYDROCARBON CAGES

Ball-and-stick models of cage molecules invite the question, "What can be put into the middle?" The cage designation implies the possibility of incarceration. But space filling models disclose a more pessimistic prospect. The "holes" in the center of cage hydrocarbons, even of dodecahedrane (~ 0.7 Å in diameter), are very small. For this reason, synthetic attempts to prepare dodecahedrane by joining triquinacene halves complexed by metal atoms were doomed to failure. "Cryptands," which do enclose metal cations, have much larger cavities. However, metal–heteroatom interactions are involved. But what about the much smaller hydrocarbon cage with the elementary particles, H^+ or the electron? Such possibilities would be difficult to examine experimentally, but they are easy to study computationally. It is simple to carry out quantum mechanical calculations on center-protonated cubane,[68] adamantane, dodecahedrane,[69] and so on. But all these species prove to be perfectly terrible energetically. From a MO viewpoint the center of molecules is the worst position for protonation. Instead, protons are attracted to positions of high electron density, that is, to the bonds. The C—H and C—C protonated isomers are more favorable energetically.

Cage hydrocarbons do not have positive electron affinities. The radical anions of saturated hydrocarbons are not bound. Even if they were, the extra electron would seek the lowest lying unoccupied molecular orbitals, and these possess lower symmetry. But a possible way to put electrons into the center of a cage system occurred to me in 1976. In our first joint paper in 1964,[32a] Olah and I had represented the stable 1-adamantyl cation with the "vacant" orbital pointing inwards (see below). Suppose *all* the bridgehead hydrogen atoms of adamantane were removed. The four *p* orbitals would now extend towards the center of the cage. Overlap might be significant. Since only one of the four possible symmetry combinations of these orbitals would be bonding (**51**), this electronic system could only accommodate two electrons as a singlet in T_d symmetry. Thus, the 1,3-dehydro-5,7-adamantdiyl dication (**52**) was conceived.

But such qualitatively attractive ideas, based only on orbital symmetry considerations, often prove to be disappointing when subjected to more quantitative theoretical (or experimental!) examinations. This was the fate of many of my other "cute" projects in the mid 1970s. I fear that many well-publicized orbital symmetry-based analyses are equally implausible. But **52** passed quantitative tests impressively. A modified intermediate neglect of differential overlap (MINDO)/3 evaluation of Eq. (1) indicated that a high degree of stabilization was to be expected for **52** due to the 4*c*-2*e* interaction.

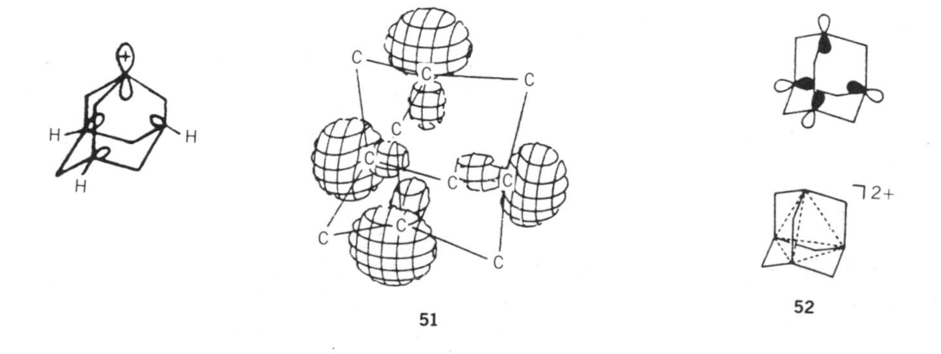

Furthermore, MINDO/3 favored tetrahedral symmetry over the more localized bonding arrangement (**53**, and the three equivalent forms).

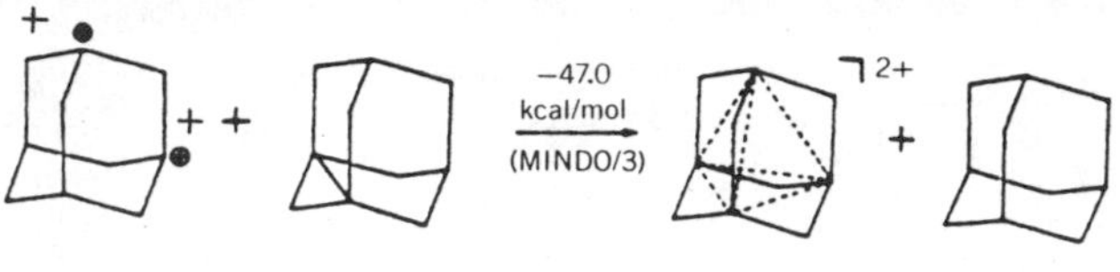

−47.0 kcal/mol (MINDO/3) (1)

Armed with this evidence, I was able to convince several Erlangen students to undertake the experimental investigation. The difficulty lay in the preparation of a suitable precursor that might give **52** with magic acid. A doubly substituted dehydroadamantane derivative like **54** was needed. This class of highly strained compounds were known to be very unstable. But even preparing a suitable unstrained tetrahalo (e.g., **53**) precursor to **54** proved to be frustrating. Tedious separations of mixed tetrahaloadamantane derivatives appeared to be necessary. German laboratory training is still thorough. The experience gives students confidence in their experimental abilities. Despite the bad reputation the project had gained, Matthias Bremer accepted the challenge for his Master's thesis. After a year's work Bremer had learned a lot about

*n*BuLi

53 **54** **55**

interconverting haloadamantanes but had only indications that **54** might have formed as a reactive intermediate. Bremer requested another subject for his Ph.D. thesis but promised that he would carry out "a few more experiments" to try to make **52**. He succeeded after all! *Angewandte Chemie* rewarded us with a "cover girl" (Fig. 11).[70]

Bremer's specific trans-halogenation reaction yielded the trifluoroiodo precursor (**53**). Butyllithium gave **54**. The latter explodes if warmed to room temperature, but Bremer learned how to isolate and to manipulate the material. Assisted by Karl Schötz, the experienced superacid chemist in my group, Bremer added **54** to a SbF_5/SO_2ClF mixture at low temperature. I watched the experiment with anticipation, but the results of first NMR Fourier transform accumulations seemed unpromising to me. I went home to dinner discouraged, but the dication (**52**) had formed without difficulty on the very first attempt. Bremer and Schötz proudly showed me the decisive spectra the next morning.

Only two ^{13}C signals were exhibited by **52**, but the chemical shifts (6.6 and 33 ppm) seemed peculiar for a carbodication since they were not deshielded. This excluded classical, rapidly equilibrating structures akin to **55**, but did not

ANGEWANDTE
CHEMIE

Herausgegeben
von der Gesellschaft
Deutscher Chemiker

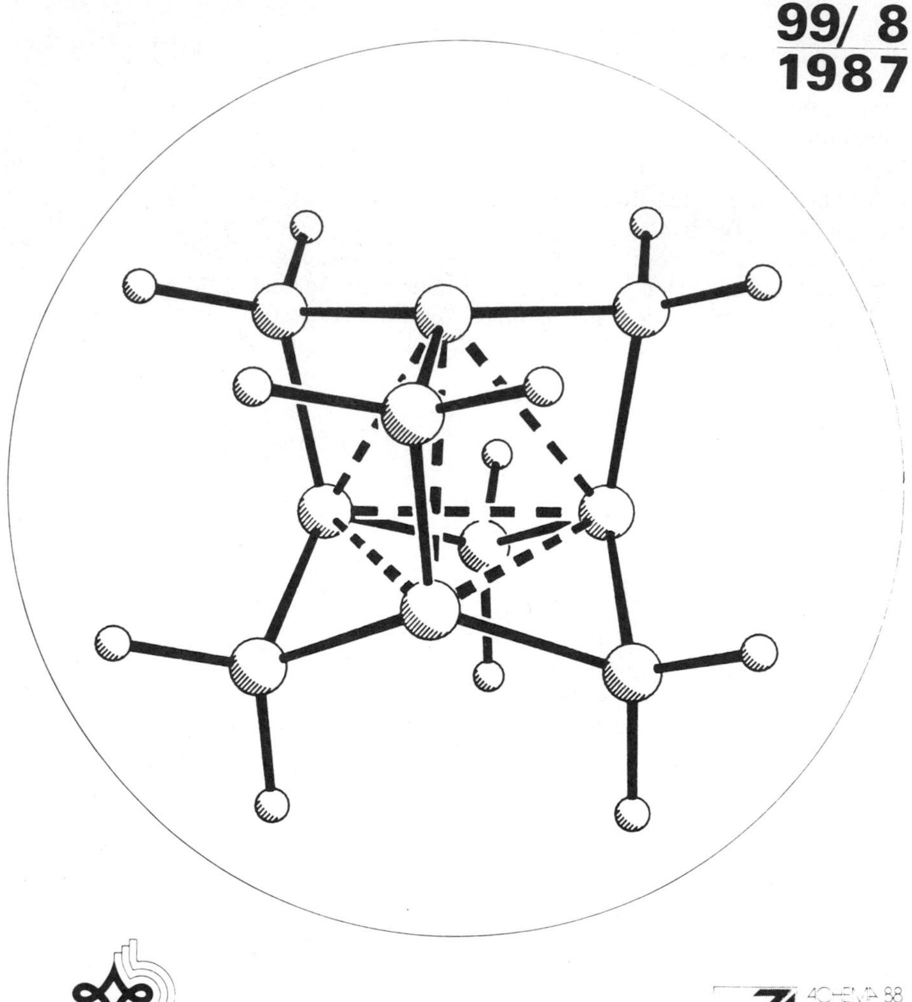

VCH
Verlagsgesellschaft

ANCEAD 99 (8) 727 - 828 (1987) · ISSN 0044 - 8249

ACHEMA 88

Fig 11. The *Angewandte Chemie* cover depicting the *4c-2a* bonding in the 1,3-dehydro-5,7-adamantdiyl dication (reproduced from reference 70).

prove nonclassical structures like **52**. Individual Gauge for Localized molecular Orbitals (IGLO) came to the rescue!

Kutzelnigg's theoretical group at Bochum had developed an ab initio method for calculating NMR chemical shifts IGLO. His associate, Michael Schindler, had applied IGLO to carbocation problems—many of the results were impressive.[71] Bremer provided Schindler with an ab initio optimized geometry of **52**, but did not reveal the experimental chemical shift values. Schindler's IGLO predictions puzzled him, too, but were in fantastically good agreement with our measurements. Structure **52** was established conclusively.[70] We have had much success in employing IGLO method for the similar solution of carbocation structural problems.[72]

Based on an electronic structural definition, **52** represents an inherently *three-dimensional* (3-D) aromatic species.[73] Pyramidal cations with six interstitial electrons (e.g., **56** and **57**)[74] are the next members of this 3-D aromatic series. Carboranes and boranes are their neutral counterparts. We are investigating these molecules now by using the IGLO/ab initio structural determination method. My fascination with cage structures continues. Carbon atoms need not even be present!

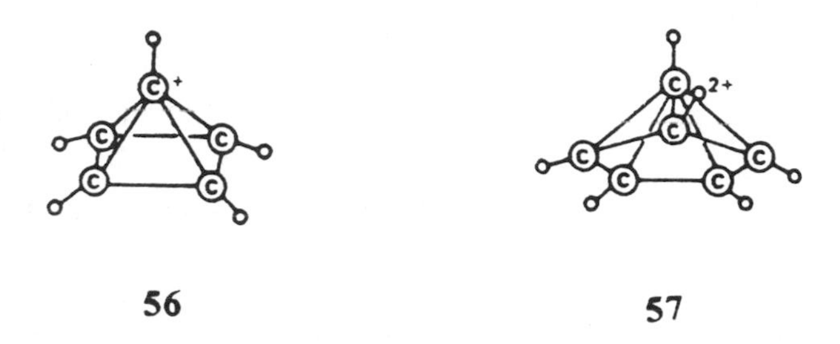

56 **57**

Children grow up, mature, and achieve their own success. I find great pleasure in encountering the ever increasing number of cage structures in the chemical literature. Adamantane chemistry has become so well established that there is little tendency anymore to cite its beginnings. Thus, I was once asked, "Are you the same von Schleyer (sic) who prepared adamantane?" I certainly am. I have thoroughly enjoyed my 30 years in cage hydrocarbons.

ACKNOWLEDGMENTS

The opportunity to share the enthusiasm for research, the excitement of discovery, and the satisfaction of good results with co-workers and collaborators is a major reward of the academic profession. I have had more than my measure of able students and stimulating scientific associates. Warmest thanks to all of you for your involvement! My special appreciation, George, for arranging this volume and our many years of friendship.

REFERENCES

1. (a) P. von R. Schleyer, *J. Am. Chem. Soc.* **1957**, *79*, 3292. (b) P. von R. Schleyer and M. M. Donaldson, *J. Am. Chem. Soc.* **1960**, *82*, 4645.
2. More effective catalysts now give adamantane quantitatively. See G. A. Olah, A. Wu, 0. Farooq, and G. K. Surya Prakash, *J. Org. Chem.* **1989**, *54*, 1450 and references cited therein.
3. P. Duden and M. Scharff, *Liebig's Ann. Chem.* **1895**, *288,* 218. These authors, noting that **3** could only be modeled in space, emphasized the advantages of three-dimensional structural representations as these "are closer to the truth than those depicting molecules in a plane."
4. W. H. Bragg and W. L. Bragg, *Nature London* **1913**, *91*, 557; *Proc. R. Soc. London, Ser. A.* **1913**, *89*, 277.
5. Reviews from Princeton: (a) R. C. Fort, Jr. and P. von R. Schleyer, *Chem. Revs.* **1964**, *64*, 222. (b) R. C. Bingham and P. von R. Schleyer, *Fortsch. Chem. Forsch.* **1971**, *18*, 1. (c) E. M. Engler and P. von R. Schleyer, *MTP International Review of Science, Organic Chemistry Series One*, Butterworth, London, **1973**, Vol. 5, Chapter 7. Also see (d) R. C. Fort, Jr., *Adamantane. The Chemistry of Diamond Molecules*, Dekker, New York, 1976. (e) E. Osawa and H. Musso, *Angew. Chem.* **1983**, *95*, 1; *Angew. Chem. Int. Ed. Engl.* **1983**, *22*, 1. (f) M. Saunders, J. Chandrasekhar, and P. von R. Schleyer in *Rearrangements in Ground and Excited States*, P. de Mayo, Ed. Academic Press, New York, **1980**, Vol. 1, pp. 1–53. (g) R. E. Leone and P. von R. Schleyer, *Angew. Chem. Int. Ed. Engl.* **1970**, *9*, 860. (h) R. E. Leone, J. Barborak, and P. von R. Schleyer in *Carbonium Ions*, G. A. Olah and P. von R. Schleyer, Eds. Wiley, New York, **1973**, Vol. IV, p. 1837. (i) P. Ahlberg, G. Jonsäll, and C. Engdahl, *Adv. Phys. Org. Chem.* **1983**, *19*, 223.
6. Early review: H. Stetter, *Angew. Chem.* **1954**, *66*, 217. Also see H. Stetter, *Angew. Chem.* **1962**, *74*, 361; *Angew. Chem. Int. Ed. Engl.* **1962**, *1*, 286.
7. H. Meerwein, *J. Prakt. Chem.* **1922**, *104*, 179. Meerwein was able to prepare the first noradamantane derivative by ring closure.
8. D. Böttger, *Chem. Ber.* **1937**, *10*, 314. This paper does not mention Landa's discovery of adamantane in 1933.
9. V. Prelog and R. Seiwerth, *Chem. Ber.* **1941**, *74*, 1644, 1769. The first of these papers describes the history of the elucidation of the constitution of adamantane. Also, V. Prelog, private communication.
10. S. Landa, *Chem. Listy* **1933**, *27*, 415, 433. S. Landa, *Chem. Ind. (Paris)* **1933**, 506. S. Landa and V. Machacek, *Collect. Czech. Chem. Commun.* **1933**, *5*, 1. S. Landa, *Petrol. Ztg.* **1934**, *30*, 1.
11. S. Landa, S. Kriebel and E. Knobloch, *Chem. Listy* **1954**, *48*, 61.
12. R. E. Eaton and T. W. Cole, Jr., *J. Am. Chem. Soc.* **1964**, *86*, 3157. See R. E. Eaton and M. Moggini, *J. Am. Chem. Soc.* **1988**, *110*, 1230. R. Gilandi, M. Moggini, and P. E. Eaton, *J. Am. Chem. Soc.* **1988**, *10*, 1232 and references cited therein. P. E. Eaton, *Chem. Eng. News*, **1988**.
13. (a) R. J. Ternansky, D. W. Balogh, and L. A. Paquette, *J. Am. Chem. Soc.* **1982**, *104*, 4503. (b) See L. A. Paquette's review in this volume, Chapter 9. L. A Paquette, *Polyquinane Chemistry* Springer-Verlag, Heidelberg, **1987**. (c) L. A. Paquette, *Top. Curr. Chem.* **1984**, *119*, 1.

14. H. W. Kroto, J. R. Heath, S. C. O'Brien, R. F. Curl, and R. E. Smalley, *Nature (London)* **1985**, *318*, 162 and many subsequent papers.

15. C. A. Cupas, P. von R. Schleyer, and D. J. Trecker, *J. Am. Chem. Soc.* **1965**, *87*, 917.

16. H. Stetter and P. Goebel, *Chem. Ber.* **1963**, *96*, 550.

17. (a) H. W. Whitlock, Jr., *J. Am. Chem. Soc.* **1962**, *84*, 3412. (b) H. W. Whitlock, Jr. and M. W. Siefken, *J. Am. Chem. Soc.* **1968**, *90*, 4929.

18. The chemistry of adamantane and of cage molecules has been reviewed many times. See references 5, 6, 13 and inter alia (a) M. A. McKervey, *Chem. Soc. Rev.* **1974**, *3*, 479; (b) M. A. McKervey, *Tetrahedron*, **1980**, *36*, 971; (c) S. Hala, *Chem. Listy* **1977**, *71*, 8; (d) C. Ganter, *Top. Curr. Chem.* **1976**, *67*, 15; (e) N. V. Averina and N. S. Zefirov, *Russ. Chem. Rev.* **1976**, *45*, 544; (f) E. I. Bagrii and A. T. Saginaev, *Russ. Chem. Rev.* **1983** *52*, 881; (g) T. Sasaki, *Adv. Heterocycl. Chem.* **1982**, *30*, 79; (h) A. Nickon and E. F. Silversmith, *Organic Chemistry: The Name Game,* Pergamon, New York, **1987**, p. 282–284; (i) A. P. Marchand, *Tetrahedron* **1988**, *44*, 2377; (j) A. deMeijere and S. Blechert, Eds., *Strain and Its Implications in Organic Chemistry* V. 273, NATO ASI, Kluwer, **1989**.

19. P. E. Eaton and G. Gastaldi, *J. Am. Chem. Soc.* **1985**, *107*, 724.

20. G. A. Olah, G. K. Surya Prakash, T. Kobayashi, and L. A. Paquette **1989**, *110*, 1304.

21. W. D. Fessner, B. A. R. C. Murty, J. Wörth, D. Hunkler, H. Fritz, H. Prinzbach, W. D. Roth, P. von R. Schleyer, A. B. McEwen, and W. F. Maier, *Angew. Chem.* **1987**, *99*, 484; *Angew. Chem. Int. Ed. Engl.* **1987**, *26*, 452.

22. R. Pinkos, J.-P. Melder, H. Fritz, and H. Prinzbach, *Angew. Chem.* **1989**, *101*, 319 and earlier papers in the same set.

23. The title of Hubert Alyea's popular end-of-the-year demonstration lecture which I enjoyed often, first as a Princeton undergraduate and then as a faculty colleague.

24. P. von R. Schleyer and R. D. Nicholas, *J. Org. Chem.* **1961**, *26*, 3740–3751.

25. P. von R. Schleyer and R. D. Nicholas, *J. Am. Chem. Soc.* **1961**, *83*, 182–187.

26. (a) C. S. Foote, *J. Am. Chem. Soc.* **1964**, *86*, 1853. (b) P. von R. Schleyer, *J. Am. Chem. Soc.* **1964**, *86*, 1854. (c) P. von R. Schleyer, *J. Am. Chem. Soc.* **1964**, *86*, 1856.

27. (a) G. J. Gleicher and P. von R. Schleyer, *J. Am. Chem. Soc.* **1967**, *89*, 698–699. (b) R. C. Bingham, and P. von R. Schleyer, *J. Am. Chem. Soc.* **1971**, *93*, 3189–3199. (c) E. M. Engler, J. D. Andose, and P. von R. Schleyer, *J. Am. Chem. Soc.* **1973**, *95*, 8005–8025. Also see (d) J. E. Williams, P. J. Stang and P. von R. Schleyer, *Ann. Rev. Phys. Chem.* **1986**, *19*, 531. (e) P. von R. Schleyer, J. E. Williams, and K. R. Blanchard, *J. Am. Chem. Soc.* **1970**, *92*, 2377. (f) P. von R. Schleyer, in *Conformational Analysis*, G. Chiurdoglu, Ed., Academic Press, New York, 1971, pp. 241–249.

28. (a) P. Müller and J. Mareda, *Helv. Chim. Acta* **1985**, *68*, 119. (b) P. Müller, J. Blanc, and J. Mareda, *Helv. Chim. Acta* **1986**, *69*, 635. (c) P. Müller and J. Mareda, *Helv. Chim. Acta* **1987**, *70*, 1017 and references cited therein. Also see (d) T. W. Bentley and K. Roberts, *J. Org. Chem.* **1985**, *50*, 5852.

29. (a) D. L. Fry, C. J. Lancelot, L. K. M. Lam, J. M. Harris, R. C. Bingham, D. J. Raber, R. E. Hall and P. von R. Schleyer, *J. Am. Chem. Soc.* **1970**, *92*, 2538–2540. (b) J. L. Fry, J. M. Harris, R. C. Bingham, and P. von R. Schleyer, *J. Am. Chem. Soc.* **1970**, *92*, 2540–2542. (c) J. L. Fry, L. K. M. Lam, C. J.

Lancelot and P. von R. Schleyer, *J. Am. Chem. Soc.* **1970**, *92*, 2542. (d) S. H. Liggero, J. J. Harper, P. von R. Schleyer, A. P. Krapcho, and D. E. Horn, *J. Am. Chem. Soc.* **1970**, *92*, 3802. (e) J. M. Harris, R. E. Hall, and P. von R. Schleyer, *J. Am. Chem. Soc.* **1971**, *93*, 2551. (f) D. J. Raber, J. M. Harris, R. E. Hall, and P. von R. Schleyer, *J. Am. Chem. Soc.* **1971**, *93*, 4821. (g) P. von R. Schleyer, *Reaction Transition States* J. E. Dubois, Ed., Gordon & Breach, New York, **1972**, Chapter 19, pp. 197–210. (h) D. J. Raber, J. M. Harris, and P. von R. Schleyer *Ions and Ion Pairs in Organic Reactions* Michael Szwarc, Ed., Wiley, New York, Vol. 2, Chapter 3, **1974**, pp. 248–366.

30. (a) T. W. Bentley and P. von R. Schleyer, *J. Am. Chem. Soc.* **1976**, *98*, 7658. (b) F. L. Schadt, T. W. Bentley, and P. von R. Schleyer, *J. Am. Chem. Soc.* **1976**, *98*, 7667. (c) T. W. Bentley and P. von R. Schleyer, *Adv. Phys. Org. Chem.* **1977**, *14*, 1. (d) T. W. Bentley, C. T. Bowen, D. H. Morten, and P. v. R Schleyer, *J. Am. Chem. Soc.* **1981**, *103*, 5466.

31. (a) Z. Majerski, S. H. Liggero, P. von R. Schleyer, and A. P. Wolf, *J. Chem. Soc. Chem.Commun.* **1970**, 1596. (b) J. Majerski, P. von R. Schleyer, and A. P. Wolf, *J. Am. Chem. Soc.* **1970**, *92*, 5731. (c) P. Vogel, M. Saunders, W. Thielecke, and P. von R. Schleyer, *Tetrahedron Lett.* **1971**, *12*, 1429.

32. (a) P. von R. Schleyer, R. C. Fort, Jr., W. E. Watts, M. B. Comisarow, and G. A. Olah, *J. Am. Chem. Soc.* **1964**, *86*, 4195. (b) G. A. Olah, G. K. Surya Prakash, J. G. Shih, V. V. Krishnamurthy, G. D. Mateescu, G. Liang, G. Sipos, V. Buss, T. Gund, and P. von R. Schleyer, *J. Am. Chem. Soc.* **1985**, *107*, 2764. (c) G. A. Olah, G. Liang, P. von R. Schleyer, W. Parker, and C. I. F. Watt, *J. Am. Chem. Soc.* **1977**, *99*, 966. (d) P. von R. Schleyer, D. Lenoir, P. Mison, G. Liang, G. K. Surya Prakash, and G. A. Olah, *J. Am. Chem. Soc.* **1980**, *102*, 683. (b) G. A. Olah, G. K. Surya Prakash, G. Liang, P. von R. Schleyer, and W. D. Graham, *J. Org. Chem.* **1982**, *47*, 1040.

33. (a) P. von R. Schleyer, P. R. Isele, and R. C. Bingham, *J. Org. Chem.* **1968**, *33*, 1239. (b) R. C. Bingham, W. F. Sliwinski, and P. von R. Schleyer, *J. Am. Chem. Soc.* **1970**, *92*, 3471. (c) R. C. Bingham, P. von R. Schleyer, Y. Lambert, and P. Deslongchamps, *Can. J. Chem.* **1970**, *48*, 2729. (d) A. Karim, M. A. McKervey, E. M. Engler, and P. von R. Schleyer, *Tetrahedron Lett.* **1971**, 3987. (e) T. M. Gund, P. von R. Schleyer, G. D. Unruh, and G. J. Gleicher, *J. Org. Chem.* **1974**, *39*, 2995. (f) W. Parker, R. L. Tranter, C. I. F. Watt, L. W. K. Chang, and P. von R. Schleyer, *J. Am. Chem. Soc.* **1974**, *96*, 7121. (g) E. Osawa, E. M. Engler, S. A. Godleski, Y. Inamoto, G. J. Kent, M. Kausch, and P. von R. Schleyer, *J. Org. Chem.* **1980**, *45*, 984.

34. P. von R. Schleyer and R. D. Nicholas, *Tetrahedron Lett.* **1961**, *2*, 305.

35. (a) For the $C_{11}H_{18}$ isomerization graph, see E. Osawa, K. Aigami, N. Takaishi, Y. Inamoto, Y. Fujikura, Z. Majerski, P. von R. Schleyer, E. M. Engler, and M. Farcasiu, *J. Am. Chem. Soc.* **1977**, *99*, 5361. (b) M. Farcasiu, K. R. Blanchard, E. M. Engler, and P. von R. Schleyer, *Chem. Lett.* **1973**, 1189. (c) N. Takaishi, Y. Inamoto, Y. Fujikura, K. Aigami, B. Golicnik, K. Mlinaric-Majerski, Z. Majerski, E. Osawa, and P. von R. Schleyer, *Chem. Lett.*, **1976**, 763.

36. A. Schneider, R. W. Warren, and E. J. Janoski, *J. Org. Chem.* **1966**, *31*, 1617.

37. (a) Review: S. A. Godleski, P. von R. Schleyer, E. Osawa, and W. T. Wipke, *Prog. Phys.Org. Chem.* **1981**, *13*, 63. (b) P. von R. Schleyer, G. J. Gleicher, and C. A. Cupas, *J. Org. Chem.* **1966**, *31*, 2014.

38. M. Nomura, P. von R. Schleyer, and A. A. Arz, *J. Am. Chem. Soc.* **1967**, *89*, 3901.

39. Also see Ref. 46.

40. I. L. Karle and J. Karle, *J. Am. Chem. Soc.* **1965**, *89*, 918.
41. V. Z. Williams, Jr., P. von R. Schleyer, G. J. Gleicher, and L. B. Rodewald, *J. Am. Chem. Soc.* **1966**, *88*, 3862.
42. Diamond hydrocarbons have been classified and named systematically, see A. Balaban and P. von R. Schleyer, *Tetrahedron* **1978**, *34*, 3599.
43. (a) P. von R. Schleyer, E. Osawa, and M. G. B. Drew, *J. Am. Chem. Soc.* **1968**, *90*, 5934. (b) E. Osawa, A. Furusaki, N. Hashiba, T. Matsumoto, V. Sing, Y. Tahara, E. Wiskott, M. Farcasiu, T. Iizuka, N. Tanaka, T. Kan, and P. von R. Schleyer, *J. Org. Chem.* **1980**, *45*, 2985.
44. (a) W. Burns, T. R. B. Mitchell, M. A. McKervey, J. J. Rooney, G. Ferguson, and P. Roberts, *J. Chem. Soc. Chem. Commun.* **1976**, 893. (b) W. Burns, M. A. McKervey, T. R. B. Mitchell, and J.J. Rooney, *J. Am. Chem. Soc.* **1978**, *100*, 906.
45. G. N. Schrauzer, B. N. Bastian, and G. A. Fosselius, *J. Am. Chem. Soc.* **1966**, *88*, 4890.
46. (a) T. M. Gund, V. Z. Williams, Jr., E. Osawa, and P. von R. Schleyer, *Tetrahedron Lett.* **1970**, 3877–3880. (b) T. M. Gund, W. Thielecke, and P. von R. Schleyer, *Org. Synth.* **1973**, *53*, 30–34. (c) T. M. Gund, E. Osawa, V. Z. Williams, Jr., and P. von R. Schleyer, *J. Org. Chem.* **1974**, *39*, 2979–2987. (d) T. Courtney, D. E. Johnson, M. A. McKervey, and J. J. Rooney, *J. Chem. Soc. Perkin Trans. 1* **1972**, 2691. (e) The structure of **34** was established by V. V. Krishnamurthy, J. G. Shih, and G. A. Olah, *J. Org. Chem.* **1985**, *50*, 3005.
47. T. M. Gund, P. von R. Schleyer, P. H. Gund, and W. T. Wipke, *J. Am. Chem. Soc.* **1975**, *97*, 743.
48. D. Farcasiu, H. Bohm, and P. von R. Schleyer, *J. Org. Chem.* **1977**, *42*, 96.
49. D. Farcasiu, E. Wiskott, E. Osawa, W. Thielecke, E. M. Engler, J. Slutsky, P. von R. Schleyer, and G. J. Kent, *J. Am. Chem. Soc.* **1974**, *94*, 4669.
50. E. M. Engler, M. Farcasiu, A. Sevin, J. M. Cense, and P. von R. Schleyer, *J. Am. Chem. Soc.* **1973**, *95*, 5769.
51. P. von R. Schleyer, P. Grubmüller, W. F. Maier, O. Vostrovsky, L. Skattebol, and K. H. Holm, *Tetrahedron Lett.* **1980**, *21*, 921.
52. M. Farcasiu, E. W. Hagaman, E. Wenkert, and P. von R. Schleyer, *Tetrahedron Lett.* **1981**, *22*, 1501.
53. (a) F.J. Jäggi and C. Ganter, *Helv. Chim. Acta* **1980**, *63*, 866. (b) A. M. Klester, F. J. Jäggi, and C. Ganter, *Helv. Chim. Acta* **1980**, *63*, 1294. (c) A. M. Klester and C. Ganter, *Helv. Chim. Acta* **1983**, *66*, 1200. (d) H. R. Känel and C. Ganter, *Helv. Chim. Acta* **1985**, *68*, 1226. (e) M. Brossi and C. Ganter, *Helv. Chim. Acta* **1987**, *70*, 1963. (f) M. Brossi, and C. Ganter, *Helv. Chim. Acta* **1988**, *71*, 848. (g) K. L. Ghatak and C. Ganter, *Helv. Chim.* **1988**, *71*, 124. (h) Also see Y. Tobe, K. Terashima, Y. Sahai, and Y. Odaira, *J. Am. Chem. Soc.* **1981**, *103*, 2307. (i) Z. Majerski, S. Djagas, and V. Vinkovic, *J. Org. Chem.* **1979**, *44*, 4064. (j) J. R. Wiseman, J. J. Vanderbilt, and W. R. Butler, *J. Org. Chem.* **1980**, *45*, 667. (k) L. A. Paquette, C. W. Docke, and G. Klein, *J. Am. Chem. Soc.* **1979**, *101*, 7599.
54. Z. Majerski, S. H. Liggero, P. von R. Schleyer, and A. P. Wolf, *J. Chem. Soc., Chem. Commun.*, **1970**, 1596.
55. D. Lenoir, R. Glaser, P. Mison, and P. von R. Schleyer, *J. Org. Chem.* **1971**, *36*, 1821.
56. (a) C. W. Woodworth, V. Buss, and P. von R. Schleyer, *Chem.Commun.* **1968**, 569. (b) Z. Majerski and P. von R. Schleyer, *Tetrahedron Lett.* **1969**, 6995.

57. (a) E. Osawa, P. von R. Schleyer, L. W. K. Chang, and V. V. Kane, *Tetrahedron Lett.* **1974**, 4189. (b) R. Stoiber and H. Musso, *Angew. Chem.* **1977**, *89*, 430; *Angew. Chem. Int. Ed. Engl.* **1977**, *16*, 415 and earlier references cited. Also see Ref. 5e.

58. (a) J. L. Fry, unpublished results, Princeton University, 1972–1973. (b) N. J. Jones, W. D. Readman, and E. Le Goff, *Tetrahedron Lett.* **1973**, 2087.

59. Z. Wiedenhofer, S. Hala, and S. Landa, *Scientific Papers of the Institute of Chemical Technology, Prague* **1971**, *D22*, 85.

60. (a) P. Grubmüller, W. F. Maier, P. von R. Schleyer, M. A. McKervey, and J. J. Rooney, *Chem. Ber.* **1980**, *113*, 1989. (b) P. Grubmüller, P. von R. Schleyer, and M. A. McKervey, *Tetrahedron Lett.* **1979**, 181.

61. (a) W. F. Maier, P. Grubmüller, I. Thies, P. M. Stein, M. A. McKervey, and P. von R. Schleyer, *Angew. Chem.* **1979**, *91*, 1004; *Angew. Chem. Int. Ed. Engl.* **1979**, *18*, 939. (b) W. F. Maier, W. Roth, I. Thies, and P. von R. Schleyer, *Chem. Ber.* **1982**, *115*, 808. (c) J. G Andrade, W. F. Maier, L. Zapf, and P. von R. Schleyer, *Synthesis* **1980**, 802.

62. (a) W. F. Maier, K. Bergmann, W. Bleicher, and P. von R. Schleyer, *Tetrahedron Lett.* **1981**, *22*, 4227. (b) W. F. Maier, I. Thies, and P. von R. Schleyer, *Z. Naturforsch.* **1982**, *37b*, 392. (c) W. Roth and P. von R. Schleyer, *Z. Naturforsch.* **1983**, *38b*, 1697.

63. P. Grubmüller, doctoral disseration, Erlangen, 1979.

64. C. Brosz, L. Lammertsma, I. Thies, and J. Vieth, unpublished research, Erlangen 1978–1983.

65. W.-D. Fessner, H. Prinzbach, and G. Rihs, *Tetrahedron Lett.* **1983**, *24*, 5857.

66. W. D. Roth, doctoral dissertation, Erlangen, 1984.

67. (a) M. Bremer, A. McEwen, P. S. Gregory, and P. von R. Schleyer, *New J. Chem.*, **1989**, *13*, 767. (b) A. B. McEwen and P. von R. Schleyer, *J. Am. Chem. Soc.* **1986**, *108*, 3951. See W. F. Maier and P. von R. Schleyer, *J. Am. Chem. Soc.* **1981**, *103*, 1891. (c) A. B. McEwen and P. von R. Schleyer, *J. Org. Chem.* **1986**, *51*, 4357.

68. G. Wipff, Ph.D. thesis, Strasbourg, 1970, J. M. Lehn, private communication.

69. R. L. Disch and J. Schulman, *J. Am. Chem. Soc.* **1981**, *103*, 3297.

70. M. Bremer, P. von R. Schleyer, K. Schötz, M. Kausch, and M. Schindler, *Angew. Chem.* **1987**, *99*, 795; *Angew. Chem. Int. Ed. Engl.* **1987**, *26*, 761.

71. W. Kutzelnigg, *Isr. J. Chem.* **1980**, *19*, 173. (b) M. Schindler and W. Kutzelnigg, *J. Chem. Phys.* **1982**, *76*, 1989. (c) M. Schindler, *J. Am. Chem. Soc.* **1987**, *109*, 1020.

72. (a) P. von R. Schleyer, K. B. Wiberg, M. Saunders, and M. Schindler, *J. Am. Chem. Soc.* **1988**, *110*, 300. (b) K. Laidig, M. Saunders, K. B. Wiberg, and P. von R. Schleyer, *J. Am. Chem. Soc.* **1988**, *110*, 7652. (c) M. Bremer and P. von R. Schleyer, *J. Am. Chem. Soc.* **1989**, *111*, 1147. (d) P. von R. Schleyer, J. W. de M. Carneiro, W. Koch, and K. Raghavachari, *J. Am. Chem. Soc.* **1989**, *111*, 5475. (e) P. von R. Schleyer, W. Koch, B. Liu, and U. Fleischer, *J. Chem. Soc. Chem. Commun.* **1989**, 1098.

73. Reviews on pyramidal carbocations: (a) H. Schwarz, *Angew. Chem.* **1981**, *93*, 1046; *Angew. Chem. Int. Ed. Engl.* **1981**, *20*, 991. (b) P. Vogel, *Carbocation Chemistry*, Elsevier, Amsterdam, 1985.

74. E. D. Jemmis and P. von R. Schleyer, *J. Am. Chem. Soc.* **1982**, *104*, 4781.

2 Catalytic Routes to Adamantane and Its Homologues

M. ANTHONY McKERVEY

Department of Chemistry
University College
Cork, Ireland

JOHN J. ROONEY

Department of Chemistry
The Queen's University
Belfast Northern Ireland

Arguably the most rewarding discovery in the chemistry of diamondoid hydrocarbons was that made by Schleyer[1] at Princeton in 1957 when he recognized adamantane among the products of the reaction of *endo*-2,3-trimethylenenorbornane (**1**) with aluminium chloride.

Although the product mixture was complex, and the yield of adamantane low, Schleyer, who had been studying the interconversion of **1** with its exo isomer **2**, quickly perceived the implications of this serendipitous discovery. Not only did the new rearrangement make adamantane accessible in three stages from cyclopentadiene, contrasting with Landa's[2] lengthy process of extracting the compound from vast quantities of Czechoslovakian petroleum (adamantane was first identified as a natural product), or Prelog's[3] multistage rational synthesis, it established a paradigm for much of the future synthetic effort directed towards the production of diamondoid hydrocarbons by rearrangement. The rearrangement itself attracted intrinsic interest, posing a challenge to mechanistic analysis, which led eventually to major developments in the application of computerized molecular mechanics calculations to the structures and energies of polycycloalkanes. It also contributed significantly to our understanding of structure–energy relationships in carbocation rearrangements.

In its original form, Schleyer's synthesis involved heating a mixture of *endo*-2,3-trimethylenenorbornane (**1**) and aluminium chloride to ~ 170°C.[4] Yields of

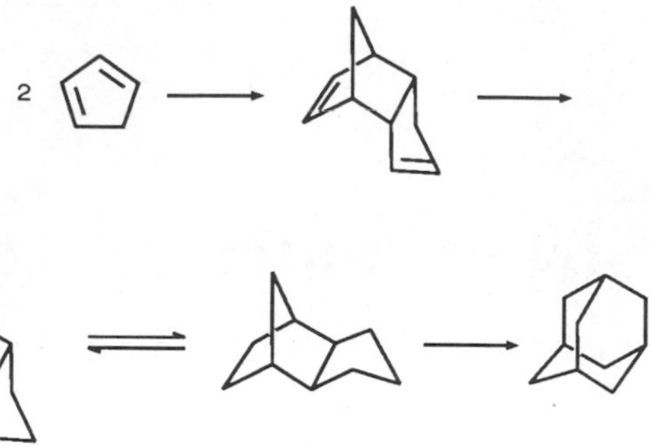

2

1 2

adamantane were poor (15–20%) due largely to the concomitant formation of hundreds of byproducts, which could be detected by gas–liquid chromatographic analysis of the mother liquors of a typical preparation. Not surprisingly, in view of the impact of this discovery on adamantane chemistry, many efforts were made to make the rearrangement more selective and efficient. At that time 2,3-trimethylenenorbornane (the hydrogenated dimer of cyclopentadiene) was (and still is) the most accessible of all the C_{10} tricyclic hydrocarbons. Consequently, efforts for improving the adamantane synthesis concentrated on the choice of Lewis acid catalyst and the operating conditions.

One of the earliest modifications, introduced by the Dupont Company,[5] called for replacement of aluminium chloride by boron fluoride–hydrogen fluoride, an acid combination capable of producing adamantane in 30% yield, with the disadvantage, however, of having to work with a very corrosive system. Yields of up to 40% were obtained with an aluminium chloride–hydrogen chloride system under a high hydrogen pressure.[6] Mixed catalysts of the aluminium halide–hydrogen halide type, often referred to as sludge catalysts, were first introduced for hydrocarbon isomerization in 1929 by Zelinsky and Turova-Pollak,[7] who described the preparation of a catalytically active liquid from aluminium bromide, hydrogen bromide, and *cis*-decalin. It is now generally accepted that the initiators of hydrocarbon rearrangement by such catalysts are species of the type $R^+AlCl_4^-$, generated from the Lewis acid and an alkyl halide formed by addition of the hydrogen halide to traces of alkene present as impurities in the alkane. One of these sludge catalysts, developed by Schneider[8] for rearrangement of various perhydroaromatics into alkyladamantanes, is prepared by mixing dimethylhexanes with aluminium bromide and hydrogen bromide. Another version due to Williams'[9] work consists of a mixture of aluminium bromide and either *t*-butyl or isobutyl bromide. Adamantane has been obtained in up to 60% yield using such catalysts. The sludge catalysts are believed to have prolonged lifetimes and although they are immiscible with alkanes or polycycloalkanes they are mobile liquids that facilitate mixing. Catalysts of the superacid type were also screened for adamantane production by Olah and his co-worker[10] and very substantial improvements were realized. Treatment of *endo*-2,3-trimethylenenorbornane (**1**) with a variety of liquid and solid superacid catalysts caused isomerization to the exo-isomer (2) followed by rearrangement into adamantane in yields of 20–98%, the yield depending on the strength of the acid system, the acid:hydrocarbon mole ratio, and the reaction temperature. The highest yields were observed with the acid systems $CF_3SO_3H + SbF_5$ (1:1) and $CF_3SO_3H + B(OSO_2CF_3)_3$ (1:1).

Gas-phase processes using catalysts on solid supports were also investigated. In the earliest of such studies Plate et al.[11] obtained adamantane in 6–13% yields from (**1**) and an aluminosilicate catalyst at 450-475°C; numerous acyclic, alicyclic and aromatic byproducts were also observed. The gas-phase approach was reexamined in Belfast in the early 1970s using new catalysts that were developed in the petroleum industry for isomerizing *n*-alkanes to their branched chain isomers at low temperatures.[12] The key feature of the catalytic activity of these solid Lewis acids is the introduction of chlorine into platinum–alumina by exposure to carbon tetrachloride or thionyl chloride. Several hydrocarbons known to yield adamantanes with conventional aluminium halide catalysis were exposed in the gas phase to chlorinated platinum–alumina in an atmosphere of hydrogen chloride. The results (Table 1) revealed a very high selectivity for adamantane formation with essentially no byproduct formation.[13] The behavior of *exo*-tetramethylenenorbornane (**3**) provides the most dramatic example: 1- and 2-methyladamantane (**4** and **5**) were the sole products in quantitative yield.

Me
Me
3 **5** **4**

The ratio of **4** to **5** depends on the operating conditions. The 1-Me isomer (**4**) is the thermodynamically more stable of the two[14] and this is reflected in the isomer composition when the gas-phase procedure is conducted at low flow rates of hydrogen chloride. The 1-Me isomer appears to arise almost entirely by cationic isomerization of the 2-Me isomer in a process that is slow relative to the rate of formation of the latter. Interestingly, labeling studies showed that this process does not involve a 1,2-methyl shift, but a skeletal reorganization probably via protoadamantyl intermediates.[15,16] It is possible with the gas-phase procedure to obtain almost pure 2-methyladamantane by conducting the rearrangement at very high flow rates; the primary product is swept from the catalyst surface before isomerization to the bridgehead isomer can occur. This gas-phase procedure has been commercialized by the Idemitsu Kosan Company of Japan using a chlorinated platinum–rhodium–alumina catalyst with an operating lifetime of several months.[17]

A very significant aspect of the rearrangement route to diamondoid hydrocarbons is the degree to which it is capable of extension.

TABLE 1. Gas-Phase Formation of Adamantanes at 165°C

Products %	Precursors (1) or (2)	Precursors (3)	Precursors (6)
Adamantane	60		
1-Methyladamantane (**4**)		98[a] 1[b]	
2-Methyladamantane (**5**)		2[a] 99[b]	
1,3-Dimethyladamantane (**7**)			86
1-Ethyladamantane (**8**)			7
1,X-Dimethyladamantane (**9**)			7
Material recovery	98–100	98–100	98–100
Starting material	39		
Byproducts	1		

[a]Equilibrium distribution at low flow rates of hydrogen chloride.
[b]Nonequilibrium distribution at high flow rates of hydrogen chloride.

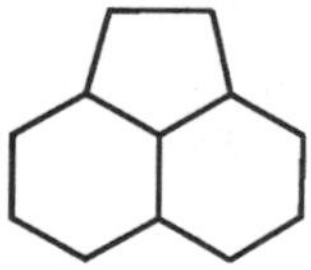

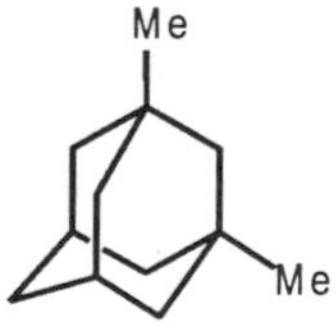

Two examples involving C_{11} and C_{12} precursors are included in Table 1. Although *endo*-trimethylenenorbornane (**1**) is the most accessible of the adamantane precursors, it is not the only C_{10} tricycle whose catalyzed rearrangement to adamantane has been observed. Indeed in mechanistic terms adamantane and **1** are quite far apart on the tricyclodecane interconversion map[18,19] with several intermediates intervening. There are several other known C_{10} tricycloalkanes: twistane (**10**),[18] tricyclo[5.2.1.$0^{4,10}$]decane (**11**),[20] protoadamantane (**12**),[21] *exo*-1,2-trimethylenenorbornane (**13**),[22] and 2,6-trimethylenenorbornane (**14**).[22]

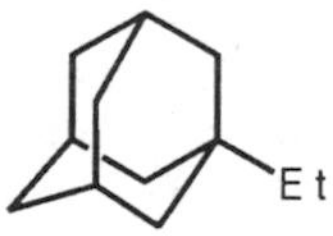

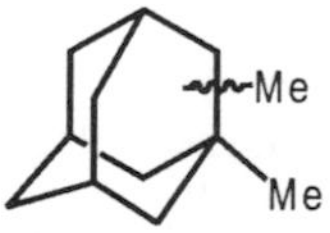

Not only do they all rearrange to adamantane under aluminium halide catalysis, they do so much more readily than either **1** or **2** suggesting that they

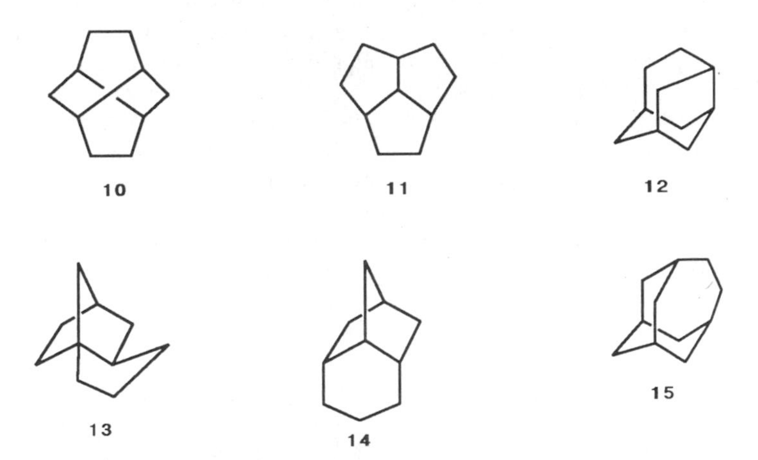

are *nearer* to adamantane on the tricyclodecane interconversion map. Protoadamantane (**12**) is in fact only one bond shift away from adamantane and, as expected, this C_{10} tricycle rearranges with remarkable ease, even in sulfuric acid or liquid bromine.[23] Rearrangement has also been observed in the gas phase on metal (0) surfaces such as palladium, platinum, rhodium, and molybdenum.[24] The mechanism of this and related surface reactions were examined in detail. An olefin metathesis-type process can be precluded by the caged structure of these bridged-ring molecules so that simple 1,2-bond shifts in alkyl *radicals* on transition metal surfaces seem to be a generally valid reaction pathway in heterogeneous interconversions of alkanes.[24,25] Homoadamantane (**15**) is a C_{11} tricycle closely related to adamantane. With Lewis acid catalysis **15** rearranges to 2-methyladamantane (**5**) within minutes.[26] In the gas phase on platinum, on the other hand, homoadamantane forms 1-methyladamantane (**4**) predominantly.[27]

Precursors much less obvious than tricycles **10–14** have also been studied. Cholesterol, nujol, abietic acid, camphene, caryophyllene, and squalene all yield small amounts of alkyladamantanes on treatment with aluminium halides.[28] None of these substances is isomeric with its diamondoid product so that disproportionation processes must necessarily occur prior to or during rearrangement. Ideally, the precursor should be isomeric with its product and it is desirable, though not essential, that it should possess a moderate degree of strain.

The use of the rearrangement route to synthesize diamantane (**16**), the second member of the polymantane series, illustrates the extension of the Schleyer methodology to the more complex pentacyclic system. It also highlights one of its more important limitations, namely, that of finding a suitable hydrocarbon to act as precursor. The choice of precursor is important. Apart from the obvious advantage of ready accessibility, it should be isomeric with the desired diamondoid product, and it should possess a moderate degree of strain to ensure an adequate thermodynamic driving force. Precursors with large components of strain energy, such as that associated with molecules containing three- or four-membered rings are generally unsatisfactory. The strongly acidic conditions required to prod the system towards equilibrium by repeatedly forming and reforming cationic intermediates, and the relatively high temperatures involved, are also the conditions under which relief of strain in the precursor can be realized by fragmentation and disproportionation processes. An additional important consideration is that the rearrangement must be mechanistically feasible, that is to say, the carbocation intermediates must be energetically accessible, and favorable bond alignments must be possible for the 1,2-bond shifts involved.

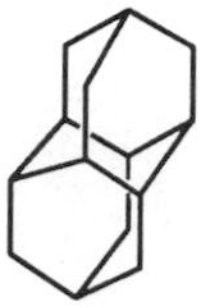

16 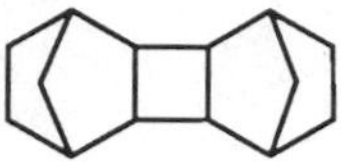 17

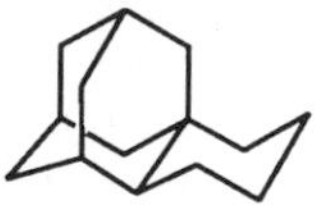

18  19 Et

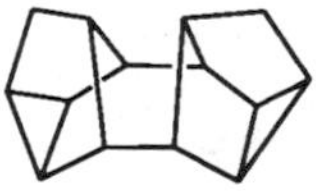

20  21 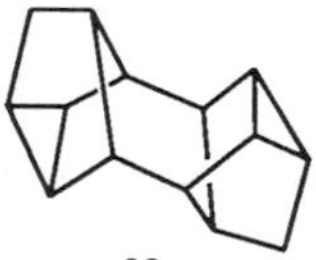 22

23 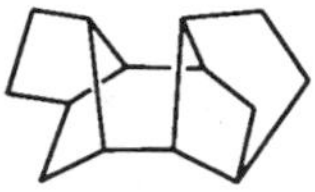 24

The importance of the choice of a precursor is nicely illustrated by the history of catalyzed rearrangements leading to diamantane (**16**). Initially,

Schleyer and his co-workers[29] used the various stereoisomers of the norbornane photodimer **17** and obtained diamantane in 1–10% yield; 1,2-tetramethylene adamantane (**18**), the major product, was the result of incomplete rearrangement and disproportionation. Diamantane was later obtained in 30% yield from precursor (**19**), but this hydrocarbon is not easily prepared.[30] Significantly greater improvements were realized independently at Belfast[31] and Princeton[32] using the norbornadiene dimers **20** and **21** as precursors. These substances, known as binor-S[33] and Katz dimer,[34] are not isomeric with diamantane but their tetrahydro derivatives are. Binor-S (**20**), along with minor amounts of its antistereoisomer binor-A (**22**), is obtained by [4+4] dimerization of norbornadiene catalyzed by cobalt(II) salts. Catalysis by rhodium on carbon, on the other hand, converts norbornadiene into the Katz dimer **21**, which also has a minor stereoisomer (**23**). Hydrogenolysis of binor-S using Adam's catalyst in acetic acid opened both cyclopropane rings producing a single tetrahydro derivative (**24**).

Hydrogenolysis of the cyclopropane ring of the Katz dimer was rather more difficult, though the tetrahydro derivative did give diamantane in good yield. Of the two, binor-S is the preferred precursor. The conversion of tetrahydrobinor-S into diamantane was observed with several catalyst systems: in the gas phase on chlorinated platinum alumina[31] in 70% yield; in solution even hot sulfuric acid is catalytically active, though inefficiently; the procedure of choice simply involves use of aluminium chloride in cyclohexane[32] or dichloromethane[31] (diamantane is quite insoluble in the latter solvent at 0°C) with yields in the range 70–90%. The superacids are the most efficient, boron tristriflate and CF_3SO_3H—SbF_5 or CF_3SO_3H—$B(OSO_2CF_3)_3$ furnishing diamantane in quantitative yield.[35] Rearrangement procedures were also developed that permit the concomitant introduction of functional groups. Thus, treatment of tetrahydrobinor-S with hot concentrated sulfuric acid produces diamantanone (**25**);[31] the combined action of aluminium chloride and acetyl chloride leads to the formation of 1- and 4-chlorodiamantanone (**26** and **27**); and use of chlorosulfonic acid as a catalyst and medium furnishes 4,9-dichlorodiamantane (**28**) as the major diamondoid product.[36]

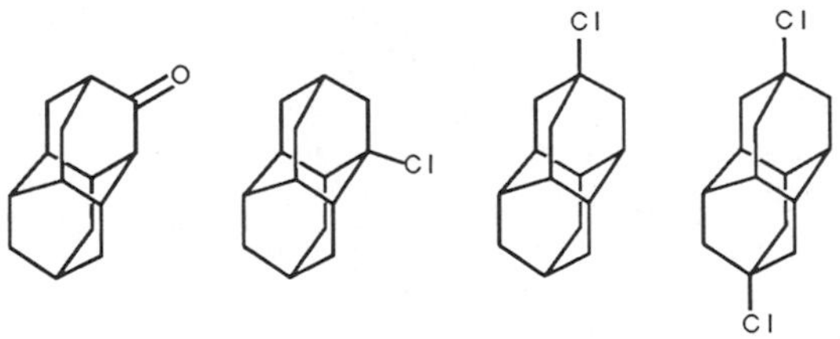

25 26 27 28

In adamantane there are two nonequivalent C—H bonds: bridgehead and nonbridgehead. Diamantane has three such positions: two bridgehead and one nonbridgehead. And just as isomerization, and ultimately equilibration, of 1- and 2-substituted adamantanes is possible (see above for the methyl derivatives),[37] so is a three-way isomerization of 1- , 3- , and 4-substituted

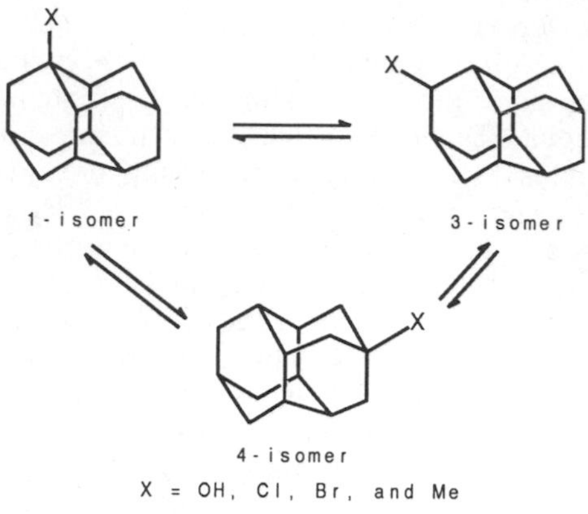

X

1-isomer

3-isomer

4-isomer

X = OH, Cl, Br, and Me

Scheme 1.

diamantanes (Scheme 1). Equilibration has been observed and quantified with methyl- ,[38] chloro- ,[39] bromo- ,[39] and hydroxydiamantanes[40] in solution and in the gas phase for the methyl derivatives[41] over wide ranges of temperature. Lewis or Brønsted catalysis is required and the various diamantyl carbocations are the likely intermediates. The thermodynamics of rearrangement have been interpreted in terms of an enthalpy change favoring the 4-substituted isomer and an entropy change, due to symmetry differences, favoring the 1-isomer; the 3-(secondary) isomer is the least stable of the three. The interconversion of the 1-methyl and 4-methyl isomers provides a very concise method of estimating the enthalpy (2.1 kcal/mol) of switching a methyl group from an axial to an equatorial position in the chair form of methylcyclohexane, a classical problem in conformation analysis.

Despite the obvious mechanistic complexity of the tetrahydrobinor-S → diamantane rearrangement (there are more than 40,000 pentacyclotetradecanes), it is possible to trace energetically and stereoelectronically reasonable pathways from precursor to product. Schleyer and his co-workers,[42] applying the graph treatment introduced by Whitlock and Siefken for the adamantane rearrangement and using enthalpies calculated by molecular mechanics, came to the conclusion that while numerous routes cannot be excluded, that shown in Scheme 2 is a very likely choice: It is energetically favorable and efficient, involving four 1,2-bond shifts, two of which are strongly exothermic and two of which are only mildly endothermic.

The evolution of a rearrangement route to triamantane $C_{18}H_{22}$ (**29**), the third member of the polymantane series, is strikingly similar to the rearrangement route to diamantane, beginning with the discovery of a very inefficient precursor followed by the development of a much better precursor. Clearly, as one progresses towards the larger polymantanes the prospect of finding "off the shelf" precursors becomes increasingly remote. Initially, Schleyer's group[43] used the cyclooctatetraene dimer (**30**) as a source of 16 carbon atoms, bringing it up to the 18 carbon atom level by Simmons–Smith cyclopropanation of both

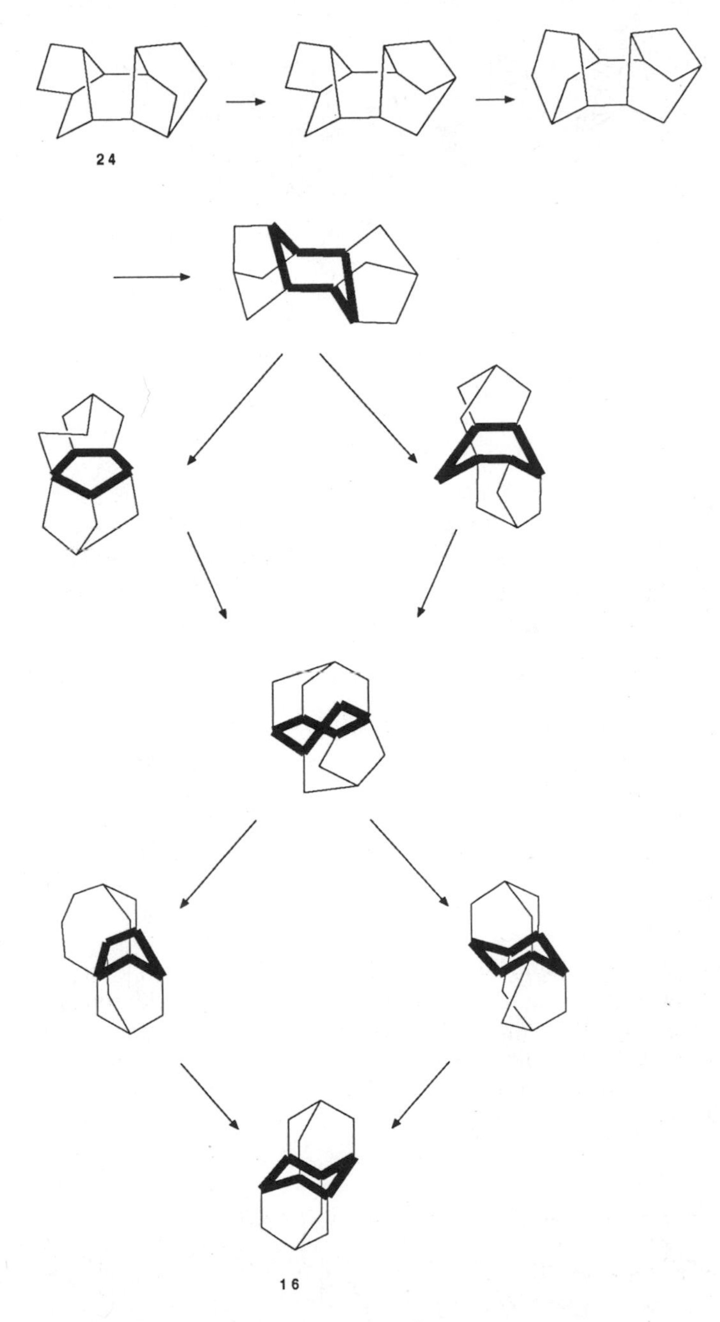

Scheme 2.

Me Me

30 31

29

olefinic bonds and hence to the correct hydrogen content in **31** by catalytic hydrogenolysis of the cyclopropane rings.

Polycycle **31** did indeed give triamantane (**29**) on exposure to aluminium bromide, but the 1% yield did not encourage routine synthesis or exploration of its properties.

In seeking to devise a better route to triamantane we were struck by the remarkable ease with which tetrahydrobinor-S rearranges into diamantane, and in Belfast in the mid-1970s we also considered the possibility of modifying the binor-S structure so as to make it a building block for triamantane.[44] Elaboration of binor-S into a triamantane precursor required the addition of four more carbon atoms with the restriction that since binor-S and triamantane are both heptacycles this change should be brought about without an increase in the total number of rings. The ideal solution would be to rearrange binor-S into a *hexacyclic* alkene, which could then be raised to the C_{18} heptacyclic level by a Diels–Alder reaction with butadiene and hence to the desire $C_{18}H_{22}$ level by hydrogenation.

Binor-S is a highly strained polycycle consisting of two nortricyclyl units conjoined in such a way that the inside edges of the cyclopropane rings are held in close proximity. We discovered that one cyclopropane ring could be opened with concomitant involvement of the other in a transannular process leading ultimately to the formation of about equal amounts of two hexacyclic alkenes (**32** and **33**).[44]

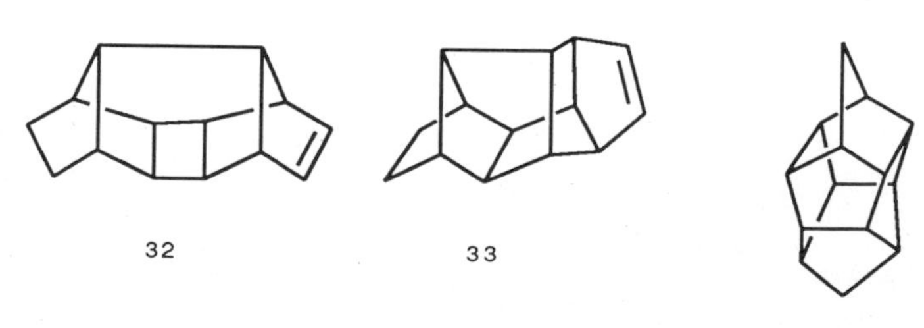

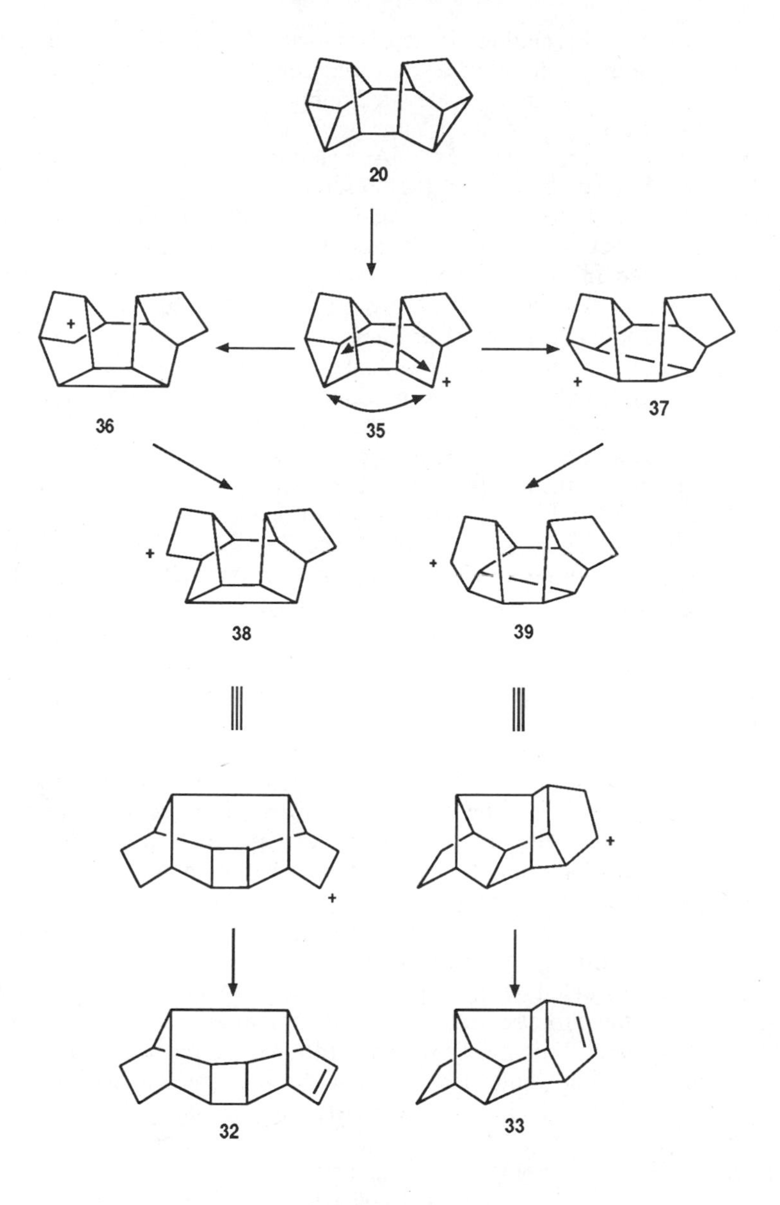

Scheme 3.

This change occurred under a variety of conditions in solution and in the gas phase. Initial studies in the gas phase were conducted under nitrogen on a platinum–silica catalyst, though we found later that acidic silica alone was catalytically active, and a fortiori silica–alumina. The first compound to emerge from the catalyst bed was the heptacycle **34** in trace amounts followed by alkenes **32** and **33**. Formation of heptacyclotetradecane (**34**) in this fashion parallels our previous observation of its formation from the Katz dimer (**21**) in the gas phase on silica–alumina at elevated temperatures.[45] In solution, binor-S behaved in the same way when treated with silver perchlorate in benzene; olefins **32** and **33** were obtained as Ag^+ adducts, which precipitated from solution as they formed. However, the presence of the silver ion turned out to be superfluous vis-a-vis rearrangement since we found later that perchloric acid in benzene also produced the two alkenes though less cleanly. Even acidic silica gel suspended in hot decalin was catalytically active. An interpretation of the mechanism of formation of alkenes **32** and **33**, based on the assumption that it is a proton-initiated process, is summarized in Scheme 3. The presence of Brønsted activity in the activated silica-based catalysts was confirmed by a visual test using triphenylmethanol as indicator. Protonation of one cyclopropane ring of binor-S produces the cation **35**, which is followed by or concerted with participation of the second cyclopropane ring in a transannular process for which there are two regiochemical possibilities. Both are apparently followed leading to cations **36** and **37**, which cannot easily eliminate a proton but which can undergo a 1,2-bond shift to form cations **38** and **39**. Deprotonation of the latter furnishes olefins **32** and **33**. One of the minor products of acid-catalyzed rearrangement of binor-S in solution was identified as its stereoisomer binor-A (**22**). Protonated binor-A is immediately formed from cation **35** by a 1,2-alkyl shift which expands the central C_6 ring. On the other hand, β-fission of the middle C—C bond of the central C_6 ring of the cation (**37**) affords the Katz minor dimer (**23**) with a protonated cyclopropane ring. An isomeric carbocation is the immediate precursor of heptacycle (**34**).

Olefins **32** and **33** proved to be excellent building blocks for triamantane synthesis. Diels–Alder addition of butadiene produced adducts **40** and **41**, which readily afforded dihydro derivatives **42** and **43** upon hydrogenation. Exposure of the **42**/**43** mixture to aluminium chloride in hot cyclohexane for several days produced triamantane (**29**) in 60% yield. Examination of the reaction mixture at intermediate times revealed numerous intermediates, none of which has been identified. If instead of butadiene, the Diels-Alder reaction of **32** and **33** was conducted with isoprene, adducts **44** and **45** were obtained. Hydrogenation of the mixture followed by treatment with aluminium chloride furnished 2-, 3-, and 9-methyltriamantane (**46**–**48**) (60% yield) in the ratio 0.1:1.0:10.0; independent studies later showed that this was the equilibrium ratio, confirming earlier predictions that 9-methyltriamantane (**48**) should be the most stable isomer.[44] Olah and his co-workers applied superacid catalysis to triamantane synthesis from precursors **42** and **43**, obtaining greater than 70% yields with $B(OSO_2CF_3)_3$ in Freon-113 and with $CF_3SO_3H—B(OSO_2CF_3)$ (1:1) neat mixtures.[35]

Collectively, adamantane, diamantane, and triamantane can be viewed as the lower polymantanes: Each exists as a unique form, now readily accessible by the catalyzed rearrangement routes described above. The higher polymantanes begin with tetramantane and, in contrast, exploration of their synthesis by any route posed new problems, since tetramantane is the first member of the series capable of existing in more than one isomeric form. There are in fact three isomers, designated isotetramantane, C_{3v} (**49**), *anti*-tetramantane, C_{2h} (**50**), and

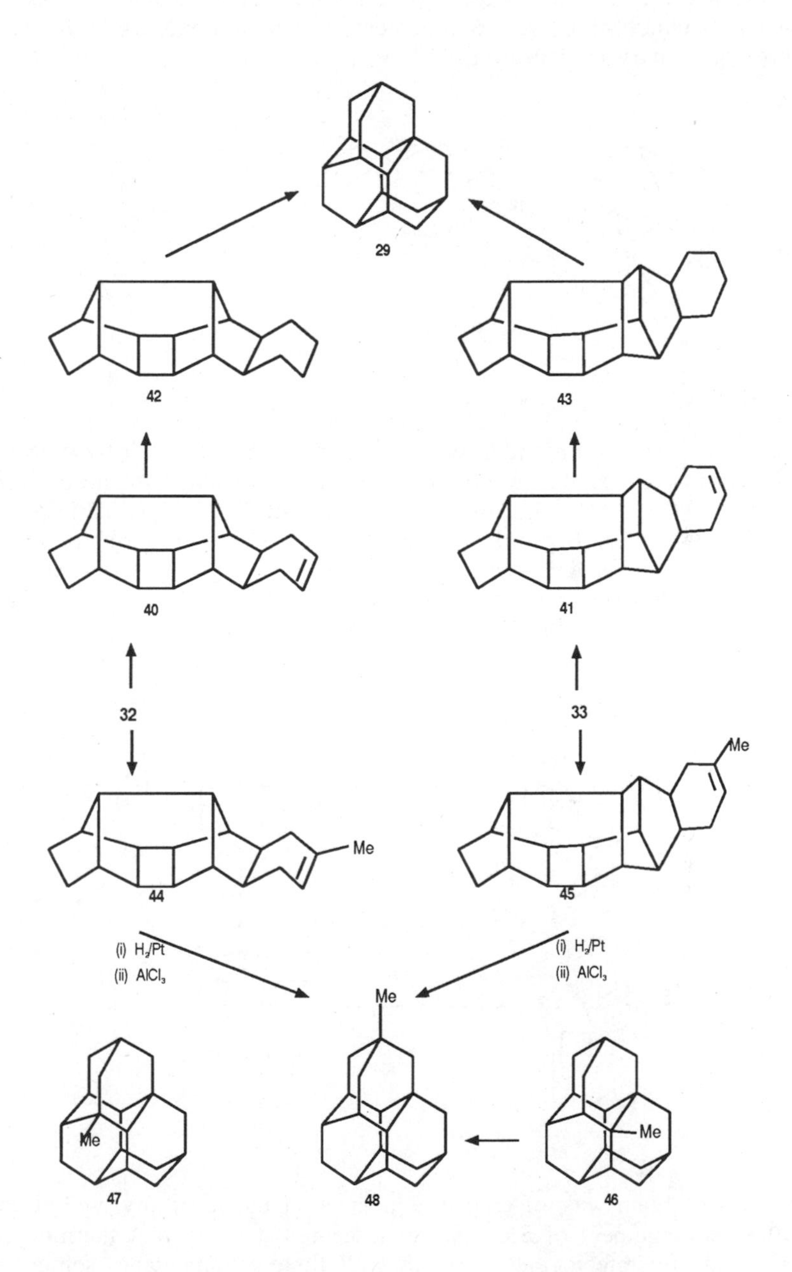
29
42
43
40
41
32
33
44
45
Me
Me
(i) H_2/Pt
(ii) $AlCl_3$
(i) H_2/Pt
(ii) $AlCl_3$
Me
Me
Me
47
48
46

skew-tetramantane, C_2 (**51**), to emphasize their topological relationship with isobutane and the anti and skew conformations of *n*-butane.[46] *Anti*- and *skew*-tetramantane each possess two quaternary carbon atoms while the iso form has three quaternary carbon atoms.

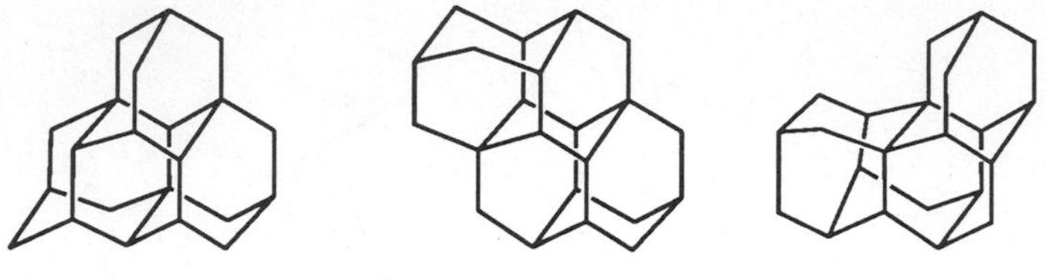

49 50 51

Again Schleyer's group[46] addressed the question as to whether the carbocation rearrangement route could be extended beyond triamantane to these large C_{22} multicyclic systems. The first problem was that of finding a suitable precursor since large polycycles of empirical formula $C_{4n+6}H_{4n+12}$ were virtually unknown. The cyclooctatetraene dimer was again deployed as the central substructure, a single Diels–Alder reaction with 1,3-cyclohexadiene furnishing adduct **52**, hydrogenation of which gave polycycle (**53**) isomeric with tetramantane.

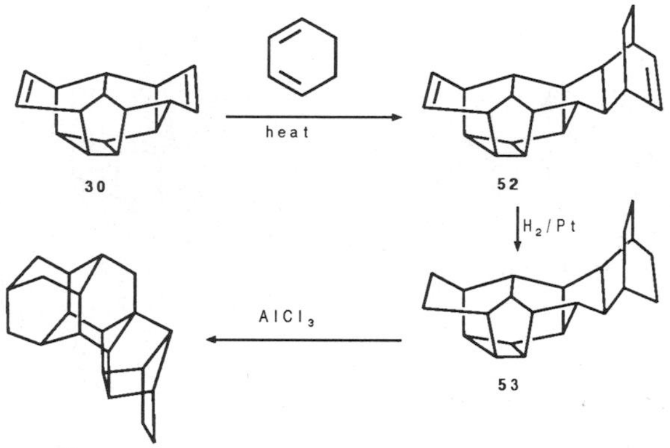

heat

30 52

H_2/Pt

$AlCl_3$

53

54

The next problem was, of course, which tetramantane, if any, would emerge from the rearrangement of **53**? Since it seemed unlikely that thermodynamic control would operate to the extent that all three tetramantane isomers could interconvert readily, the possibility of producing one or other of the isomers unpredictably, or complex mixtures containing all three, could become the limiting factor in the cation rearrangement route. In the event, aluminium chloride treatment of **53** produced none of the tetramantanes, rearrangement stopped short at the diamantane stage and produced a hydrocarbon (**54**) christened bastardane, the name having been chosen more to express its less regular structure than convey the chagrin of the investigators. Since all of the tetramantane isomers are almost certainly more thermodynamically stable than bastardane (whose structure incorporates the highly strained norbornane unit), the reasons for the failure of the rearrangement to go to completion must be kinetic (mechanistic), not thermodynamic. Prolonged exposure of bastardane to aluminium halides at elevated temperatures did not further the rearrangement, rather the molecule was destroyed by fragmentation.

This failure to produce tetramantane by multiple cationic rearrangement, and the accumulating difficulties inherent in the whole approach, led inevitably to a search for alternative approaches for constructing the higher polymantanes. Schleyer's group[47] and ours,[48] working independently, devised two new approaches that share the common proposition that it should be possible to elaborate any member of the polymantane series into the next higher homologue by adding four more carbon atoms and two rings. Both approaches are necessarily multistep and both terminate in a catalyzed rearrangement. The Schleyer synthesis (Scheme 4)[47] involved stepwise attachment of two rings to the adamantane nucleus using two ketocarbenoid C—H insertion reactions followed by Wolff–Kishner reduction. This sequence furnished a mixture of two hydrocarbons **55** and **56**, isomeric with diamantane. Hydrocarbon **55** is a protodiamantane and was expected to rearrange easily to diamantane via a 1,2-bond shift (see above the protoadamantane to adamantane rearrangement). Hydrocarbon **56**, though not a protodiamantane, is closely related to it, and in fact, brief treatment of the two with aluminium chloride caused a quantitative conversion to diamantane.

In Belfast[47] we first investigated the single homologation of diamantane into triamantane using reactions that later formed the basis of a successful double homologation of diamantane into one of the tetramantane isomers. The reactions (Scheme 5) used in the early stages of the synthesis were very similar to those employed by Schleyer's group. 1-Diamantanecarboxylic acid was transformed into cyclopentanone (**57**) using a diazoketone C—H insertion reaction, and was further elaborated into alkene **58** by standard procedures. The next, and final, transformation was brought about by vaporizing **58** in hydrogen at 430°C and passing it over a platinum–silica catalyst whereupon triamantane was produced in 21% yield. This unusual transformation is in fact a composite example in a multicyclic system of two reactions well known in catalytic reforming of petroleum-range hydrocarbons, one of the basic refining processes. Reforming catalysts are usually platinum or platinum alloys on a solid support, and reactions include isomerization of *n*-alkanes to their branched isomers, reversible ring enlargement/contraction of cycloalkanes, and dehydrocyclization of acyclic alkanes, the most important examples being the reversible conversion of methyl- and ethylcyclopentane into cyclohexane and methylcyclohexane, respectively, and the cyclization of *n*-hexane to cyclohexane and methylcyclopentane and of *n*-heptane to methylcyclohexane and ethylcyclopentane.[49]

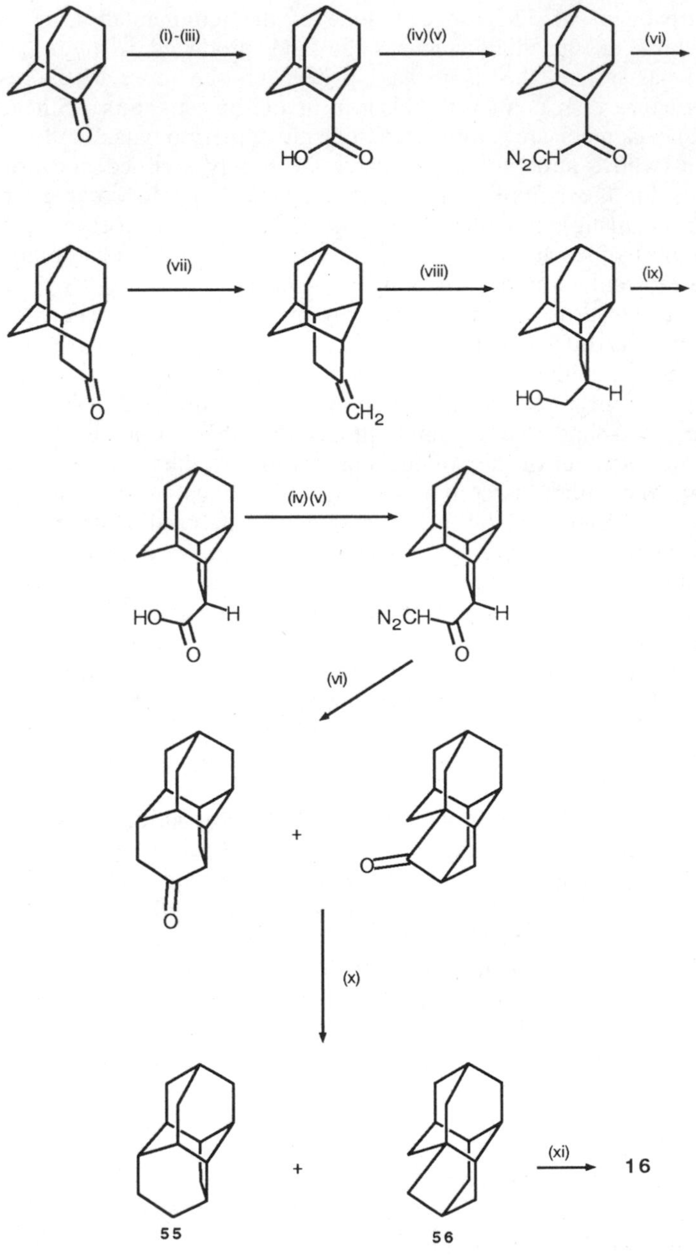

(i) $Me_2SO{=}CH_2$; (ii) $Et_2O\ BF_3$; (iii) CrO_3; (iv) $SOCl_2$; (v) CH_2N_2; (vi) $CuSO_4$/toluene; (vii) $Ph_3P{=}CH_2$; (viii) $B_2H_6/H_2O/NaOH$; (ix) CrO_3; (x) N_2O_4/KOH; (xi) $AlBr_3/CS_2$.

Scheme 4.

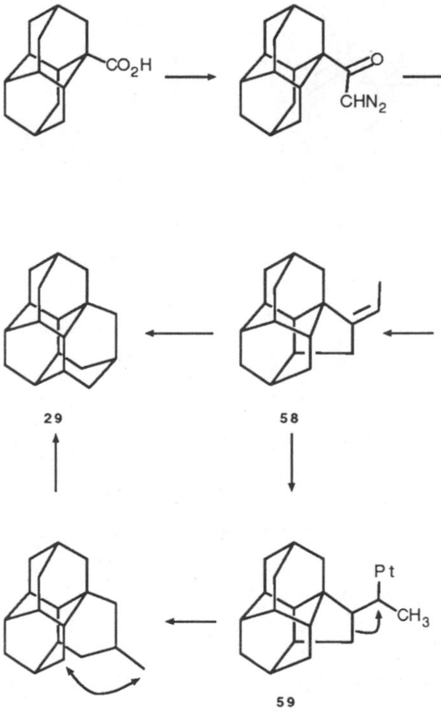

Scheme 5.

Conversion of alkene **58** into triamantane contains the essential elements of these two processes, the overall change involving ring expansion of an ethylcyclopentyl system to a methylcyclohexyl system (which in this example has a strong driving force due to release of ring strain), and a C—C bond formation with loss of hydrogen (dehydrocyclization) between the methyl group and the transannular methylene group indicated in the scheme. The ring expansion reaction may proceed via a 1,2-bond shift to platinum in a surface alkyl radical such as **59**.[25,50,51] A second metal-mediated process removes two hydrogen atoms with concomitant formation of the new C—C bond.

The success of this new route to triamantane from diamantane encouraged us to attempt a synthesis of one of the tetramantane isomers by double homologation. We were confident that the methodology would not be impeded by the sheer size of the substrate hydrocarbon since we had previously observed the clean vapor phase rearrangement of the [2+2] dimers (head–head, and head–tail) of adamantene in hydrogen over palladium and platinum into two new $C_{20}H_{28}$ isomers, even at temperatures as low as 180°C.[52] Although double homologation of diamantane is inherently more problematic than single homologation, we were able at the outset to differentiate between the structural foundations necessary for each of the tetra isomers. If one compares the three isomers with the diamantane precursor (**60**), it can be seen that the C_8 extension needed for

R^2 R^1 R^3

60

R^1 = H
R^2 = R^3 = C_4 unit

Isotetramantane

R^2 = H
R^1 = R^3 = C_4 unit

anti-Tetramantane

R^3 = H
R^2 = R^1 = C_4 unit

skew-Tetramantane

isotetramantane should be anchored at positions R^2 and R^3, at positions R^1 and R^3 for *anti*-tetramantane, and at positions R^1 and R^3 for *skew*-tetramantane.

Our efforts were directed towards the anti isomer. Koch–Haaf carboxylation of diamantane-1,6-diol (**61**) produced the 1,6-diacid, which was transformed into the 1,6-bisdiazoketone (**62**) (Scheme 6). Copper catalyzed decomposition of **62** produced a biscyclopentanone (**63**), which was probably a mixture of the two possible C—H insertion products although it behaved as a single substance. Four more carbon atoms were added to **63** via Grignard addition and the resulting tertiary diols were dehydrated to dienes (**64**). Finally, using the gas-phase procedure and a platinum–silica catalyst at 360°C, dienes (**64**) furnished a mixture of products from which *anti*-tetramantane (**50**) could be isolated in 10% yield by crystallization from acetone.[53] The diamondoid structure was confirmed by X-ray diffraction.[54]

Apart from demonstrating the feasibility of the single and double homologation route to the higher polymantanes, these syntheses show how quite complex reactions with large organic molecules can be brought about in the gas phase. Gas-phase catalytic procedures of the type described above may be applicable to a range of reactions for which liquid-phase methodology is limited, particularly for molecules not possessing functional groups. Prior to our studies these reactions have been limited largely to petroleum range molecules. There are, however, more recent dramatic demonstrations of the power of these catalytic reactions in dodecahedrane synthesis. Paquette and his group[55] used catalytic dehydrocyclization in hydrogen of C_{20} polycyclic precursors (Scheme 7) to complete the final C—C bond formation in dodecahedrane and various methyl derivatives. A double hydrogenolysis of the C_4 ring in pagodane (Scheme 7) followed by a double 1,5-dehydrocyclization of the product on a metal surface is evidently the mechanism of the recent spectacular synthesis of dodecahedrane by Schleyer, Prinsbach, and co-workers.[56] The latter approach again demonstrates the importance of choice of precursor for multiple hydrocarbon rearrangement. These dodecahedrane syntheses are discussed in more detail in Chapters 9 and 10.

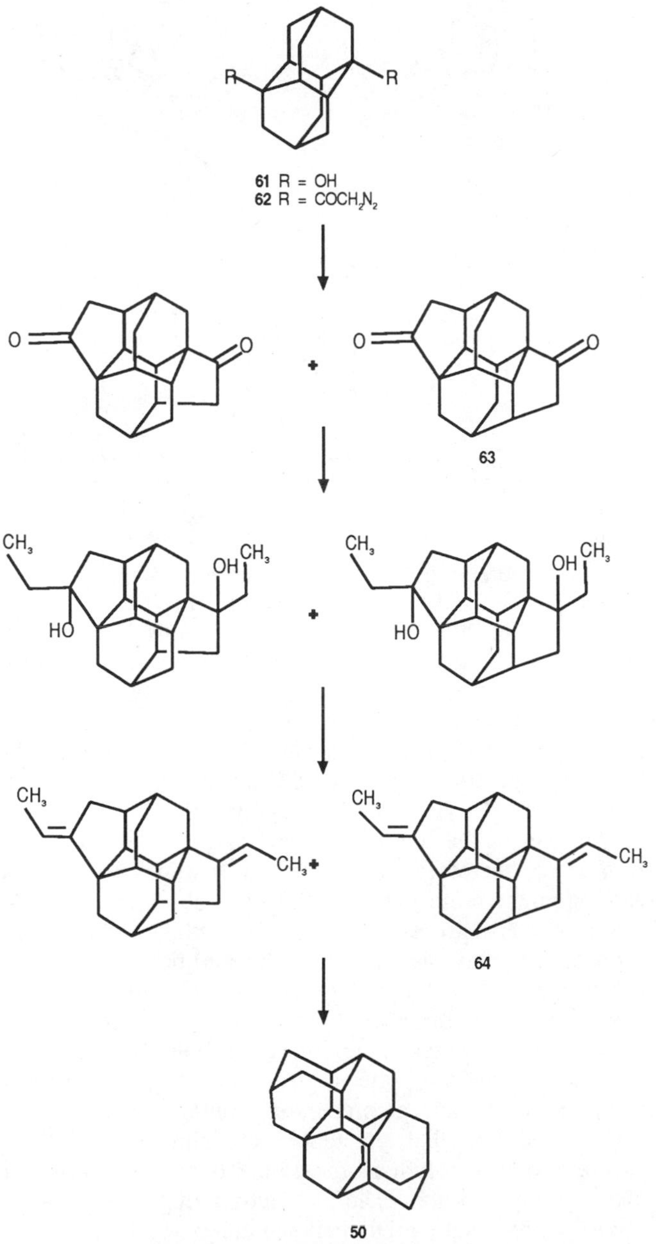
R
R
61 R = OH
62 R = COCH$_2$N$_2$
O
O
O
O
63
CH$_3$
OH
CH$_3$
HO
CH$_3$
OH
CH$_3$
HO
CH$_3$
CH$_3$
CH$_3$
CH$_3$
64
50

Scheme 6.

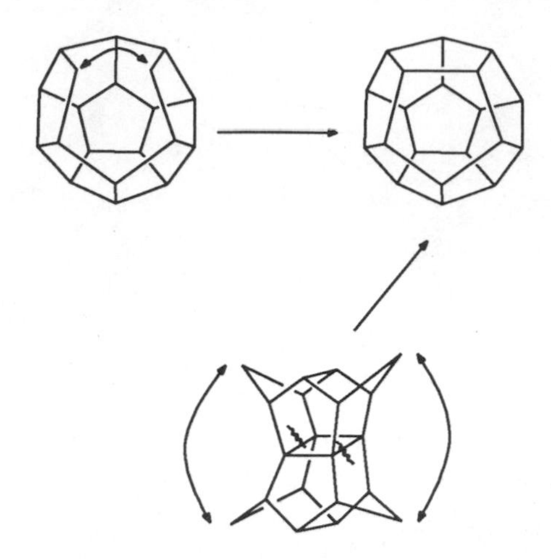

Scheme 7.

To return to the larger polymantanes, triamantane is now readily accessible, and could in principle be used to produce one or more of the pentamantanes by homologation. However, the main limitation of this approach is that it is necessarily multistep. The great attraction of the catalyzed rearrangement route involving carbocation intermediates is that if a suitable precursor can be located easily, the synthesis becomes effectively a one-step operation. Our experiences with diamantane and triamantane confirmed the importance of the choice of precursor, and although triamantane could be obtained from an elaborated cyclooctatetraene dimer, the yield compares very unfavorably with that obtained from precursors based on the binor-S derived alkenes **32** and **33**.

The question arose, therefore, as to whether it might be possible to extend the cation rearrangement route even further by elaborating alkenes **32** and **33** into potential tetramantane precursors, taking advantage, where possible, of the Diels– Alder reaction. This plan required the addition of eight carbon atoms and three more rings since tetramantane is a nonacyclodocosane. Sequential addition of two butadiene moieties to **32** and **33** is inadequate since the resulting adducts, while possessing the correct carbon content, contain one ring too few. However, a three-ring extension can be realized via the Diels–Alder route if one is prepared to accept precursors containing one carbon atom too many. It will be recalled that precursors containing "extra" carbon atoms invariably rearrange to alkylpolymantanes rather than less regular structures not possessing alkyl groups. For example, tetramethylenenorbornane (**3**) and polycycles **44/45** produced methyladamantanes and methyltriamantanes, respectively. Our expectation was, therefore, that a C_{23} precursor might rearrange to one or more of the methyltetramantanes that could then be demethylated catalytically back into the tetramantane manifold.

The precursors were assembled in a straightforward manner: sequential Diels–Alder additions (Scheme 8) of cyclopentadiene and of butadiene to

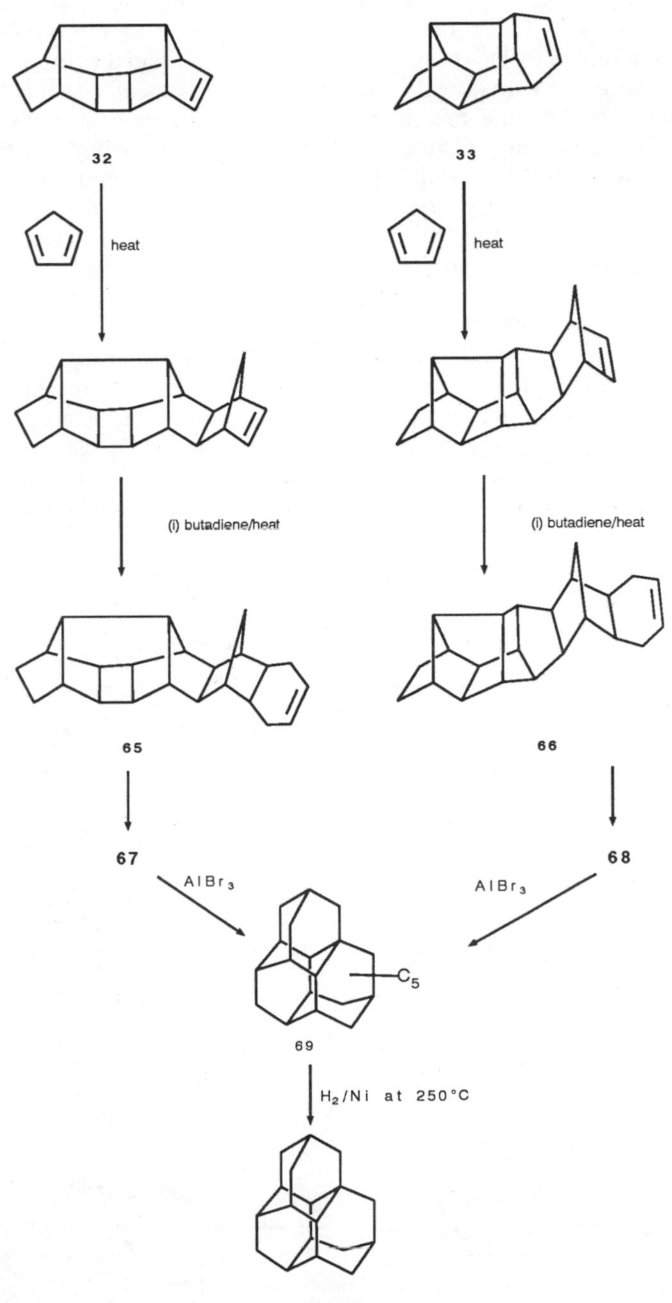
32
33
heat
heat
(i) butadiene/heat
(i) butadiene/heat
65
66
67
68
AlBr3
AlBr3
C5
69
H2/Ni at 250°C

Scheme 8.

alkenes **32** and **33** produced nonacyclic alkenes **65** and **66**, which were then hydrogenated to C_{23} hydrocarbons **67** and **68**. Exposure of the mixture of **67** and **68** to aluminium chloride in cyclohexane or to chlorinated platinum alumina in the gas phase gave a mixture of numerous new hydrocarbons.[57] There are many possible methyltetramantane isomers and the complexity of the rearrangement mixture made isolation of individual components exceedingly difficult. It was decided, therefore, to simplify the mixture by controlled catalytic hydrogenolysis in the gas phase, the intention being to demethylate any methyltetramantanes present to tetramantanes and methane and at the same time fragment any rearrangement products containing strained rings. We already knew that methyl and other alkyl groups could be cleaved from the lower polymantanes with high efficiency and selectivity using gas-phase hydrogenolysis over nickel.[58] Thus methyladamantanes, methyldiamantanes, and methyltriamantanes can all be demethylated in hydrogen at temperatures in the range 150–250°C. The parent polymantanes are much more resistant to catalytic cleavage in hydrogen, although complete degradation does occur at higher temperatures at 280°C, for example, 2-methyltriamantane is successively degraded to triamantane, diamantane, adamantane, and ultimately methane.[58]

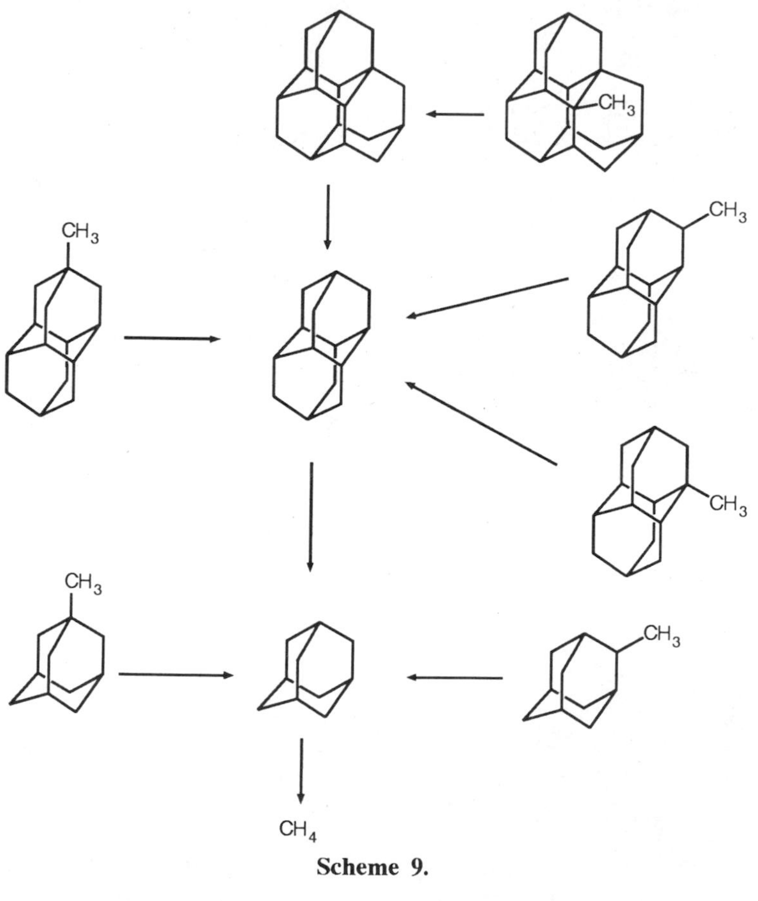

Scheme 9.

When the product mixture obtained by aluminium chloride treatment of **67** and **68** was subjected to gas-phase catalytic hydrogenolysis over nickel the major product detected was *triamantane* (Scheme 8); tetramantanes were not present in detectable amounts. If we make the assumption that nickel promotes hydrogenolytic cleavage only, and not further rearrangement, we may conclude that the aluminium chloride catalyzed rearrangement of **67** and **68** proceeds along the polymantane route as far as triamantane and then encounters a mechanistic block that prevents it from reaching the tetramantane stage. The product, or products, are, therefore, structures of type **69** consisting of a regular triamantane portion to which is attached an irregular C_5 unit. Catalytic hydrogenolysis cleaves off the C_5 unit leaving triamantane intact.

This failure to produce one or more of the tetramantanes from what appeared to be very promising precursors, while not exhausting all the possibilities, does suggest that the carbocation rearrangement route employing Lewis acid catalysts is limited to the lower polymantanes. It is, however, quite possible that new methods of C—C bond formation and reorganization in multicyclic systems involving catalytic procedures in the gas phase will emerge. As we have demonstrated, some catalytic reforming reactions can be used to make an effect in this area. The production of dodecahedrane from pagodane by rearrangement provides encouragement for the future.

ACKNOWLEDGMENTS

We thank our many former co-workers whose names appear in the references cited for their contributions to our research projects in catalyzed reactions of hydrocarbons.

REFERENCES

1. P. von R. Schleyer, *J. Am. Chem. Soc.* **1957**, *79*, 3292; P. von R. Schleyer and M. M. Donaldson, *J. Am. Chem. Soc.* **1960**, *82*, 4645.
2. S. Landa, *Chem. Listy* **1933**, *27*, 415.
3. V. Prelog and R. Seiwerth, *Ber.* **1941**, *74*, 1644.
4. P. von R. Schleyer, M. M. Donaldson, R. D. Nicholas, and C. Cupas, *Org. Synth.* **1963**, *42*, 8.
5. R.E. Ludwig, US Patent 2937211 (*Chem. Abs.* **1960**, *54*, 19540c).
6. H. Koch and J. Franken, *Brennst. Chem.* **1961**, *42*, 90 (*Chem. Abs.* **1961**, *55*, 21059i).

7. N. D. Zelinsky and M. B. Turova-Pollak, *Chem. Ber.* **1929**, *62B*, 1658.
8. A. Schneider, US Patent 3128316 (Chem. Abs. **1964**, *61*, 4244a); A. Schneider, R. W. Warren, and E. J. Janoski, *J. Am. Chem. Soc.* **1964**, *86*, 5365.
9. V. Z. Williams, jun., A.B. thesis, Princeton University, 1965. See also M. J. T. Robinson and H. J. F. Tarratt, *Tetrahedron Lett.* **1968**, 5.
10. G. A. Olah and 0. Farooq, *J. Org. Chem.* **1986**, *51*, 5410.
11. A. F. Plate, Z. K. Nikitina, and T. A. Burtseva, *Neftekhimiya* **1961**, *1*, 599 (*Chem. Abs.* **1962**, *57*, 4938a).
12. J. P. Giannetti, H. G. McIlvreid, and R. T. Sebulsky, *Ind. and Eng. Chem. (Proc. Res. Dev.)*, **1970**, *9*, 473.
13. D. E. Johnston, M. A. McKervey, and J. J. Rooney, *J. Am. Chem. Soc.* **1971**, *93*, 2798.
14. E. M. Engler, K. R. Blanchard, and P. von R. Schleyer, *J. Chem. Soc. Chem. Commun.* **1972**, 1210.
15. Z. Majerski, P. von R. Schleyer, and A. P. Wolf, *J. Am. Chem. Soc.* **1970**, *92*, 5731.
16. Z. Majerski, S. H. Liggero, P. von R. Schleyer, and A. P. Wolf, *J. Chem. Soc. Chem. Commun.* **1970**, 1596.
17. Personal communication from Dr. K. Ito, Idemitsu Kosan Co. Ltd.
18. H. W. Whitlock, Jr., and M. W. Siefken, *J. Am. Chem. Soc.* **1968**, *90*, 4929.
19. For more detailed reviews of mechanistic and thermochemical aspects of polymantane rearrangements see M. A. McKervey, *Chem. Soc. Rev.*, **1974**, *3*, 479 and *Tetrahedron* **1980**, *36*, 971.
20. L. A. Paquette, G. V. Meehan, and S. J. Marshall, *J. Am. Chem. Soc.* **1968**, *90*, 6779.
21. B. R. Vogt, *Tetrahedron Lett.* **1968**, 1575; J. E. Baldwin and W. D. Fogelsong, *J. Am. Chem. Soc.* **1968**, *90*, 4303.
22. E. M. Engler, M. Farcasiu, A. Sevin, J. M. Cense, and P. von R. Schleyer, *J. Am. Chem. Soc.* **1973**, *45*, 5769.
23. D. Lenoir and P. von R. Schleyer, *J. Chem. Soc. Chem. Commun.* **1970**, 941; D. Lenoir, R. Glaser, P. Mison, and P. von R. Schleyer, *J. Org. Chem.* **1971**, *36*, 1821; B. D. Cuddy, D. Grant, and M. A. McKervey, *J. Chem. Soc. (C)* **1971**, 3173; A. Karim and M. A. McKervey, *J. Chem. Soc. (C)*, **1974**, 2475.
24. H. A. Quinn, J. H. V. Graham, M. A. McKervey, and J. J. Rooney, *J. Catal.* **1972**, *26*, 333; M. A. McKervey, J. J. Rooney, and N. G. Samman, *J. Catal.* **1973**, 330.
25. V. Amir-Ebrahimi and J. J. Rooney, *J. Chem. Soc. Chem. Commun.* **1988**, 260.
26. Z. Majerski and K. Mlinaric, *J. Chem. Soc. Chem. Commun.* **1972**, 1030.
27. J. J. Rooney, R. W. Dale, and M. A. McKervey, unpublished results.
28. M. Nomura, P. von R. Schleyer, and A. A. Arz, *J. Am. Chem. Soc.* **1967**, *89*, 3657.
29. C. A. Cupas, V. Z. Williams, Jr., P. von R. Schleyer, and D. J. Trecker, *J. Am. Chem. Soc.* **1965**, *87*, 617.
30. T. M. Gund, V. Z. Williams, Jr., E. Osawa, and P. von R. Schleyer, *Tetrahedron Lett.* **1970**, 3877.

31. D. Faulkner, R. A. Glendinning, D. E. Johnston, and M. A. McKervey, *Tetrahedron Lett.* **1971**, 1671; T. Courtney, D. E. Johnston, M. A. McKervey, and J. J. Rooney, *J. Chem. Soc. Perkin Trans. 1*, **1972**, 2691.

32. T. M. Gund, E. Osawa, V. Z. Williams, and P. von R. Schleyer, *J. Org. Chem.* **1974**, *39*, 2979.

33. G. N. Schrauser, B. N. Bastian, and G. A. Fosselius, *J. Am. Chem. Soc.* **1966**, *88*, 4890.

34. N. Acton, R. J. Roth, T. J. Katz, J. K. Frank, C. A. Maier, and I. C. Paul, *J. Am. Chem. Soc.*, **1972**, *94*, 5446.

35. O. Farooq, S. M. F. Farnia, M. Stephenson, and G. Olah, *J. Org. Chem.* **1988**, *53*, 2840.

36. F. Blaney, D. E. Johnston, M. A. McKervey, and J. J. Rooney, *Tetrahedron Lett.* **1975**, 99.

37. For an analysis of the thermodynamics of this isomerization in terms of enthalpy and entropy contributions see Ref. 14.

38. R. Hamilton, D. E. Johnston, M. A. McKervey, and J. J. Rooney, unpublished observations.

39. M. A. McKervey, D. E. Johnston, and J. J. Rooney, *Tetrahedron Lett.* **1972**, 1547.

40. D. E. Johnston, M. A. McKervey, and J. J. Rooney, *J. Chem. Soc. Chem. Commun.* **1972**, 29.

41. R. Hamilton, D. E. Johnston, M. A. McKervey, and J. J. Rooney, *J. Chem. Soc. Chem. Commun.* **1972**, 1209.

42. T. M. Gund, P. von R. Schleyer, P. H. Gund, and W. T. Wipke, *J. Am. Chem. Soc.* **1975**, *97*, 743.

43. V. Z. Williams, Jr., P. von R. Schleyer, G. J. Gleicher, and L. B. Rodewald, *J. Am. Chem. Soc.* **1966**, *88*, 3862.

44. R. Hamilton, M. A. McKervey, J. J. Rooney, and J. F. Malone, *J. Chem. Soc. Chem. Commun.* **1976**, 1027; R. Hamilton, F. S. Hollowood, M. A. McKervey, and J. J. Rooney, *J. Org. Chem.* **1980**, *45*, 4954.

45. M. A. McKervey, J. J. Rooney, and N.G. Samman, *J. Chem. Soc. Chem. Commun.* **1972**, 1185.

46. P. von R. Schleyer, E. Osawa, and M. G. B. Drew, *J. Am. Chem. Soc.* **1968**, *90*, 5034.

47. D. Farcasiu, H. Bohm, and P. von R. Schleyer, *J. Org. Chem.* **1977**, *42*, 96.

48. W. Burns, M. A. McKervey, and J. J. Rooney, *J. Chem. Soc. Chem. Commun.* **1975**, 965.

49. For a general discussion of the chemistry involved in catalytic reforming see B. C. Gates, J. R. Katzer, and G. C. A. Schuit, *Chemistry of Catalytic Processes*, McGraw-Hill, New York, 1979, and P. Wiseman, *Introduction to Industrial Organic Chemistry*, Applied Science, 1972.

50. The mechanisms proposed for skeletal isomerization and cyclization of alkanes have been reviewed by J. K. A. Clarke, *Chem. Rev.* **1975**, *75*, 291; J. K. A. Clarke and J. J. Rooney, *Adv. Catal.* **1976**, *25*, 125.

51. For recent examples of ring expansion/contraction in some related bridged ring hydrocarbons in the gas phase on noble metals see H. A. Quinn, J. H. Graham, J. J. Rooney, and M. A. McKervey, *J. Catal.* **1972**, *26*, 333; M. A. McKervey,

N. G. Samman, and J. J. Rooney, *J. Catal.* **1973**, *30*, 330; D. Grant, M. A. McKervey, J. J. Rooney, N. G. Samman, and G. Step, *J. Chem. Soc. Chem. Commun.* **1972**, 1186.

52. W. Burns, M. A. McKervey, J. J. Rooney, N. G. Samman, J. Collins, P. von R. Schleyer, and E. Osawa, *J. Chem. Soc. Chem. Commun.* **1977**, 95.
53. W. Burns, T. R. B. Mitchell, M. A. McKervey, J. J. Rooney, G. Ferguson, and P. Roberts, *J. Chem. Soc. Chem. Commun.* **1976**, 893; W. Burns, M. A. McKervey, T. R. B. Mitchell, and J. J. Rooney, *J. Am. Chem. Soc.* **1978**, *100*, 906.
54. P.J. Roberts and G. Ferguson, *Acta Crystallogr.* **1977**, 33B, 2335.
55. L. A. Paquette, R. J. Ternansky, D. W. Balogh, and W. J. Taylor, *J. Am. Chem. Soc.* **1983**, *105*, 5441; L. A. Paquette, R. J. Ternansky, D. W. Balogh, and G. Kentgen, *J. Am. Chem. Soc.*, **1983**, *105*, 5446.
56. W. D. Fessner, B. A. R. C. Murthy, J. Worth, D. Hunkler, H. Fritz, H. Prinzbach, W. D. Roth, P. von R. Schleyer, A. B. McEwen, and W. F. Maier, *Angew. Chem. Int. Ed. Engl.* **1987**, *26*, 452.
57. Unpublished work of F. S. Hollowood, M. A. McKervey, and J. J. Rooney.
58. P. Grubmuller, P. von R. Schleyer, and M. A. McKervey, *Tetrahedron Lett.* **1979**, 181; P. Grubmuller, W. F. Maier, P. von R. Schleyer, M. A. McKervey, and J. J. Rooney, *Chem. Ber.*, **1980**, *113*, 1989.

3 The Superacid Route to 1-Adamantyl Cation

TED S. SORENSEN and STEVEN M. WHITWORTH

Department of Chemistry
University of Calgary
Calgary, Alberta, Canada

INTRODUCTION

In 1957, Schleyer[1] observed that tricyclo[5.2.1.0^{2,6}]decane (**1**, tetrahydrodicyclopentadiene) could be converted to tricyclo[3.3.1.1^{3,7}]decane (**2**, adamantane) using strong Lewis acids (Scheme 1), and more recently, Olah and co-worker[2] showed that this reaction proceeds much more efficiently in the presence of strong acids such as a 1:1 mixture of trifluoromethanesulfonic acid and antimony pentafluoride. That this conversion involved carbocation intermediates has never been in doubt, and in 1964, Schleyer, Olah, and co-workers[3] reported that the observable tetrahydrodicyclopentadiene (THDPD) cation (**3**) was indeed converted, in good yields, to the 1-adamantyl cation (**4**) (Scheme 2). This chapter is concerned with the latter reaction and related examples where other $C_{10}H_{15}^{+}$ cations were also shown to give the 1-adamantyl

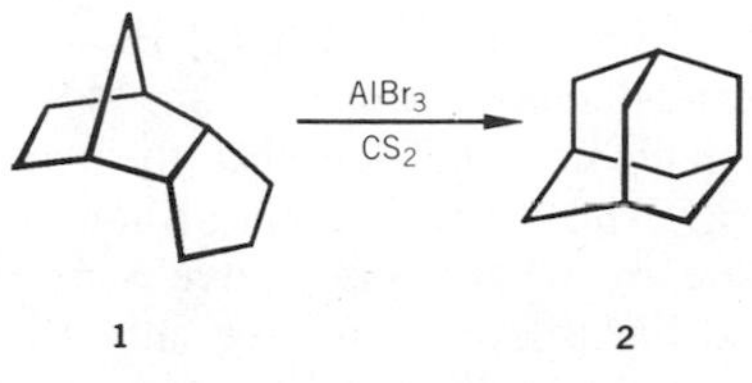

$AlBr_3$ / CS_2

1 → 2

Scheme 1.

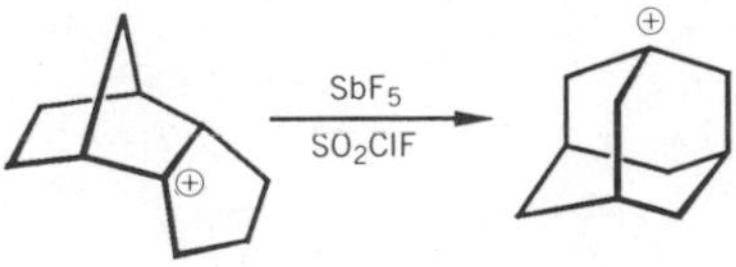

$\xrightarrow[SO_2ClF]{SbF_5}$

3 4

5

Scheme 2.

cation under stable ion conditions. The important differences between this process and the Lewis acid catalyzed isomerization of neutral $C_{10}H_{16}$ hydrocarbons will be expounded later. Since the 1-adamantyl cation can be efficiently trapped with a variety of nucleophiles, the end product is a functionalized 1-adamantyl derivative, for example, 1-adamantanol from aqueous quenching. Both methods provide a direct route to the adamantane framework from readily available starting materials. Previously, the adamantane framework had only been prepared by either inefficient syntheses,[4] or by its isolation from crude hydrocarbon fractions.[5] The rearrangement is quite general in that the higher homologs of THDPD rearrange under similar conditions to give the appropriate methyl substituted adamantanes.[6]

Before discussing the isomerization under stable ion conditions, it is appropriate at this point to describe briefly the Lewis acid catalyzed isomerization process. This is schematically shown in Figure 1, which depicts the isomerization of several different carbon frameworks (1, 2, 3, . . .). At any given time, most of the organic material is present as hydrocarbon (HC1, HC2, . . .), but hydride transfers are frequent and rapid, so that there may be several carbocations with a common carbon framework (1a, 1b, . . .). The obvious way to progress to a new hydrocarbon framework (e.g., 1d—2a, Fig. 1) is via a carbocation to carbocation skeletal rearrangement (e.g., a Wagner—Meerwein shift), with subsequent rearrangements requiring that the carbocation center "move" within a given framework. The dynamic equilibrium between cation and hydrocarbon provides one route by which this "move" can occur, and is illustrated in Figure 1 by the dashed lines. An alternative, albeit more selective, route can be achieved by intramolecular hydride shifts, and these are illustrated as thin solid lines in Figure 1. It is experimentally difficult in practice to distinguish

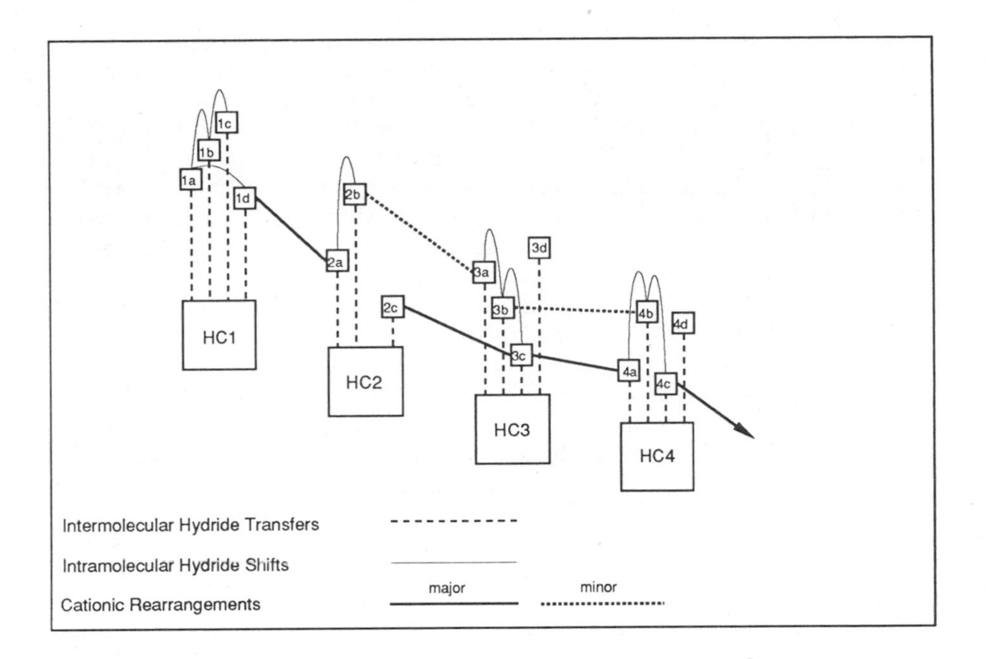

Fig. 1. Schematic representation of the Lewis acid catalyzed isomerization.

between these two possibilities, but some elegant labeling studies have been carried out to probe this aspect of the reaction.[7]

In the Lewis acid catalyzed isomerization of *endo*-tetrahydrodicyclopentadiene (*endo*-THDPD), it is important to note that only the geometric isomer **5** (*exo*-THDPD) and adamantane[8] are observed, that is, the concentrations of the intermediate hydrocarbons are extremely small, and the Figure 1 schematic is misleading in this respect. A number of other hydrocarbons display similar reactivity, rearranging directly to adamantane, with no observable intermediates.[9] There are, however, a number of $C_{10}H_{16}$ tricyclic hydrocarbons that exhibit a more complex behavior on Lewis acid treatment,[10] in the sense that many detectable intermediates are present during the isomerization. There are of course an enormous number of theoretically possible tricyclic $C_{10}H_{16}$ isomers, but in looking at the isomerization pathway for THDPD, one can fairly confidently exclude from serious consideration as intermediates, those structures with three- and four-membered rings, inverted bridgeheads, and so on, because of the high ring strain involved (thermodynamics). The resultant set of 19 tricyclic $C_{10}H_{16}$ hydrocarbons have been described as residing in "adamantaneland,"[9a] and their structures are shown in Figure 2. All of the 19 adamantaneland hydrocarbons have now been synthesized, a number of them in the last 10 years. Many of these compounds have been shown to isomerize to adamantane under Lewis acid catalysis, and this aspect of the rearrangement is discussed elsewhere in this volume.

As already indicated, the hydrocarbon rearrangement proceeds through a series of cationic intermediates, and graph theory[11] approaches can be used to enumerate the routes from any $C_{10}H_{16}$ hydrocarbon to any other. This does not, however, provide much insight into the actual energetics of the process. In 1960, Schleyer and Donaldson[8] suggested that the THDPD isomerization to adamantane might proceed by a very direct route involving only three steps, but one of these steps was equivalent to a 1,3-alkyl shift in a carbocation system, a rearrangement for which there are scant precedents. Later, Whitlock and Siefkin,[9a] in a notable paper, discussed the problem further and showed that one can go from THDPD to adamantane in a series of five steps, all involving only traditional Wagner—Meerwein shifts as well as processes that move the positive center within any given structure. The Whitlock—Siefkin route is accepted today as the major pathway from THDPD to adamantane, although it is recognized that myriad minor routes will inevitably be present (simply from kinetic–thermodynamic considerations, if nothing else). All of the hydrocarbon intermediates involved in the Whitlock–Siefkin scheme have been separately isomerized under Lewis acid conditions and the results generally are not inconsistent with the reaction scheme. Some of this work involves some rather elegant labeling experiments, for example, with specific ^{13}C enrichment.[12]

For the Lewis acid catalyzed process, the driving force for the reaction is the relief of strain that accompanies the formation of adamantane from its more strained tricyclic $C_{10}H_{16}$ precursors. That adamantane is the "bottomless pit"[9a] into which everything falls is easily understandable. Adamantane consists entirely of cyclohexyl "chairs," fused so that the torsional interactions are minimal, and a tetrahedral geometry can be adopted. This is not to say that the structure is "strain-free",[13] since the resultant tetrahedral geometry is not necessarily the most stable conformation of an unequally substituted carbon atom. The 1-adamantyl cation on the other hand, while having the same skeletal structure, has the disadvantage (energywise) that it has a tertiary cationic center that cannot achieve planarity. That stable ion solutions of a variety of precursors eventually result in the formation of the 1-adamantyl cation demonstrates that the unfavorable nonplanarity of the cation is more than compensated by the relief in skeletal strain arising from the formation of the adamantyl framework. Additional features affecting the stability of the 1-adamantyl cation will be discussed later.

The mediation of the Lewis acid catalyzed and stable ion processes by the respective hydrocarbon and cation stabilities is one way in which the two processes differ. Another difference arises from the absence of any hydrocarbon material in the cationic rearrangement, thus preventing intermolecular hydride transfers from scrambling the positive charge among the different positions, a process that is often required to allow access to the next framework. If there is no corresponding intramolecular process to migrate the positive charge to the position required for the skeletal rearrangement, then that particular rearrangement route may not be operative in the stable ion process, even though it is in the Lewis acid catalyzed process. Several examples where this is the case are discussed later.

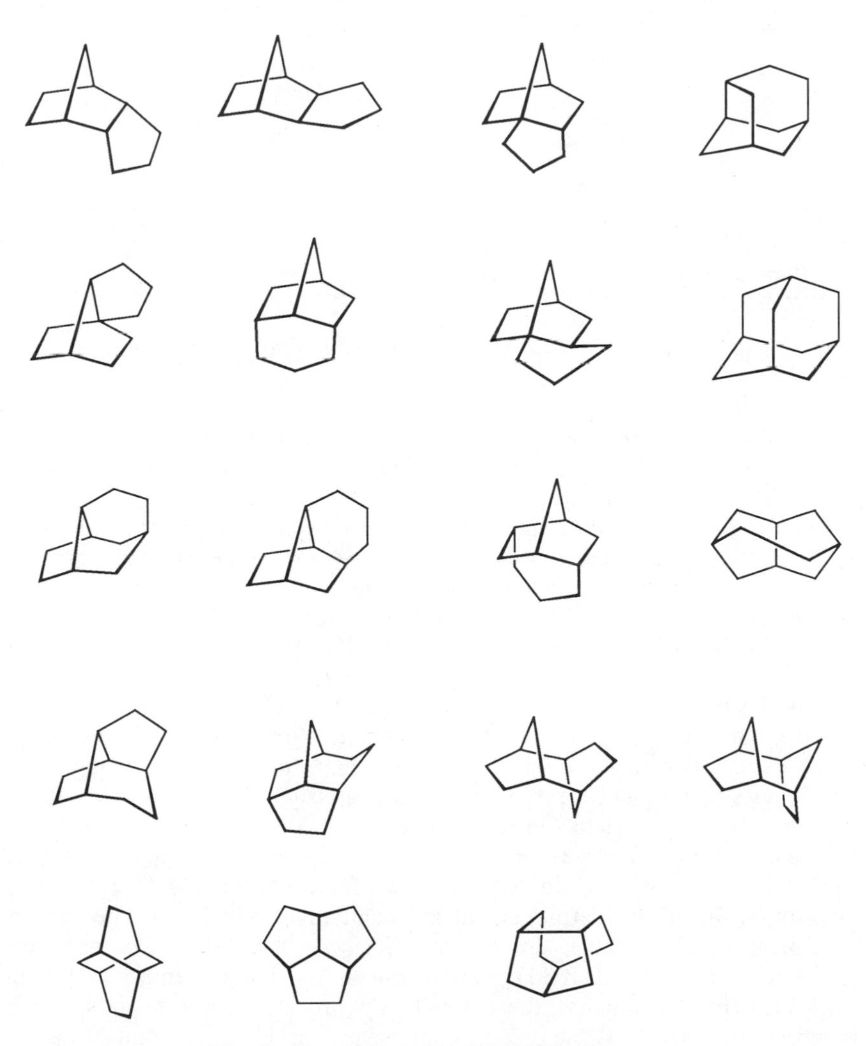

Fig. 2. The 19 "adamantaneland" structures.

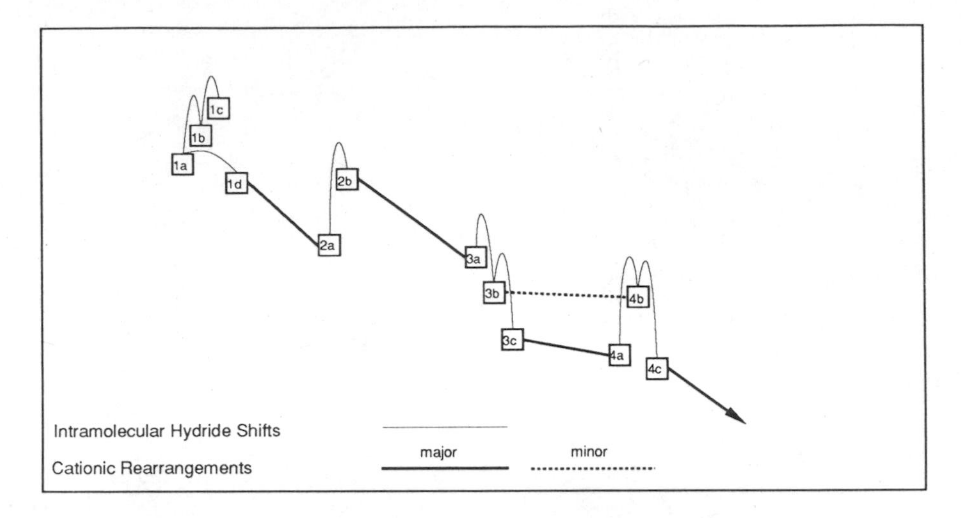

Fig. 3. Schematic representation of the stable ion isomerization.

The cationic rearrangement is illustrated schematically in Figure 3, in which the intermolecular hydride transfers have been removed from Figure 1, and as can be seen, an overall simplification is observed. Despite the differences, the rearrangement of the THDPD cation to the 1-adamantyl cation is expected to follow the same general pathway as that proposed for the Lewis acid catalyzed isomerization, that is, involving the same skeletal intermediates, although their interconversion may occur by slightly different routes. Although the major features of the reaction pathway have been delineated for the Lewis acid catalyzed process,[9a] the problem is far from closed. Even though we probably know how the transformation occurs in a strict structural sense, we have very little feeling for what the individual transition state barriers are for many of the reactions steps. The paper by Whitlock–Siefkin,[9a] for example, contains the correct analysis of the "how" problem, but is seriously flawed in the discussion of the rearrangement barriers (discussed later).

One can of course measure the kinetics of various Lewis acid catalyzed hydrocarbon isomerizations. In reality, most of the literature data concerning the rates of hydrocarbon rearrangement are qualitative in nature. The ease of rearrangement of different hydrocarbons (i.e., k_{obs} and ΔG_{obs}) is generally compared to that of the THDPD rearrangement, so that the more facile the rearrangement (to adamantane), the closer that hydrocarbon is to adamantane in the reaction sequence. These barriers, however, rarely correspond to single steps and are not necessarily involved in the major reaction pathway in going from **1** to **2**. Even worse, a meaningful estimate of the barriers involved in carbocation to carbocation rearrangements would require an estimate of either the reactive cation concentrations or the intermolecular hydride shift barriers. These are virtually impossible to obtain. This is one of the major advantages of working with fully formed cations, since any rearrangement observed under stable ion conditions will correspond to a cation to cation rearrangement. It is also possible to gain some insight into the barriers of intramolecular hydride shifts, since the removal of the hydrocarbon component from the reaction formally eliminates the intermolecular hydride transfer pathway from consideration, although it will be shown that some intermolecular hydride transfers occur in the latter stages of the rearrangement under stable ion conditions. It is also very probable, as discussed later, that the stable ion route is less branched (i.e., "contaminated" with minor pathways) because intramolecular hydride shifts are more selective since they are governed by dihedral angle and steric (distance) considerations.

In the process of probing the barriers involved in the cation to cation rearrangements, one also has the opportunity to study the chemistry of any intermediate cations that may be observed under stable ion conditions. While this may not be of direct relevance to the rearrangement under investigation, we will cover some of the interesting chemistry of these individual structures or of any notable degenerate rearrangements that may occur.

GENERAL OVERVIEW

The disadvantage in studying fully formed $C_{10}H_{15}^+$ cations is apparent when one acknowledges that observable carbocations have been prepared from only 3 of the 19 possible skeletons shown in Figure 2. In addition, one can infer the formation of two further carbocations as minor equilibrium partners to two of the above. Fortunately, all five of these cations relate directly to the most probable path by which THDPD is transformed to adamantane.

In this chapter we endeavor to discuss all of the relevant literature dealing with the solvolytic and/or acid catalysis generation of relevant carbocations. However, the Lewis acid catalyzed rearrangement of various hydrocarbons will not be discussed in any great detail because these studies do not allow one to generate a site-specific carbocation.

There is also a small amount of literature relating to the attempted superacid generation of some $C_{10}H_{15}^+$ cations, which are thought to be on the direct route from the THDPD cation to the adamantyl cation. The formation of these cations and their subsequent rearrangement to the 1-adamantyl cation will therefore be reviewed. The attempted preparation of many intermediate $C_{10}H_{15}^+$ cations, in most cases, results in the immediate (under the reaction conditions) rearrangement of the cation to other, more stable carbocations. This does not necessarily imply that the original cation is particularly unstable, just that there is a relatively low barrier to an even more stable cation. In these circumstances, the size of the barrier can be estimated at less than 10 kcal/mol for NMR samples of cations made up carefully at the lowest temperatures possible (-150°C) and analyzed within 20 min.

Rather than allow the energy profile to develop in parallel with the discussion of the individual steps, we propose to present the generally accepted cationic pathway for the overall rearrangement of the THDPD cation to the 1-adamantyl

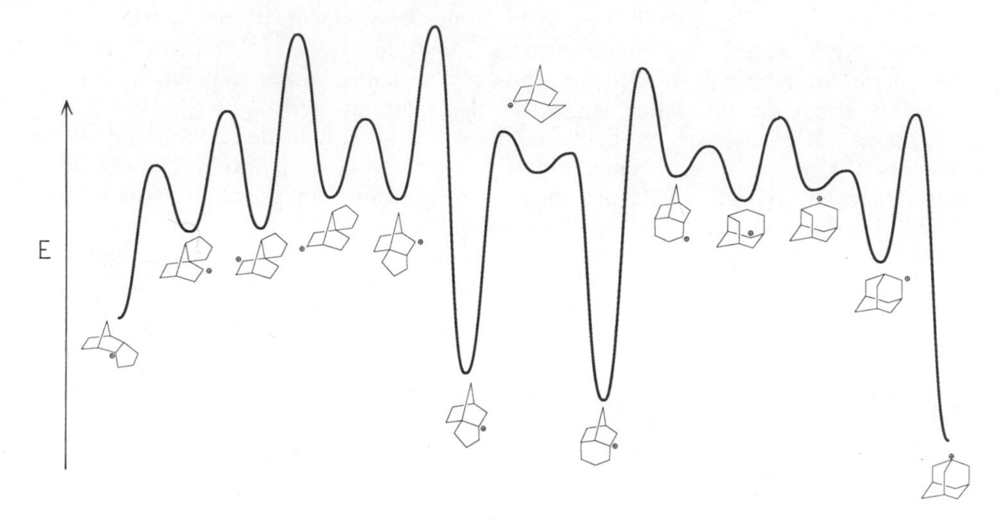

Fig. 4. Potential energy diagram for the rearrangement of the THDPD cation to the 1-adamantyl cation.

cation at the outset. This includes the estimated transition state barriers for the various steps and is shown in Figure 4. Using this as a guide, one can now discuss in detail what is known about the individual steps, including details about the observable ions in this sequence and the way in which transition state barriers were estimated.

Before doing this, however, there are several general points that we should comment on. First, the structures of most carbocations are *written* in a "classical" form, simply to make for a clearer presentation. In the following detailed discussion, questions concerning the actual "structures" are discussed. Second, we can legitimately question the ground-state existence of some of the cationic structures shown in Figure 4, particularly those where a 1-alkyl substituted secondary ion is present; for example, the partial structure (**6**).

6

Such a structure would be expected to be "bridged" across the C-2–C-6 carbon atoms, and this bridging might simply collapse to the tertiary ion without any activation barrier, that is, formulations such as (**6**) would not correspond to minima on the potential energy surface. There is, however, some reasonable possibility that these secondary ions could correspond to a very shallow energy minimum and we propose to use this formalism. Whitlock and Ganter, among others, have both used this approach. This point is not too important for the transition state barrier estimates, since we use the behavior of closely related model systems to derive estimates of various activation barriers. Although the detailed interpretation of structures might therefore be in some doubt, one can still have complete faith in the actual energy values themselves.

There are reports of some miscellaneous $C_{10}H_{15}^+$ cations not involved in the major reaction pathway outlined in this chapter, but nevertheless having some interesting chemical features, including their ability to rearrange to the 1-adamantyl cation. We therefore propose to include a brief description of the chemistry of these cations, including any relevant solvolysis work, in a separate section in the final stages of this chapter.

THDPD CATION

The rearrangement of the THDPD cation to the 1-adamantyl cation closely parallels the commercial route to adamantane, originally discovered by Schleyer and Donaldson.[1] In 1964, Olah et al.[3] prepared the tertiary THDPD cation, but left open the question of structure (exo or endo) because the ^{1}H NMR spectrum is rather complex at 60 MHz. They did note, however, that one eventually gets the 1-adamantyl cation, albeit in a slow reaction requiring temperatures near 25°C. Our laboratory undertook a reexamination of this system in 1976,[14] using ^{13}C NMR spectroscopy, and Olah also recently published a ^{13}C NMR study.[15] Very recently we obtained the 400 MHz ^{1}H NMR spectrum (1-D and 2-D) of the ion (**3**),[14] and the complete assignments at this field, with modern pulse methods, becomes quite straightforward, however, the endo/exo assignment relies on the following indirect evidence. First, the rapid 3,2-hydride shift observed in the THDPD cation (discussed later) is characteristic of an exo hydride shift rather than an endo hydride shift.[16] Second, the methyl group in 3-methyl-2-norbornanone has been equilibrated in both neutral and acidic conditions; in all cases, both isomers were formed, but the more acidic the medium was, the higher the proportion of the endo methyl over exo methyl.[17] Since the protonated carbonyl is expected to behave as a partial carbocation, it appears that there is some preferential stabilization of the endo substituent over the exo substituent. Finally, quenching studies of the cation show that the final product is the exo alcohol **7**[15,16] by comparison with authentic material prepared according to the procedure of Takaishi.[18]

The first three cations shown in Figure 4 exist within a potential energy surface containing one additional cation, and all four cations are shown in Figure 5. The tertiary cations **3** and **8** are interconverted through the three-step sequence of a Wagner–Meerwein shift, a 2,6-hydride shift, and finally, another Wagner–Meerwein shift. High external barriers surround these compounds, and in fact surmounting this barrier corresponds to the k_{obs} activation barrier for the overall reaction of **3** to the 1-adamantyl cation **4**. There are several interesting points concerning these ions.

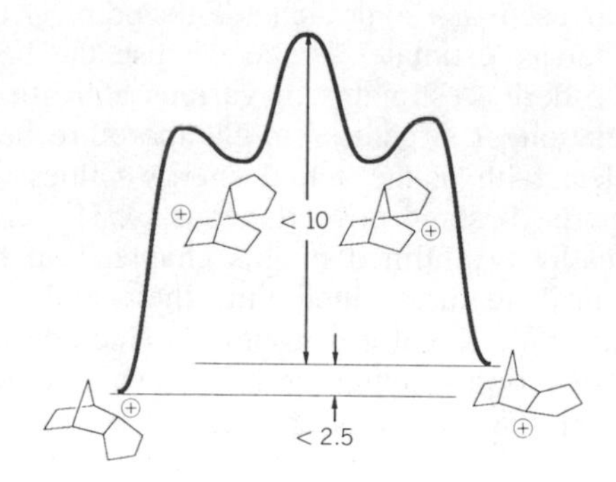

Fig. 5. Potential energy diagram for the THDPD manifold. All energies in kilocalories per mole.

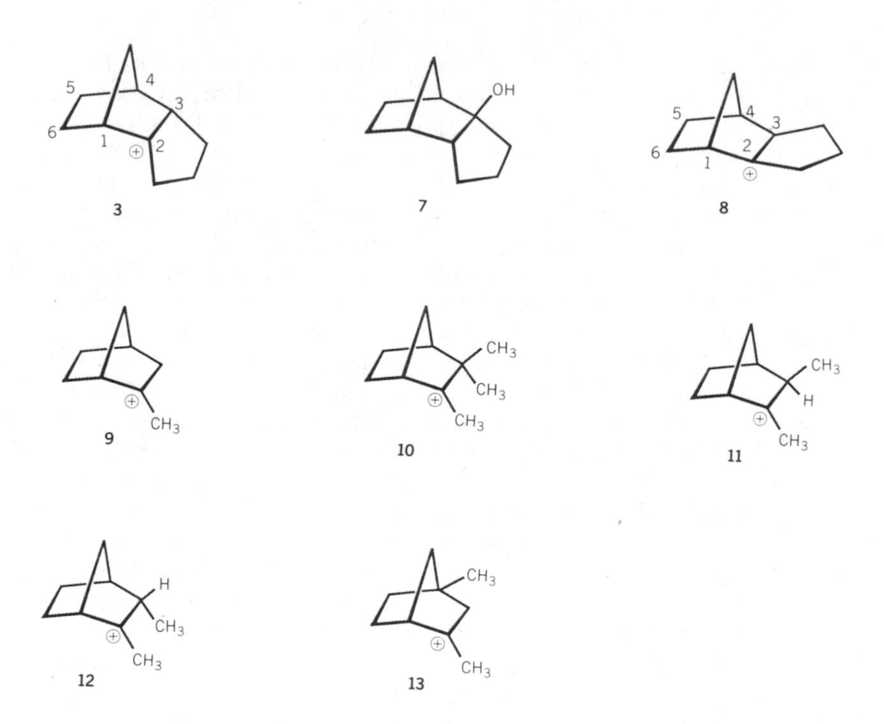

1. Of the two possible tertiary ions, **3** or **8**, only **3** is observed, that is, where the trimethylene ring is endo. The same ion is produced from alcohol precursors containing either the endo or exo trimethylene ring, so that the reaction **8** to **3** must be quite fast on a conventional kinetic time scale. The interconversion between **3** and **8** involves the transient formation of two secondary cations (Fig. 5). The details of this process are discussed in the next section since they are involved in the route out of the endo–exo THDPD

manifold. Attempts to prepare the putative exo cation (**8**) at –130°C and within an observation time of ~ 10–15 min showed no evidence of **8** in the NMR spectrum, which corresponds to a maximum barrier of ~ 10 kcal/mol for the **8** to **3** transformation. A limit on the equilibrium constant **8/3** was also estimated by measuring the NMR spectrum of **3** at a high signal-to-noise level. We noted no peaks ascribable to **8**, this measurement being carried out at temperatures where it was fairly certain that equilibrium conditions would exist, and under these circumstances, we can confidently exclude the presence of more than 1% of the corresponding exo isomer.[14] A second more sensitive approach was also investigated, the NMR technique known as "exchange with a hidden partner."[19] For two exchanging signals of very different intensities, such that the minor peak is not observable for signal-to-noise reasons, the chemical shift differences of the major peak in the slow exchange limit and the average peak in the fast exchange limit will be negligible. However, in going from the slow to the fast exchange limit, considerable line broadening of the major peak may occur, depending on the chemical shift separation of the two peaks and the actual population of the minor isomer. To a good approximation the maximum line broadening is given by $p\ \Delta\nu$ (p=minor population, $\Delta\nu$=chemical shift difference). In conditions in which the endo isomer (**3**) undergoes a rapid 3,2-hydride shift, the ^{13}C peak widths were measured as a function of temperature (between 210 and 250 K). There was some sharpening of the peaks involved in the 3,2-hydride shift process, but the ^{13}C peaks of those carbon atoms in the symmetry plane remained sharp. Assuming that a broadening of 2 Hz could be detected (peak widths at half-height were ~ 2 Hz) and a chemical shift difference of 5 ppm (500 Hz at 100 MHz), the minor population, that is **8**, can be estimated at less than 0.4% or >2.5 kcal/mol less stable[14] than the major population (Fig.5).

2. The observed barrier of $\Delta G^{\ddagger} < 10$ kcal/mol for the interconversion of **8** to **3** can be compared to a number of model systems where the same interconversion mechanism is thought to be operating. This reaction can be degenerate, as in the case of the 2-methyl-2-norbornyl cation **9**,[20] the camphenehydro cation **10**,[21] or the 2,4-dimethyl-2-norbornyl cation **13**[20c], where $\Delta G^{\ddagger}$ = 11.3 kcal/mol, 10.7 kcal/mol and 11.1 kcal/mol, respectively. Nondegenerate systems have also been studied, such as the very closely related *exo*-2,3-dimethyl- (**11**) to *endo*-2,3-dimethyl-2-norbornyl cation (**12**),[17] where $\Delta G^{\ddagger}$ = 11.9 kcal/mol. Although the barrier for **8** to **3** is observed to be less than 10 kcal/mol, there is reason to believe from these model systems that the actual barrier will not be much less.

Why is the endo cation **3** more stable than the exo cation **8**? In the hydrocarbon rearrangement, equilibration studies show just the opposite preference for the hydrocarbons, the exo hydrocarbon being favored over the endo by 2.7 kcal/mol (25°C).[8] Thus the exo skeleton is more stable than the endo. On forming the cation at the 2 position, there will be a concomitant change in geometry, but the associated changes in strain are unlikely to account for the reversing of cation stabilities. A more likely explanation is that there is preferential stabilization of the endo cation **3** over the exo cation **8**, resulting

TABLE 1. Chemical Shifts of the H-6 Protons in Selected Tertiary 2-Norbornyl Cations

Cation	δ_{exo}/ ppm	δ_{endo}/ppm	Reference
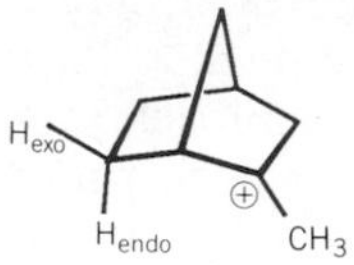 H_{exo}, H_{endo}, ⊕, CH_3 **9**	3.56	1.71	20
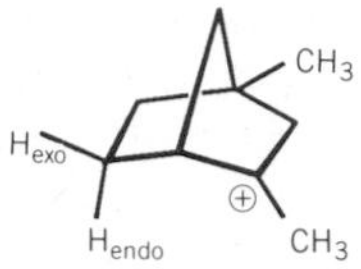 H_{exo}, H_{endo}, ⊕, CH_3, CH_3 **13**	3.68	1.70	23
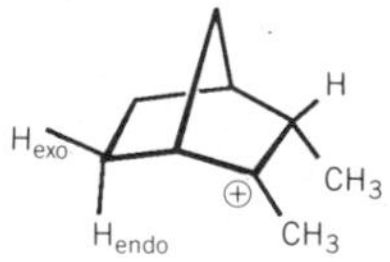 H_{exo}, H_{endo}, ⊕, H, CH_3, CH_3 **12**	3.7	1.93	21

from a combination of delocalization effects. The *endo*-2,3-dimethyl-2-norbornyl cation (**12**) is somewhat more stable than the exo isomer (**11**),[17] and this can be attributed to the better C—H hyperconjugative overlap of the C^+ orbital when the hydrogen at C-3 is exo. A similar result would be expected in the case of **3** (exo hydrogen at C-3). In addition, the endo ring in **3** appears to allow better partial delocalization of the C-2 positive charge into the exo C-6—H bond. There is considerable evidence for partial delocalization in this kind of tertiary norbornyl structure, including the recent X-ray evidence of Laube.[22] In terms of ^1H NMR spectroscopy, the major result of such delocalization is an abnormally low field chemical shift of the exo H-6 proton. In Table 1, we compare the ^1H NMR shifts for both the exo and endo protons at C-6 in several tertiary 2-norbornyl cations. The results for **3** compare quite closely with those for the 2-methyl-2-norbornyl cation (**9**), which is widely postulated to be a "partially bridged" structure. The minimum energy difference between **3** and **8**, that is, 2.5 kcal/mol, is in essence at least a partial measure of the stabilization that results from this type of charge delocalization.

Cation **3** also undergoes a degenerate rearrangement involving an exo 3,2-hydride shift. The barrier is 7.2 kcal/mol[15] and is similar to that found in the *endo*-2,3-dimethyl-2-norbornyl cation (**12**)[17] (6.6 kcal/mol at –117°C). This degenerate process is of no direct concern in considering the overall rearrangement to the 1-adamantyl cation, except that isotopic labels would be "scrambled" by such a process. In fact, this degenerate rearrangement is only the first of several of this class, which are likely to cause such "scrambling."

THDPD TO HOMOBRENDYL CATION

Further rearrangement of the *endo*-THDPD (**3**) and related cations requiresa combination of several high energy steps. The first step is part of the *endo*-THDPD to *exo*-THDPD manifold described in the previous section and is shown in more detail in Figure 6. Thus the secondary cation **14** is formed from **3** by a Wagner–Meerwein shift. The next step involves a 3,2- hydride shift that converts the secondary cation **14** into **16**, which then rearranges via a Wagner–Meerwein shift to give the secondary cation **17**. An exo 3,2-hydride shift finally gives the tertiary 1,2-trimethylene-2-norbornyl cation (**18**). There are two possible parallel pathways to cation **18**, and both these are illustrated in Figure 6. The upper pathway (previously described) is expected to dominate, since the final step is an exo 3,2-hydride shift (in a 2-norbornyl cation), rather than the higher energy endo 3,2-hydride shift in the lower example (Fig. 6). The tertiary to secondary barrier from **3** to **14** has been estimated at ~ 11–12 kcal/mol from several different experiments,[17,20,23] as well as from the arguments presented in the previous section. The 3,2-hydride shift in a secondary 2-norbornyl cation is experimentally measurable from NMR line broadening studies, and is ~ 10-11 kcal/mol.[24] The combined barrier from **3** to **16** is therefore estimated from these model systems to be ~ 21-22 kcal/mol, which is in good agreement with our experimentally measured E_a of 22 kcal/mol[14] for the overall rearrangement of **3** to the 1-adamantyl cation **4**.

Although there is good correspondence between the experimental and estimated barriers for the overall rearrangement, a more direct result is provided by ionizing the iodide **19** in $SbF_5/SO_2ClF—SO_2F_2$. Only two cations are produced, the THDPD cation (**3**) and 2,6-trimethylene-2-norbornyl (homobrendyl) cation (**20**).[14] This partitioning shows that there are in fact *two* very similar high points on the potential energy surface (Fig. 7). The reverse exo 3,2-hydride shift in **16** leads ultimately to **3**, while a combination Wagner–Meerwein shift and exo 3,2-hydride shift in **16** leads forward to cation **18** and eventually, cation **20**. The actual partitioning gives nearly equal amounts of both **3** and **20** (actually in 55:45 ratio), suggesting that the barriers out of the secondary cation manifold shown in Figure 7 are very nearly equal. The inability to observe cations **16** ⇌ **17**, even at the lowest temperatures attainable, means that both the barriers to the forward and reverse reactions are less than 10 kcal/mol, although they are not expected to be much less.[24] Since both high points in Fig. 7 correspond to exo 3,2-hydride shifts, it seems reasonable that the barriers should be similar. At this point it should perhaps be reemphasized that secondary

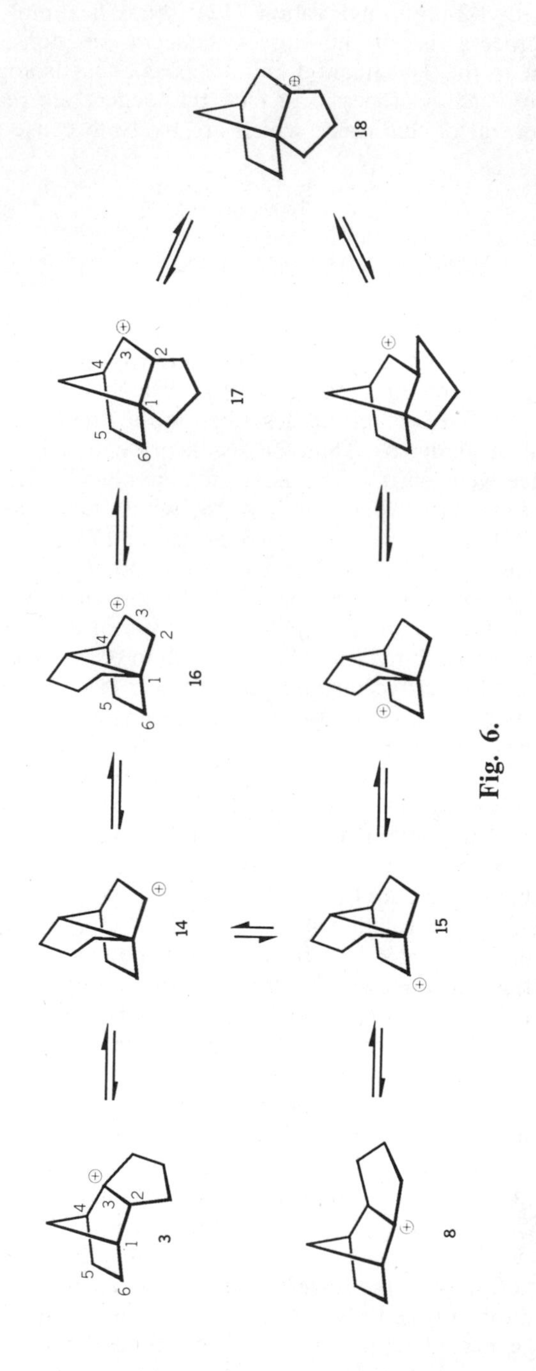

Fig. 6.

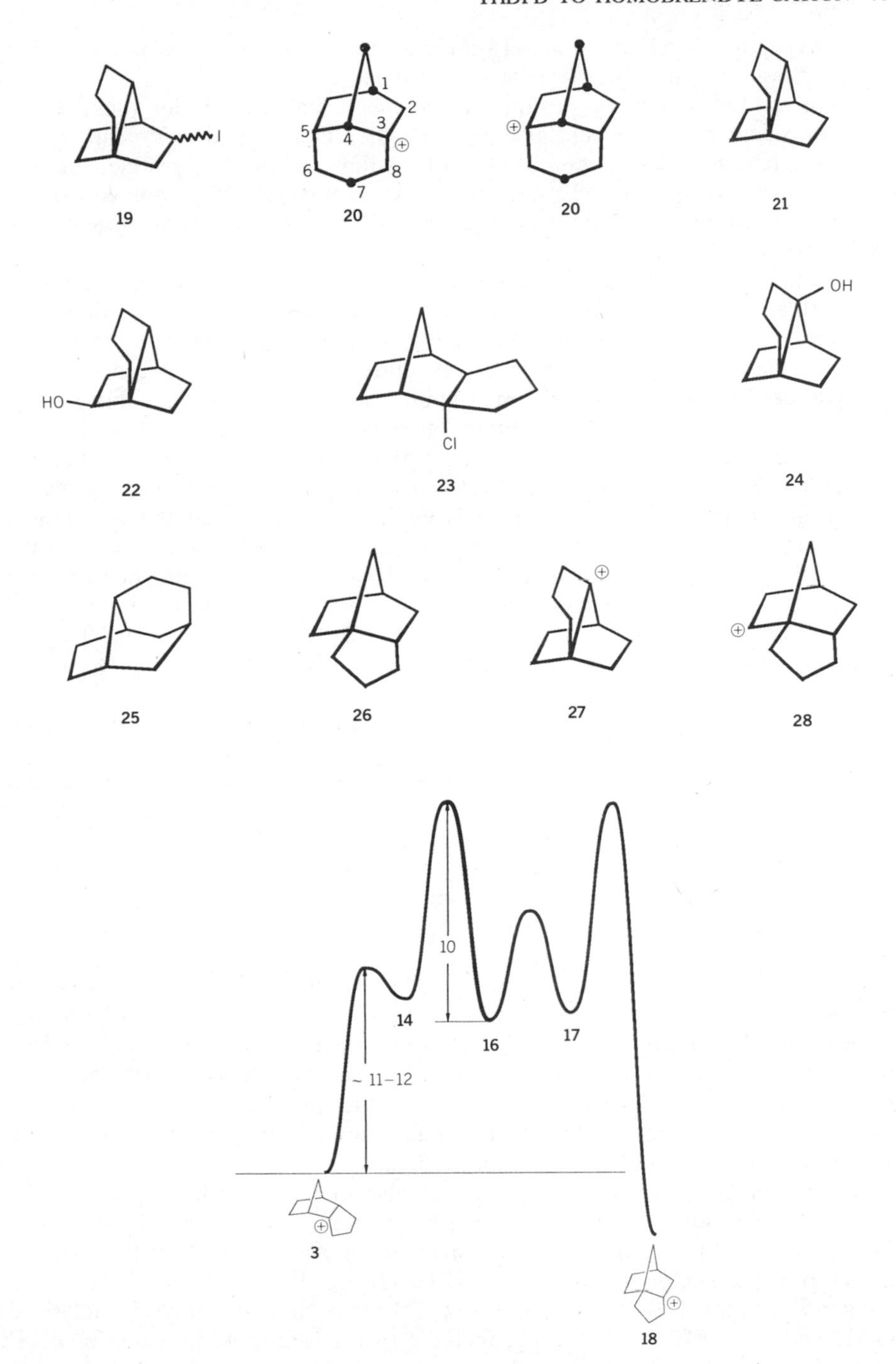

Fig. 7. Potential energy diagram for the route out of the THDPD manifold. All energies in kilocalories per mole.

norbornyl cations such as **14** and **15** are expected to lie in very shallow energy minima, assuming that they are indeed minima.

Holm et al.[25] previously postulated that the overall barrier for the THDPD to adamantane rearrangement occurs with the intermediate hydrocarbon **21**. This is consistent with the "backward" thionyl chloride–pyridine induced rearrangement of alcohol **22** to the endo chloride **23**,[26] compared to the "forward" rearrangement of alcohol **24** to hydrocarbon **25** on treatment with 98% sulfuric acid-pentane.[27]

A comparison of the Lewis acid catalyzed results with the superacid results is informative. In the first place, hydrocarbon **21** can proceed along the "adamantanization" route via a bimolecular hydride transfer to give cation **16**, and subsequently via a Wagner–Meerwein shift, cation **17**, which can then exit as hydrocarbon **26**. The high energy internal exo 3,2-hydride shift in the formation of cation **18** can therefore be avoided by a bimolecular hydride transfer from hydrocarbon **26**. A similar argument can be used to show that the barrier between **14** and **16** (see Fig. 7) can also be reduced by the presence of bimolecular hydride transfers. There is evidence, in fact, that the Lewis acid plus promoter catalyzed rearrangements are as rapid as, or more so, than the superacid rearrangements, and the major reason is likely the avoidance of these high internal hydride shift barriers. Hydrocarbon **21** represents a watershed because the random formation of cation **14** or **15** via bimolecular hydride transfers will naturally yield THDPD, while the formation of **16** would eventually yield adamantane.

Holm and co-workers[25] also noted that hydrocarbon **21**, on rearrangement, produced yet other $C_{10}H_{16}$ intermediates in small amounts, indicative of a second (or alternate) route to adamantane, and the cationic rearrangement observed by Tobe, et al.[27] is consistent with this observation. These routes likely involve the initial formation of cations such as **27**. The carbocation route to adamantane is in fact likely to be much less diverging since such "aberrant" cations are not likely to be formed in superacid media because of "mechanistic" considerations, that is, steric factors preventing 1,3 or 1,4 intramolecular-hydride shifts.

The Whitlock–Siefkin route from **3** to **4**, as outlined in Figure 4, suggests that the next observable carbocation after **3** might be the tertiary 1,2-trimethylene-2-norbornyl cation **18**. However, starting from the high energy intermediate represented by cation **16**, one passes right through **18** and stops only at the homobrendyl cation (**20**). This latter cation in fact represents the only observable cationic intermediate on the whole pathway, a statement that will be amplified in the following discussion.

The homobrendyl cation sits in a reasonably deep valley and only rearranges to the 1-adamantyl cation (no detectable intermediates observed) at about -20°C, $\Delta G^{\ddagger}$ = 18 kcal/mol. However, since this barrier is less than the THDPD barrier of ~ 22 kcal/mol, this cation is not observed as an intermediate when one starts with the THDPD cation. Cation **20** has been completely characterized by ^{1}H and ^{13}C NMR spectroscopy, including various 2-D NMR experiments.[14] The precursor homobrendyl alcohol was originally prepared by Corey and Glass[28] using an acid catalyzed rearrangement of 1,2-trimethylene-5-norbornene via the secondary 6-ylium cation (**28**).

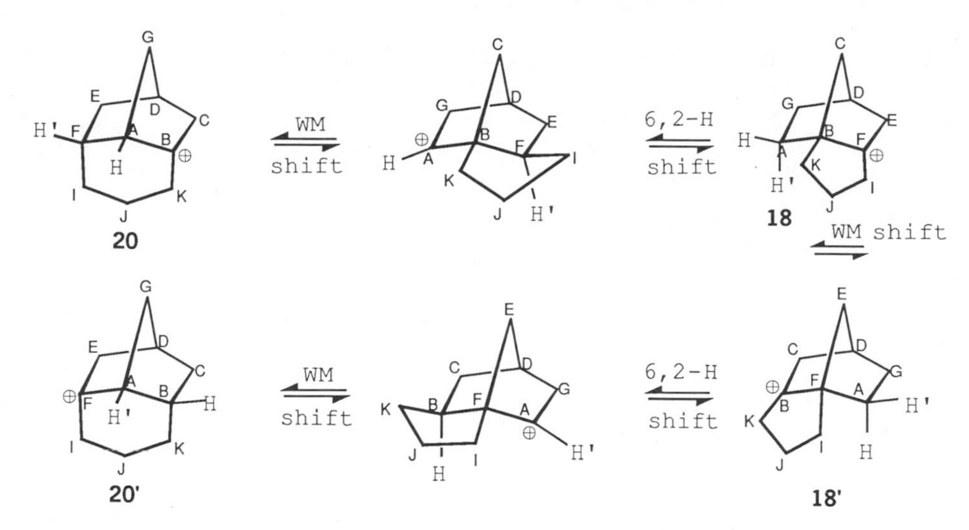

Fig. 8. The degenerate rearrangement of the homobrendyl cation.

At a temperature near where the irreversible rearrangement to the 1-adamantyl cation is occurring, one sees the beginning of a degenerate rearrangement in **20** (line broadening of selected NMR peaks). This degenerate rearrangement has been characterized by both ^{13}C and ^{1}H NMR spectroscopy, the former by noting initial line broadening of 6 of the 10 carbon atoms, the latter by both spin saturation transfer techniques and by 2D nuclear Overhauser effect correlation spectroscopy (NOESY) experiments.[14]

Formally, the line broadening observed in the ^{13}C NMR spectrum represents an operation in which the cation **20** equilibrates with its mirror image **20′**. The carbon atoms shown in solid circles are on the "symmetry plane" and would be equivalent in both structures. Hence, they are not broadened by the rearrangement. In theory, such a process might be thought to involve a simple hydrogen migration from C-5 to C-3 (or vice versa). Sterically, however, this is not possible for an internal hydride shift and there is no evidence that intermolecular hydride shifts are occurring in this cation. The ^{1}H NMR results provide the answer. They show that the H4 and H5 protons are exchanged at the same rate (k_{obs} = 6.3 s^{-1} at 230 K) as the carbon equivalence process, and the only scheme that allows for all of these occurrences is shown in Figure 8.

Experimentally, this degenerate rearrangement has a barrier of 12.5 kcal/mol.[14] At a solution temperature of –80°C, cation **18** itself was not visible as a minor equilibrium partner of **20**, and a minimum K of 50 for **20/18** was therefore estimated, $\Delta G^{\ddagger} > 1.5$ kcal/mol. This means that the reverse rearrangement from **18** to **20** has a barrier of less than 11 kcal/mol. However, since the cation (**18**) cannot be observed directly (because of its immediate rearrangement to **20**, the barrier for this process must actually be less than 10 kcal/mol, making the cation (**20**) at least 2.5 kcal/mol more stable than **18** (Fig. 9). In effect then, one does not stop at cation **18** when proceeding from the high energy plateau represented by cation **16**, even though the latter was generated at about –150°C.

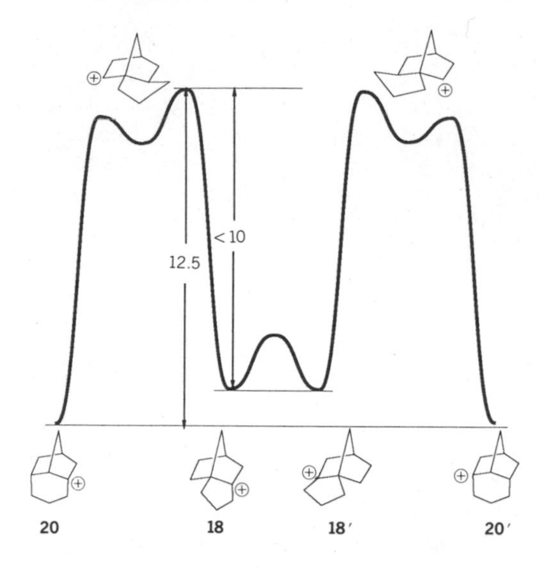

Fig. 9. Potential energy diagram for the homobrendyl cation. All energies in kilocalories per mole.

All of the information contained in this degenerate rearrangement is highly relevant to the overall rearrangement picture, and the various ground-state and activation energies have been incorporated into Figure 9. In essence the degenerate rearrangement process allows one to probe "backwards" from the homobrendyl cation. The rapid Wagner–Meerwein shift in **18** is completely expected. For example, the corresponding 1,2-dimethyl-2-norbornyl cation has a barrier for this process of less than 3 kcal/mol.[29] This degenerate process represents yet another way in which isotopic labels can become rapidly scrambled. Indeed this same degenerate process has been observed using the neutral hydrocarbon **26** and mild Lewis acids.[7]

The existence of the homobrendyl cation (**20**) is of interest with regard to the original rearrangement mechanistic proposals of Whitlock and Siefkin.[9a] Although their cation mechanism is very probably the correct one, their ideas about reaction barriers are quite incorrect. They saw cation **20** as a possible candidate for the overall reaction "bottleneck" (or highest barrier), basically because the expected nonplanarity of this cation was equated with a high potential energy. One sees now that cation **20** represents a very low point on the potential energy surface, and is kinetically quite stable.

3-HOMOBRENDYL TO 2-ADAMANTYL CATION

This section describes the rearrangement of the 3-homobrendyl cation (**20**) to the 2-adamantyl cation (**30**). The overall process is depicted in Figure 10, which includes the final step to the 1-adamantyl cation (**4**). Movement of **20** out of the relatively deep "well" previously described requires the formation of the homobrendyl cation (**31**), which in turn, allows access to the penultimate carbon framework—protoadamantane, as shown in Figure 11.

The conversion of **20** to **31** can be accomplished in one of two ways. First, a 1,2-hydride shift from the 2-position, followed by a 1,3-hydride shift from the 8-position can occur. Second, a simple 1,2-hydride shift may occur. The unfavorable overlap between the migrating hydride and the cation center means that the latter process is not expected to occur from the ground-state conformation of the 3-homobrendyl cation (**20**), but rather from the conformation in which the cyclohexyl ring has undergone a chair (**32**) to boat (**33**) ring flip (Fig. 12).

Once the intermediate secondary homobrendyl cation (**31**) has been formed, ring expansion to the protoadamantyl system can occur in one of two ways. First, a single 1,2-alkyl shift can give the 7-protoadamantyl cation (**34**) (route a, Fig. 11). Second, a series of sequential 1,2-alkyl shifts can give **34** (route b, Fig. 11), via the intermediacy of **35**. Solvolysis of *exo*-2-homobrendyl brosylate gave small amounts of 7-protoadamantyl derivatives,[30] not inconsistent with this picture.

The attempted preparation of the secondary ion **31**, from the corresponding mesylate, produced a clean spectrum of only the 1-adamantyl cation.[14] The total absence of any observable 3-homobrendyl cation, in conditions in which it would be stable, indicates that the activation barrier for the return to **20** is significantly larger than are any subsequent barriers in the rearrangement to the 1-adamantyl cation.

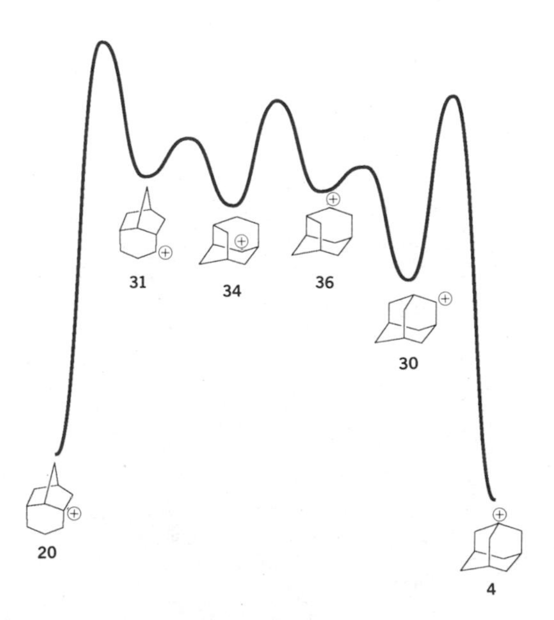

Fig. 10. Potential energy diagram for the homobrendyl to the 1-adamantyl transformation.

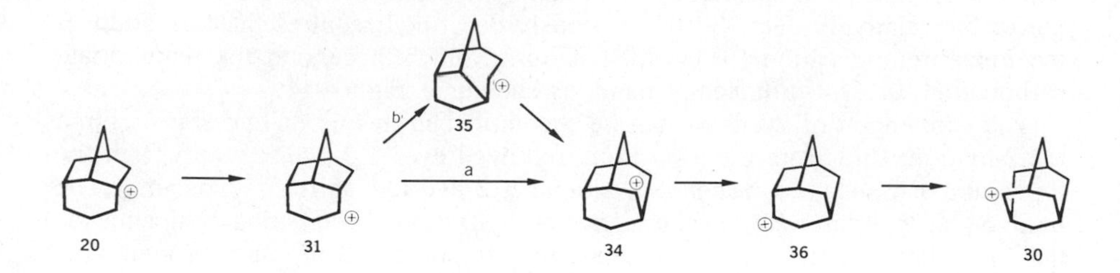
35
b
a
20
31
34
36
30

Fig. 11.

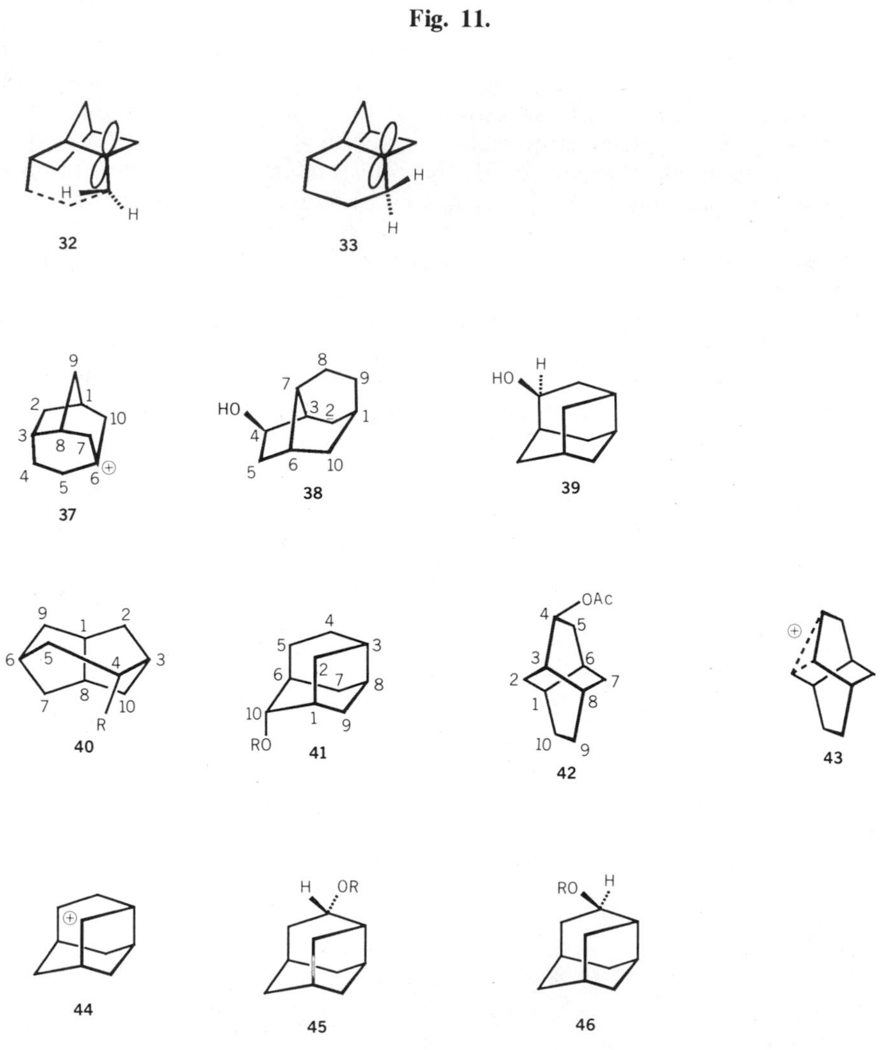
H
H
32
H
H
33
37
HO
38
HO
H
39
40
R
41
RO
OAc
42
43
44
H
OR
45
RO
H
46

Fig. 12.

Disregarding the uncertainty in the exact route that is followed to give the protoadamantyl framework, the positive charge is initially formed in the 7-position. It is generally accepted that the immediate precursor to the 2-adamantyl cation is the 4-protoadamantyl cation (**36**), with a simple 1,2-alkyl shift facilitating this interconversion. The exact route converting **34** to **36** (assuming that only one route is involved) remains unclear. There is no solvolysis work directly relevant. The interconversion may occur directly, via a single 1,4-hydride shift. Such intramolecular 1,4-hydride shifts between the apex positions of a cyclohexyl "boat" are well documented and can occur with considerable ease.[31] A two-step process involving a 1,3- followed by a 1,2-hydride shift appears to be less favorable on steric grounds. Before discussing the rearrangement of **36** to **30**, the chemistry of some of the protoadamantyl cations will be discussed.

The tertiary 6-protoadamantyl cation **37** might be formed as a temporary "resting place" in the overall reaction. An intramolecular 1,2-hydride shift from the 7-position is unlikely to be responsible for this transformation because of unfavorable orbital overlap, but the possibility of intermolecular hydride transfers being involved in its formation are discussed later. Experimentally, the attempted preparation of the 6-protoadamantyl cation,[14] under stable ion conditions, gave only the 1-adamantyl cation, so that the rearrangement barrier to the 1-adamantyl cation must be less than 10 kcal/mol.

Despite not being able to observe the 6-protoadamantyl cation directly, there is indirect evidence of its stability (relative to other protoadamantyl cations). Ionic bromination reacts preferentially at those carbon centers that are best able to stabilize a positive charge. It has been shown that under these conditions, the major product arises from bromination at the 6-position, with small amounts of the 3-bromo compound also being reported.[32,33] Similar selectivities have been reported for the reaction of protoadamantane with $Pb(OAc)_4$, CrO_3, and O_3 on silica.[34] These results are in agreement with the predictions of molecular mechanics calculations.[32] Reactions that presumably involve the intermediacy of the 6-protoadamantyl cation (Koch–Haaf carbonylation and the Ritter reaction) have been reported to occur without rearrangement of the carbon framework.[33] 6-Bromoprotoadamantane, however, rearranged to 1-adamantanol on treatment with 96% H_2SO_4.[33] That the 6-protoadamantyl cation appears to be the most stable cation within that framework is not too surprising, since the 6-position is the closest of all protoadamantyl bridgehead positions to that of the tertiary position in adamantane. Interestingly, there is evidence that the 6-protoadamantyl cation is of comparable stability to the 1-adamantyl cation. 6-Protoadamantane is brominated faster than adamantane (after taking into account statistical effects), and also in support of this, is the fact that solvolysis of 6-bromoprotoadamantane is twice as fast as in 1-bromoadamantane.[32]

The solvolysis of several secondary protoadamantyl derivatives has been reported. Spurlock and Clark[35] solvolyzed both the *exo*- and *endo*-2-protoadamantyl brosylates under a variety of conditions and found that the major product was tricyclo[4.3.1.$0^{3,7}$]decan-4-ol (**38**), from a simple 1,2-alkyl shift. A large number of minor products, including 5–30% of 2-adamantyl derivatives, were also recovered. The reaction of *endo*-5-protoadamantanol (**39**) under chlorination conditions (PCl_5 or $SOCl_2$) produced significant amounts of

4-chlorobicyclo[4.2.1.1^{3,8}]decane (**40**, R=Cl) as well as the unrearranged chloride.[36] The subsequent reaction of both these chlorides with silver acetate–acetic acid gave identical product mixtures, which included small amounts of the 2-adamantyl acetate. Majerski[37] observed the same result in the solvolysis of tosylate **40** (R=OTs), but the alcohol **40** (R=OH) rearranged to adamantane on treatment with 98% sulfuric acid-pentane. The solvolysis of *exo*-10-protoadamantyl tosylate (**41**, R=OTs) results in a mixture of *exo*-10-protoadamantyl acetate (**41**, R=OAc) and *exo*-4-tricyclo[4.4.0.0^{3,8}]decyl acetate (*exo*-4-twistyl acetate) (**42**).[38] The same product distribution is observed in the solvolysis of *exo*-4-twistyl acetate, and, combined with the results of deuterium labeling reactions, Kniezo and co-workers[38] proposed a bridged intermediate (**43**). Although the various twistyl cations are not on the direct route to the 1-adamantyl cation, their chemistry is discussed here. The attempted preparation of the 2-twistyl cation under superacid conditions lead directly to the 1-adamantyl cation in high yield.[9a] The tertiary 1-twistyl cation, as expected, appears to be more stable than either of the secondary cations, since solvolysis occurs without rearrangement.[39] No rearrangements were observed in reactions that proceed through the intermediacy of the 1-twistyl cation (Koch–Haaf carbonylation or the Ritter reaction.[39a]

The wide range of products obtained on solvolysis of the different protoadamantyl derivatives serves once again to illustrate the possible differences between the rearrangement under stable ion conditions and under Lewis acid catalysis. The formation of the 2-protoadamantyl cation (**44**) can easily be accomplished under Lewis acid catalysis (via intermolecular hydride transfers), however, it is more difficult to see how it could occur as an intramolecular process from the 7-protoadamantyl cation (**34**). Assuming that the 2-protoadamantyl cation[35] can be transformed to the adamantyl cation (as observed under solvolysis conditions), then this process may be a "real" route under Lewis acid catalysis, but an "imaginary" route under stable ion conditions.

The solvolysis of 4-protoadamantyl derivatives was studied in detail by several workers. The rates of solvolysis of *exo*-4-protoadamantyl (**45**) derivatives are ~ 10^4 times as fast as are those of the corresponding *endo* isomer (**46**).[40] This differential reactivity was also observed in the epimeric alcohols on treatment with aqueous sulfuric acid.[41] Acetolysis of *exo*-4-protoadamantyl tosylate gives 2-adamantyl acetate, while under the same conditions, the *endo* tosylate gives a mixture of 2-adamantyl acetate and *endo*-4-protoadamantyl acetate.[40,41] It will become clear that the 4-protoadamantyl and 2-adamantyl cations are inextricably linked, and so the interconversion between these cations is discussed in conjunction with their respective structures and stabilities.

Despite repeated attempts,[14,42] the 2-adamantyl cation has not been characterized under stable ion conditions. Consequently, details of the stability and structure of this important intermediate have been obtained indirectly, and solvolysis work has provided much of this information. In general, the solvolysis of most secondary derivatives occurs with some degree of solvent participation (k_s), but it was conclusively shown that secondary 2-adamantyl derivatives solvolyze without solvent participation (k_c limiting).[43] Despite this, the role of the solvent is important, and as we will illustrate, the course of a

Fig. 13.

reaction or product ratios can be controlled to some extent by the appropriate choice of solvent. We should always bear in mind that the conclusions arrived at from solvolysis arguments may not always be applicable to stable ion work.

Isotopic labeling studies have been used to probe the 4-protoadamantyl to 2-adamantyl cationic rearrangement. Solvolysis of *endo*-4-protoadamantyl-4-d_1-dinitrobenzoate (**46**, R = ODNB) gave 2-adamantyl acetate and 4-*endo*-protoadamantyl acetate, in which 50% of the deuterium was in the 4-protoadamantyl and 2-adamantyl positions, and the remaining deuterium was in the 6-protoadamantyl and 1-adamantyl positions.[40,41] These results are consistent with the formation of a weakly bridged structure as the initial solvolysis product (Fig. 13). This may either be a true energy minimum or a degenerate interconversion of the two localized cations. That the initial bridged structure is a reactive species, is illustrated by: first, the large amount of "leakage" to the 2-adamantyl cation and second, by the small amount of deuterium scrambling observed in recovered starting material (8%) in conditions that favor return.[40]

Solvolysis of *exo*-4-protoadamantyl-4-d_1-tosylate (**45**, R = ODNB), shows that, in the resulting 2-adamantyl acetate, the deuterium occupies solely the 1-position and not the 2-position (within detectable limits).[40,41] The greater reactivity of the exo isomer is presumed to be the result of favorable alignment between the 2,3 bond and the leaving group (tosylate), allowing bond migration to assist in the ionization, resulting in the direct formation of the 2-adamantyl cation. The solvolysis results of 2-adamantyl tosylate are complementary to those of *exo*-4-protoadamantyl tosylate. On careful product analysis, small amounts (typically 0.5%) of 4-*exo*-protoadamantyl derivatives are found, but none of the corresponding endo isomer.[40] The rearrangement of the protoadamantyl framework to the adamantyl framework was also observed under more vigorous conditions.[44] Both the formation of a bridged structure (Fig. 14) or alternatively, the collapse of an external ion pair,[40] have been invoked to rationalize these results. It is expected that, if bridged, such bridging will be weak and unsymmetric since the 4-protoadamantyl cation is expected to be more strained than the 2-adamantyl cation.

The deuterium label in *exo*-4-protoadamantyl-5-d_1-dinitrobenzoate established that solvolysis to 2-adamantyl acetate proceeded with retention of configuration of the oxygen functionality,[45] consistent with ionization to the bridged cation, followed by solvent attack from the opposite face to the migrating C—C bond. If a discrete 2-adamantyl cation were involved, attack from either face should be equally efficient.

The protonation of protoadamantene (**47**) with trifluoroacetic acid (TFA) was reported by Nordlander to give mainly 2-adamantyl trifluoroacetate and small

Fig. 14.

amounts of *exo*-4-protoadamantyl trifluoroacetate.[46] The initial formation and subsequent reaction of trifluoroacetate esters was tested for and discounted. With deuterated TFA, four diasteriomeric alcohols were produced and characterized. The product distributions were interpreted as involving initial acid–π-complexation, preferentially from the exo face, with subsequent rearrangement to 2-adamantyl acetate. The possible existence of a bridged intermediate was raised to account for the marked preference for nucleophilic attack on the opposite face to the migrating C—C bond (Fig. 15). Recently, the results of Nordlander were questioned, and an alternative route involving the rate-limiting protonation of the alkene was put forward by Tidwell.[47]

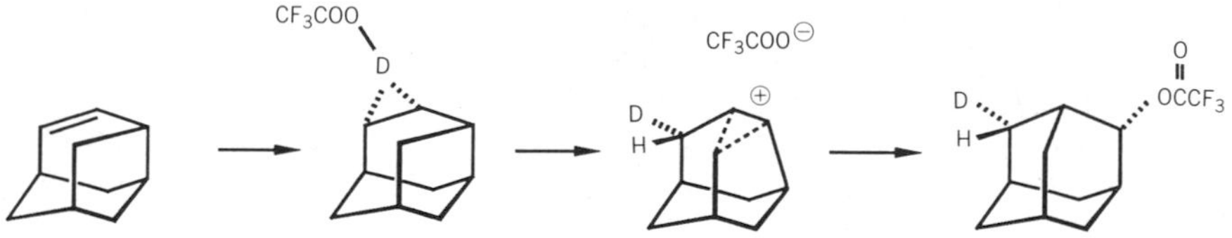

Fig. 15.

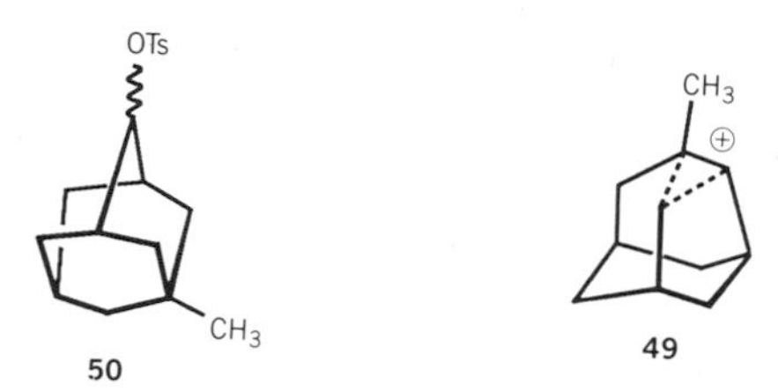

The unsymmetric nature of the bridge structure **48** can potentially be overcome, to some extent, by the addition of a methyl group to the 4-protoadamantyl position to give **49**. The strain of the protoadamantyl framework is then offset by the more stable tertiary cation, compared to the secondary adamantyl cation. This system has been studied extensively, and it has been concluded that the cation is indeed a bridged structure.[48]

Bone and Whiting[49] solvolyzed 5-methyl-2-adamantyl tosylates (**50**) under a variety of solvent conditions (changing the leaving group to picrate or nisylate had no dramatic effect). It was found that in all solvents, except dimethyl acetamide, solvolysis occurred with retention of configuration. Superimposed on this behavior was a lesser preference for nucleophilic attack syn to the 5-methyl group, such that the observed ratios of retention/inversion for the (*Z*) isomer were between 9:1 in aqueous acetone and 3.8:1 in TFA; and 3.2:1 (aq. acetone) and 0.85:1 (TFA) for the (*E*) isomer.[49a] Combined with small amounts of protoadamantyl derived material found in the reaction products, σ-bridging was invoked to account for these observations (although it was conceded that it was less important than in the 2-norbornyl system). 2,5-Dimethyl-2-adamantyl tosylates were also solvolyzed under comparable conditions, but considerably smaller retention/inversion ratios were found, indicating the more "classical" nature of tertiary cations.[49b]

It may be argued that steric effects of the 5-methyl group could, in some way, account for the observed reactivities (in fact, Whiting argued that steric effects were responsible for the preferential attack syn to the 5-methyl group). In 1986, le Noble[50] conclusively showed that this was not the case. Solvolysis of 2-adamantyl-5-d_1-tosylate gave retention to inversion ratios of 92:8 in aqueous acetone, and 65:35 in hexafluoro-2-propanol. The solvent dependence of the retention/inversion ratios is consistent with the following:

1. *Anchimeric Assistance.* The more polar the solvent is, the less delocalization is required to stabilize the positive charge.
2. *The Involvement of Ion Pairing in the Solvolysis Process.* Less ion pairing is required in more polar solvents, so less overall retention is observed.

It should be noted that the only solvent in which Whiting observed overall inversion was the polar, aprotic dimethyl acetamide, and Schleyer[40] suggested that this may be significant, in that it indicates the collapse of an external ion pair.

2-Adamantyl tosylates with (inductively) electron withdrawing groups in the 5-position, showed a marked preference for retention of configuration,[50] superimposed on a lesser preference for nucleophilic attack syn to the electron withdrawing group. In the corresponding tertiary compounds (e.g., 5-fluoro-2-methyl-2-adamantanol), the preference for retention is now dominated by attack syn to the electron withdrawing substituent.

The process was rationalized by σ-participation. Stronger σ-bridging, or a shorter lifetime, accounts for the net retention in the secondary ions. A weaker σ bridge allows equilibration of the more stable tertiary cation, so that the more electron rich β,γ bonds are involved in the σ-bridge (Fig. 16).[50]

Grob[51] has studied the solvolysis of 4-substituted 2-adamantyl nisylates. Plots of solvolysis rates against the inductive substituent constants (σ_I^q) of the 4-substituents show good straight lines. The ρ_I values obtained for the two epimers **51** and **52** are indicative of borderline bridging. Product analysis of **51**

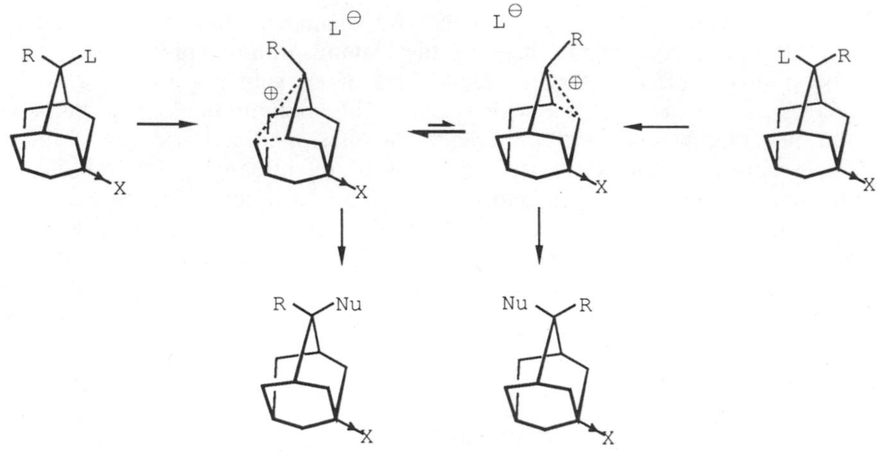

Fig. 16.

shows predominantly retention of configuration for electron donating groups that falls off as the electron demand increases. Product analysis of **52** again shows a net preference for retention, but the ratios are independent of the 4-substituent. Grob concluded that bridging was operative in both **51** and **52**, but that the description of that bridging was different.[51] The two proposed bridged structures are shown in Fig. 17.

These results show that there is a considerable amount of evidence suggesting that bridging does occur, at least in secondary 2-adamantyl cations. The evidence for this is solvolytic in nature, and it has been pointed out by several workers that the results are not inconsistent with solvent effects. At the beginning of this section it was pointed out that 2-adamantyl cations solvolyzed without direct solvent participation (i.e., a k_c rather than a k_s process). There is however, evidence that solvolysis occurs with significant return of the leaving group,[52] and this adds support to the notion that solvent effects may be involved (at least to some extent) in the observed preference for retention.

It is perhaps gratifying to conclude this section with some observations made under stable ion conditions. Schleyer, Olah and co-workers[53] prepared the 1,3,5,7-tetramethyl-2-adamantyl cation (**53**). The four tertiary methyl groups prevent the positive charge from migrating to a bridgehead position. The

Fig. 17.

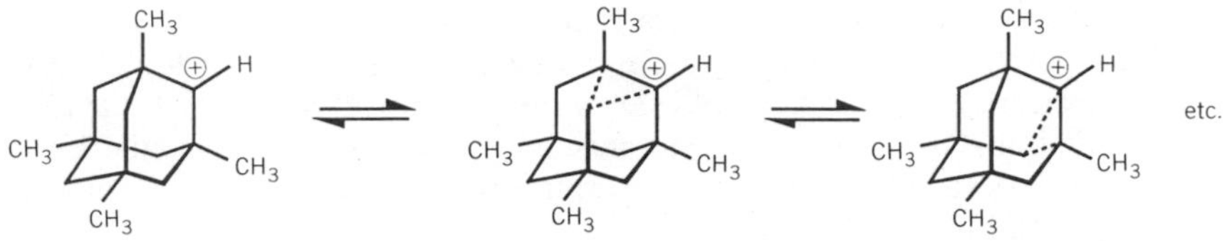

Fig. 18.

chemical shift of the proton attached to the cationic center is ~ 8 ppm upfield from the position expected for a localized secondary cation. Combined with the NMR symmetry of the molecule, an equilibrium between a series of degenerate structures, all involving σ-bridging (Fig. 18), was invoked. More recently, Sorensen[54] suggested, on the basis of stable ion work, that the results of Whiting[49] may be explained by a preferential hyperconjugative stabilization of the 2-adamantyl cation that resulted in the nonplanarity of the cation center. This differentiation between the two faces of the cation could then account for the observed results of Whiting. These results were obtained in tertiary 2-adamantyl cations, and so may not be of direct relevance, although it is to be expected that any such stabilization will be more important in secondary rather than tertiary cations.

2-ADAMANTYL TO 1-ADAMANTYL CATION

The final step in the rearrangement from THDPD is the facile interconversion of the 2-adamantyl cation to the 1-adamantyl cation. Solvolysis reactions of 1-adamantyl derivatives show no evidence for the formation of any 2-adamantyl derivatives.[55] That the reverse process was shown to be, at most, insignificant,[55] illustrates the integrity of these two systems under solvolytic conditions. This is not entirely unexpected, since an intramolecular 1,2-hydride shift, because of the near orthogonality between the migrating hydrogen and vacant *p* orbital, is predicted to be very unfavorable. The possibility of intramolecular 1,3-hydride shifts occurring in the solvolysis of 2-adamantyl derivatives has been tested for,[55] but found not to occur within detectable limits. These observations, combined with the facile reaction under stable ion conditions, suggests that the mechanism for the interconversion under stable ion conditions may not be operative under solvolysis conditions.

Under Koch–Haaf carbonylation conditions, the rearrangement of the 2-adamantyl cation to the 1-adamantyl cation was shown to follow second-order kinetics with respect to the adamantyl precursor.[56] Thus the product distribution could be controlled by the substrate concentration. In dilute solutions, 2-adamantyl carboxylate was the major product and 1-adamantyl carboxylate the minor product. In more concentrated solutions, the product distribution was reversed. Despite this work, the details of the mechanism remain unclear. It should be pointed out that second-order kinetics do not apply to the earlier stages of rearrangement of the THDPD to the 1-adamantyl cation,[14] that is, the rate-determining steps. However, one must conclude that intermolecular hydride transfers are responsible for the rapid rearrangement from the 2- to 1-adamantyl cation under stable ion conditions.

In 1968, Whitlock[9a] looked at the 3,5,7-d_3-1-adamantyl cation (**54**) under stable ion conditions. After 30 min at –78°C, mass spectral analysis of the quenched products showed that the deuterium incorporation was the same as that of the starting material. In a second experiment, an ~ 1:1 mixture of d_3 and d_0 material was employed, and product analysis showed that, under comparable conditions to his original experiment, there was complete deuterium randomization at the combined bridgehead positions. Assuming the 30 min is greater than five half-lives, this gives an activation barrier of less than 14 kcal/mol for the exchange of bridgehead hydrogen atoms in the 1-adamantyl cation. In 1971, Schleyer[57] repeated Whitlock's first experiment under more forcing conditions using a deuterium label in the secondary adamantyl position. After 90 min at 105°C, the ^{1}H NMR signal of the bridgehead tertiary hydrogen atoms showed less than 10% loss of intensity. This places a lower limit on the exchange between secondary and tertiary hydrogen atoms of 29–30.5 kcal/mol.

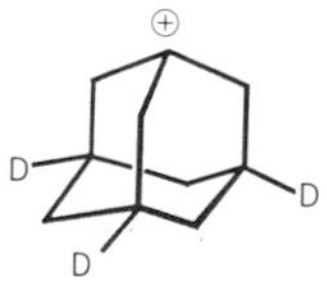

54

Considerable caution has to be exercised in the analysis of the activation barriers for the 2- to 1-adamantyl rearrangement since this is clearly an intermolecular process, and the barrier heights may be radically altered by changes in concentration. However, assuming that the results of Whitlock and Schleyer were carried out at similar concentrations (likely for stable ion work), then the following (abstract) system may be arrived at (Fig. 19). The salient features of this scheme are the following:

1. The barrier for the A to B transformation is less than 11 kcal/mol, allowing for the facile transformation of the 2-adamantyl cation to the 1-adamantyl cation under stable ion conditions.
2. The barrier C to B to C is less than 14 kcal/mol (and may be considerably less), and allows for the exchange of bridgehead protons in the 1-adamantyl cation.

3. The large barrier of ~ 30 kcal/mol in going from C to A accounts for the lack of exchange between the secondary and tertiary protons in the 1-adamantyl cation.

Assuming that other tertiary and secondary cations behave in a similar way at comparable concentrations allows the following generalization to be made. Intermolecular hydride transfers may convert secondary cations to tertiary cations with relative ease, but the reverse process is expected to be a high energy process *for typical concentrations of carbocations under stable ion conditions.* This analysis takes no account of the mechanism by which these intermolecular hydride shifts occur. Since the rate-determining barriers in the early stages of the reaction of the THDPD cation to the 1-adamantyl cation involve secondary intermediates (that undergo subsequent framework rearrangements or intramolecular hydride shifts to give other secondary cations), it is possible that intermolecular hydride transfers may raise the effective barriers to the rearrangement by forcing the reaction "backwards" to more stable tertiary cations.

The mechanism of the intermolecular rearrangement can be accounted for in one of two ways:

1. A small (catalytic) amount of extraneous hydrocarbon can act as a hydride donor to the 2-adamantyl cation, thus forming adamantane and a new cationic intermediate. Recombination of these two species, again by an intermolecular hydride transfer, could then give the more stable 1-adamantyl

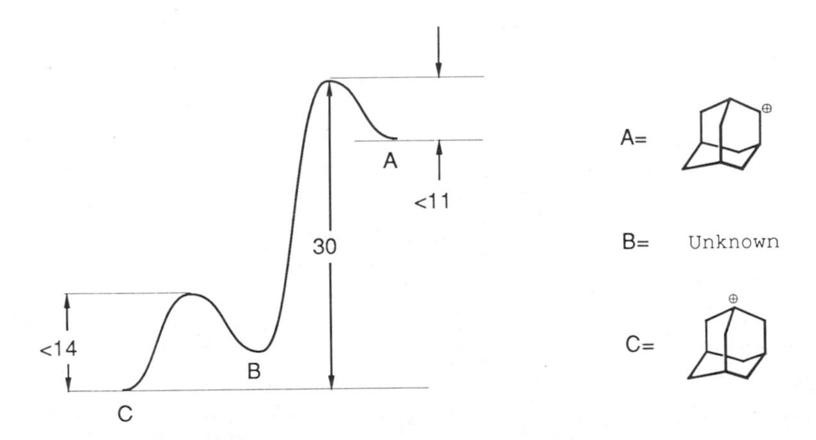

Fig. 19. Schematic for intermolecular hydride transfers in the adamantane framework. All energies in kilocalories per mole.

+ RH ⇌ + $R^{\oplus}$ ⇌ + RH

(A) (B) (C)

Fig. 20. Bimolecular hydride transfer via a hydrocarbon.

cation and regenerate the original hydrocarbon (Fig. 20). The same process acting on the 1-adamantyl cation would also account for the isomerization of the bridgehead positions in that cation. The high barrier to the 2-adamantyl cation from the 1-adamantyl cation implies that the catalytic cationic species (R^+, Fig. 20) would be much more stable than the 2-adamantyl cation itself. Whitlock[9a] found rigorous solvent and reagent purification had no effect on the product distribution in a deuterium labeling experiment, in which the observed scrambling was thought to be a result of intermolecular hydride transfers.

2. Disproportionation of the 2-adamantyl cation, to give an adamantyl dication and adamantane, could occur. The subsequent recombination of these, to give two 1-adamantyl cations, would be strongly favored from a thermodynamic point of view. However, the dication would have to be a high energy species since attempts to directly observe these have been unsuccessful.[58] A more likely disproportionation species is a monocation-monoester (or halide, depending on the superacid medium used). In any case, the other disproportionation product, adamantane, could serve as the RH species in Fig. 20 and as B in Fig. 19, and need be present only in minute concentrations.

The stability of the 1-adamantyl cation has been considered, especially concerning the possibility of C—C hyperconjugative stabilization of the cation. The solvolysis rates of 1-adamantyl and 1-bicyclo[2.2.2]octyl (**55**) derivatives are not expected to be much different on the basis of ring strain. However, 1-adamantyl bromide solvolyzes ~ 10^3 times as fast as that of **55** (X=Br and R=H), and the possibility of hyperconjugative stabilization was put forward to account for these observations.[59] Substitution of the bridgehead hydrogen for methyl groups in 1-adamantyl bromide would be expected to increase the hyperconjugative stabilization, and therefore increase the solvolysis rates. The opposite effect is in fact observed, and Schleyer and co-workers[59] raised the possibility that methyl groups were in fact electron withdrawing when attached to a saturated carbon center. The anomalous behavior of hydrogen relative to other substituents was later expanded by Schleyer and co-workers.[60] The solvolysis of 3-substituted 1-adamantyl bromides (**56**, X=Br) and 4-substituted 1-bicyclo[2.2.2]octyl brosylates (**55**, X=OBs) were reported. Plots of the rates of solvolysis (relative to the rates of their respective unsubstituted derivatives) of **56** (X=Br) versus **55** (X=OBs) showed excellent correlation except when the substituent was hydrogen. It was concluded that, on changing a bridgehead group from hydrogen to alkyl, there were subtle changes in the cyclic geometry, which depended on the nature of the bicyclic (or tricyclic) structure.[60]

A similar approach was subsequently followed by Grob.[61] Plots of solvolysis rates of **56** (X = Br, OTs, or ONs) versus the inductive substituent constants (σ_I^q) showed good correlation for most substituents. However, strongly electron donating groups showed anomalous (enhanced) rates, as well as large amounts of rearranged products. The conclusion reached at the time was that C—C hyperconjugation was only important in those cases where the 3-substituent was electron donating.

Some of the strongest evidence for hyperconjugative stabilization in the 1-adamantyl cation comes from the stable ion work of Olah and Schleyer.[3,15] Under stable ion conditions, the chemical shifts of the γ carbon and proton signals are at lower field than are the corresponding β signals, the opposite effect is expected based only on proximity to the positive charge. The C—C hyperconjugation or the backside interaction between the cation *p* orbital with the back lobes of the bridgehead C—H bonds was put forward to account for the unusual chemical shifts observed.

The lengthened β,γ bond expected to be associated with C—C hyperconjugation was observed by X-ray crystallography.[62] The crystal structure of the 3,5,7-trimethyl-1-adamantyl cation (as its SbF_6^- salt) showed a β,γ bond length of 1.62 Å, compared to bond lengths of 1.44 Å for the α,β bonds. The out of plane deformation at the cationic center was 0.21 Å.

CONCLUSION

At first sight, the Lewis acid catalyzed rearrangement of THDPD to adamantane appears to be a remarkably complex transformation that contains an embarrassing lack of information concerning both the mechanism and the energetics of the process (apart from the obvious fact that the rearrangement occurs through the intermediacy of carbocations). By careful study of those hydrocarbons that are believed to be intermediates in the reaction, the mechanistic problem was eventually solved. This chapter has attempted to complete the analysis by estimating the barriers for each of the individual steps in the cationic process, using both those compounds actually involved in the transformation as well as related model compounds. In this way, a detailed potential energy surface has been constructed from its individual components, and the picture thus obtained agrees very well with the observed overall kinetics of the process, confirming the validity of this approach.

Although only three stable cations were observed in the cationic THDPD rearrangement, this does not preclude the observation of other stable carbocations that are not directly involved in the rearrangement itself. The attempted observation of such cations is now described. Paquette[63] obtained the 1-adamantyl cation by ionizing either the tricyclic alcohol (**57**) or the epimeric alcohols (**58**) in fluorosulfonic acid at –78°C. No intermediate cations were observed, and the initial skeletal rearrangement probably involves a 1,2-alkyl shift to a secondary cation.[63] The final stages of the proposed rearrangement are identical to those described in this chapter. Cation **59** was not prepared directly, but in light of the high barriers of 1,2-hydride shifts (>15 kcal/mol) observed in the structurally related dodecahedral cation;[64] cation **59** might be observable if such high barriers remain. Alcohol **60** (R = OH) was ionized in fluorosulfonic acid at –78°C, but extensive decomposition occurred.[65] However, 4% of an epimeric alcohol was obtained, but its structure was not determined. We only know that it was not 1-adamantanol. Acetolysis of the corresponding tosylate (**60**, R = OTs) gave only the stereoisomer **61**. Although the cation produced by both the solvolysis and superacid experiments is secondary, there appears to be no easy way for it to rearrange, except for an interconversion with its stereoisomer. Indeed Whitlock[9a] commented that this framework "lies on the periphery" of the THDPD rearrangement. The regioselective generation of cation **62** was reported under ionic type hydrogenation conditions[66] in which the cation (or rearranged cation) is subsequently trapped as the hydrocarbon. Under these conditions, a 4:1 mixture of adamantane and protoadamantane was recovered, providing little insight into the rearrangement.

Since only three observable adamantaneland $C_{10}H_{15}^+$ cations are presently known, one can wonder about the reason for this, and in the broader context of Fig. 2, whether these three are the *only* possible candidates. The hyperconjugative stabilization of both the 1-adamantyl cation and the THDPD

OH

OH

57 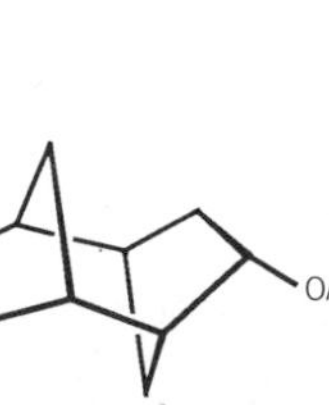 58 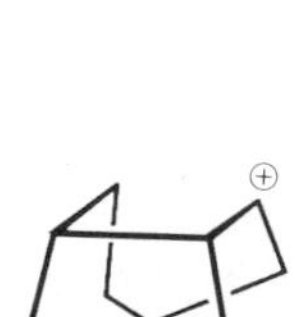 59

OAc

OR

60 61 62

cation was already commented on, and appears to play a significant role in cation stabilization. On the basis of chemical shifts, some hyperconjugative interactions appear to be operative in the 3-homobrendyl cation. Thus, the 4- and 5-protons of **20** have chemical shifts of 5.32 and 5.24 ppm, respectively;[14] and the C7 chemical shift appears at abnormally low field (53.8 ppm).[14] In addition to good cation stability, the observation of a cation requires high energy barriers out of the energy minimum, so as to prevent the rearrangement of the cation. On the basis of the known chemistry of the THDPD rearrangement, large barriers will probably involve the unfavorable rearrangement of a secondary cation "partner." Finally, a moderate degree of nonplanarity at the cationic center does not, on its own, appear to preclude the observation of a stable carbocation, as evidenced by the observation of both the 1-adamantyl and 3-homobrendyl cations. The enforcement of more severe nonplanarity, such as at a [2.2.1] bridgehead, however does prevent the formation of a stable carbocation.

Taking into account the above considerations, we conclude this chapter with a brief analysis of any other possible "adamantaneland" cations that may be stable enough to be observable. The seven remaining frameworks that have not already been discussed in detail are shown in Figure 21, in which the possible sites of stable carbocations are shown with solid circles. The degree to which both the skeletal strain and the nonplanarity of the cation contribute to the overall stability is probably best estimated using semiempirical techniques.[67] However, in the absence of such information, we will consider those cations that seemingly have some possibility of being directly observable.

Cation **64** can be converted to cation **63** by a sequence of rearrangements involving exo 3,2-hydride shifts and Wagner–Meerwein shifts; and **63** in turn can rearrange to the homobrendyl cation via these two processes and that of a 2,6-hydride shift (the numbering being used here is that for a bicyclo-[2.2.1]heptane unit). Since none of these processes are expected to be prohibitively large it is unlikely that either of these cations will be observable. Cation **65** contains two sites at the bridgehead positions of a bicyclo[2.2.2] system, and since the bicyclo[2.2.2]octyl system is known to accommodate a bridgehead cation,[68] then **65** may be an observable cation. Cation **66** has two sites that are not too distorted, but both of these cations have excellent orbital overlap with adjacent hydrogen atoms, so the possibility of 1,2-hydride shifts (these numbers are used generically) occurring is great, and the resultant cations can both rearrange to the protoadamantyl system without the need to postulate a high energy process. The least strained tertiary center in cation **67** has poor orbital overlap with adjacent hydrogen atoms, but if a 1,2-hydride shift does occur, the subsequent rearrangement to the protoadamantyl system is expected to be rapid.[66] Cation **68** also has poor orbital overlap, but again, if a 1,2-hydride shift does occur, then the rearrangement to the 2-twistyl cation is expected to be rapid. Since both the protadamantyl[14] and the twistyl[9a] cations are known to rearrange immediately to the 1-adamantyl cation, then neither cation **67** or **68** is expected to be observable if the 1,2-hydride shifts have barriers of < 10 kcal/mol. Cation **69** is contained within a [3.3.2]decyl system in which some additional strain has been introduced by joining the two bridges together. Like cations **67** and **68**, an unfavorable 1,2-hydride shift is all that

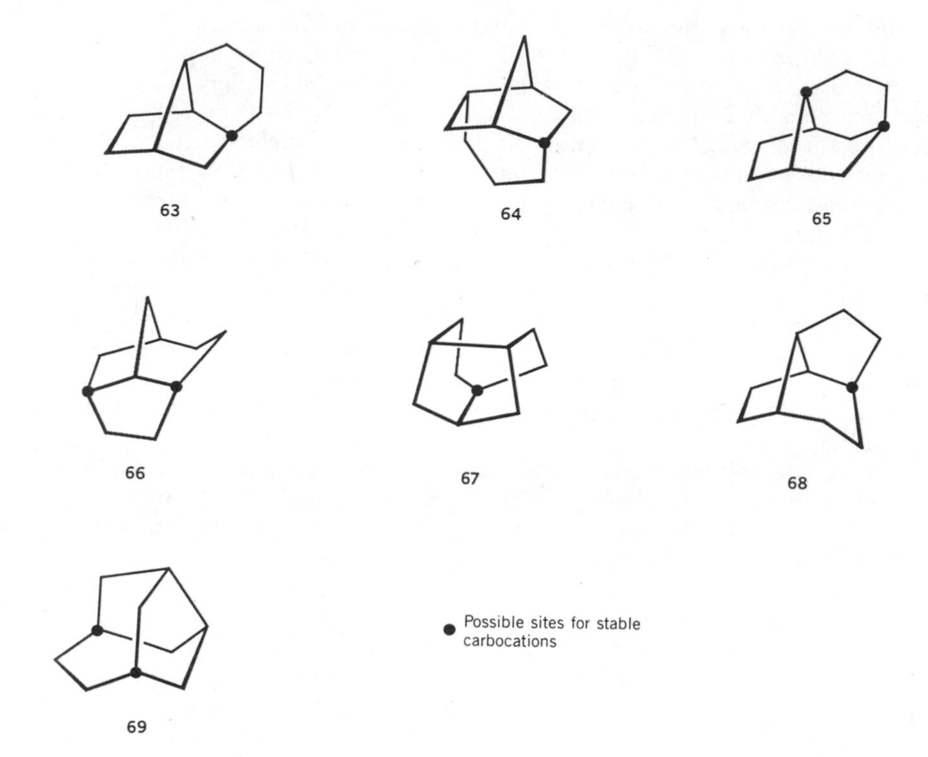

Fig. 21. Possible carbon frameworks that are capable of supporting observable carbocations.

prevents the cascade to the 1-adamantyl cation. It appears that these cations are in fact the best chances one has of observing other stable adamantaneland cations.

REFERENCES

1. P. von R. Schleyer, *J. Am. Chem. Soc.* **1957**, *79*, 3292.
2. G. A. Olah and O. Farooq, *J. Org. Chem.* **1986**, *51*, 5410.
3. P . v. R. Schleyer, R. C. Fort, Jr., W. E. Watts, M. B. Comisarow, and G. A. Olah, *J. Am. Chem. Soc.* **1964**, *86*, 4195.
4. (a) V. Prelog and R. Seiwerth, *Chem. Ber.* **1941**, *74*, 1644. (b) V. Prelog and R. Seiwerth, *Chem. Ber.* **1941**, *74*, 1769.
5. (a) S. Hala and S. Landa, *Erdoel Kohle*, **1966**, *19*, 727. (b) S. Landa and V. Machacek, *Collect. Czech. Chem. Commun.* **1933**, *5*, 1.

6. (a) A. Schneider, R. W. Warren, and E. J. Janoski, *J. Org. Chem.* **1966**, *31*, 1617. (b) S. Hala, S. Landa, and V. Hanus, *Angew. Chem.* **1966**, *78*, 1060. (c) H. Hamill, A. Karim, and M. A. McKervey, *Tetrahedron* **1971**, *27*, 4317.
7. (a) A. M. Klester, and C. Ganter, *Helv. Chim. Acta*, **1985**, *68*, 734. (b) A. M. Klester and C. Ganter, *Helv. Chim. Acta*, **1985**, *68*, 104.
8. P. von R. Schleyer and M. M. Donaldson, *J. Am. Chem. Soc.* **1960**, *82*, 4645.
9. (a) H. W. Whitlock, Jr. and M. W. Siefkin, *J. Am. Chem. Soc.* **1968**, *90*, 4929. (b) L. A. Paquette, G. V.Meeham, S. J. Marshall, *J. Am. Chem. Soc.* **1969**, *91*, 6779. (c) D. Lenoir and P. von R. Schleyer, *Chem. Commun.* **1970**, 941.
10. (a) E. M. Engler, M. Farcasiu, A. Sevin, J. M. Cense, and P. von R. Schleyer, *J. Am. Chem. Soc.* **1973**, *95*, 5769. (b) F. J. Jaggi and C. Ganter, *Helv. Chim. Acta* **1980**, *63*, 866. (c) A. M. Klester, F. J. Jaggi, and C. Ganter, *Helv. Chim. Acta* **1980**, *63*, 1294.
11. N. Tanaka, T. Iizuka, and T. Kan, *Chem. Lett.* **1974**, 539.
12. (a) Z. Majerski, S. H. Liggero, P. von R. Schleyer, and A. P. Wolf, *Chem. Commun.* **1970**, 1596. (b) A. M. Klester and C. Ganter, *Helv. Chim. Acta* **1983**, *66*, 1200. (c) M. Farcasiu, E. W. Hagaman, E. Wenkert, and P. von R. Schleyer, *Tetrahedron Lett.* **1981**, *22*, 1501.
13. P. von R. Schleyer, J. E. Williams, and K. R. Blanchard, *J. Am. Chem. Soc.* **1970**, *92*, 2377.
14. R. P. Kirchen, T. S. Sorensen, and S. M. Whitworth, unpublished results.
15. G. A. Olah, G. K. S. Prakash, J. G. Shih, V. V. Krishnamurthy, G. D. Mateescu, G. Liang, G. Sipos, V. Buss, T. M. Gund, and P. von R. Schleyer, *J. Am. Chem. Soc.* **1985**, *107*, 2764.
16. L. Huang, K. Ranganayakulu, and T. S. Sorensen, *J. Am. Chem. Soc.* **1973**, *95*, 1936.
17. A. J. Jones, E. Huang, R. Haseltine, and T. S. Sorensen, *J. Am. Chem. Soc.* **1975**, *97*, 1133.
18. N. Takaishi, Y. Fujikura, and Y. Inamoto, *Synth. Commun.* **1983**, 293.
19. (a) F. A. L. Anet and V. J. Basus, *J. Magn. Reson.* **1978**, *32*, 339. (b) N. Okazawa and T. S. Sorensen, *Can. J. Chem.* **1978**, *56*, 2737.
20. (a) R. Haseltine, E. Huang, K. Ranganayakulu, T. S. Sorensen, and N. Wong, *Can. J. Chem.* **1975**, *53*, 1876. (b) K. Ranganayakulu and T. S. Sorensen, *J. Org. Chem.* **1984**, *49*, 4310. (c) R. Haseltine, N. Wong, T. S. Sorensen, and A. J. Jones, *Can. J. Chem.* **1975**, *53*, 1891.
21. R. Haseltine, E. Huang, K. Ranganayakulu, and T. S. Sorensen, *Can. J. Chem.* **1975**, *53*, 1056.
22. T. Laube, *Angew. Chem. Int. Ed. Eng.* **1987**, *26*, 560.
23. E. Huang, K. Ranganayakulu, and T. S. Sorensen, *J. Am. Chem. Soc.* **1972**, *94*, 1780.
24. (a) G. A. Olah, A. M. White, J. R. DeMember, A. Commeyras, and C. Y. Lui, *J. Am. Chem. Soc.* **1970**, *92*, 4627. (b) G. A. Olah, G. Liang, G. D. Mateescu, and J. L. Riemenschneider, *J. Am. Chem. Soc.* **1973**, *95*, 8698.
25. P. von R. Schleyer, P. Grubmuller, W. F. Maier, O. Vostrowsky, L. Skattebol, and K. H. Holm, *Tetrahedron Lett.* **1980**, *21*, 921.
26. F. J. Jaggi and C. Ganter, *Helv. Chim. Acta*, **1980**, *63*, 214.

27. Y. Tobe, K. Terashima, Y. Sakai, and Y. Odaira, *J. Am. Chem. Soc.* **1981**, *103*, 2307.

28. E. J. Corey and R. S. Glass, *J. Am. Chem. Soc.* **1967**, *89*, 2600.

29. (a) G. A. Olah and G. Liang, *J. Am. Chem. Soc.* **1974**, *96*, 189. (b) G. A. Olah, G. Liang, *J. Am. Chem. Soc.* **1974**, *96*, 195.

30. J. G. Henkel and L. A. Spurlock, *J. Am. Chem. Soc.* **1973**, *95*, 8339.

31. G. A.-Craze and I. Watt, *J. Chem. Soc. Perkin Trans. 2* **1981**, 175.

32. A. Karim, M. A. McKervey, E. M. Engler, and P. von R. Schleyer, *Tetrahedron Lett.* **1971**, 3987.

33. A. Karim and M. A. McKervey, *J. Chem. Soc. Perkin Trans. 1,* **1974,** 2475.

34. J. J. Sosnowski, E. B. Danaher, and R. K. Murry, Jr., *J. Org. Chem.* **1985**, *50*, 2759.

35. L. A. Spurlock and K. P. Clark, *J. Am. Chem. Soc.* **1972**, *94*, 5349.

36. E. J. Jaggi and C. Ganter, *Helv. Chim. Acta,* **1980**, *63*, 2087.

37. Z. Majerski, S. Djigas, and V. Vinkovic, *J. Org. Chem.* **1979**, *44*, 4064.

38. (a) M. Tichy, L. Kniezo, and J. Hapala, *Tetrahedron Lett.* **1972**, 699. (b) M. Tichy, L. Kniezo, and J. Hapala, *Collect. Czech. Chem. Commun.* **1975**, *40*, 3862.

39. (a) A. Belanger, Y. Lambert, and P. Deslongchamps, *Can. J. Chem.* **1969**, *47*, 795. (b) R. C. Bingham, P. von R. Schleyer, Y. Lambert, and P. Deslonchamps, *Can. J. Chem.* **1970**, *48*, 3739.

40. D. Lenoir, R. E. Hall, and P. von R. Schleyer, *J. Am. Chem. Soc.* **1974**, *96*, 2138.

41. D. Lenoir and P. von R. Schleyer, *Chem. Commun.* **1970**, 941.

42. G. A. Olah, G. Liang, and G. D. Mateescu, *J. Org. Chem.* **1974**, *39*, 3750.

43. J. M. Harris, D. J. Raber, R. E. Hall, and P. von R. Schleyer, *J. Am. Chem. Soc.* **1970**, *92*, 5729.

44. (a) J. Boyd and K. H. Overton, *J. Chem. Soc. Perkin Trans. 1*, **1972**, 2533. (b) H. J. Storesund and M. C. Whiting, *J. Chem. Soc. Perkin Trans. 2*, **1975**, 1452.

45. J. E. Nordlander and J. E. Haky, *J. Org. Chem.* **1980**, *45*, 4780.

46. J. E. Nordlander, J. E. Haky, and J. P. Landino, *J. Am. Chem. Soc.* **1980**, *102*, 7487.

47. A. D. Allen and T. T. Tidwell, *J. Am. Chem. Soc.* **1982**, *104*, 3145.

48. (a) D. Lenoir, D. J. Raber, and P. von R. Schleyer, *J. Am. Chem. Soc.* **1974**, *96*, 2149. (b) D. Kovacevic, B. Goricnik, and Z. Majerski, *J. Org. Chem.* **1978**, *43*, 4008. (c) J. E. Nordlander and J. E. Haky, *J. Am. Chem. Soc.* **1981**, *103*, 1518.

49. (a) J. A. Bone and M. C. Whiting, *Chem. Commun.* **1970**, 115. (b) J. A. Bone, J. R. Pritt, and M. C. Whiting, *J. Chem. Soc. Perkin Trans 2*, **1975**, 1447.

50. C. K. Cheung, L. T. Tseng, M.-H. Lin, S. Srivastava, and W. J. le Noble, *J. Am. Chem. Soc.* **1986**, *108*, 1598.

51. C. A. Grob, G. Wittwer, and K. R. Rao, *Helv. Chim. Acta*, **1985**, *68*, 651.

52. C. Paradisi, and J. F. Bunnett, *J. Am. Chem. Soc.* **1985**, *107*, 8223.

53. (a) D. Lenoir, P. Mison, E. Hyson, P. von R. Schleyer, M. Saunders, P. Vogel, and L. A. Telkowski, *J. Am. Chem. Soc.* **1974**, *96*, 2157. (b) P. von R. Schleyer, D. Lenoir, P. Mison, G. Liang, G. K. S. Prakash, G. A. Olah, *J. Am. Chem. Soc.* **1980**, *102*, 683.

54. E. S. Finne, J. R. Gunn, and T. S. Sorensen, *J. Am. Chem. Soc.* **1987**, *109*, 7816.

55. M. L. Sinnott and M. C. Whiting, *J. Chem. Soc. Perkin Trans. 2,* **1975**, 1446.

56. (a) P. von R. Schleyer, *Angew. Chem. Int. Ed. Eng.* **1969**, *8*, 529. (b) P. von R. Schleyer, L. K. M. Lam, D. J. Raber, J. L. Fry, M. A. McKervey, J. R. Alford, B. D. Cuddy, V. G. Keizer, H. W. Geluk, and J. L. M. A. Schlatmann, *J. Am. Chem. Soc.* **1970**, *92*, 5246. (c) J. R. Alford, B. D. Cuddy, D. Grant, and M. A. McKervey, *J. Chem. Soc. Perkin Trans.* I **1972**, 2707.

57. P. Vogel, M. Saunders, W. Thielecke, and P. von R. Schleyer, *Tetrahedron Lett.* **1971**, 1429.

58. G. K. S. Prakash, V. V. Krishnamurthy, A. Arvanaghi, and G. A. Olah, *J. Org. Chem.* **1985**, *50*, 3985.

59. R. C. Fort, Jr. and P. von R. Schleyer, *J. Am. Chem. Soc.* **1964**, *86*, 4194.

60. P. von R. Schleyer and C. W. Woodworth, *J. Am. Chem. Soc.* **1968**, *90*, 6528.

61. (a) C. A. Grob, W. Fischer, and H. Katayama, *Tetrahedron Lett.* **1976**, 2183. (b) C. A. Grob and B. Schaub, *Helv. Chim. Acta* **1982**, *65*, 1720. (c) C. A. Grob, G. Wang, and C. Yang, *Tetrahedron Lett.* **1987**, *28*, 1247.

62. T. Laube, *Angew. Chem. Int. Ed. Eng.* **1986**, *25*, 349.

63. L. A. Paquette, G. V. Meehan, and S. J. Marshall, *J. Am. Chem. Soc.* **1969**, *91*, 6779.

64. G. A. Olah, G. K. S. Prakash, W.-D. Fessner, T. Kobayashi, and L. A. Paquette, *J. Am. Chem. Soc.* **1988**, *110*, 8599.

65. L. A. Paquette, C. W. Doecke, and G. Klein, *J. Am. Chem. Soc.* **1979**, *101*, 7599.

66. H.-R. Kanel and C. Ganter, *Helv. Chim. Acta* **1985**, *68*, 1226.

67. (a) R. C. Bingham, M. J. S. Dewar, and D. H. Lo, *J. Am. Chem. Soc.* **1975,** *97*, 1285. (b) M. J. S. Dewar and W. Thiel, *J. Am. Chem. Soc.* **1977**, *99*, 4899. (c) M. J. S. Dewar, E. G. Zoebisch, E. F. Healy, and J. J. P. Stewart, *J. Am. Chem. Soc.* **1985**, *107*, 3902.

68. A. de Meijere, O. Schallner, P. Golitz, W. Weber, and P. von R. Schleyer, *J. Org. Chem.* **1985**, *50*, 5255.

4 Carbocations and Electrophilic Reactions of Cage Hydrocarbons

GEORGE A. OLAH

Loker Hydrocarbon Research Institute and
Department of Chemistry
University of Southern California
Los Angeles, California

EARLY SYNTHESES OF ADAMANTANE AND SCHLEYER'S CARBOCATIONIC ISOMERIZATION ROUTE

Diamondoid hydrocarbons fascinated chemists for a long time.[1] Interest goes back to the 1920s when organic chemists considering the perfect structure of diamond speculated on the possibility of someday obtaining hydrocarbons with such a structure. To them, of course, six-membered rings represented cyclohexanes.[2]

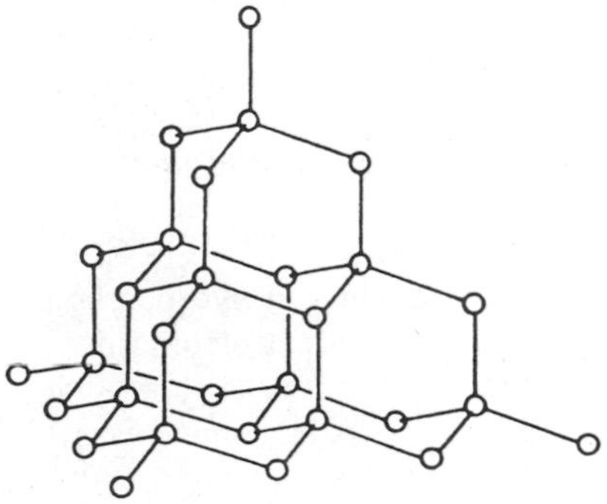

Diamond

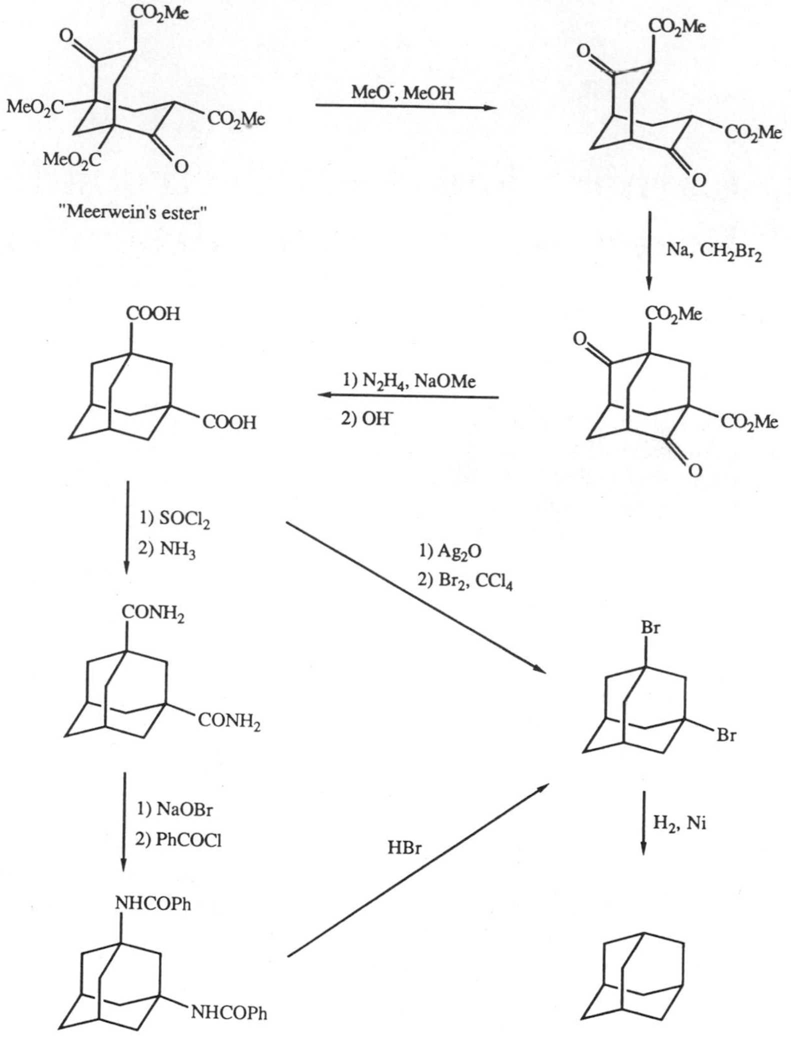

Scheme 1. Prelog's adamantane synthesis.[4]

As it turned out, the parent diamandoid hydrocarbon, adamantane, is present in nature and Landa isolated it from petroleum in Czechoslovakia in 1933.[3] Many at that time questioned Landa's structure, which, however, subsequently was proven by Prelog's synthesis.[4] Adamantane is a C_{10} hydrocarbon, that is, tricyclodecane, comprising the smallest repeating unit of the diamond lattice.

Diamantane and triamantane (vide infra) are the next higher homologues. Prelog's multistep adamantane synthesis (Scheme 1) in 1941 achieved only a 0.3% overall yield.

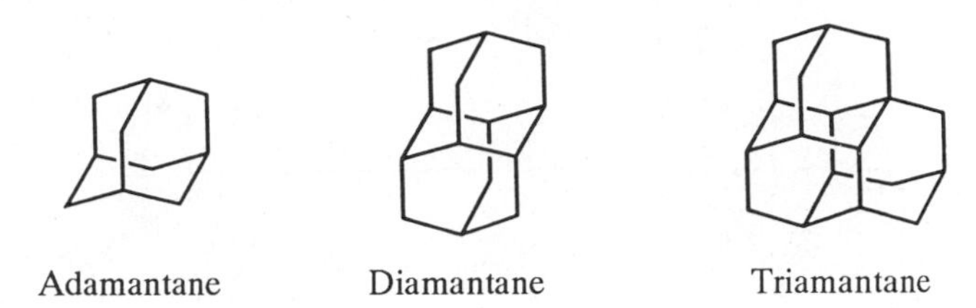

Stetter's subsequent adamantane synthesis[5] (Scheme 2) improved the yield to 6%, but still the field would never have opened up for extensive study without a much more convenient way to make adamantane.

CO_2Me O MeO_2C CO_2Me MeO_2C O

1) H_2/Pt
2) PCl_5

CO_2Me Cl MeO_2C CO_2Me MeO_2C Cl

H_2/Ni
OH^-

COOH HOOC COOH HOOC

1) Ag_2O
2) Br_2, CCl_4

Br Br Br Br

H_2, Ni

Scheme 2. Stetter's adamantane synthesis.[5]

This is the background against which Schleyer's breakthrough must be viewed.[6] In 1957 he found that *endo*-trimethylene norbornane, the hydrogenated dicyclopentadiene dimer, isomerizes readily to the exo isomer using $AlCl_3$ as catalyst (Scheme 3). In the course of the $AlCl_3$ catalyzed isomerization studies he unexpectedly also observed formation of a crystalline, high melting compound, which he identified as adamantane.

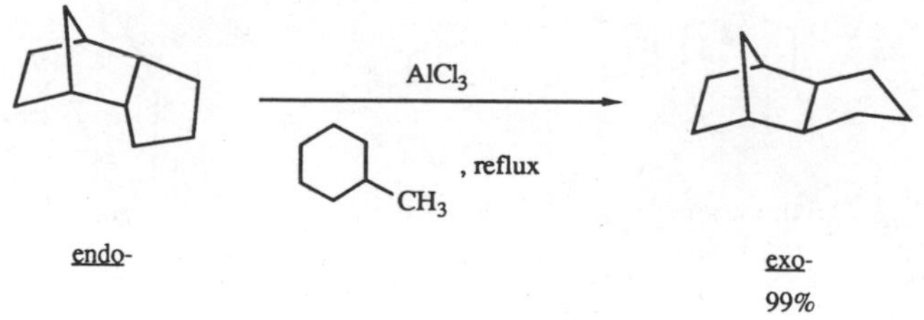

$AlCl_3$

, reflux

CH_3

endo-

exo-

99%

Scheme 3.

The readily available and cheap *endo*-trimethylenenorbornane gave 10% yield of adamantane, a remarkable result at the time considering that up to this point adamantane chemistry was limited to the two previously mentioned pioneering but low yield multistep syntheses. Schleyer subsequently improved the yield of his isomerization process using $AlCl_3$ sludge catalysts to ~ 19% (Scheme 4).[7,8] The yield of adamantane was low because of the concomitant formation of

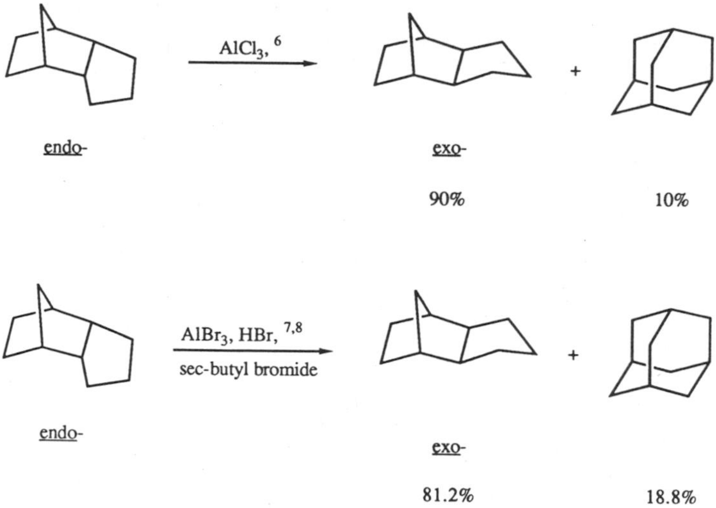

$AlCl_3$, [6]

+

endo-

exo-

90%

10%

$AlBr_3$, HBr, [7,8]

sec-butyl bromide

+

endo-

exo-

81.2%

18.8%

Scheme 4. Schleyer's isomerization of trimethylenenorbornane.

virtually hundreds of byproducts as shown by the gas–liquid chromatography of the mother liquor from the isomerization reaction.

This breakthrough was the starting point of the spectacular development and growth of what we now recognize as cage hydrocarbon chemistry. In addition, some of the higher homologues of adamantane, such as diamantane and triamantane, later become available by similar isomerization methodology.

Subsequent progress in the preparation of adamantane was achieved by Plate et al.[9] using heterogeneous, gas-phase isomerization over aluminosilicate

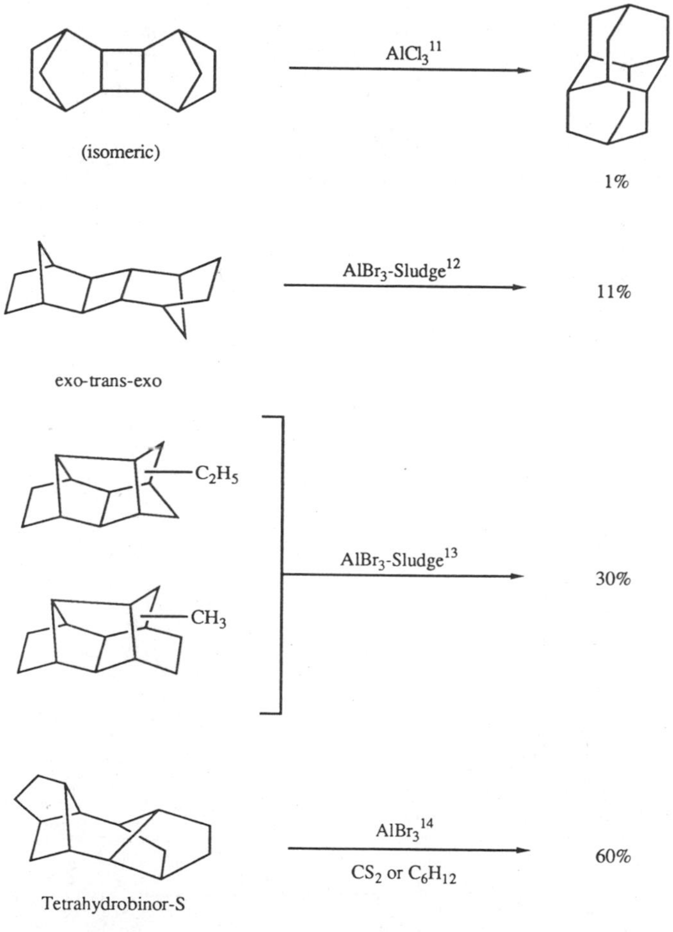

(isomeric) — $AlCl_3$[11] → 1%

exo-trans-exo — $AlBr_3$-Sludge[12] → 11%

C_2H_5; CH_3 — $AlBr_3$-Sludge[13] → 30%

Tetrahydrobinor-S — $AlBr_3$[14], CS_2 or C_6H_{12} → 60%

Scheme 5. Preparation of Diamantane from various C_{14}-precursors.

catalysts at 450–475°C and notably by McKervey et al.[10] using a chlorinated Pt on aluminum catalyst. Whereas the former gave only a 6–13% yield, the latter achieved a 60% yield, which in subsequent work was even further improved.

The next higher homologue, diamantane, is a particularly interesting molecule. I remember attending an IUPAC conference in London in 1963 at which the emblem of the conference was this intriguing, and at the time yet unknown, C_{14} hydrocarbon. On the inside cover of the Congress program there was a challenge to the chemical community to attempt to prepare this hydrocarbon and it was suggested that it be named "congressane." It took Schleyer and Cupas only 2 years to make congressane, albeit in 1% yield.[11] Subsequent research improved the yield to 11%, to 30%, and finally to 60% (Scheme 5).[12–14]

In 1966 Schleyer and co-workers reported the next homologue, that is, triamantane.[15] Initially the yield was again low (2–5%), but McKervey and co-workers found more suitable precursors to raise it to 60%,[16] a quite remarkable achievement considering the difficulties which otherwise would be involved in synthesizing a molecule of this size (Scheme 6).

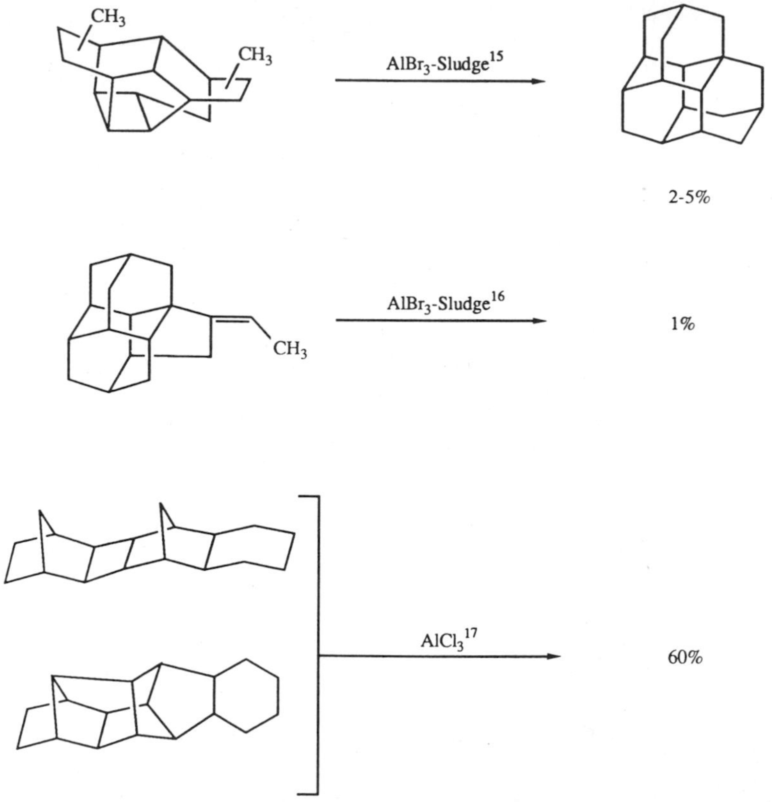

CH3
CH3
$AlBr_3$-Sludge[15]
2-5%
$AlBr_3$-Sludge[16]
CH3
1%
$AlCl_3$[17]
60%

Scheme 6. Preparation of triamantane from various C_{18} precursors.

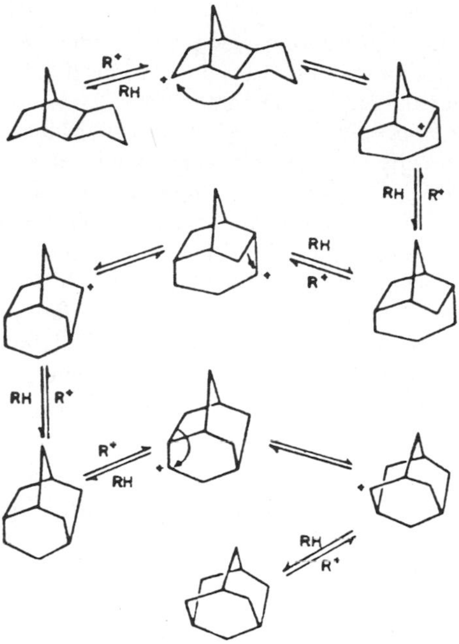

Scheme 7. Carbocationic pathway of the adamantane rearrangement.[18]

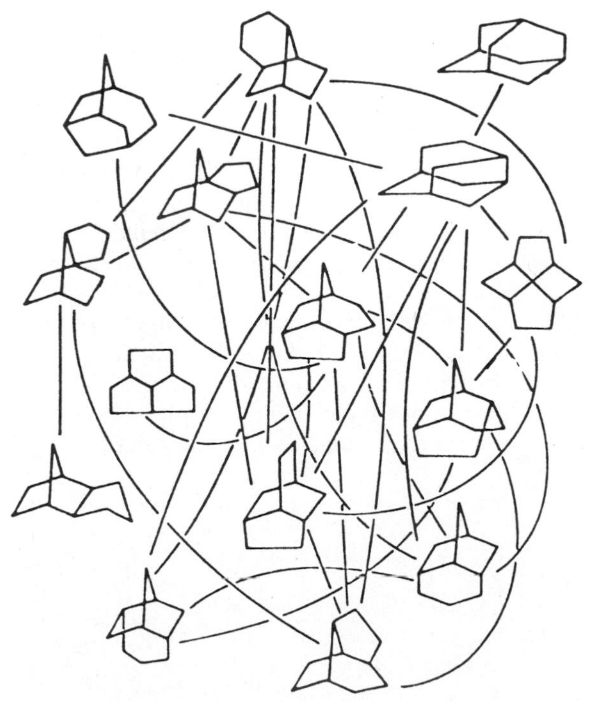

Scheme 8. Partial map of adamantaneland.[17]

The acid catalyzed isomerization reactions of precursor polycyclic aromatics to adamantanoid cage hydrocarbons involve carbocationic processes.[18] Scheme 7 illustrates this adamantane rearrangement.

It is necessary to emphasize that the isomerization of possible $C_{10}H_{16}$ isomers to the thermodynamically most stable adamantane is a very complex system, which literary can involve thousands of steps (and intermediate ions).[17] Very few of these were *de facto* observed. The map of "adamantaneland" (Scheme 8) is very complex, as discussed in the interconversion possibilities of all possible $C_{10}H_{16}$ hydrocarbons to adamantane by Whitlock and Siefkin.[19] The key to our understanding of the acid catalyzed carbocationic isomerization processes is that if allowed to react long enough, all involved ions eventually convert to the thermodynamically favored adamantane system.

SUPERACID CATALYZED ISOMERIZATION TO ADAMANDOID HYDROCARBONS

Over the years, my group has developed and studied superacids and their chemistry.[20] In the course of this work we have applied superacids to the isomerization of hydrocarbons and compared them with the conventional $AlCl_3$ and related Friedel–Crafts systems. Table 1 shows a comparison of reported results of the isomerization of *endo*-trimethylenenorbornane to adamantane with varying superacid catalysts including $HF—BF_3$, $HF—SbF_5$, $FSO_3H—SbF_5$, $CF_3SO_3H—SbF_5$, and the recently developed $CF_3SO_3H—B(O_3SCF_3)_3$ giving 40–98% yields.

The $CF_3SO_3H—SbF_5$ and $CF_3SO_3H—B(O_3SCF_3)_3$ systems with a 1:1 acid to hydrocarbon molar ratio gave practically quantitative isomerization to adamantane. When starting with the endo isomer rapid initial isomerization to the exo isomer takes place, followed by slower skeletal isomerization to adamantane.

The superacidic systems are able to achieve highly efficient isomerization to adamantane because in the course of the carbocationic conversion process of isomeric C_{10} hydrocarbons, the 1-adamantyl cation is the thermodynamically most stable carbocation, into which eventually all intermediate ions convert. Under stable ion conditions in superacids the bridgehead 1-adamantyl cation is readily obtained.[25,26]

TABLE I. Superacid Isomerization of *endo*- and *exo*-Trimethylenenorbornane to Adamantane

Acid Systems	Acid:Hydrocarbon Molar Ratio	Solvent	Temperature (°C)	Time (hr)	*exo*-Trimethylene-norbornane (% Yield)	Adamantane isolated (% Yield)
$AlCl_3$—HCl (H_2 40 atm)[21a]						40.0
HF—BF_3 (Ref. 21b)						30.0
HF + BF_3 (800 psi)	10:1 (*endo-*)		100	6	67.0	32.0
	20:1 (*endo-*)		100	5	67.0	33.0
FSO_3H + SbF_5, 1 : 1	1:3 (*endo-*)[a]		0[b]	18	59.5	38.0
HF—SbF_5 (Refs. 22,33), 1 : 1			100			47.0
	1:1 (*endo-*)[a]		0[b]	18	71.0	16.0
	1:1 (*exo-*)[a]		0[b]	18	66.0	20.0
CF_3SO_3H + HF + BF_3, 1 : 1	10:1 (*endo-*)[c]		100	7	66.0	32.7
(500 psi)	5:1 (*endo-*)		*b*	7	27.0	27.0
CF_3SO_3H + SbF_5 (Ref. 23), 1 : 1	1:1 (*endo-*)[c,d]		0[b]	10	0.4	95.4
			0[b]	7	0.1	98.0
	1:1 (*exo-*)[d]		0[b]	12		94.0
	1:1 (*endo-*)[a]	Freon-113	0[b]	16	41.0	50.0
		Freon-113	0[b]	50	30.0	64.3
	1:3 (*exo-*)[a]		0[b]	16	28.3	58.2
			0[b]	50	27.0	68.0
CF_3SO_3H + $B(OSO_2CF_3)_3$ (Ref. 24), 1 : 1	1:1 (*endo-*)		0[b]	12		98.0

[a] A yet unidentified product was also obtained: (6–13%) in CF_3SO_3H—SbF_5, (~1%) in FSO_3H—SbF_5, (13–14%) in HF—SbF_5.
[b] Room temperature.
[c] 1-Adamantanol was obtained (<2%) after aqueous workup.
[d] A yet unidentified product was obtained: (3–6%) in CF_3SO_3H—SbF_5, <2% in FSO_3H—SbF_5.

The near quantitative superacid catalyzed isomerization to adamantane was subsequently extended to the preparation of di- and triadamantane from their isomeric precursors (Scheme 9). Again with new generation of superacidic catalysts, such as boron tris-triflate in Freon solution or its conjugate acid with triflic acid or with CF_3SO_3H—SbF_5, nearly quantitative isomerization of discussed C_{14} precursors to diamantane and 70–75% yield of isomerization of C_{18} precursors to triamantane was achieved (Table 2).

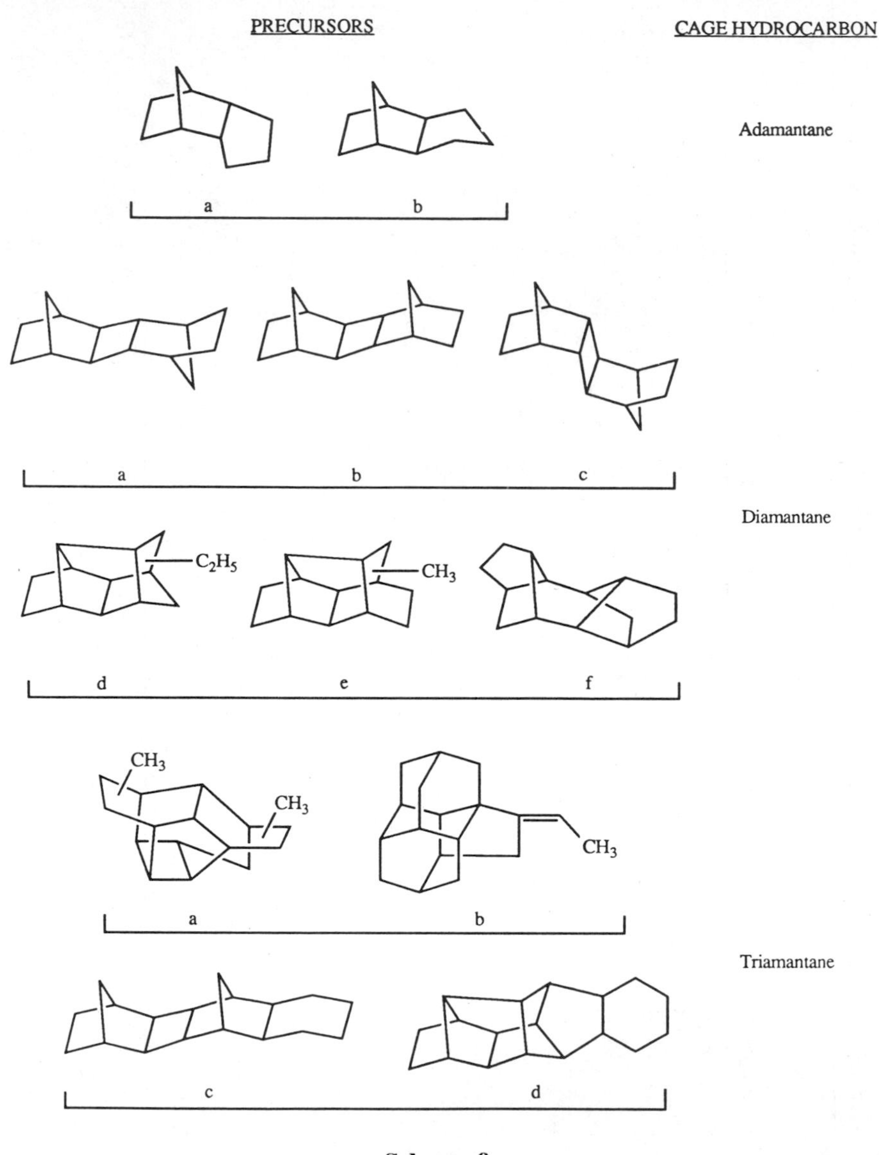

Scheme 9.

TABLE 2. Superacid Catalyzed Isomerization[a] of Polycycloalkanes to Adamantanoid Hydrocarbons

Precursor Hydrocarbon	Catalyst[b]	Time (h)	Adamantanoids (%) Ada	Dia	Tri
C_{10}	(i)	0.5	98		
	(ii)	10	98		
		(1.5)[c]	(98)[c]		
	(iii)	12	98		
C_{14}	(i)	0.25		99	
	(ii)	10		98	
		(1.75)[c]		(98)[c]	
	(iii)	11		98	
C_{18}	(i)	74			70
	(ii)	74			71
		(8)[c]			(74)[c]
	(iii)[d]	74			69

[a]Catalyst:hydrocarbon, 1:1; temperature, 0°C to ambient.
[b](i) $B(OSO_2CF_3)_3$ in Freon-113; (ii) $CF_3SO_3H:SbF_3$ (1:1) neat; (iii) $CF_3SO_3H\text{-}B(OSO_2CF_3)_3$ neat.
[c]Values in parentheses are for sonicated mixtures in Freon 113.
[d]Catalyst:hydrocarbon 1:2.

The yield of isomerization of heptacyclooctadecanes to triamantane was eventually increased to 95% (isolated yield) in the isomerization with boron tris-triflate in Freon solution when a small amount of 1-bromoadamantane was added to act as a carbocationic initiator. Such carbocationic initiation in alkane isomerization was used previously by Kramer.[27]

+

$B(OSO_2CF_3)_3$, 1-Ad-Br

Freon 113
72 hrs

95%

Sonication also facilitates the reactions by speeding them up. The new generation of superacidic catalyst systems allows the preparation of three diamondoid hydrocarbons, that is, adamantane, diamantane, and triamantane in nearly quantitative yield.

Adamantane gained significance in many areas of chemistry, which include the use of its 1-amino derivative as an antiviral agent.[28] It is being used in the Soviet Union on a large scale as an anti-influenza drug. Its effectiveness, however, is still not well known, although the DuPont Company, which started its development as a potential antiviral drug, has done extensive studies on it. Other pharmaceutical companies also showed significant interest in cage hydrocarbon derivatives. Commercial scale production of adamantane is carried out in Japan, although no known major industrial use has yet been announced. Adamantane and its analogues certainly offer many possibilities, and because of their ready availability by Schleyer's method, practical applications are expected.

ADAMANTYL CATIONS

After discussing the superacid catalyzed preparation of diamondoid hydrocarbons, we now should give further consideration to the carbocationic

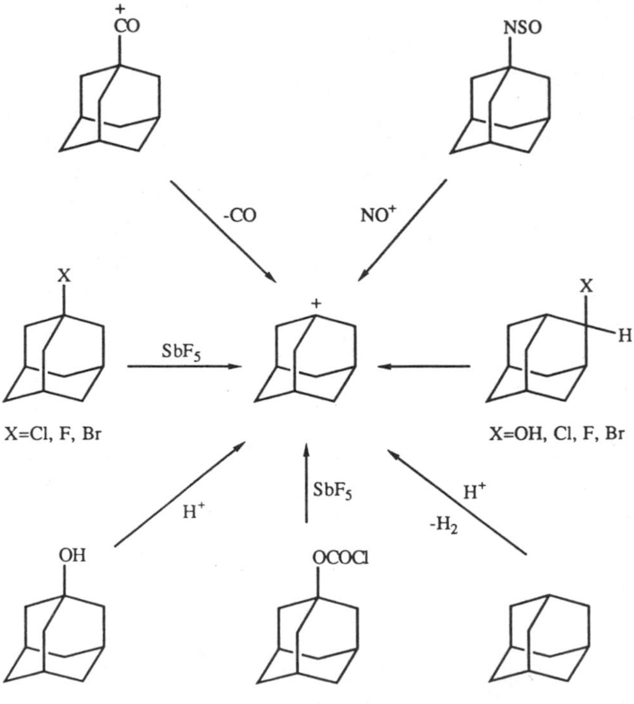

Scheme 10.

intermediates involved. In 1964[25] we published our initial study of the bridgehead 1-adamantyl cation jointly with Schleyer, followed later by a more detailed study.[26] We prepared the ion using stable ion methodology from adamantanol, 1-adamantyl halides, and related precursors in superacidic media (Scheme 10).

The ^1H NMR spectra of the 1-adamantyl cation obtained at successively higher fields (60, 100, and 250 MHz) are shown in Figure 1. At higher fields the proton coupling patterns became evident. Proton decoupled and coupled ^{13}C NMR spectra are shown in Figure 2.

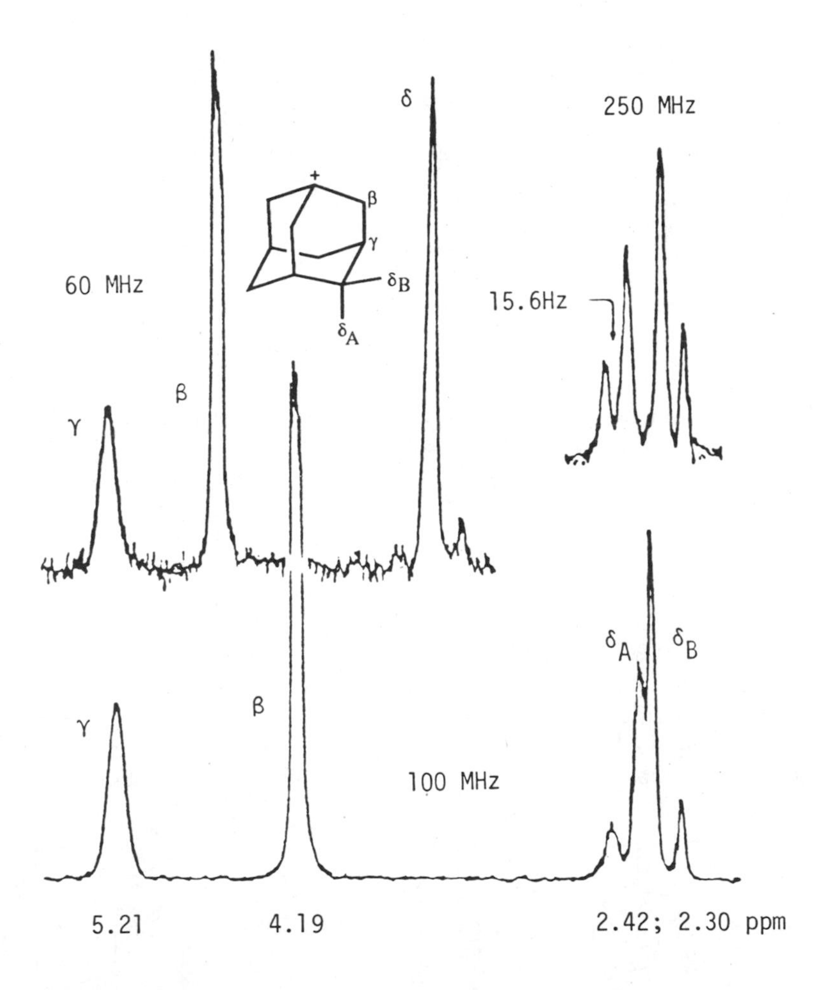

Fig. 1. ^1H NMR spectrum of the 1-adamantyl cation at (*a*) 14.09 kG (60 MHz), (*b*) 23.48 kG (100 MHz), and (*c*) 58.70 kG (250 MHz).

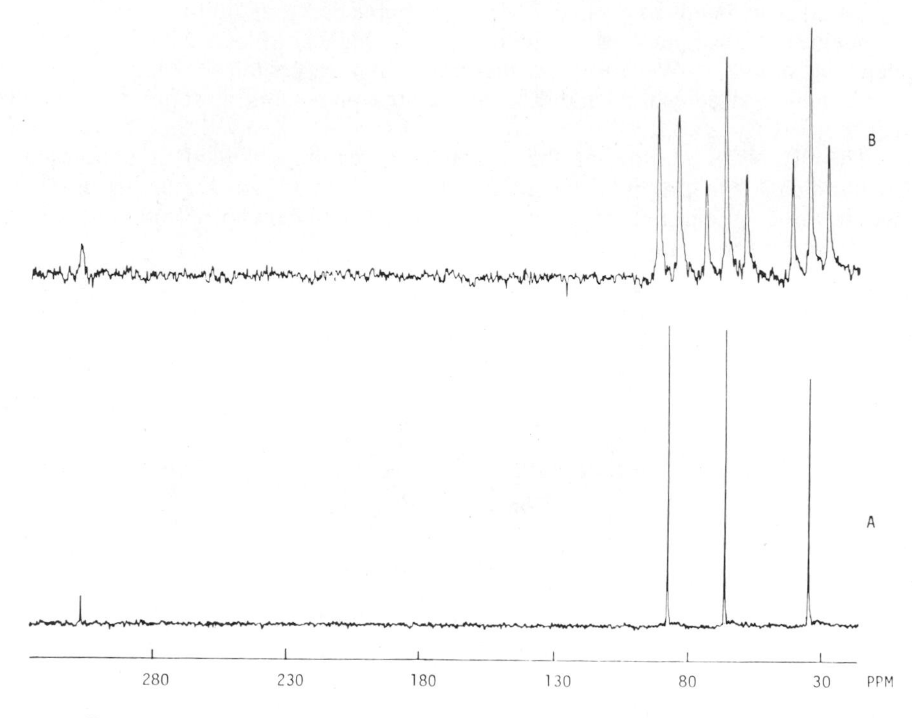

Fig. 2. ^{13}C NMR spectra (20 MHz) of the 1-adamantyl cation (**3a**) in SbF_5/SO_2ClF solution at –80°C; (*A*) proton decoupled; (*B*) proton coupled.

The carbocationic $^{+}C_1$ center is at $\delta^{13}C$ 300, which is indicative of a trivalent carbocation. At the same time the C-3,5,7 bridgehead positions also show increased deshielding, indicative of hyperconjugative C—C bond delocalization. This effect is also reflected by the ^{1}H spectra. Comparison with the 1-dodecahedryl cation (vide infra), where such effects are not operative because of unfavorable geometry, is particularly interesting.

Similar ^{1}H and ^{13}C NMR studies were also carried out on diamantyl[26,29] and triamantyl cations.[30] In the former case two bridgehead cations, 1-diamantyl

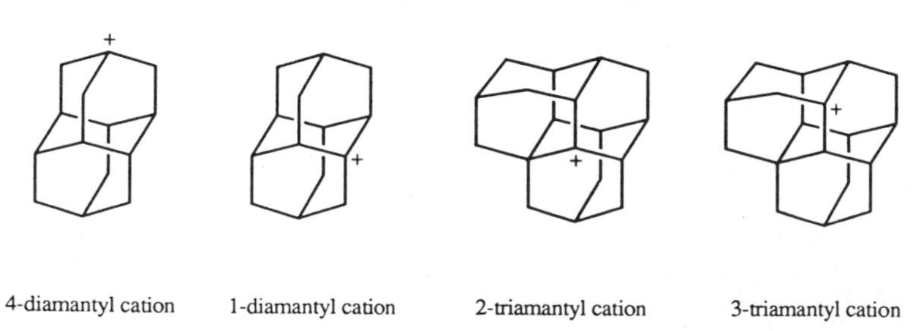

and 4-diamantyl, were observed and of the two, belt 1-diamantyl cation was very stable. In the case of bridgehead triamantyl cations only 2- and 3-bridgehead cations were observed. Attempted generation of the 9-triamantyl cation only led to the rearranged 3-derivative through intermolecular hydride shifts. All these polycyclic bridgehead cations are extensively stabilized by C—C hyperconjugation.

Table 3 summarizes the ^{1}H and ^{13}C NMR chemical shifts of some of the adamantanoid cations.

Figure 3 shows the ^{13}C NMR spectrum of the 1-diamantyl cation.

Study of coupling constants, particularly of C-C couplings, is also of interest when compared with that in the *t*-butyl cation.

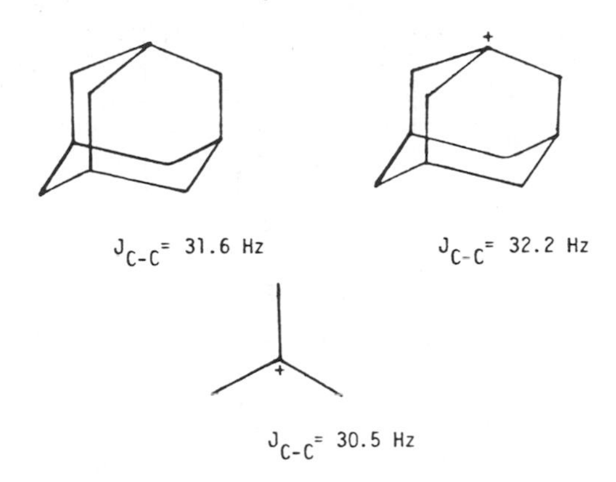

Two dimensional ^{1}H-^{13}C connectivity NMR studies were also carried out on the 1-adamantyl cation (Fig. 4) and served to demonstrate that the technique can be well adapted to the study of stable carbocations, which in more complex systems is of substantial use.

In the 2-D NMR study of the 1-diamantyl cations, the ^{1}H-^{13}C shift correlation (Fig. 5) offers a good demonstration of this point.

Once stable, long lived adamantanoid cations could be prepared in superacidic

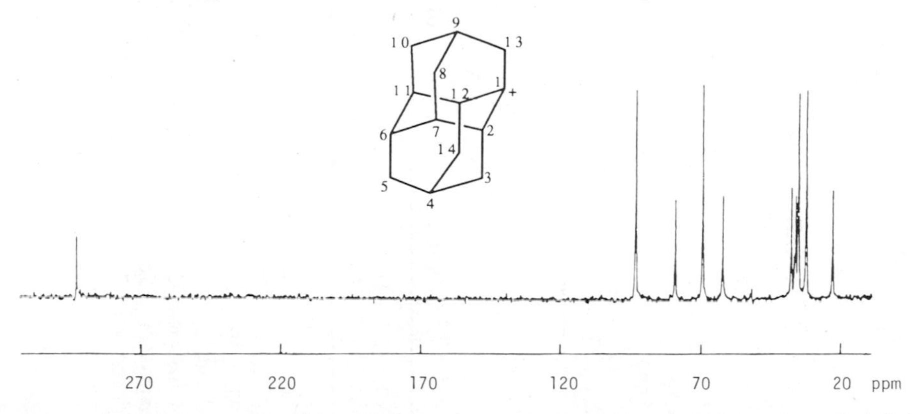

Fig. 3. Proton-decoupled ^{13}C NMR spectrum (20 MHz) of 1-diamantyl cation in SbF_5/SO_2ClF at –80°C.

TABLE 3. ^1H NMR Chemical Shifts of Adamantyl, Diamantyl, and Triamantyl Cations

Cation	Temp. (°C)	H-1	H-2	H-3	H-4	H-5	H-6	H-7	H-8	H-9	H-10	H-11	H-12	H-13	H-14	H1-5	H-16	H-17	H-18
1-Ad$^+$	-80	—	4.19	5.19	2.3	5.19	2.3	5.19	4.19	2.3	4.19	—	—	—	—	—	—	—	—
1-Diad$^+$	-80	—	4.1	2.0	2.0	1.9	2.0	4.67	2.3	4.4	2.3	4.67	4.1	3.5	2.0	—	—	—	—
2-Triad$^+$	-80	—	—	3.2	3.7	1.4	3.7	3.2	1.5	1.0	1.2	1.6	4.0	1.6	1.2	1.0	1.7	1.7	1.5
3-Triad$^+$	-80	—	3.0	—	2.9	1.0	1.0	1.0	1.5	1.0	1.0	1.0	1.2	3.5	1.4	3.4	1.4	1.0	2.0

^{13}C NMR Chemical Shifts of Adamantyl, Diamantyl, and Triamantyl Cations

Cation	Temp. (°C)	C-1	C-2	C-3	C-4	C-5	C-6	C-7	C-8	C-9	C-10	C-11	C-12	C-13	C-14	C-15	C-16	C-17	C-18
1-Ad$^+$	-80	300	65.9	86.8	34.5	86.8	34.5	86.8	65.7	34.5	65.7	—	—	—	—	—	—	—	—
1-Diad$^+$	-80	297.9	70.5	35.1	22.7	36.1	37.7	95.3	32.3	80.6	32.3	95.3	70.5	63.1	35.1	—	—	—	—
2-Triad$^+$	-80	75.1	296.3	68.3	90.9	35.9	90.9	68.3	36.5	25.5	32.9	38.8	106.8	38.8	32.9	25.5	39.4	39.4	36.5
3-Triad$^+$	-80	100.8	77.6	294.3	67.1	33.9	34.3	45.8	33.2	24.9	32.6	31.4	36.7	91.1	41.3	78.8	43.4	34.9	62.3
Adamantane	25	28.5	37.8	28.5	37.8	28.5	37.8	28.5	37.8	37.8	37.8	—	—	—	—	—	—	—	—
Diamantane	25	37.7	37.7	38.4	26.0	38.4	37.7	37.7	38.4	26.0	38.4	37.7	37.7	38.4	38.4	—	—	—	—
Triamantane	25	33.6	46.9	38.1	35.5	38.6	35.3	38.1	38.1	27.8	38.1	38.1	46.9	38.1	38.1	27.8	45.3	45.3	38.1

solution, their investigation was not limited to spectroscopic studies (primarily NMR). It was possible, for example, to also isolate crystalline salts of the 1-adamantyl and related cations.[31] This allowed us to extend our studies to the solid state ^{13}C NMR investigation of the 1-adamantyl cation. As Figure 6 shows the cross polarization magic angle spinning (CPMAS) spectrum is similar to the solution ^{13}C NMR spectrum, thus indicating no particular solid state effects.

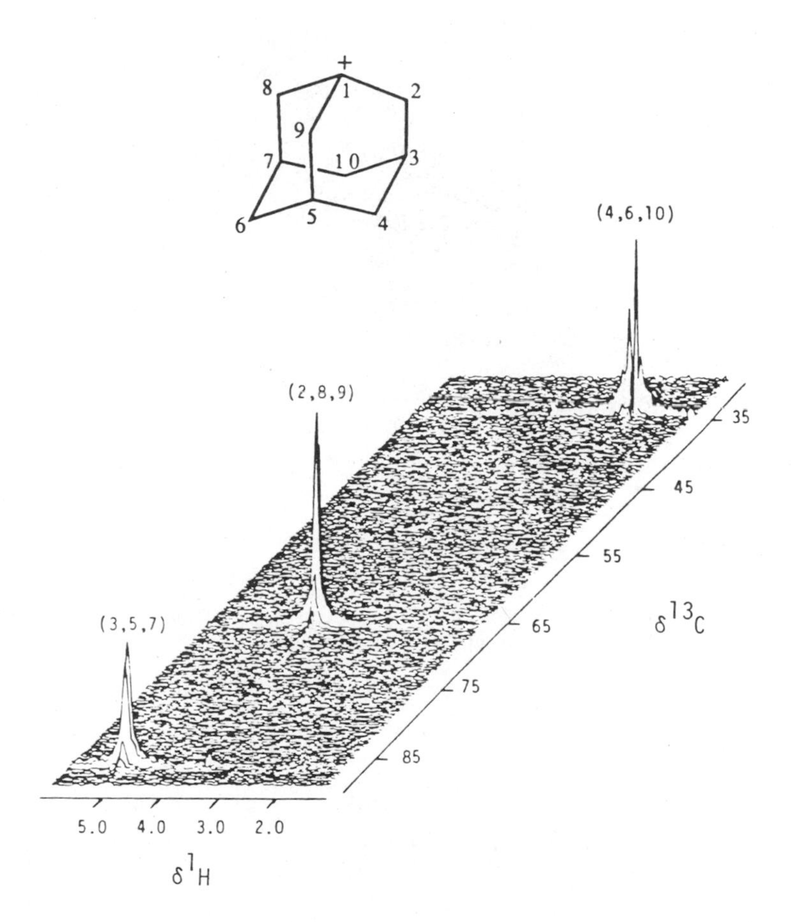

Fig. 4. ^{1}H-^{13}C chemical shift correlated 2-D NMR spectrum of the 1-adamantyl cation in SbF_5/SO_2 at –30°C; (*a*) contour plot; (*b*) stacked plot.

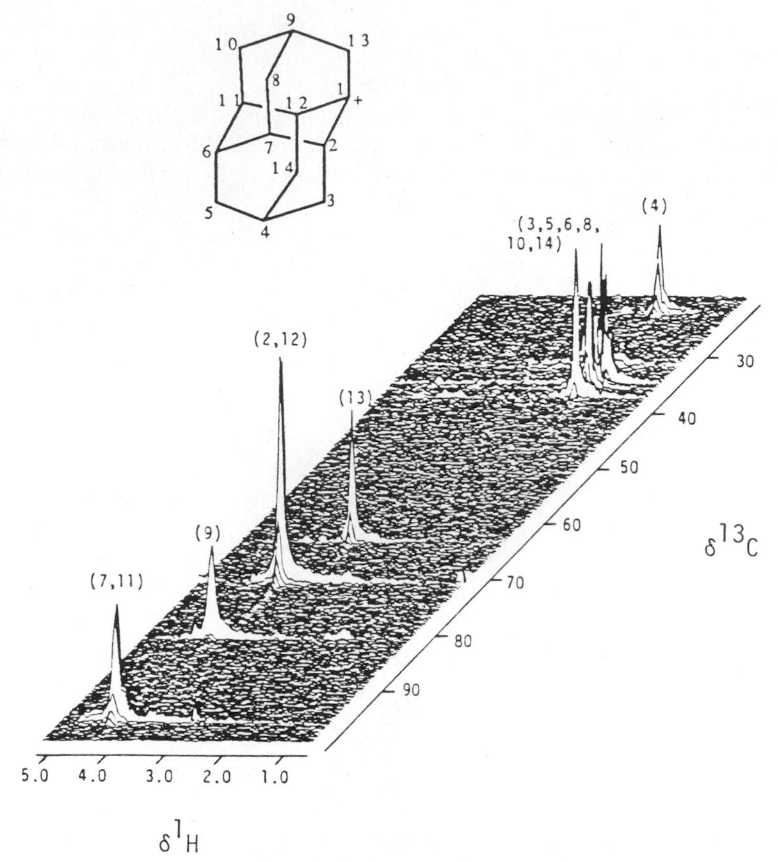

Fig. 5. ^{1}H-^{13}C chemical shift correlated 2-D NMR spectrum of the 1-adamantyl cation in SbF_5/SO_2 at –30°C; (*a*) contour plot; (*b*) stacked plot.

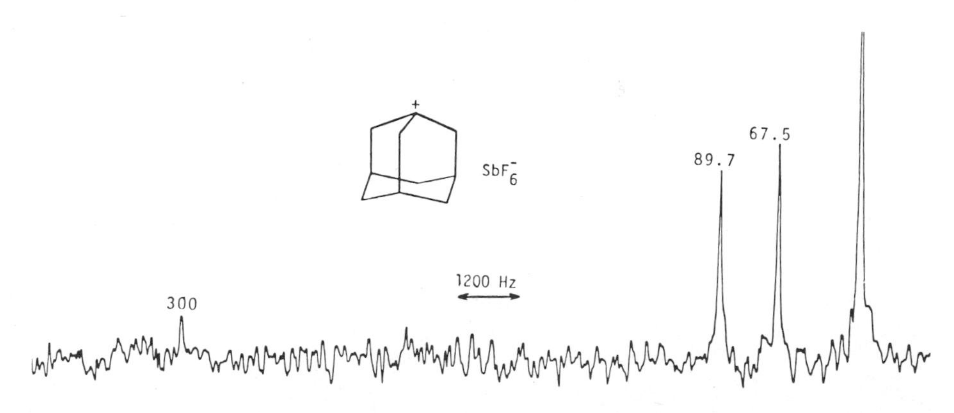

Fig. 6. 50 MHz solid state ^{13}C CPMAS NMR spectrum of the 1-adamantyl cation (as the hexafluoroantimonate salt). Chemical shifts are in parts per million (ppm) from external tetramethylsilane.

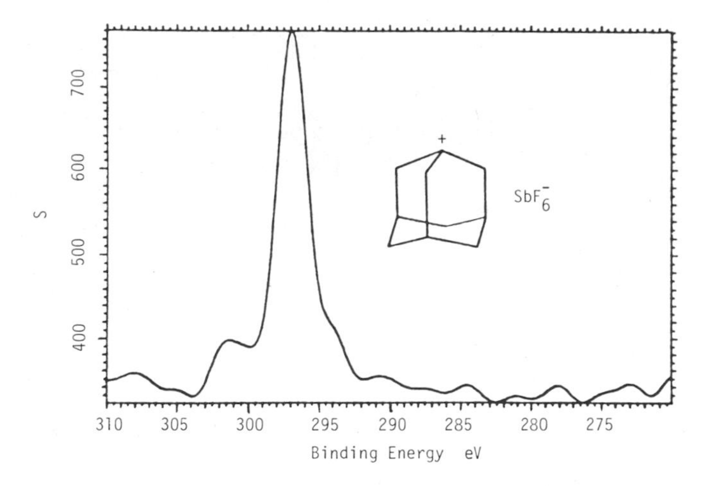

Fig. 7. Carbon 1*s* electron spectrum of the 1-adamantyl cation.

Isolated 1-adamantyl cation salts also made it possible to carry out a core electron spectroscopic (ESCA) study of the solid ion. Figure 7 shows the carbon 1*s* electron spectrum of the 1-adamantyl cation.

It is significant to note that the internal 1*s* binding energy separation (ΔE_b) is 3.8 eV, characteristic of a typical trivalent carbocation.

The isolation of adamantyl cation salts also made possible their single crystal X-ray crystallographic structural study, notably carried out by Laube.[32]

As mentioned previously, the bridgehead 1-adamantyl cation is strongly stabilized by C—C hyperconjugation depicted by canonical structures:

This results in significant delocalization of the positive charge effecting all bridgehead positions. Definitive proof for the existence of such C—C hyperconjugation comes from single crystal X-ray crystallographic studies. In a pioneering effort Laube has obtained[32] the structure of the 3,5,7-trimethyl-1-adamantyl cation as the $Sb_2F_{11}^-$ salt, with an excellent R factor (0.070) at low temperatures.

The crystallographic data is revealing. The C-2–C-3, C-9–C-5, and C-8–C-7 bonds are consistently longer than the C-1–C-2, C-1–C-9, and C-1–C-8 bonds (by ~ 0.15–0.2 Å) giving clear evidence for the β-C–C hyperconjugative effect. Furthermore, the carbocationic center at the C-1 position is quite flat (average bond angle 118°).

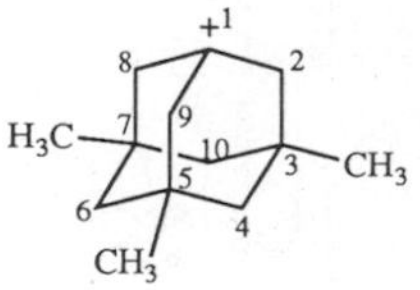

Important Bond Lengths (Å)

C-1–C-2 = 1.43	C-2–C-3 = 1.61
C-1–C-9 = 1.42	C-9–C-5 = 1.62
C-1–C-8 = 1.47	C-8–C-7 = 1.62

Important Bond Angles (°)

C-2–C-1–C-9 = 120
C-8–C-1–C-9 = 116
C-2–C-1–C-8 = 118

Bau et al.[33] studied the structure of the parent bridgehead 1-adamantyl cation as the $Sb_2F_{11}^-$ salt, cocrystallizing with a molecule of $H_3O^+Sb_2F_{11}^-$. Because of severe disorder problems, it was not possible to obtain an accurate data set for the ion salt. The standard deviations of the observed bond lengths and angles of the 1-adamantyl cation were too high for an accurate structure determination.

The study of long lived stable cations also provided additional information for the mechanism of the formation of adamantane and its homologues in acid catalyzed isomerization of suitable precursors. However, as these rearrangements take place under thermodynamic conditions, generally only the most stable carbocations can be observed. In the low temperature (–78°C) study of the ionization of 2-*exo*-chlorotrimethylenenorbornane, an intermediate bridgehead carbocation was observed, which at –30°C rearranged to the 1-adamantyl cation.[15]

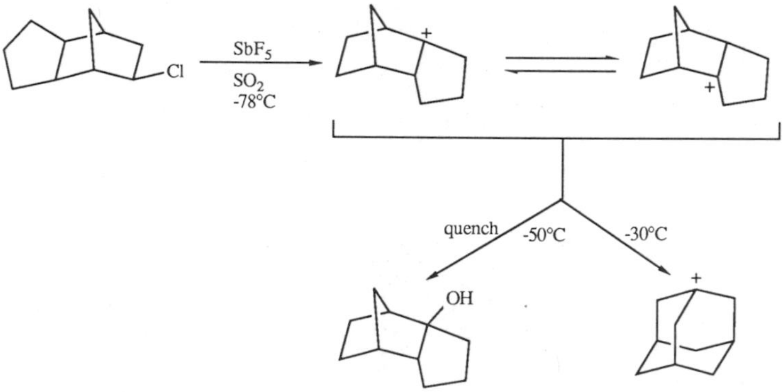

Attempts to prepare the stable 1,3-adamantyl dication failed, and only a monocation–monodonor–monoacceptor complex was observed.

F

F $\xrightarrow{SbF_5/SO_2ClF}$ + δ+ F → δ- SbF_5

Based on a theoretical prediction, Schleyer and co-workers[34] were able to prepare, by ionization of 1,3-difluorodehydroadamantane, the remarkable $C_{10}H_{12}^{2+}$ adamantane dication, a tetra-trishomoaromatic tetrahedral system containing 2*e*-4*c* bonding. This dication is discussed in more detail in Chapter 2.

F

F $\xrightarrow[-80°C]{SbF_5/SO_2ClF}$ 2+

Olah and co-workers succeeded in preparing the 3,3-(1,1-biadamantyl)dication.[25]

SbF_5/SO_2ClF

+ +

Br Br

SbF_5/SO_2ClF

Similarly, the 4,9-apical, apical diadamantyl dication was also prepared and studied,[25] but attempts to obtain the belt,belt 1,6-dication were unsuccessful. Charge–charge repulsion effects may be responsible.

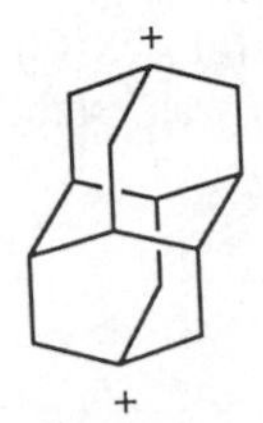

PAGODANE CATIONS AND DICATION

Whereas the study of long lived cage hydrocarbon cations centered heavily on diamondoid systems, in recent years it was also extended to include the fascinating $C_{20}H_{20}$ pagodane and dodecahedrane systems.

The highly symmetrical undecacyclic $C_{20}H_{20}$ hydrocarbon pagodane (undecacyclo$[9.9.0.0^{1,5}.0^{2,12}.0^{2,18}.0^{3,7}.0^{8,12}.0^{13,17}0^{16,20}]$eicosane has been synthesized by Prinzbach et al.[35] as a potential precursor for its structurally related isomer dodecahedrane.[36]

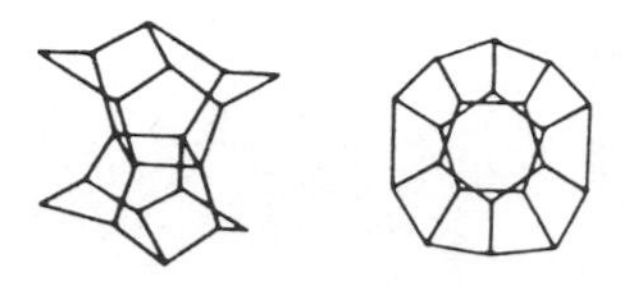

Dissolution of pagodane in five-fold excess of freshly distilled SbF_5/SO_2ClF solution at –78°C resulted in the formation of an ionic species. The 200 MHz ^{1}H NMR spectrum indicated a complex spectral pattern in the aliphatic region. However, after several hours at –80°C, the solution showed a very clean ^{1}H NMR spectrum (Fig. 8*a*); δ^1H 3.37 (br, 8H), 3.68, and 2.72 (AX doublets, $J_{H—H}$ = 13.2 Hz, 8H), and 2.39 (br, 4H). The 50-MHz ^{13}C NMR spectrum of the same solution at –80°C (Fig. 8*b and c*) showed only four peaks at δ ^{13}C:251.0 (singlet), 65.3 (triplet, $J_{C—H}$ = 141.9 Hz), 57.2 (doublet, $J_{C—H}$ = 148.0 Hz), and 52.3 (doublet, $J_{C—H}$ = 152.8 Hz). The observed symmetry and the extent of deshielding in both the ^{1}H and ^{13}C NMR spectra of the species in SbF_5/SO_2ClF solution when compared to the progenitor pagodane (Table 4) indicated that the ionic species has the D_{2h} symmetry of the parent pagodane itself.

The ion solution was found to be surprisingly stable. Quenching the ion solution with an excess of cold methanol (at –78°C) provided a white crystalline dimethoxy product, the structure of which was established by NMR spectroscopy and by X-ray crystallography.

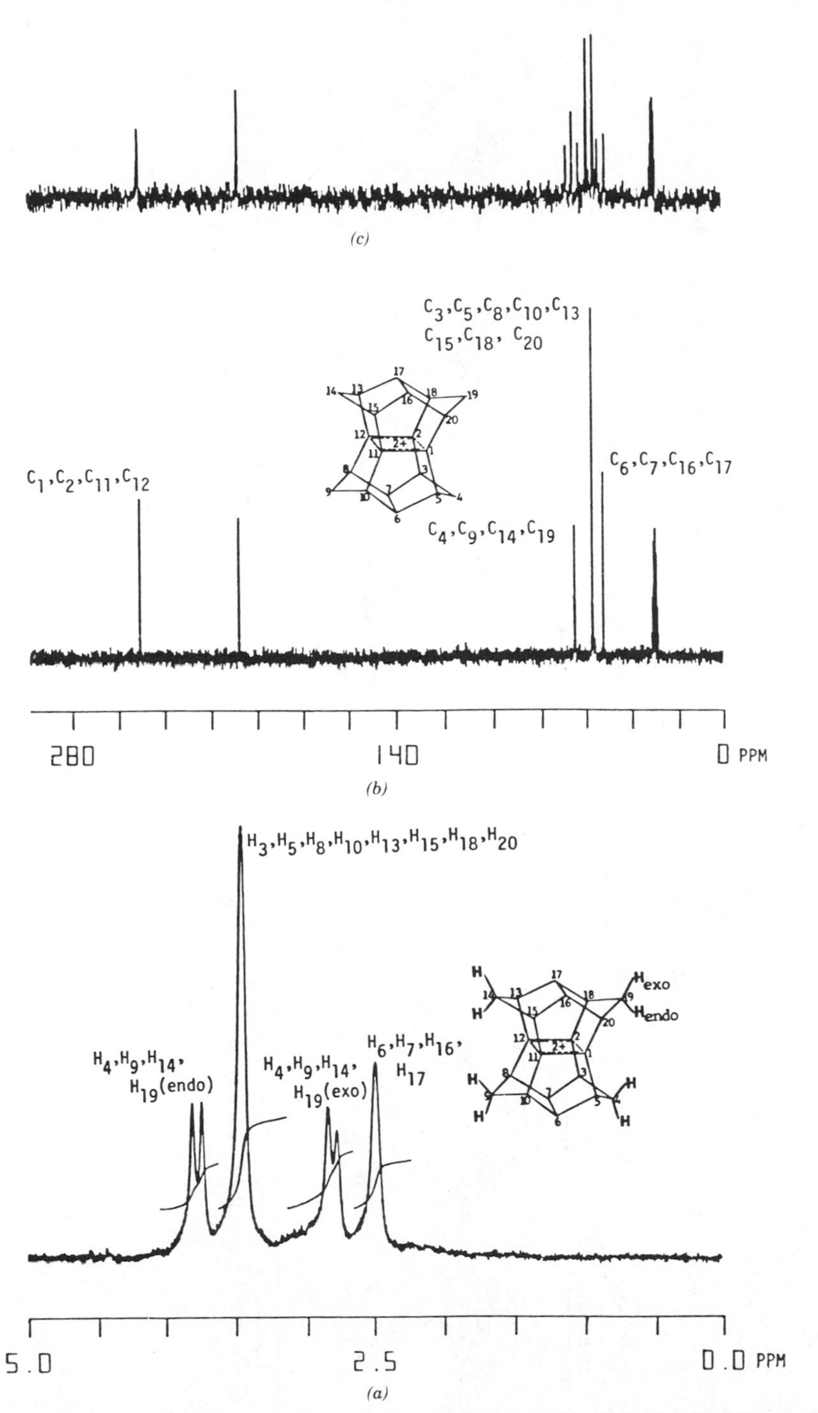

Fig. 8. (*a*) 200-MHz ^{1}H NMR spectrum of pagodane dication in SbF_5/SO_2ClF solution at –80°C. (*b*) 50-MHz proton-decoupled ^{13}C NMR spectrum. (*c*) Proton-coupled spectrum [* peaks due to lock solvent $(CD_3)_2CO$].

TABLE 4. ^{1}H and ^{13}C NMR Spectroscopic Data of Pagodane and Pagodane Dication

structure	^1H, δ^a	^{13}C, δ^a
14, 13, 17, 18, 19, 15, 16, 12, 2, 20, 8, 11, 7, 3, 1, 9, 4, 10, 6, 5 [b]	H_4,H_9,H_{14},H_{19} = 1.58 (AB, 8 H); $H_3,H_5,H_8,H_{10},H_{13},H_{15},H_{18},H_{20}$ = 2.24 (br, 8 H); H_6,H_7,H_{16},H_{17} = 2.60 (br, 4 H)	C_1,C_2,C_{11},C_{12} = 62.8 (s); C_6,C_7,C_{16},C_{17} = 59.6 (d, $J_{C\text{-}H}$ = 137.8 Hz);$C_3,C_5,C_8,C_{10},C_{13},C_{15},C_{18},C_{20}$ = 42.6 (d,$J_{C\text{-}H}$ = 140.6 Hz); C_4,C_9,C_{14},C_{19} = 41.9 (t, $J_{C\text{-}H}$ = 129.8 Hz)
14, 13, 17, 18, 19, 15, 16, 12, 2, 20, 8, 11, 7, 3, 1, 9, 4, 10, 5, 6 [c]	H_4,H_9,H_{14},H_{19} = AX spin system; H_{endo} = 3.68 (d, $J_{H\text{-}H}$ = 13.2 Hz, 4 H);[d] H_{exo} = 2.72 (d, $J_{H\text{-}H}$ = 13.2 Hz, 4 H);[d] H_6,H_7,H_{16},H_{17} = 2.39 (br, 4 H); $H_3,H_5,H_8,H_{10},H_{13},H_{15},H_{18},H_2O$ = 3.37 (br, 8 H)	C_1,C_2,C_{11},C_{12} = 251.0 (s); C_4,C_9,C_{14},C_{19} = 65.3 (t, $J_{C\text{-}H}$ = 141.9 Hz); $C_3,C_5,C_8,C_{10},C_{13},C_{15},C_{18},C_{20}$ = 57.2 (d, $J_{C\text{-}H}$ = 148.0 Hz); C_6,C_7,C_{16},C_{17} = 52.3 (d, $J_{C\text{-}H}$ = 152.8 Hz)

[a] Chemical shifts are in ppm from external capillary tetramethylsilane. [b] In $CDCl_3$ solution at 25 °C. [c] In SbF_5/SO_2ClF solution at −80 °C. [d] Endo and exo assignments are only tentative. Multiplicities: s = singlet, d = doublet, t = triplet, br = broad.

The ionic product derived from pagodane is thus the pagodane dication that is formed by the 2*e* oxidation of the strained cyclobutane ring. Its structure can be depicted by either of the two D_{2h} forms. The first is given preference in view of the X-ray structure of the obtained dimethoxy quench product.[37]

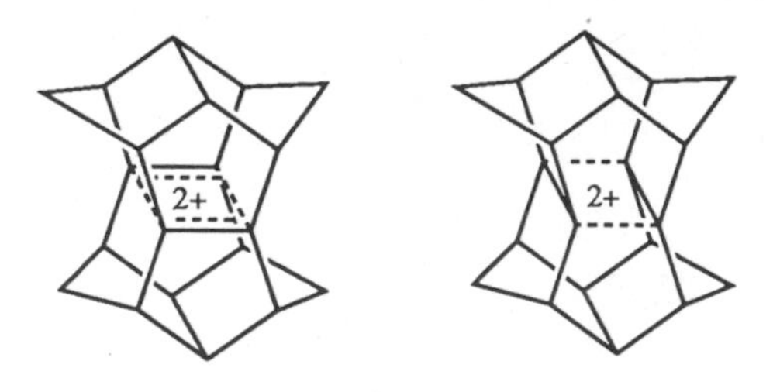

The remarkable stability of the pagodane dication can be rationalized as that of a unique 2π-aromatic system. In fact, the dication can be considered to be topologically equivalent to the cyclobutadiene dication. The situation is analogous to the transition state for the allowed cycloaddition of ethylene with an ethylene dication (mode 2 and mode 0 interaction).[38]

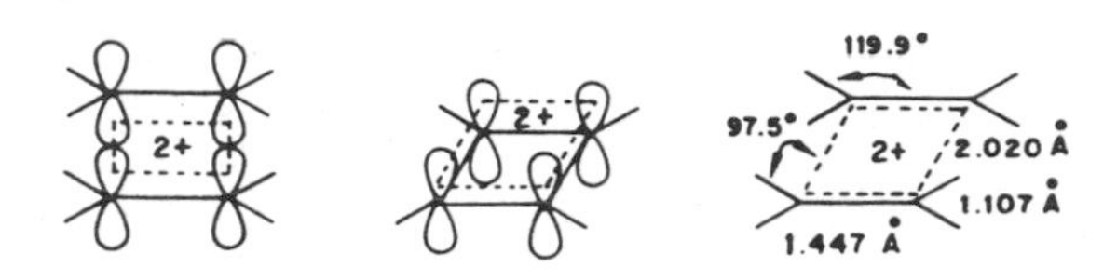

An experimental test for aromatic systems is the detection of a ring current. A ring current effect in the pagodane dication seems to be indicated at positions C-6, C-7 and C-16, C-17. Both ^{1}H and ^{13}C NMR shifts at these positions are shielded compared to those in pagodane (by 7.3 ppm for carbon atoms and 0.21 ppm for the protons). It is noted that the cyclobutanoid carbon atoms in the dication are much more deshielded than those observed in the isoconjugate tetramethylcyclobutadiene dication (by 41.3 ppm).[39] However, it is improper to compare these chemical shifts in view of the different types of π-bonding.

All attempts to protonate pagodane to the 2-*seco* pagodyl cation with superacids, including the newly developed only mildly oxidizing $CF_3SO_3H{:}B(O_3SCF_3)_3$ system, were unsuccessful. Under all conditions, the dication was exclusively formed by 2*e* oxidation. This prompted us to investigate its preparation from other, better suited precursors. Ionization of 2-chloro-*seco*[1.1.1.1]pagodane in fivefold excess of SbF_5 in SO_2ClF at –78°C resulted in a light yellow colored solution. The 50 MHz ^{13}C NMR spectrum showed twelve ^{13}C resonances with proper multiplicities similar to the progenitor. This is in accord with the formation of stable 2-*seco*[1.1.1.1] monocation with retained C_s symmetry.[40] The same secocation was obtained by the protonation of bis-*seco*-dodecahedradiene in the $CF_3SO_3H{:}B(O_3SCF_3)_3/SO_2ClF$ system. Furthermore, protolytic ionization of

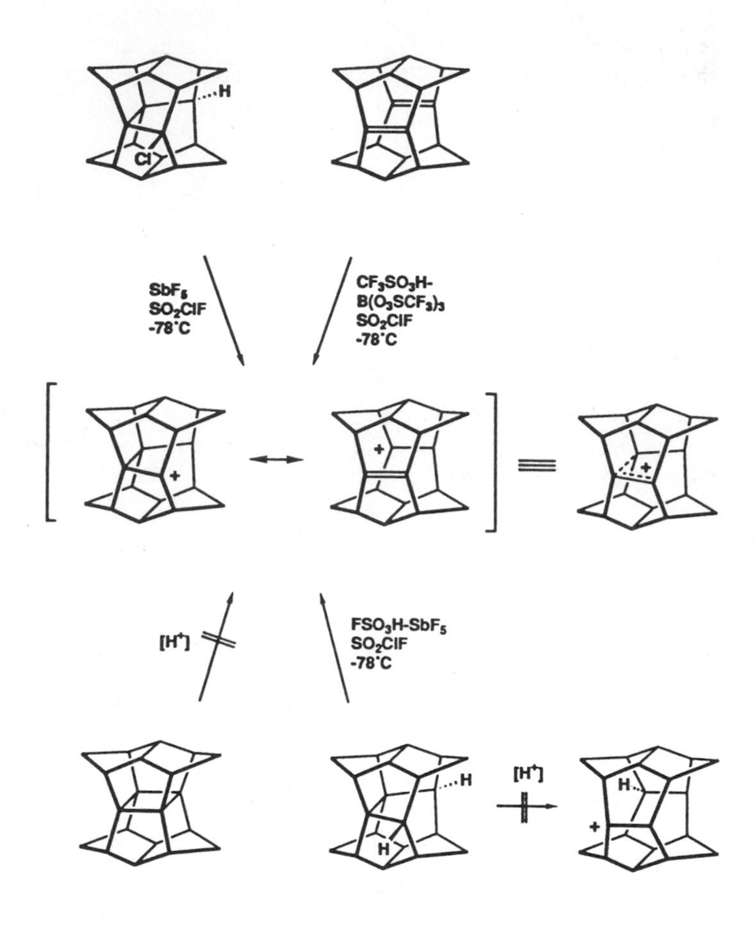

Scheme 11.

seco-pagodane in the $FSO_3H:SbF_5$ (Magic Acid®)/SO_2ClF system also resulted in the *seco*-pagodyl cation. See Scheme 11.

The significant ^{13}C NMR shifts in the *seco*-pagodyl cation of the cationic center C-2 at $\delta^{13}C$ 279.2 as well as that of adjoining C-1 at 129.1 and C-11 at 140.6 indicate that the electron deficiency at C-2 is strongly stabilized by the C-1—C-11 σ bond through C—C hyperconjugation involving related resonance structures (C-11 attracts more positive charge than C-1). Such phenomenon was discussed in the case of the bridgehead 1-adamantyl cation, wherein all the bridgehead carbon atoms (γ-carbon atoms) become much more deshielded than the β-methylene carbon atoms. It appears, however, that the C-1—C-11 bond in the *seco*-pagodyl cation interacts with the C-2 cationic center much more strongly.

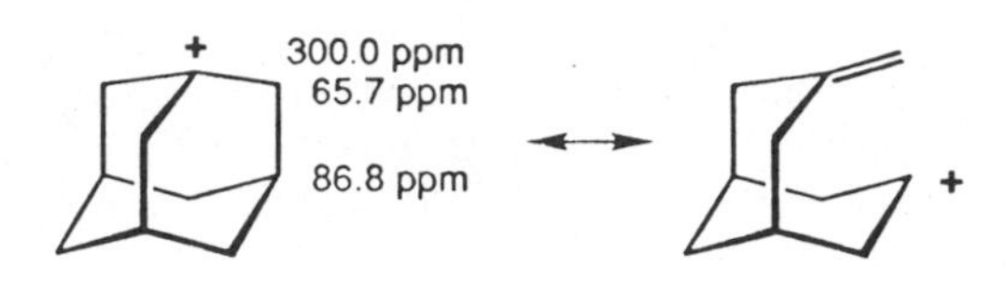

The ease of formation of the *seco*-pagodyl cation by both protonation of bis-*seco*-dodecahedradiene and regiospecific protolytic ionization of *seco*-pagodane attests to the stability of the ion. Thus, the facile formation of the pagodane dication, as well as *seco*-pagodyl cation, from various pagodane derivatives indicate a thermodynamic hurdle to superacid catalyzed isomerization of pagodane to dodecahedrane.

DODECAHEDRYL CATION AND DICATION

The challenge of synthesizing the platonic hydrocarbon dodecahedrane has been met by Paquette et al.[41] and by Prinzbach et al.[42] Paquette also prepared a wide range of monofunctionalized dodecahedranes. Practical exploitation of this chemistry rests heavily (although not exclusively) on the transient generation and efficient trapping of the dodecahedryl monocation. For this reason, in a joint effort, Olah and Paquette prepared and studied the ion under long life conditions.[43]

Careful dissolution of chlorododecahedrane in a solution of SO_2ClF/SbF_5 at –78°C yielded a pale yellow colored solution (Scheme 12). The 200 MHz ^1H NMR spectrum of this solution at –70°C revealed a set of three absorptions at δ 4.64 (br, 3H), 3.05 (br, 7H), and 2.59 (br, 9H) as shown in Figure 9. The peak did not split further at 500 MHz. The 50 MHz ^{13}C NMR spectrum consisted of six absorptions at 363.9 (s), 81.1 (d), 64.4 (d), 64.1 (d), 63.0 (d), and 60.9 (d) ppm, clearly indicating formation of the static dodecahedryl cation. The ion was similarly obtained from 1-dodecahedranol and dodecahedrane, but it took longer to form. The limited solubility of dodecahedrane necessitated that the superacidic medium be warmed to 0°C for dissolution to occur.

It was thought that the dodecahedryl cation would undergo rapid hydrogen scrambling (through 1,2-hydride shifts), similar to that observed in the cyclopentenyl cation, a process that would render all the carbon and hydrogen atoms equivalent. However, no such degenerate process was observed as is indicated by the lack of change in the ^1H NMR line shapes, even when solutions of the cation were allowed to warm to 0°C. Thus, the barrier for such degenerate rearrangement is at least 15 kcal/mol. The orientation of the empty π-orbital with the C—H bonds would seem to make such rearrangement difficult.

Upon standing in the superacidic solution for 6–7 h at –50°C, the monocation is slowly and irreversibly transformed into a new ionic species displaying a more simplified NMR spectra. Only two absorptions are seen at δ 4.74 (br, 6H) and 3.23 (br, 12H) in the ^{1}H spectrum at –70°C. The three signals at 379.2 (s), 78.8 (d), and 59.8 (d) ppm that constitute the ^{13}C spectrum of the dodecahedryl dication are shown in Figure 10.

The NMR data allow the dication to be assigned as the symmetrical apical,apical 1,16-dodecahedryl dication. The formation of the dication can be rationalized by protolytic ionization of the C—H bond at position 16 through a five-coordinate species. Independent generation of the dication was accomplished by ionization of an isomeric mixture of dibromododecahedranes that is produced when dodecahedrane is brominated in the presence of $AlBr_3$. This mixture consists of three $C_{20}H_{18}Br_2$ components (GC–MS analysis) in the ratio of 5:6:2. Since the only observable ion is the 1,16-dication, 1,2-hydride shifts occur readily once the system is charged or intermolecular hydride transfer is involved.

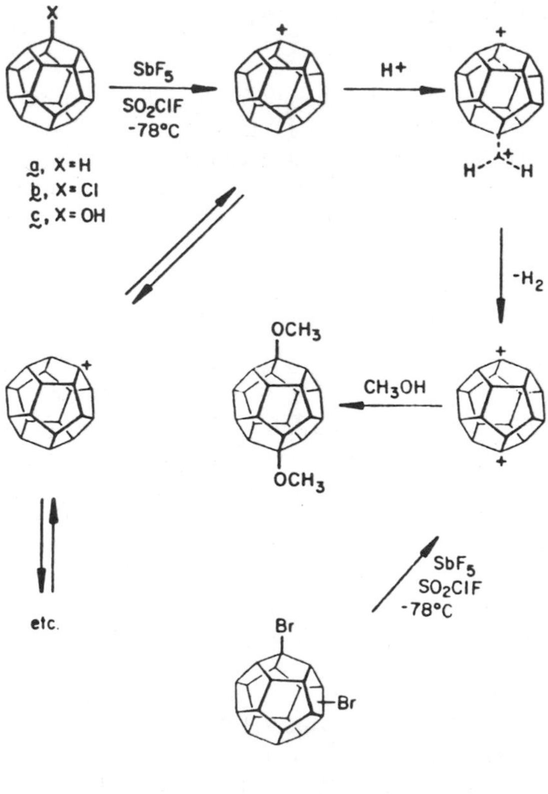

Scheme 12.

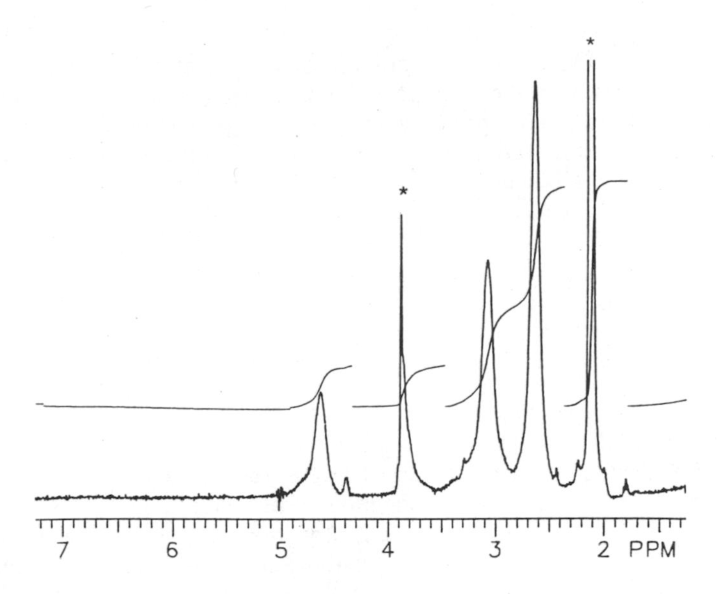

Fig. 9. 200 MHz ^{1}H spectrum of the 1-dodecahedryl cation in SbF_5/SO_2ClF at –70°C. *, peaks due to lock solvent (acetone-d_6 containing some water).

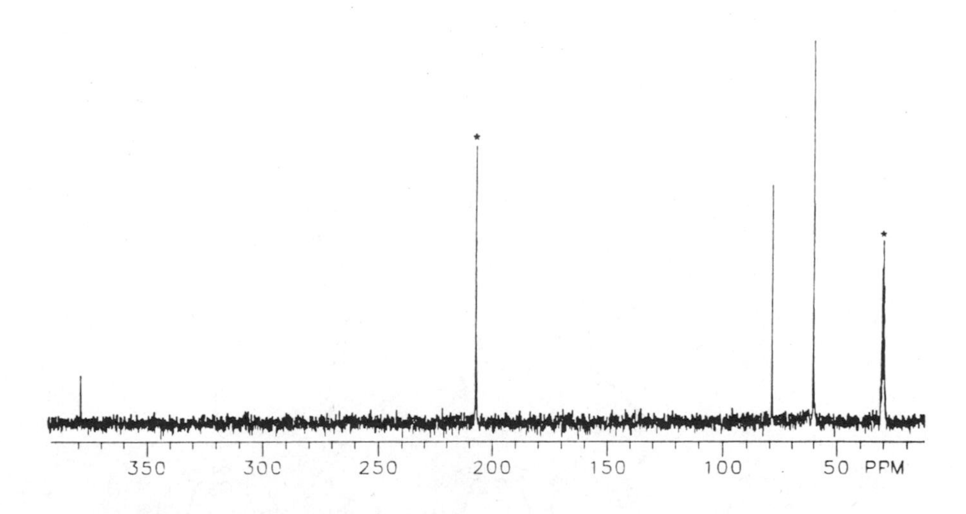

Fig. 10. 50-MHz proton decoupled ^{13}C NMR spectrum of 1,16-dodecahedryl dication in SbF_5/SO_2ClF at –70°C. *, peaks due to lock solvent (acetone-d_6 containing some water).

The ^{13}C chemical shifts of the positively charged centers in the dodecahedryl mono (363.9 ppm) and dication (379.2 ppm) are the most deshielded ever observed for carbocationic species. The previously recorded most deshielded carbocationic shifts were those of the 1-bicyclo[3.3.3]undecyl cation and the 1,5-bicyclo[3.3.3]undecyldiyl dication.[44]

+ 356.3 ppm

+ 346.2 ppm

Application of the ^{13}C chemical shift additivity criterion[45] reveals a net ^{13}C chemical shift deshielding of 283 and 610 ppm, respectively in accordance with the progression from a mono- to a dication. The magnitude of deshielding per unit positive charge in both the dodecahedryl mono- and dication is less than that observed in a typical tertiary carbocation such as the 1-methyl-1-cyclopentyl cation ($\Sigma\Delta = 374$).[45] Schleyer[46] analyzed these experimental NMR ^{13}C shift data in comparison with those calculated by the theoretical IGLO method. He concluded that the C-1^{+} δ 363.9 value in the dodecahedryl cation is in excellent agreement with the theoretical value. It thus represents the value for a trivalent carbocation center where geometrical restrictions do not allow C—C or C—H delocalization. Usually in trivalent carbocations, even in the *t*-butyl cation, partial delocalization (bridging) affects the C^{+} chemical shifts.

The 1,16-dodecahedryl dication was found to be rather stable at 0°C for several days. Quenching solutions of the ion in methanol gave 1,16-dimethoxy-dodecahedrane in $\geq$ 85% yield.

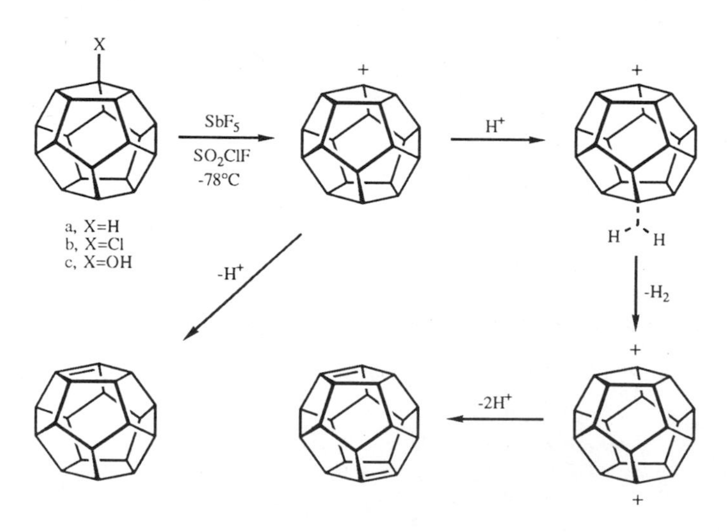

Scheme 13.

The possibility exists that the mono- and dication can be deprotonated to form dodecahedrene and dodecahedradiene, respectively. These unsaturated molecules may be involved in observed condensation or decomposition processes of the ions and may even prove isolable upon controlled deprotonation with hindered bases.

ELECTROPHILIC SUBSTITUTION OF ADAMANTANE

The study and understanding of cage hydrocarbon ions opened up the way for the electrophilic chemistry of cage hydrocarbon systems.[47] To demonstrate this emerging and fascinating chemistry our discussion will center on the case of adamantane.

Electrophilic reactions of saturated hydrocarbons, such as adamantane, in contrast to S_N1-type reactions (involving intermediate formation of the related trivalent carbocations, which then rapidly react with various nucleophiles acting as alkylating agents) are S_E2-type reactions in which a reactive electrophile affects substitution of a C—H bond via a five-coordinate carbocation intermediate.

As already discussed, the bridgehead adamantane cation can be very readily generated from various precursors, including protolysis of the parent hydrocarbon itself. When the ion formed is allowed to react with various

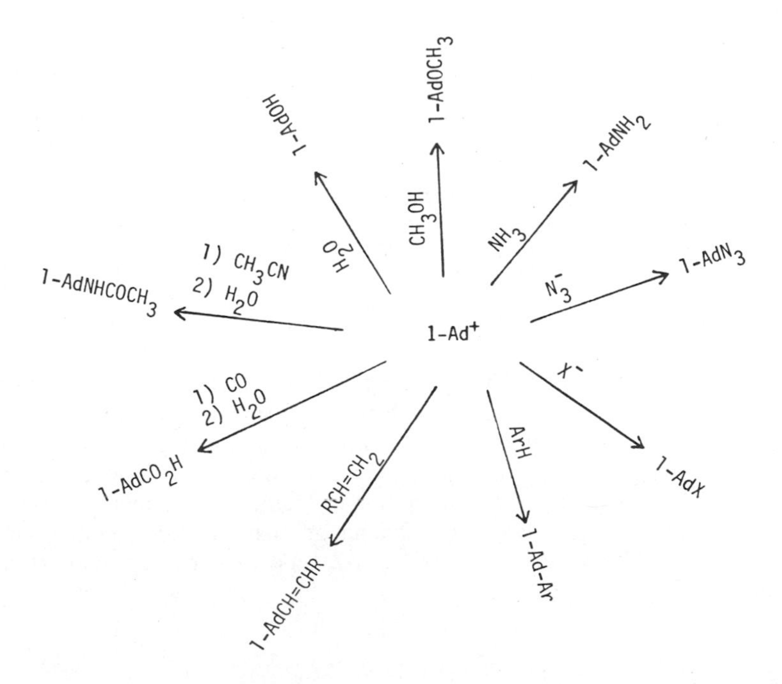

nucleophiles, the corresponding adamantylated products are obtained in what amounts to S_N1-type reactions.[29]

Examples of S_N1 reactions involving the 1-adamantyl cation include hydrolysis to 1-adamantanol, various solvolysis reactions, preparation with halide ions of 1-haloadamantanes, ammonolysis to 1-aminoadamantane, reaction with azides to give 1-azaadamantane, the Ritter reaction with nitriles, the Koch–Haaf reaction to 1-adamantanecarboxylic acid, and adamantylation of olefins and aromatics.

The reactions follow the well-established pathway of related open chain alkyl systems, such as those of *t*-butyl, *t*-cumyl, or cycloalkyl systems. The kinetics of the solvolysis reactions of 1-adamantyl systems were extensively studied, as well as those of 2-adamantyl derivatives proceeding through S_N2 pathways.[48] Their further discussion is not considered in the scope of this chapter.

Study of the superacidic chemistry of adamantane opened up the possibility of the electrophilic substitution of adamantane (and related cage hydrocarbons).

Hydrogen-Deuterium Exchange

When adamantane was treated with deuterated superacids, such as $DF—SbF_5$ or $DSO_3F—SbF_5$, or 1,3,5,7-tetradeuterioadamantane with $HF\text{-}SbF_5$ or $HSO_3F—SbF_5$, some hydrogen–deuterium exchange is observed to accompany the protolytic ionization to the 1-adamantyl cation.[49] This reaction is indicative of the protolytic (deuterolytic) attack on the C—H (or C—D) bond forming the five-coordinate adamantyl cation.

H DF-SbF$_5$ H D + SbF$_6^-$ -H$^+$ D + +HD

In reactions carried out at atmospheric pressure HD formed in the ionization reaction primarily escapes or reacts with the acid system. Its solubility in the liquid reaction medium is anyhow very limited. Competing quenching of the 1-Ad^+ by HD can therefore hardly account for the observed H–D exchange, which thus indicates intermediacy of a five-coordinate adamantane cation.

Electrophilic hydrogen–deuterium exchange can be considered as the prototype reaction for other electrophilic S_E2 type of substitution reactions of adamantane.

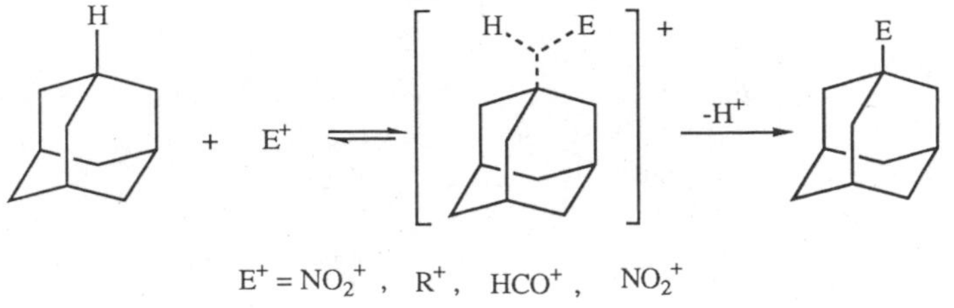

$E^+ = NO_2^+$, R^+, HCO^+, NO_2^+

Nitration

The nitration of adamantane with nitronium salts, such as $NO_2^+BF_4^-$ or $NO_2^+PF_6^-$, is illustrative of electrophilic substitution.[50] The major reaction product is 1-nitroadamantane (with small amounts of 2-nitroadamantane formed in a much slower reaction of the secondary C—H bonds). Some 1-adamantyl nitrite and 1-adamantanol is also formed in the reaction, indicative of the hydrolysis of the 1-adamantyl cation, as well as of a limited nucleophilic reaction by NO_2^-, formed from HNO_2 in the competing ionization reaction.

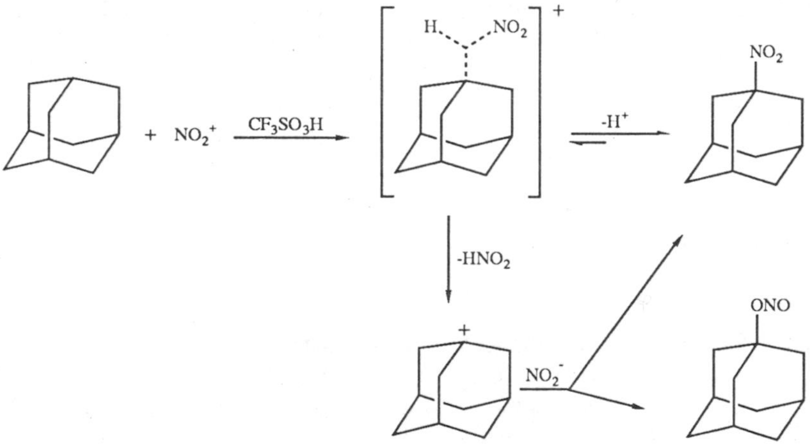

The corresponding nucleophilic nitration reaction of 1-bromoadamantane with nitriles was also studied. Conventional S_N2-type reactions fail, because the nucleophile is unable to approach the reaction center from the backside (i.e., through the cage). Thus alkali metal nitrites fail to give 1-nitroadamantane with 1-haloadamantanes. Silver nitrite, however, gives 1-adamantyl nitrite (and 1-adamantanol by its hydrolysis) as well as a very low yield of 1-nitroadamantane in an S_N1-type reaction due to the ambident nature of the nitrite anion.

Halogenation

The Lewis acid halide catalyzed *halogenation* of adamantane with halogens takes place with great ease and the reaction can give not only 1-haloadamantanes, but can be carried out to proceed through di- and trihalogenation all the way to the preparation of 1,3,5,7-tetrahaloadamantane.[51]
The electrophilic halogenation of adamantane in mechanistic terms can be again considered as an S_E2-type reaction.

$$\text{adamantane} \xrightarrow{Cl_2,\ AlCl_3} \left[\text{adamantane}\cdots H\cdots Cl\right]^+ \bar{AlCl_4} \xrightarrow[-AlCl_3]{-HCl} \text{1-chloroadamantane}$$

Alternatively, however, collapse of the 1-adamantyl tetrahaloaluminate complex, formed via protolytic ionization of adamantane by inevitable protic impurities or HCl formed in the chlorination reaction, can also explain the reaction in terms of an S_N1 substitution

$$\text{adamantane} \xrightarrow[AlCl_3]{HCl(H_2O)} \text{1-adamantyl}^+ \ \bar{AlCl_4} \longrightarrow \text{1-chloroadamantane}$$

The metathetic chlorination reaction of adamantane with $Cl_2 + AgSbF_6$,[49] however, would seem to favor the S_E2 route over the S_N1-type chlorination.

Alkylation

In the *Friedel–Crafts alkylation* of adamantane with olefins (ethylene, propylene, and butylenes) alkyladamantanes are again obtained by two possible pathways: (a) adamantylation of olefins by the adamantyl cation formed through hydride abstraction from adamantane by alkyl cations (generated by the protonation of the olefins) and (b) S_E2-type σ alkylation of adamantane by alkyl cations via insertion into the bridgehead C—H bonds of adamantane through a pentacoordinate carbonium ion.[51]
Based on Schmerling's[52] as well as Bartlett's and Nentizescu's[53] pioneering work, the conventional acid catalyzed alkylation of isoalkanes by alkenes, from

a mechanistic point of view, must be considered as alkylation of alkenes by a trivalent alkyl cation, produced via hydride abstraction from the isoalkane by the initial carbocation formed via protonation of the alkene.

$$RCH{=}CH_2 + H^+ \rightleftharpoons R\overset{+}{C}HCH_3$$

$$R_3CH + R\overset{+}{C}HCH_3 \rightleftharpoons R_3C^+ + RCH_2CH_3$$

$$RCH{=}CH_2 + R_3C^+ \rightleftharpoons R\overset{+}{C}HCH_2CR_3 \xrightarrow{-H^+} RCH_2CH_2CR_3$$

This alkylation path is fundamentally different from that wherein an alkyl cation reacts directly with the alkane via a 2*e*-3*c*-bonded five-coordinate carbocation (σ alkylation).

$$RH + R'^+ \rightleftharpoons \left[R \cdots \begin{smallmatrix} H \\ R' \end{smallmatrix} \right]^+ \xrightarrow{-H^+} R\text{-}R'$$

To study the direct σ-alkylation reaction of alkanes, Olah et al.[54,55] carried out experiments reacting lower alkanes with stable alkyl cations generated under superacidic stable ion conditions. Typical alkylation reactions were those of propane, isobutane, and *n*-butane with *t*-butyl or *s*-butyl cations. As intermolecular hydride transfer between tertiary and secondary alkyl cations and alkanes is generally much faster than the alkylation reactions, the products obtained also included those derived from the alkanes and alkyl cations formed via the hydride transfer reactions.

An interesting example of σ alkylation is the reaction of the *t*-butyl cation with isobutane.[53] Despite the highly unfavorable, sterically crowded interaction, formation of a small amount of 2,2,3,3-tetramethylbutane was found. The fast intermolecular Bartlett–Nentizescu–Schmerling type hydrogen transfer reaction predominates but ~ 2% of 2,2,3,3-tetramethylbutane was also obtained, indicative of a nonlinear (although obviously highly distorted) five-coordinate carbocation intermediate (or transition state).

$$(CH_3)_3C^+ + (CH_3)_3CH \rightleftharpoons \left[\begin{smallmatrix} & H & \\ (CH_3)_3C & \cdots & C(CH_3)_3 \end{smallmatrix} \right] \xrightarrow{-H^+} (CH_3)_3CC(CH_3)_3$$

The superacid catalyzed alkylation of adamantane with olefins was studied by Olah et al.[51]

The mechanistic problem in the acid catalyzed alkylation of adamantane by olefins is again to differentiate the direct σ alkylation of adamantane by alkyl cations (generated by the protonation of olefins) from the conventional π-adamantylation of olefins by 1-adamantyl cation (formed via the hydride abstraction from adamantane by the initially generated alkyl cation). The rigid cage framework of adamantane does not allow the ready formation of a bridgehead olefin and no backside (nucleophilic or electrophilic) attack.

The CF_3SO_3H or $CF_3SO_3H{:}B(OSO_2CF_3)_3$ catalyzed alkylation of adamantane with lower olefins (ethene, propene, 1-butene, *trans*-2-butene, *cis*-2-butene, and 2-methylpropene) was carried out in carbon tetrachloride solution at 0°C.

When adamantane was treated with 1 equivalent or less of CF_3SO_3H in the absence of olefin, quenched, and analyzed, only trace amounts of 1-adamantanol (≤0.1%) were observed. However, the formation of 1-adamantanol increased up to 2% with a higher (10:1) acid to adamantane molar ratio. This indicates the existence of a limited equilibrium between adamantane and the 1-adamantyl cation (and possibly 1-adamantyl triflate) in the acid medium. In order to suppress the formation of the 1-adamantyl cation by protolysis in these acid systems during the alkylation reactions (and the subsequent adamantylation of olefins), reactions were carried out with 1 equivalent or less of superacid.

H^+

Reaction of adamantane with ethene in CF_3SO_3H gave 1-ethyladamantane. Polyethylated adamantanes were also formed along with ethene oligomers.

$$\text{Ad} + CH_2{=}CH_2 \xrightarrow[CCl_4]{CF_3SO_3H} 1(C_2H_5)\text{Ad} + \text{polyethyladamantanes}$$

Similar reaction of propene with CF_3SO_3H as the acid catalyst gave monopropyladamantane [both 1-*n*-propyl and 1-(2′-propyl)], along with polypropylated adamantanes.

$$\text{Ad} + CH_3CH{=}CH_2 \xrightarrow[CCl_4]{CF_3SO_3H}$$

$$1\text{-}(n\text{-}C_3H_7)\text{Ad} + 1\text{-}(i\text{-}C_3H_7)\text{Ad} + \text{polypropyladamantanes}$$

When a stronger acid system (1:1 triflic acid/boron tristriflate) was used the oligomerization of the olefin became more predominant, but the yields of alkyl- and polyalkyladamantanes were also significantly higher. However, with more prolonged reaction time only polyalkyladamantanes were obtained.

The formation of alkyladamantanes, in accordance with the previous discus-

$$\text{olefin} + H^+ \rightleftharpoons R^+$$

$$\text{Ad} + R^+ \rightleftharpoons \text{1-Ad}^+ + \text{RH}$$

$$\text{1-Ad}^+ + \text{>C=C<} \longrightarrow \text{1-Ad—C—}\overset{+}{\text{C}}\text{<}$$

$$\text{1-Ad—C—}\overset{+}{\text{C}}\text{<} + \text{Ad} \longrightarrow \text{1-(R)Ad} + \text{1-Ad}^+$$

Scheme 14. π Adamantylation of olefins.

$$\text{olefin} + H^+ \rightleftharpoons R^+$$

$$\text{Ad} + R^+ \longrightarrow [\text{Ad}\cdots\text{H}\cdots\text{R}]^+ \longrightarrow \text{1-(R)Ad} + H^+$$

Scheme 15. σ Alkylation of adamantane.

sion, can be depicted to take place either by π-adamantylation of olefins (Scheme 14) or by σ-alkylation of adamantanes (Scheme 15).

Alkyl cations formed by protonation of olefins can σ alkylate adamantane by insertion into the tertiary (bridgehead) C—H bond through five-coordinated carbonium ion intermediates. Alternatively, the alkyl cation can be hydride abstracted from adamantane to generate the more stable 1-adamantyl cation, which then adamantylates the olefin. The 1-adamantylalkyl cation formed in the adamantylation of olefin can then abstract hydrogen from another molecule of adamantane to give the alkyladamantane product. Adamantane thus serves as a source for the adamantyl cation for alkylating the olefin.

The formation of 1-ethyladamantane in the reaction with ethene can be explained by either route, and no differentiation between the two pathways is possible. However, more insight into the mechanism can be obtained considering the results of the reaction with higher olefins such as propene and butenes (see Schemes 15 and 16). The adamantylation of these olefins should proceed by Markovnikov addition and thus give the more substituted 1-adamantylalkyl cations. Hydride transfer then leads to the corresponding alkylated products, that is, 1-(*n*-propyl)adamantane in the case of propene. However, direct σ alkylation of adamantane with alkyl cations will give the anti-Markovnikov product, that is, in case of propylation with propene 1-(2′-propyl)adamantane. Indeed, both 1-(*n*-propyl)- and 1-(2′-propyl)adamantane were observed indicative of σ alkylation competing with π adamantylation of propene. To further study the σ alkylation of adamantane, the reaction of adamantane with 2-propanol in CCl_4/CF_3SO_3H as well as with the isopropyl cation, prepared from a solution of isopropyl chloride in SbF_5—SO_2ClF and CH_2Cl_2 at −78°C was carried out. In both reactions 1-(*n*-propyl)- and 1-(2′-

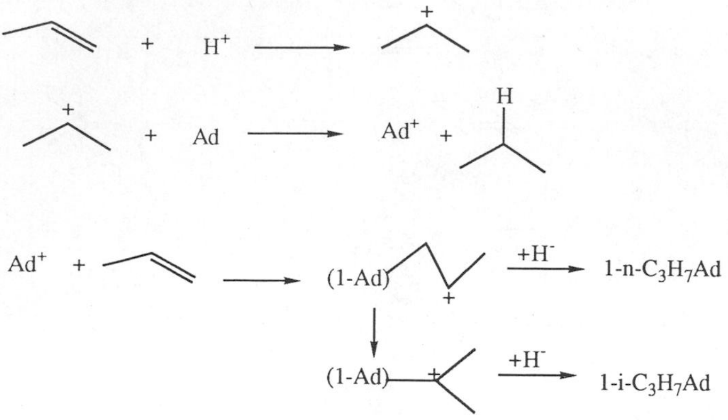

+ H^+ ⟶ +

+ + Ad ⟶ Ad^+ + H

Ad^+ + ⟶ (1-Ad) + $\xrightarrow{+H^-}$ 1-n-C_3H_7Ad

(1-Ad)+ $\xrightarrow{+H^-}$ 1-i-C_3H_7Ad

Scheme 16.

propyl)adamantane were formed in 5:1 and 7:1 ratio, respectively. The formation of 1-*n*-propyladamantane in the reactions is due to the formation of propene by deprotonation of the isopropyl cation under these reaction conditions and its subsequent 1-adamantylation. These results are in accord with the results of the propylation of propane by the isopropyl cation.

1-(2′-Propyl)adamantane is considered to be formed by C—H insertion of the isopropyl cation into the tertiary C—H bond of adamantane, through the corresponding pentacoordinate carbonium ion (Scheme 16). Alternatively, one could argue, however, that the 1-(2′-propyl)adamantane can also be formed by the rearrangement of the initially formed secondary cation from the Markovnikov adamantylation of propene. Thus, both 1-(*n*-propyl)- and 1-(2′-propyl)-adamantane could be products of the adamantylation of propene. On the basis of the obtained propylation data, however, no clear differentiation is possible.

Consequently, in order to gain further understanding of the mechanism of the formation of alkyladamantanes, the reaction of adamantane with butenes was studied.

Butenes (*n*-butene, *trans*-2-butene, *cis*-2-butene, and 2-methylpropene) were reacted with adamantane in CCl_4/CF_3SO_3H with 10:1 adamantane to acid ratio (Scheme 17). Reactions with 2-butenes gave mostly 1-*n*-butyladamantene, 1-*s*-butyladamantane, and 1-isobutyladamantane. Occasionally, trace amounts of 1-*t*-butyladamantane were also formed. Isobutylene (2-methylpropene), however, consistently gave relatively good yield of 1-*t*-C_4H_9Ad along with other isomeric 1-butyladamantanes. 1-Butene gave only the isomeric butyladamantanes with only trace amounts of 1-*t*-C_4H_9Ad.

The formation of 1-*n*-C_4H_9Ad, 1-*s*-C_4H_9Ad, and 1-*i*-C_4H_9Ad in these reactions can be explained by adamantylation of olefins. In a control reaction, when 1-butene was passed through CCl_4/CF_3SO_3H (under the usual reaction conditions) in the absence of adamantane, apart from large amounts of oligomeric products, both 2-butene and 2-methylpropene were formed. Similarly 2-butene was found to isomerize 2-methylpropene in a control experiment. Even 2-methylpropene was isomerized to 2-butene under the reaction conditions. The formation of 1-*s*- and 1-*i*-C_4H_9Ad in the reaction with 1-butene and 1-*i*-C_4H_9Ad in the reaction with 2-butene is readily explained. The formation of small amounts of 1-*s*-C_4H_9Ad in the reaction with 2-methylpropene can be explained by an intramolecular rearrangement of the intermediate 1-(1-adamantyl)-2-methyl-2-propyl

Ad + (1-butene) $\xrightarrow[CCl_4]{CF_3SO_3H}$ 1-(n-C_4H_9)Ad + 1-(s-C_4H_9)Ad + 1-(i-C_4H_9)Ad + polybutyl adamantanes

Ad + (trans-2-butene) ⟶ same

Ad + (cis-2-butene) ⟶ same

Ad + (2-methylpropene) ⟶ same + tert-C_4H_9Ad

Scheme 17.

cation (formed by adamantylation of 2-methylpropene) to the 2-(1-adamantyl)-2-butyl cation. Such a rearrangement in all probability involves a "protonated cyclopropane" intermediate, similar to that suggested in butane isomerization.

H^+ CH_3 (1-Ad)

The formation of *t*-butyladamantane in the studied butylation reactions is significant. Since in control experiments acid catalyzed isomerization of other isomeric butyladamantanes did not give even trace amounts of 1-*t*-butyladamantane, the tertiary isomer must be formed in the direct σ-*t*-butylation of adamantane by the *t*-butyl cation through a pentacoordinate carbocation ion. The same intermediate is involved in the concomitant formation of the 1-adamantyl cation via intermolecular hydrogen transfer (the indicated major reaction). The formation of even low yields of 1-*t*-C_4H_9Ad in the reaction is a clear indication that the pentacoordinate carbocation does not attain a linear geometry ⋜C---H-⋞(which could result only in hydrogen transfer), despite the unfavorable steric crowding. This reaction is similar to the earlier discussed reaction between the *t*-butyl cation and isobutane to form 2,2,3,3-tetramethylbutane.

$$Ad + (CH_3)_3C^+ \longrightarrow [\,H\cdots C(CH_3)_3\text{–}Ad\,]^+ \xrightarrow{-H^+} 1\text{-}t\text{-}C_4H_9Ad$$

$$\Updownarrow$$

$$Ad^+ + (CH_3)_3CH$$

The alternate pathway for the formation of *t*-butyladamantane through hydride abstraction of an intermediate 1-adamantylalkyl cation would necessitate involvement of an energetic "primary" cation or highly distorted "protonated cyclopropane," which is not likely under the reaction conditions.

$$(1\text{-Ad})\text{–}C(CH_3)=CH_2 \not\longrightarrow [\,(1\text{-Ad})\text{–}C(CH_3)_2\text{–}CH_2^+\,] \xrightarrow{+H^-} 1\text{-}t\text{-}C_4H_9Ad$$

The reaction of adamantane with *t*-butyl alcohol in CCl_4/CF_3SO_3H as well as with the *t*-butyl cation under stable ion conditions, was also carried out. In both cases 1-*t*-C_4H_9Ad was formed together with isomeric butyladamantanes.

These results are in accord with studies on the alkylation of alkanes with the *t*-butyl cation under stable ion conditions.

The observation of 1-*t*-butyladamantane in the superacid catalyzed alkylation of adamantane provides evidence for the σ alkylation of adamantane by the *t*-butyl cation. As this involves a most unfavorable sterically crowded tertiary–tertiary interaction, it is reasonable to suggest that similar σ alkylation can also be involved in less strained interactions with secondary and primary alkyl systems, although superacid catalyzed alkylation of adamantane with olefins under usual Friedel–Crafts conditions predominantly occurs via adamantylation of olefins. As the adamantane cage allows attack by the alkyl group only from the front side, the alkylation studies under superacidic conditions provide significant insight into the mechanism of electrophilic reactions at saturated hydrocarbons and the nature of their carbocationic intermediates.

Koch Reaction and Electrophilic Formylation

The reaction of an alkyl or cycloalkyl cation with carbon monoxide yielding acyl cations constitutes the key steps in the well-known Koch–Haaf reaction[56] used for the preparation of carboxylic acids from alkenes, CO, and H_2O. Synthetic aspects of this reaction have been studied extensively and reviewed by Falbe.[57a] Kinetic and thermodynamic aspects of the various reaction steps have been investigated by Hogeveen under stable ion conditions.[57b] Generation of acyl cations by reaction of CO with the corresponding alkyl cations leads to the formation of a new C—C bond. Whether this process occurs by alkylation at carbon or alternatively at oxygen followed by a C→O alkyl shift is still uncertain.

$$R^+ + CO \rightleftharpoons R{-}\overset{+}{C}{\equiv}O \quad (1)$$

$$R^+ + CO \rightleftharpoons R{-}\overset{+}{O}{\equiv}C \quad (2a)$$

$$R{-}\overset{+}{O}{\equiv}C \longrightarrow R{-}\overset{+}{C}{\equiv}O \quad (2b)$$

While a similar O-alkylated intermediate has been postulated[58] in the formation of carbocations from alkoxides and carbenes, justification for the mechanism was obtained so far only from nonempirical calculations[59] of possible structures of a model system, for example, protonated carbon monoxide.[60]

The parent of the acyl cations (R = H) is the formyl cation. In the course of acid catalyzed formylation of aromatics with CO, the formyl cation (protonated CO) is suggested to be the reactive electrophile. Electrophilic formylation of aromatics with CO in superacid media was explained by formylation by the protosolvated formyl cation.[61–65]

Whereas electrophilic formylation of aromatics with CO is well studied under both the Gatterman–Koch condition and with superacid catalysis,[60–65] electrophilic formylation of saturated aliphatics remained till recently virtually unrecognized.

The formation of C_6 or C_7 acids along with some ketones was reported by Paatz and Weisberger in the reaction of isopentane, along with methylcyclopentane and cyclohexane, with CO in $HF{:}SbF_5$ at ambient temperatures and atmospheric pressure.[66a] Yoneda et al.[66b] found that other alkanes can also be carboxylated with CO in $HF{:}SbF_5$. Tertiary alkyl cations, which are produced by the protolysis of C—H bonds of branched alkanes in $HF{:}SbF_5$, undergo skeletal isomerization and disproportionation before reacting with CO. Reactions of the acyclic hydrocarbons of various skeletal structures with CO in superacid media were also recently studied by Yoneda et al.[66c] The products obtained were only isomeric carboxylic acids with a lower number of carbon atoms than the starting alkanes. Formation of the carboxylic acids were accounted for by the reactions of parent, isomerized, and fragmented alkyl cations with CO to form the corresponding acyl cation intermediates (Koch–Haaf reaction), followed by their quenching with water. No formylated products were identified in these reactions.

Okamoto et al.[67a] obtained 3-hydroxy-4-homoadamantyl-1-adamantanecarboxylate by the triflic acid catalyzed reaction of 1-adamantyl triflate with CO at atmospheric pressure in the presence of adamantane. They considered that the initial step involves the reaction of the 1-adamantyl cation with CO giving the 1-adamantanoyl cation, which then abstracts a hydride ion from adamantane yielding 1-adamantanecarboxaldehyde. The transient aldehyde, however, could not be observed. It was suggested that this is due to its fast further reaction under these conditions. In the patent literature[67c] a claim has been made for the preparation of 1-adamantanecarboxaldehyde by the $AlCl_3$ catalyzed reaction of adamantane with CO in CH_2Cl_2 at $\leq$ 25°C. However, no details were given.

Olah and co-workers investigated the superacid-catalyzed reaction of CO with adamantane under mild reaction conditions.[68]

When adamantane was allowed to react with CO under pressure (1200 psi) in trifluoromethanesulfonic (triflic) acid with an acid/adamantane molar ratio of 10:1 at room temperature in 1,1,2-trichlorotrifluoroethane (Freon-113) solution, 1-adamantanecarboxaldehyde was obtained in only 0.2% yield upon quenching the reaction mixture in ice–bicarbonate. 1-Adamantanecarboxylic acid, the usual Koch–Haaf product, is the major product. When the reaction is carried out in the much stronger superacid system $B(OSO_2CF_3)_3$—CF_3SO_3H by using an acid/adamantane molar ratio of 3:1, the yield of 1-adamantanecarboxaldehyde increased to 3.4%. When the same reaction was repeated under identical reaction conditions and CO pressure in the SbF_5—CF_3SO_3H superacid system, the yield of 1-adamantanecarboxaldehyde was further improved to 8.2%. Just as the reaction in the CF_3SO_3H system, reactions in a higher acidity superacid system also gave 1-adamantanecarboxylic acid as the major product (60–75%), together with an increased yield of 1-adamantanol (2–6%).

+ superacid —(1) CO (1200 psi), room temperature; (2) H_2O→

CHO + OH + COOH

When the superacid catalyzed reactions of adamantane with CO were repeated under solvent-free condition by using only the excess of acid as the reaction medium, the yield of 1-adamantanecarboxaldehyde further increased (9.1, 14.5, and 21% in CF_3SO_3H,$B(OSO_2CF_3)_3$—HSO_3, and SbF_5—HSO_3CF_3 system, respectively).

Formation of 1-adamantanol and 1-adamantanecarboxylic acid in the reactions is indicative of the initial protolytic ionization of the adamantane to the 1-adamantyl cation.

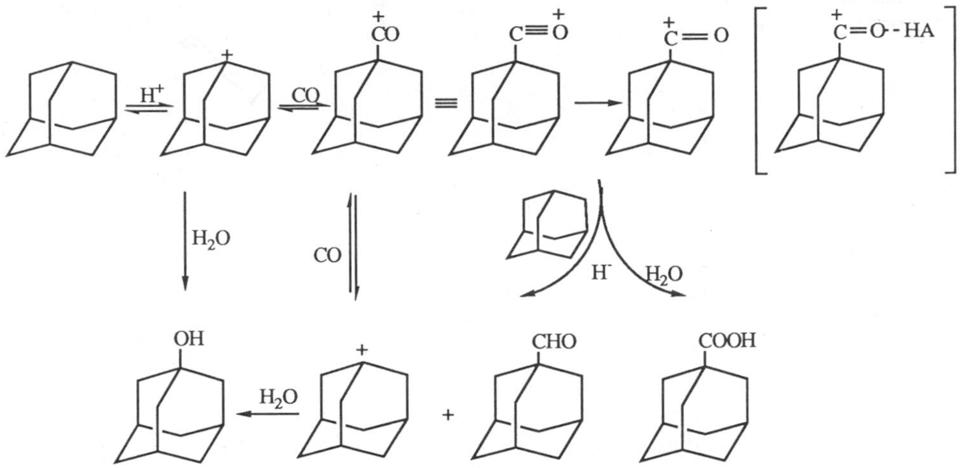

As adamantane is in equilibrium with the 1-adamantyl cation (or possibly with the 1-adamantyl triflate in the case of the triflic acid catalyzed reaction) quenching of the 1-adamantyl cation gives 1-adamantanol.

Kinetics of reversible carbonylation–decarbonylation as applied to acyclic and cyclic systems[69] were investigated by Hogeveen,[57b] Olah,[70,71] Brouwer,[72] and others. It was found that both carbonylation and decarbonylation processes are very rapid. The somewhat off-planar *t*-adamantyl cation was found to be 2.1 kcal less stable than tertiary acyclic cations, but the carbonylation equilibrium constant for the former system was found to be 30 times larger than those for acyclic systems.[57] The stabilization of positive charge in alkanoyl cations has been realized as mainly due to the resonance $RC^{+}{=}O \leftrightarrow RC{\equiv}O^{+}$. The effect of the R group on this stabilization is only of lesser importance. Nevertheless, the magnitude of carbonylation–decarbonylation equilibrium constants provides a quantitative measure of the stabilization of the alkanoyl cations in solution. In superacid systems 1-adamantanoyl cation formed from the 1-adamantyl cation and CO remains in significant concentration under CO pressure and upon quenching gives the 1-adamantanecarboxylic acid as the major product.

An alternative route by which the 1-adamantanoyl cation or its protosolvated form can lead to 1-adamantanecarboxyaldehyde is by hydride abstraction from excess adamantane. The latter then would form the 1-adamantyl cation that can reenter the reaction processes. To prove or disprove the feasibility of the formation of 1-adamantanecarboxaldehyde by this route, 1-adamantanoyl cation was prepared from the reaction of 1-adamantanecarbonyl chloride with either SbF_5—CF_3SO_3H or SbF_5.[70] The prepared adamantanoyl cation was then allowed to react with adamantane (serving as the hydride source) for comparable lengths of time as in the reaction of adamantane with CO. Quenching thc reaction mixture in ice–bicarbonate gave 0.2 and 1.1% 1-adamantanecarboxyldehyde, respectively, (based on the formed 1-adamantanecarboxylic acid from the 1-adamantanoyl cation in SbF_5—SO_3H and SbF_5, respectively).

$^{+}$CO   +   ⟶   CHO   +   OH   +   COOH

(0.2-1.1%)

Formation of 1-adamantanecarboxaldehyde, albeit in very low yield, shows that whereas the 1-adamantanoyl cation can hydride abstract from adamantane, it is not a favorable reaction. Hydride abstraction by alkanoyl cations to give aldehydes consequently is seldom observed.

Brouwer[73] reported hydride abstraction by the acetyl cation from isobutane. Olah et al.[74] showed that in aprotic media the reaction does not take place. In superacidic media it is protosolvation of the acetyl cation that increases its reactivity allowing hydride abstraction from isobutane.

$$CH_3CO^{+}\text{----}HA + HC(CH_3)_3 \longrightarrow CH_3CHO^{+}H + (CH_3)_3C^{+}$$

The yields of 1-adamantanecarboxaldehyde obtained via hydride abstraction by the adamantanoyl cation from adamantane are much lower than those obtained in the reaction of adamantane with CO in the superacid media. It is, therefore, clear that hydride abstraction by the 1-adamantanoyl cation (generated from the 1-adamantyl cation with CO in superacids) is not the major route responsible for the formation of 1-adamantane carboxaldehyde in the former reaction.

To further study the validity of this conclusion Olah et al.[74] also investigated the reaction of 1,3,5,7-tetradeuterioadamantane with CO under comparable reaction conditions. 1,3,5,7-Tetradeuterioadamantane was allowed to react with CO in the presence of CF_3SO_3H. Quenching of the reaction mixture with ice–bicarbonate and extraction in CH_2Cl_2 gave, along with the usual 1-adamantanol and 1-adamantanecarboxylic acid, 3,5,7-trideuterio-1-adamantanecarboxaldehyde-H and 3,5,7-trideuterio-1-adamantanecarboxaldehyde-*D* in a 94:6 ratio (based on ^1H and ^2H NMR spectra with use of an internal standard). In a control reaction of 1,3,5,7-tetradeuterioadamantane with triflic acid under the reaction conditions, no significant hydrogen–deuterium exchange was observed.

These results conclusively show the direct insertion of the formyl cation into the C—H bonds of adamantane, that is, σ formylation is the major pathway leading to the formation of 1-adamantanecarboxaldehyde. Intermolecular hydride (deuteride) abstraction by the adamantanoyl cation appears to be only a minor reaction.

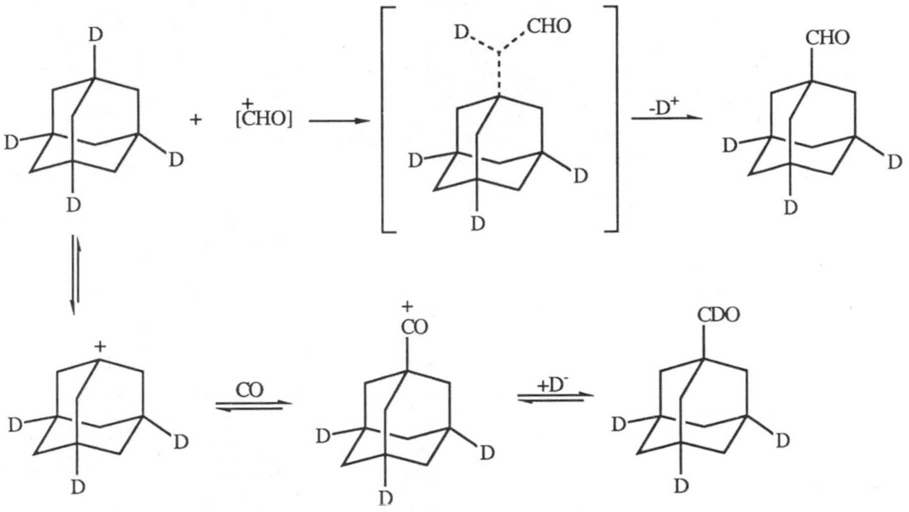

The transition state for the reaction is considered to be a *2e-3c* bonded pentacoordinate carbocation, which via proton elimination gives the aldehyde.

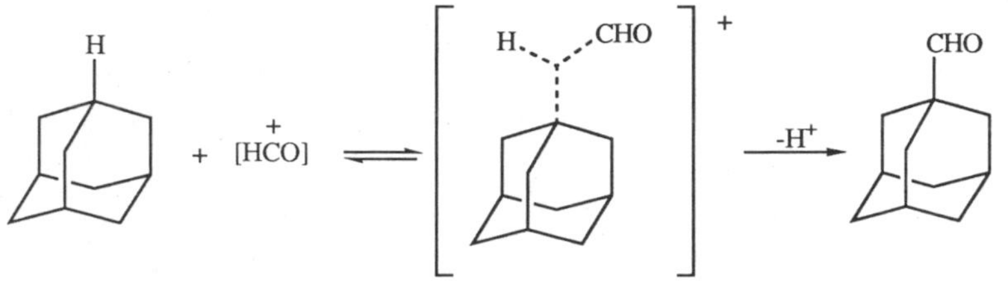

Oxygenation

S_N1-type oxygenation of adamantane to 1-hydroxyadamantane is readily achieved with hydrolyzing solutions of the 1-adamantyl cation. When adamantane is heated in sulfuric acid solution, 2-adamantanone is formed. The reaction is assumed to involve intermolecular hydrogen transfer of the initially formed 1-adamantyl cation, not only with the tertiary C—H, but also secondary C—H bonds of excess adamantane, with subsequent ready oxidation of the 2-adamantyl system.

Electrophilic σ hydroxylation of alkanes has been achieved using either hydrogen peroxide or ozone under superacid catalysis involving $H_3O_2^+$ or HO_3^+, respectively.[75] Handling of anhydrous hydrogen peroxide or ozone, however, is frequently inconvenient. Ricci introduced[76] bis(trimethylsilyl) peroxide as an anhydrous hydrogen peroxide equivalent and used it effectively for the

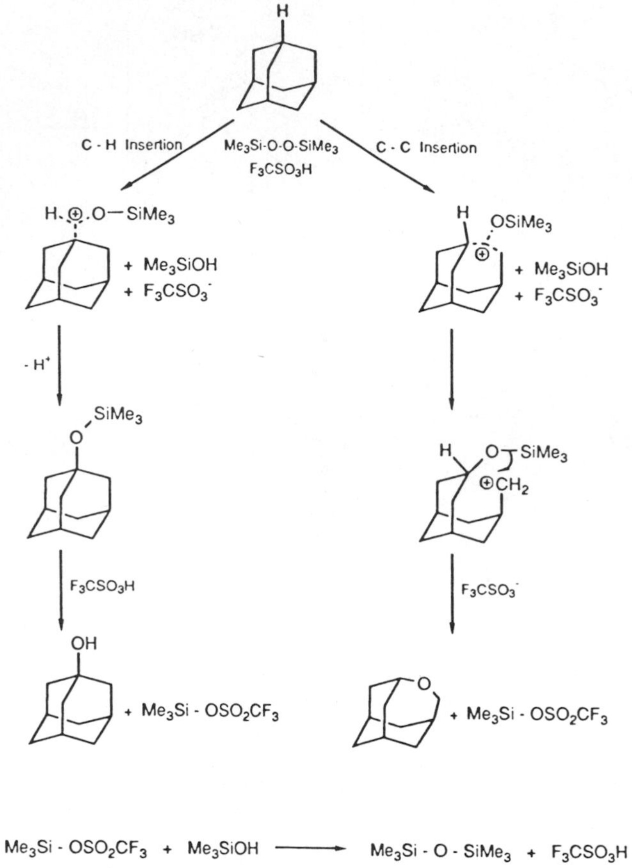

$$Me_3Si\text{-}OSO_2CF_3 + Me_3SiOH \longrightarrow Me_3Si\text{-}O\text{-}SiMe_3 + F_3CSO_3H$$

Scheme 18.

hydroxylation of organolithium compounds. Olah and Ernst employed[77a] this reagent in conjunction with trifluoromethanesulfonic acid as an electrophilic hydroxylating system for aromatics. It was subsequently also found by Olah, Ernst, and Prakash[77b] to be an effective system for the electrophilic oxygenation of adamantane and diamantane.

Adamantane dissolved in dichloromethane at 0°C reacts smoothly with bis(trimethylsilyl)peroxide in the presence of an excess of triflic acid. The main reaction product (79%) is 4-oxahomoadamantane. The minor product (besides unreacted adamantane) is 1-adamantanol. See Scheme 18.

The mechanism of the reaction can be rationalized by initial insertion of trimethylsiloxy cation [formed by the protonation of bis(trimethylsilyl)peroxide] into the C—C σ-bond insertion. The observed selectivity appears to be due to the better electron donor nature of the C—C over the C—H σ bond. The alternative reaction pathway involving initial adamantyl cation formation

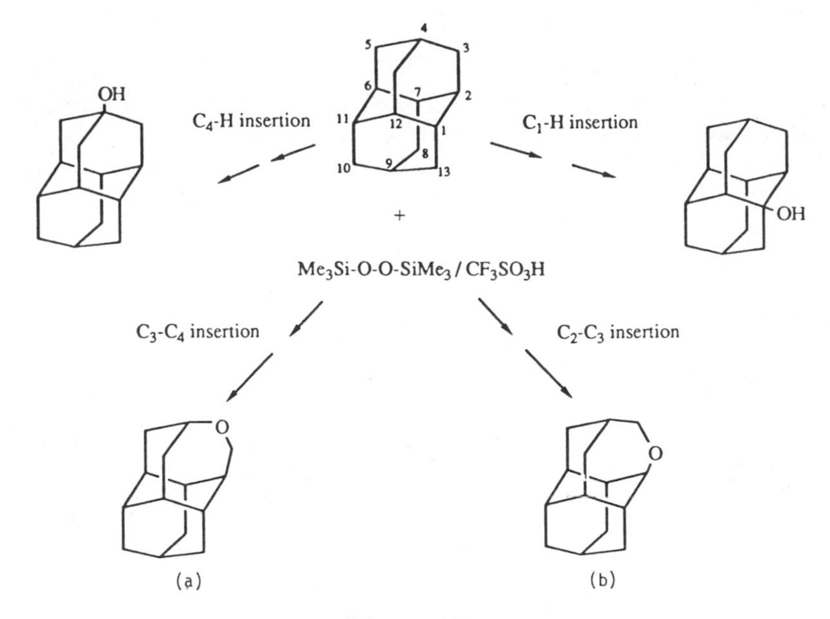

Scheme 19.

followed by reaction with bis(trimethylsilyl)peroxide appears to be unlikely since this should lead predominantly to formation of 1-adamantanol.

Diamantane reacts similarly with $[(CH_3)_3SiO]_2/CF_3SO_3H$ (Scheme 19). In this case, however, four products were obtained due to the presence of two different kind of bridgehead C—H and related C—C σ bonds. The two isomeric oxahomodiamantanes were obtained in 32% isolated yield (based on diamantane used). Their relative ratio in the reaction mixture is 1:3.

In addition, two isomeric bridgehead diamantanols were also isolated and characterized in 24% yield, which is much higher than the observed hydroxylation in the case of adamantane, indicating more ready kinetic accessibility of the C_4—H bond. (The C_1—H bond insertion, although preferred thermodynamically, is more sterically hindered.) Such effects have been previously observed in other electrophilic reactions of diamantane.[78]

CONCLUSION

Study of carbocationic intermediates of cage hydrocarbons led to subsequent development of their superacid catalyzed electrophilic substitution reactions. These studies so far centered mainly on adamantane. Besides helping a better understanding of the mechanistic aspects, the studies also gain preparative utility.

REFERENCES

1. For a review see: R. C. Fort, Jr., *Adamantane, the Chemistry of Diamond Molecules,* Marcel-Decker, New York, 1976.
2. Diamond does not contain hydrogen, all the carbon atoms are bound to outer carbon atoms. This is also the case with the more recently discovered carbon clusters such as C_{60} (Buckminster Fullevane or footballane).
3. S. Landa, *Chem. Listy.* **1933**, *27*, 415, 443.
4. V. Prelog and R. Seiwerth, *Ber.* **1941**, *74*, 1644, 1769.
5. H. Stetter, O. E. Bender, and W. Newmann, *Ber.* **1956**, *89*, 1922.
6. P. v. R. Schleyer, *J. Am. Chem. Soc.* **1957**, *79*, 3292.
7. P. v. R. Schleyer and M. M. Donaldson, *J. Am. Chem. Soc.* **1960**, *82*, 4645.
8. P. v. R. Schleyer, M. M. Donaldson, R. D. Nicholas, and C. Cupas, *Org. Synth.* **1963**, *42*, 16.
9. A. F. Plate, Z. K. Nikitina, and T. A. Burtseva, *Neftekhimiya*, **1961**, *1*, 599; *Chem. Abst.* **1962**, *57*, 4938a.
10. D. A. Johnston, M. A. McKervey, and J. J. Rooney, *J. Am. Chem. Soc.* **1971**, *93*, 2798.
11. C. Cupas, P. v. R. Schleyer, and D. J. Trecker, *J. Am. Chem. Soc.* **1965**, *87*, 917.
12. V. Z. Williams, Jr., A.B. thesis, Princeton University, 1965.
13. T. M. Gund, V. Z. Williams, Jr., E. Osawa, and P. v. R. Schleyer, *Tetrahedron Lett.* **1970**, 3877.
14. T. M. Gund, W. Thilecke, and P. v. R. Schleyer, *Org. Synth.* **1973**, *53*, 30.
15. V. Z. Williams, Jr., P. v. R. Schleyer, G. J. Gleicher, L. B. Rodewald, *J. Am. Chem. Soc.* **1966**, *88*, 3862.
16. W. Burns, M. A. McKervey, and J. J. Rooney, *Chem. Commun.* **1975**, 965.
17. F. S. Hollowood and M. A. McKervey, *J. Org. Chem.* **1980**, *45*, 4954.
18. P. v. R. Schleyer and M. M. Donaldson, *J. Am. Chem. Soc.* **1960**, *82*, 4645.
19. H. W. Whitlock and M. W. Siefkin, *J. Am. Chem. Soc.* **1968** *90,* 4929.
20. G. A. Olah, G. K. S. Prakash, and J. Sommer, *Superacids*, Wiley, New York, 1985.
21. (a) H. Koch and J. Franken, *Brennst-Chem.* **1961**, *42*, 90; *Chem. Abst.* **1961**, *55*, 21059i. (b) R. E. Ludwig, U.S. Patent 2,937,211 (1960); *Chem. Abst.* **1960**, *54*, 1954C.
22. J. A. Olah, and G. A. Olah, *Synthesis* **1973**, 488.
23. O. Farooq, S. M. F. Farnia, M. Stephenson, and G. A. Olah, *J. Org. Chem.* **1988,** *53*, 2840; O. Farooq, Ph.D. thesis, University of Southern California, 1984.
24. G. A. Olah, K. Laali, and O. Farooq, *J. Org. Chem.* **1984**, *49*, 4591.
25. P. v. R. Schleyer, R. C. Fort, Jr., W. E. Watts, M. B. Comisarow, and G. A. Olah, *J. Am. Chem. Soc.* **1964**, *86*, 4195.
26. G. A. Olah, G. K. S. Prakash, J. G. Shih, V. V. Krishnamurthy, G. D. Mateescu, G. Liang, G. Sipos, V. Buss, T. M. Gund, and P. v. R. Schleyer, *J. Am. Chem. Soc.* **1985**, *109*, 2764.

27. G. Kramer, Tetrahedron, **1986**, *42*, 1071.

28. W. L. Davies, R. R. Grunert, R. F. Haat, J. Paulshock, J. C. Watts, T. R. Word, E. C. Hermann, and C. E. Hoffmann, *Science* **1964**, *144*, 862.

29. G. A. Olah and J. Lukas, *J. Am. Chem. Soc.* **1968**, *90*, 933.

30. M. Stephenson, Ph.D. thesis, University of Southern California, **1987**.

31. G. A. Olah, J. J. Svoboda, and A. T. Ku, *Synthesis* **1973**, 492.

32. T. Laube, *Angew. Chem. Int. Ed. Engl.* **1986**, 349.

33. R. Bau, G. A. Olah, and G. K. S. Prakash, unpublished results.

34. M. Bremer, P. v. R. Schleyer, K. Schötz, M. Kausch, and M. Schindler, *Angew. Chem.* **1987**, *99*, 795.

35. (a) W.-D. Fessner, H. Prinzbach, and G. Rhis, *Tetrahedron Lett.* **1983**, 5857. (b) W.-D. Fessner, Ph.D. thesis, University of Freiburg, 1986.

36. L. A. Paquette, R. J. Ternansky, D. W. Balogh, and G. Kentgen, *J. Am. Chem. Soc.* **1983**, *105*, 5446.

37. (a) G. K. S. Prakash, V. V. Krishnamurthy, R. Herges, R. Bau, H. Yuan, G. A. Olah, W.-D. Fessner, and H. Prinzbach, *J. Am. Chem. Soc.* **1986**, *108*, 836. (b) **1988**, *110*, 7764.

38. M. J. Goldstein and R. Hoffmann. *J. Am. Chem. Soc.* **1971**, *93*, 6193.

39. G. A. Olah and J. S. Staral, *J. Am. Chem. Soc.* **1976**, *98*, 6290 and references cited therein.

40. G. K. S. Prakash, W.-D. Fessner, G. A. Olah, G. Lutz, and H. Prinzbach, *J. Am. Chem. Soc.* **1989**, *111*, 746.

41. R. J. Ternansky, D. W. Balogh, and L. A. Paquette, *J. Am. Chem. Soc.* **1982**, *104*, 5446.

42. W.-D. Fessner, B. A. R. C. Murty, J. Worth, D. Hunkler, H. Fritz, H. Prinzbach, W. D. Roth, P. v. R. Schleyer, A. B. McWewn, and W. F. Maier, *Angew. Chem. Int. Ed. Engl.* **1987**, *26*, 452.

43. (a) G. A. Olah, G. K. S. Prakash, T. Kobayashi, and L. A. Paquette, *J. Am. Chem. Soc.* **1988**, *110*, 1304. (b) G. A. Olah, G. K. S. Prakash, W.-D. Fessner, T. Kobayashi, and L. A. Paquette, *J. Am. Chem. Soc.* **1988**, *110*, 8599.

44. G. A. Olah, G. Liang, P. v. R. Schleyer, W. Parker, and C. I. F. Watt, *J. Am. Chem. Soc.* **1977**, *99*, 966.

45. P. v. R. Schleyer, D. Lenoir, P. Mison, G. Liang, G. K. S. Prakash, and G. A. Olah, *J. Am. Chem. Soc.* **1980**, *102*, 683.

46. P. v. R. Schleyer, lecture presented at the *Electron Deficient Clusters* Symposium, University of Southern California, Los Angeles, CA, January 1989.

47. For a review of the electrophilic reactions of alkanes see: G. A. Olah, G. K. S. Prakash, R. E. Williams, L. D. Field, and K. Wade, *Hypercarbon Chemistry*, Wiley-Interscience, New York, 1987, pp. 215–287.

48. T. H. Lowry and K. S. Richardson, *Mechanism and Theory in Organic Chemistry*, 3rd ed. Harper and Row, New York, 1987 and references therein.

49. G. A. Olah, et al. unpublished results.

50. G. A. Olah and H. C. Lin, *J. Am. Chem. Soc.* **1971**, *93*, 1239.

51. G. A. Olah, O. Farooq, V. V. Krishnamurthy, G. K. S. Prakash, and K. Laali, *J. Am. Chem. Soc.* **1985**, *107*, 7541.

52. L. Schmerling., *J. Am. Chem. Soc.* **1944**, *66*, 1422; **1945**, *67*, 1778; **1946**, *68*, 153.

53. (a) P. D. Bartlett, R. E. Condon, and A. Schneider, *J. Am. Chem. Soc.* **1944**, *66*, 1531. (b) C. D. Nenitzescu, M. Avram, and E. Sliam, *Bull. Soc. Chim. Fr.* **1955**, 1266 and references therein.

54. G. A. Olah, Y. K. Mo, J. A. Olah, *J. Am. Chem. Soc.* **1973**, *95*, 4939.

55. G. A. Olah, J. R. DeMember, and J. Shen, *J. Am. Chem. Soc.* **1973**, *95*, 4952.

56. H. Koch and W. Haaf, *Org. Synth.* **1964**, *44*, 1.

57. (a) J. Falbe, *Carbon Monoxide in Organic Synthesis*, Verlag: Berlin, 1970; Chapter III. (b) H. Hogeveen, *Advances in Physical Chemistry,* V. Gold, Ed., 1973, Vol. 10, p.29.

58. P. S. Skell, and I. Starer, *J. Am. Chem. Soc.* **1959**, *81*, 4117.

59. H. B. Jansen and P. Ros, *Chem. Phys. Lett.* **1969**, *3*, 140.

60. G. A. Olah and S. J. Kuhn, in *Friedel–Crafts and Related Reactions,* G. A. Olah, Ed, Wiley-Interscience, New York, 1964, Vol. III, Part 2, p. 1153.

61. G. A. Olah, F. Pelizza, S. Kobayashi, and J. A. Olah, *J. Am. Chem. Soc.* **1976**, *98*, 296.

62. G. A. Olah, K. Laali, and O. Farooq, *J. Org. Chem.* **1985**, *50*, 1483.

63. S. Fujiyama, M. Takagawa, and S. Kajiyama, Japan Patent 2,425,691, **1974**.

64. J. M. Deldricyh and A. Kwantes, British Patent 1,123,966, **1968**.

65. (a) B. L. Booth, T. A. El-Fekky, and G. F. M. Noori, *J. Chem. Soc. Perkin Trans. 2* **1980**, 181. (b) N. N. Crounse, *Org. React.*, **1949**, *5*, 290 and references therein. (c) W. E. Truce, *Org. React.* **1957**, *9*, 37 and references therein. (d) G. A. Olah, J. A. Olah, and T. Ohyama, J. Am. Chem. Soc. **1984**, *106*, 5284.

66. (a) R. Paatz, and G. Weisberger, *Chem. Ber.* **1967**, *100*, 984. (b) N. Yoneda, T. Fukuhara, Y. Takahishi, and A. Suzuki, *Chem. Lett.* **1983**, 17. (c) N. Yoneda, Y. Takahashi, T. Fukuhara, and A. Suzuki, *Bull. Chem. Soc. Jpn.* **1986**, *57*, 2819.

67. (a) K. Takeuchi, T. Miyazaki, I. Kitagawa, and K. Okamoto, *Tetrahedron Lett.* **1985**, 661. (b) K. Takeuchi, F. Akiyama, T. Miyazaki, I. Kitagawa, and K. Okamoto, *Tetrahedron* **1987**, *43*, 701. (c) J. Polis, B. P. Raguel, and E. Liepins (Institute of Organic Synthesis, Academy of Sciences, Latvian, S.S.R.), USSR 4 328 249; *Chem. Abstr.* **1977**, *86*, 43298W.

68. O. Farooq, M. Marcelli, G. K. S. Prakash, and G. A. Olah, *J. Am. Chem. Soc.* **1988**, *110*, 864.

69. H. Stetter, M. Schwarz, Mand A. Hirschhorn, *Berichte* **1959**, *92*, 1628.

70. G. A. Olah, and M. B. Comissarow, *J. Am. Chem. Soc.* **1966**, *88*, 4442.

71. G. A. Olah, A. M. White, J. R. DeMember, A. Commeyras, and C. Y. Lui, *J. Am. Chem. Soc.* **1970**, *92*, 4627.

72. D. M. Brouwer, *Recl. Trav. Chim. Pays-Bas* **1968**, *87*, 210,1435.

73. D. M. Brouwer and A. A. Kiffen, *Recl. Trav. Chim. Pays-Bas.* **1973,** *92*, 689, 809, 906.

74. G. A. Olah, A. Germain, H. C. Lin, and D. A. Forsyth, *J. Am. Chem. Soc.* **1975**, *97*, 2928.

75. For a review see: G. A. Olah, D. G. Parker, and N. Yoneda, *Angew. Chem. Int. Ed. Engl.* **1978**, *17*, 909.

76. M. Taddei and A. Ricci, *Synthesis* **1986**, 633.
77. (a) G. A. Olah and T. D. Ernst, *J. Org. Chem.* **1989**, *54*, 1204. (b) G. A. Olah, T. D. Ernst, C. B. Rao, and G. K. S. Prakash, *New J. Chem.* **1989**, *13*, 791.
78. T. Courtney, D. E. Johnston, M. A. McKervey, and J. J. Rooney, *J. Chem. Soc. Perkin Trans.* I **1972**, 2691.

5 Fragmentation and Transannular Cyclization Routes to Cage Hydrocarbons

A. G. YURCHENKO

Department of Chemical Technology
Kiev Polytechnic Institute
Kiev, USSR

INTRODUCTION

For more than 30 years after its discovery adamantane remained a kind of chemical curiosity, inaccessible and mysterious. Two incidental events attracted the attention of researchers to it, which later on developed into the starting point for a new branch of modern organic chemistry. The first was the discovery of an elegant and remarkably simple method for the preparation of adamantane by Schleyer.[1] After his report adamantane became readily available for studies and allowed the creation of the field, initially named adamantane chemistry, to grow into the chemistry of cage hydrocarbons. There are cases in the history of organic chemistry when very interesting fields of research fell into decay because no practical application of its results were found. A typical example is the chemistry of azulenes, which stopped its development in the early 1960s. That is why the second event, the discovery of the antiviral activity of adamantane derivatives,[2] was also very important. This has stimulated the continued synthesis of adamantane and other cage hydrocarbon derivatives for biological screening purposes.

The elaboration of synthetic approaches to compounds with desired skeletal structure and substituents, as well as methods of substitution of the bridge and bridgehead positions of cage nuclei have become important problems in cage hydrocarbon chemistry. One of the fruitful approaches to the solution of these synthetic problems is the transannular closure of mono- and bicyclic compounds,

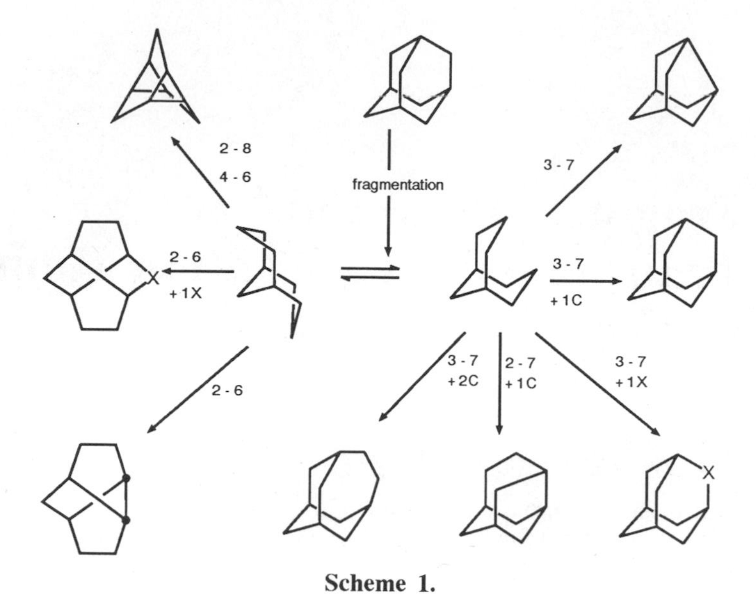

Scheme 1.

bicyclo[3.3.1]nonane derivatives being among the most important. The bicyclo-[3.3.1]nonane skeleton can be recognized in a series of tri- or tetracyclic cage systems and their structural entity is of interest for the construction of such molecules (see Scheme 1).

Scheme 1 summarizes some possible transformations of the bicyclo[3.3.1]-nonane nucleus into different tricyclic cage compounds. Carbon–carbon bond formation between the C-3 and C-7 positions of the bicyclo[3.3.1]nonane skeleton results in noradamantane derivatives. This approach was used by Meerwein[3] and Mori.[4] The introduction of one carbon atom between the C-3 and C-7 positions gives the adamantane nucleus, whereas inclusion of a heteroatom in the same position gave 2-heteroadamantane. Numerous examples of such transformations were collected by Fort in his monograph[5] and Sasaki in his review[6] (for some later papers see Refs. 7,8). The introduction of a carbon atom between the C-2 and C-7 positions leads to protoadamantanes. The most suitable methods of protoadamantan-4-one preparation utilized just this approach.[9–11] Inclusion of a two carbon atom bridge between the C-3 and C-7 positions gives homoadamantane derivatives.[12] Formation of a bond between the C-2 and C-6 positions yields brexane derivatives[13] and inclusion of a heteroatom in these positions allows the preparation of heterotwistanes.[14] These transformations become possible when bicyclo[3.3.1]nonane possesses a double boat conformation. Simultaneous formation of 2-8 and 4-6 bonds in the same conformation gives the triasterane derivative.[15,16] The synthesis of a large variety of variously substituted cage compounds can be achieved by application of the aforementioned principles to bicyclo[3.3.1]nonane derivatives containing different substituents in the secondary and tertiary positions of the skeleton. In this vast field of synthetic organic chemistry so far only a few landmarks are in place and future efforts are needed to elaborate on the details of the synthetic methods and to determine the limits of their applications.

This chapter primarily summarizes the results of our efforts in studying synthetic approaches, stereochemistry, and mechanistic aspects of the following two processes: transannular cyclization reactions of unsaturated bicyclo[3.3.1]nonane derivatives and fragmentation reactions of adamantane dibromides. The latter reaction was usually employed for synthesis of the unsaturated bicyclo[3.3.1]nonanes. In total these two reactions represent the general synthetic approach to the preparation of substituted cage compounds with different functional groups from adamantane derivatives.

TRANSANNULAR REACTIONS OF BICYCLO[3.3.1]NONANE LEADING TO CAGE COMPOUNDS

Electrophilic Addition Reactions to Unsaturated Bicyclo[3.3.1]nonanes

Stetter was the first to demonstrate the transformation of unsaturated bicyclo[3.3.1]nonanes into adamantane derivatives. This electrophilic addition to 3-methylenebicyclo[3.3.1]nonan-7-one (**1**)[17] and to 3,7-dimethylenebicyclo-[3.3.1]nonane (**2**)[18] formed the corresponding 3-substituted 1-hydroxy- and 1-methyladamantanes. These reactions are examples of the electrophilic addition to carbon–oxygen (as in **1**) or carbon–carbon (as in **2**) double bonds involving internal nucleophile participation. In these cases the carbon–carbon double bond plays the role of the nucleophile. Carboxyl,[19] nitrile,[20] hydroxyl,[21] and carbonyl[22]

3 – 13 **14 – 24**

3, 14 $R=CH_3$, $R_1=R_2=H$, $R_3=COOCH_3$
4, 15 $R=R_1=R_2=H$, $R_3=COOCH_3$
5, 16 $R=R_1=R_2=H$, $R_3=CH_2OH$
6, 17 $R=R_3=CH_3$, $R_1=R_2=H$
7, 18 $R=R_1=R_3=H$, $R_2=CH_3$
8, 19 $R=R_1=R_3=H$, $R_2=C_6H_5$
9, 20 $R=R_1=R_3=H$, $R_2=4\text{-}C_6H_4OCH_3$
10, 21 $R=R_3=H$, $R_1=R_2=CH_3$
11, 22 $R=R_1=R_3=CH_3$, $R_2=H$
12, 23 $R=R_1=R_3=H$, R
13, 24 $R=R_3=CH_3$, $R_1=H$, $R_2=C_6H_5$

groups were also reported as internal nucleophiles in the bicyclo[3.3.1]nonane to adamantane transformations.

Compounds (**1**) and (**2**) attracted our attention because they were readily accessible from bridgehead substituted adamantane precursors.[23] These compounds, after several improvements in their methods of preparation were made, became the basis for our extended studies. Compound **1** was prepared in two steps from adamantane involving simple dibromination[24] and fragmentation by alkali in aqueous dioxane,[25] whereas compound **2** was synthesized by zinc fragmentation of 1-bromo-3-bromomethyladamantane.[26] The latter method proved to be the reaction applicable to a wide variety of δ-dibromides and allowed the synthesis of a series of derivatives with substituents both at the bridgeheads of the bicyclo[3.3.1]nonane nucleus and in the exo-methylene group.[27–32]

3-Methylenebicyclo[3.3.1]non-6-ene (**25**) was prepared by this method from 1,4-dibromoadamantane in excellent yield[3] (cf. Ref. 34). The double fragmentation of the adamantane nucleus was also carried out to yield 1,3,5,7-tetramethylenecyclooctane (**26**).[35]

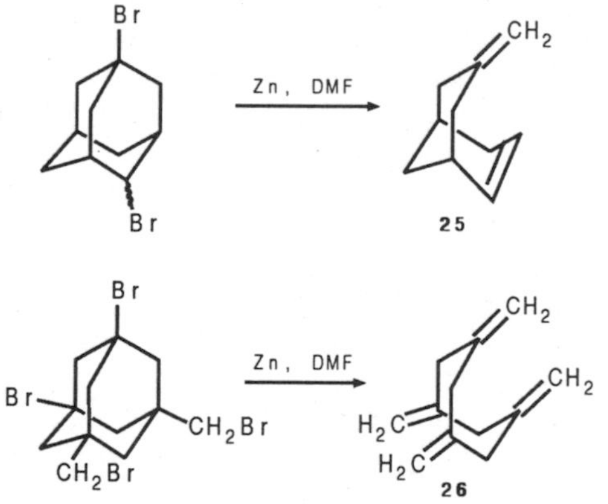

Br, Br — Zn, DMF → CH_2 **25**

Br, Br, CH_2Br, CH_2Br — Zn, DMF → CH_2, CH_2, H_2C, H_2C **26**

The structure **26** was formally attributed to the cyclic allene tetramer,[36] and its ability to give a Diels–Alder adduct with tetracyanoethylene was considered as an unusual example of a [2+2+2] cycloaddition.[37] Actually, compound **26** does not form adducts either with maleic anhydride or with tetracyanoethylene.

Under electrophilic addition conditions the above mentioned fragmentation products and compounds resulting from their chemical transformations usually regenerate the adamantane nucleus. In total these two reactions (fragmentation and electrophilic addition) can be considered as a replacement of a substituent with another, as demonstrated by the following scheme:

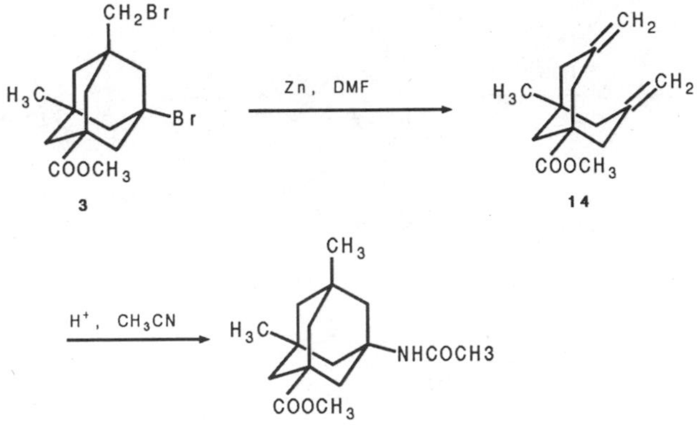

In the course of two consecutive reactions the bromine atom of the bridgehead is substituted for an acetamido group and that in the bromomethyl group for a hydrogen atom. In some cases the combination of the two reactions allowed the transfer of the substiteunt from the side chain to the cage skeleton.

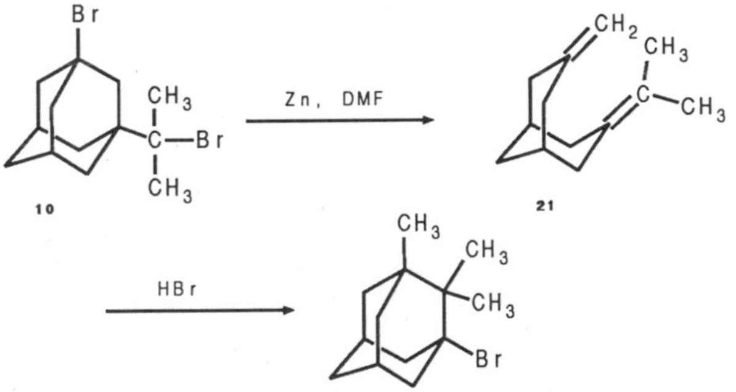

Two methyl groups are thus at once transferred from the side chain to the nucleus.

A very important feature of unsaturated bicyclo[3.3.1]nonanes is their ability to undergo a variety of reactions leading to cage compounds directly via transannular interaction or producing new synthons for construction of cage compounds. The compound **1** upon peracid oxidation yields an epoxide (**27**) that can be converted into different cage compounds under treatment with both electrophilic and nucleophilic reagents.[38,39]

1 —(RCO_3H)→ 27 —(H^+, HX)→ 2-oxaadamantane (X, CH_2OH)

27 —(NH_3)→ (NH_2, CH_2OH); 27 —(RO^-, ROH)→ (OR, CH_2OH); 27 —(t-BuOK, DMSO)→ (O=, OH)

Upon peracid oxidation of compound **2** neither mono- nor diepoxide was isolated, as transannular cyclization into the adamantane derivative takes place in the course of the reaction.

2 —(RCO_3H)→ (OCOR, CH_2OH)

Compound **1** is easily isomerized in the presence of Raney nickel,[40] allowing a series of 1,3,4-trisubstituted 2-oxaadamantane derivatives to be produced upon electrophilic or nucleophilic cyclization of the resulting ketone (**28**).[41,42]

1 —(Raney Ni)→ 28 —(H^+, HX)→ (X, CH_3)

28 —(Br_2, ROH)→ (OR, CH_3, Br); 28 —(RCO_3H)→ (OH, CH_3, OH)

Unfortunately, the isomerization of compound **2** gives a complex mixture of hydrocarbons from which the hydrocarbon **29** was isolated with considerable difficulties. As expected its reactivity[43] was found to be close to that of compound **25**, yielding 1,4-disubstituted adamantanes upon electrophilic addition.[33]

CH_2 CH_2 Raney Ni CH_2 CH_3 I_2 I CH_3 I

2 29

Another interesting set of synthons suitable for the preparation of substituted adamantanes is offered by the conversion of compound **1** and **2** into bicyclic mono- and diallenes and cumulenes. 3-Methylenebicyclo[3.3.1]nonan-7-one (**1**) with dichlorocarbene form an adduct (**30**) in 68% yield. The corresponding dibromocarbene adduct was separated in 32% yield. Upon treatment with *n*-butyllithium the carbonyl group of the adduct remains unaffected and ketallene (**31**) was produced in 80% yield. When treated with electrophilic reagents the allene (**31**) undergoes transannular cyclization to form 3-substituted 1-hydroxy-2-methyleneadamantanes.[44]

O CH_2 :CX_2 O X X n-BuLi

1 30

X = Cl, 68%
X = Br, 32%

O C CH_2 HY OH CH_2 Y

31

Y = OH, 76%
Y = OCH_3, 78%

Addition of an equimolar amount of dihalocarbene to 3,7-dimethylenebicyclo-[3.3.1]nonane leads to formation of a mixture of mono-and diadducts from which the monoadduct was separated in 25–36% yield. With twofold excess of the carbene the diadduct was isolated in 76–78% yield. The dibromocarbene adducts upon treatment with *n*-butyllithium were converted into the corresponding 3-methylene-7-vinylidenebicyclo[3.3.1]nonane (**32**) and 3,7-divinylidenebicyclo[3.3.1]nonane (**33**), compounds with unique spatial arrangement of the double bond and allene system.[45,46]

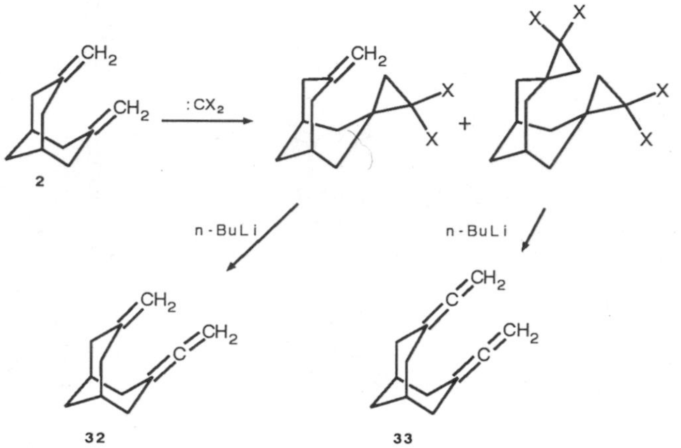

The adducts with dichlorocarbene did not react with *n*-butyllithium under these conditions, but compounds **32** and **33** were formed in greater than 90% yield from the corresponding bromo derivatives. The depicted scheme demonstrates the type of reactivity of compound **32**.

Hard acids, such as protic acids, attack compound **32** on the methylene group, but soft acids, such as iodine, attack the allene system. At a low concentration of sulfuric acid (0.1–0.5 mol/L) adamantane derivatives are formed, but at acid concentrations greater than 10 mol/L the primary unsaturated cyclization products are rearranged into protoadamantane derivatives.[45]

An X-ray structural study of compound **33** confirmed its double chair conformation with the C3—C7 distance equal to 2.953 Å. Structural parameters of the molecule are in good agreement with those obtained by INDO (intermediate neglect of differential overlap) calculations (see Fig. 1). The unique mutual arrangement of the allene system in the molecule allowed the study of their transannular interaction in the course of addition reactions. It proved that the electrophilic cyclization of compound **32** gave unsaturated adamantane and protoadamantane derivatives in good yields.[46] It should be pointed out that neither **32** nor **33** form 1,2-addition products in the course of the reactions, even in trace quantities. The nature of the species formed upon electrophilic addition to **33** as well as **32** depends considerably upon the nature of the reagent and the

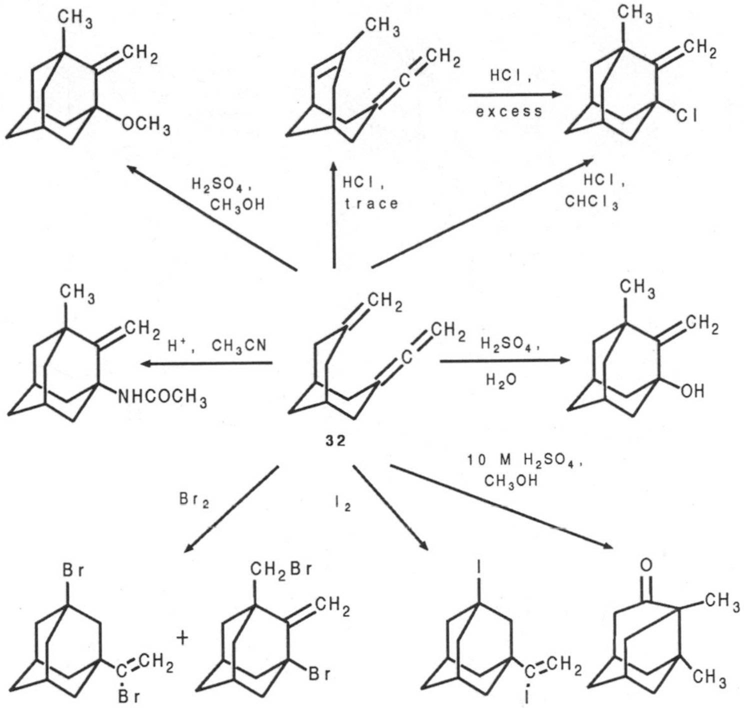

Scheme 2.

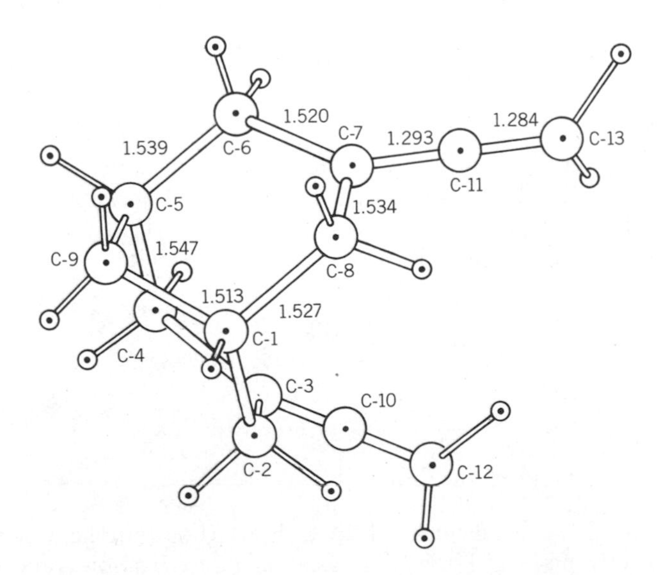

Fig. 1. X-ray structure of **33** (bond lengths in angstroms, Å).

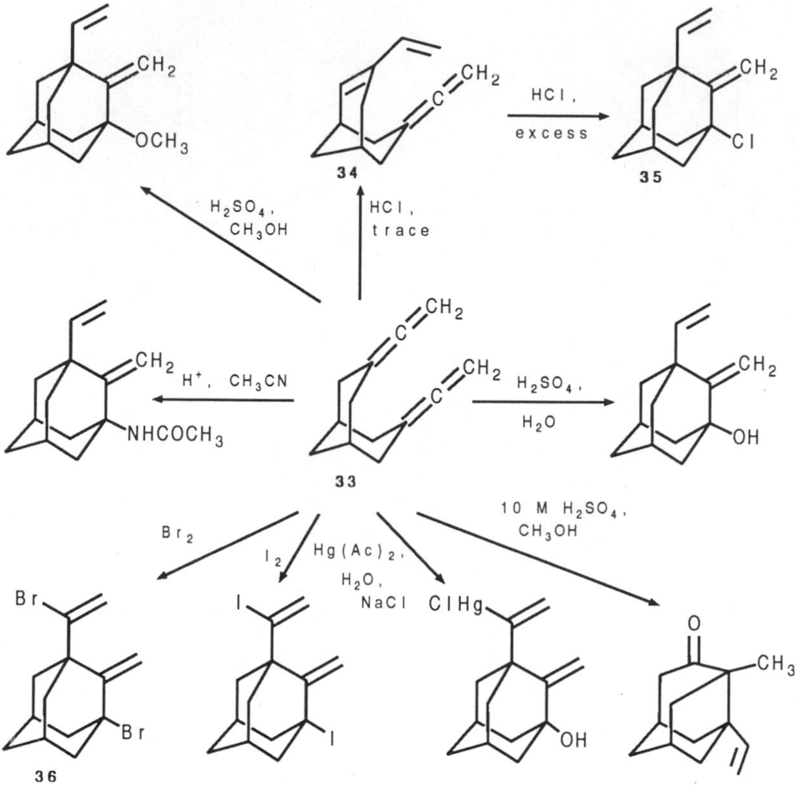

Scheme 3.

solvent. A solution of 0.1–0.5 *M* sulfuric acid in water, methanol, or acetonitrile converts **33** into 3-hydroxy- , 3-methoxy- or 3-acetamido derivatives of 1-vinyl-2-methyleneadamantane. In nonpolar solvents (hexane or CCl_4) with a deficiency of the reagent, isomerization products are mainly formed.

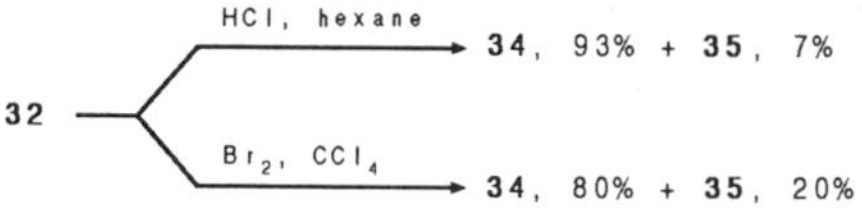

32 — HCl, hexane → **34**, 93% + **35**, 7%

32 — Br_2, CCl_4 → **34**, 80% + **35**, 20%

As in the case of **22**, upon treatment of **33** with 10 *M* sulfuric acid or a strong acid in polar solvents, the primary unsaturated cyclization products are rearranged into the corresponding protoadamantanes.[46]

Bicyclic allenes **32** and **33** on treatment with dibromocarbene form the diadducts **37** and **39**. Their spectral properties support the structures and

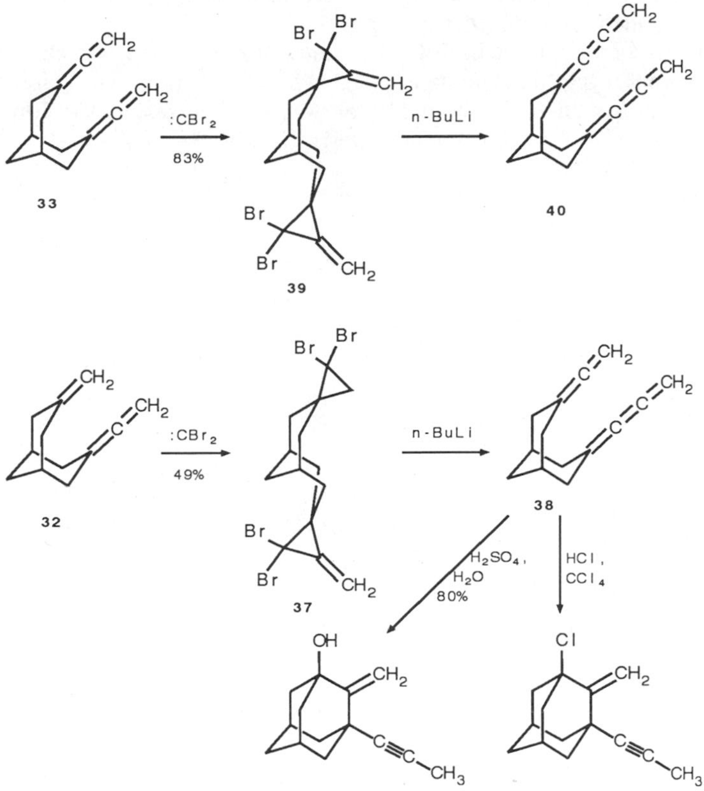

Scheme 4.

conformations depicted in Scheme 4. Butyllithium converts the adducts into the cumulenes **38** and **40**. Though these cumulenes are unstable substances and can be maintained only in dilute solutions,[47] nevertheless the possibility of synthesis of substructures hardly accessible by other methods makes them attractive for further studies.

Two reagents (dilute H_2SO_4 and HCl in CCl_4) convert the allene (**38**) into adamantane derivatives containing methylene and methylacetylene groups.

Free Radical Addition Reactions

Concerning the addition reactions discussed previously, much evidence indicates that in ionic transannular cyclization reactions of bicyclo[3.3.1]nonane derivatives only slightly strained compounds such as adamantane, 2-oxaadamantane, or protoadamantane derivatives can be formed.

Upon transforming the bicyclo[3.3.1]nonane system to the noradamantane skeleton by using free radical addition reactions, it became possible to overcome much larger transition state energy barriers.

Addition of carbon tetrachloride, bromotrichloromethane, or chloroform to 3,7-dimethylenebicyclo[3.3.1]nonane (**2**) whose reaction was initiated by benzoylperoxide, azobisisobutyronitrile, or γ-irradiation, resulted in formation of noradamantane as well as adamantane derivatives.[48–50]

CH_2
CH_2
2
$BrCCl_3$, γ - hν
CH_2CCl_3
CH_2
41
CH_2CCl_3
$CH_2\cdot$
42
CH_2CCl_3
CH_2CCl_3
CH_2Br
43
CH_2CCl_3
CH_2CCl_3
Br
44

The noradamantane/adamantane ratio was found to be temperature dependent. On elevating the temperature, the yield of the thermodynamically more stable tetrahalide with an adamantane structure increased (see Table 1).

This observation allowed us to suggest and experimentally prove the interconversion of intermediate radicals **41** and **42**.[51] Free radical dehalogenation with the tributyltin hydride of 3-bromomethyl-7-ethylnoradamantane gave a mixture of 3-methyl-7-ethylnoradamantane and 1-ethyladamantane. The temperature dependence of the relative yield of the two reaction products was about the same as that shown in Table 1. Free radical addition of CCl_4 to 2,6-dimethylenebicyclo[3.3.1]nonane gave the tetrachloride

TABLE 1. Product Distribution in the Addition of Bromotrichloromethane to 2

Temperature (°C)		22	40	77	100	150	200
Yield (%)	**43**	100	99.9	97.7	95.4	90.3	44.8
	44		0.1	2.3	4.6	9.7	55.2

with a brexane structure,[52] thus allowing us to carry out transannular cyclization using the C-2 and C-6 positions of bicyclo[3.3.1]nonane.[53]

CH_2 CH_2 → (CCl_4, hν) CH_2CCl_3 CH_2Cl ≡ CH_2CCl_3 CH_2Cl

Photochemical Reactions

Ultraviolet (UV) irradiation of 3,7-dimethylenebicyclo[3.3.1]nonane **2** and its derivatives containing substituents in the C-1 and C-5 positions and in the exo-

CH_2 R_1 R R_2 R_3 → (hν, Cu_2Cl_2) R R_1 R_2 R_3

R = H, CH_3; R_1 = H, CH_3; R_2 = H, CH_3, C_6H_5, and $COOCH_3$; R_3 = H, CH_3, CH_2OH

methylene group, converts these compounds in the presence of Cu_2Cl_2 into substituted propellanes (3,6-dehydrohomoadamantane derivatives).[54,55] In this case Cu_2Cl_2 serves as a sensitizer and reveals a kind of matrix effect. It was found that from all the unsaturated bicyclo[3.3.1]nonanes studied, only those compounds that are able to from stable π complexes with silver or cuprous ions undergo cyclization.[56] According to an X-ray determination[57] of the π complex of **2** with silver nitrate, the silver ion is located between the two exocyclic double bonds of the molecule and thus strengthens the double chair conformation (Fig. 2).

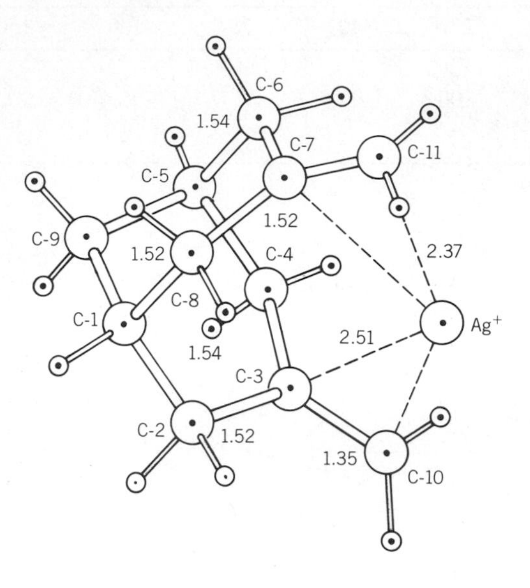

Fig. 2. X-ray structure of the **2**-Ag^+ cation. Interatomic distances are in angstroms (Å).

Contrary to the endo-complexing bicyclo[3.3.1]nonane olefins, the unsaturated bicyclo[3.3.1]nonanes that form π complexes with the exo location of the metal ion[58] do not undergo photochemical cyclization. This explains why methylene-allene (**32**) and diallene (**33**) remain unaffected by UV irradiation. Attempted photochemical cyclization of 2,6-dimethylenebicyclo[3.3.1]nonane was also unsuccessful.[53]

CH_2 CH_2 hν CH_2 CH_2

33

CH_2 CH_2 hν

32

Chemical properties of propellanes of this kind (substitution, isomerization, etc.), except bromination of 3,6-dehydrobromoadamantane,[54] were not studied.

Stereochemistry of Transannular Addition Reactions

Upon electrophilic addition to the exo methylene derivatives of bicyclo[3.3.1]-nonane, only 3,7-dimethylenebicyclo[3.3.1]nonane (**2**) forms the same reaction product irrespective of which exo methylene group is attacked by the electrophile. If one of the exo methylene groups is substituted, the energy barrier becomes different for electrophilic attack on the two possible reaction centers.[26,28,30]

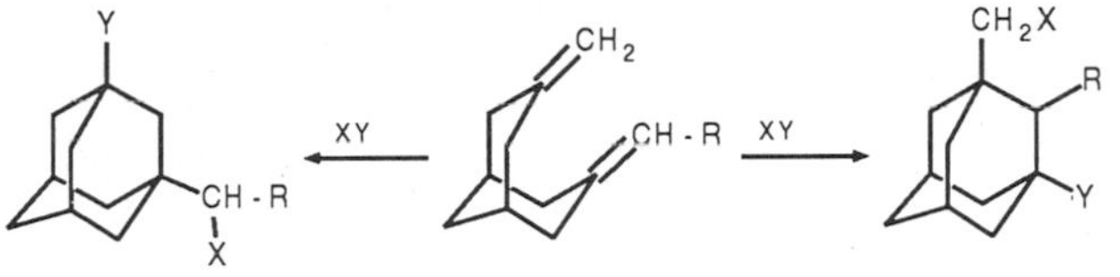

It turned out that only the unsubstituted exo methylene group was attacked by reagents of low and moderate reactivity, whereas highly reactive electrophiles led to both reaction pathways. Compound **28**, reacts with bromine in CCl_4 at room temperature to form a mixture of 1-bromoethyl-2-methyl-3-bromoadamantane and 1-(α-bromoethyl)-3-bromoadamantane (**7**) in the ratio 78:22.[27] The relative yields of dibromides are temperature dependent. Addition of bromine to compound **22** resulted in a mixture of **11** and 1-bromomethyl-2,5,7-trimethyl-3-bromoadamantane in the ratio 7:93 at 36°C and 59:42 at –80°C. Both dienes upon addition of iodine in CCl_4 give only the products corresponding to the reaction that takes place on the unsubstituted methylene group.

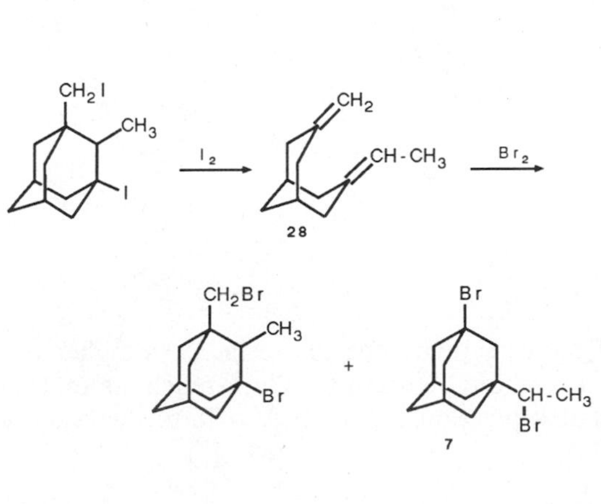

One explanation of how adamantyldimethylcarbinol undergoes rearrangement to form 1-isopropyladamantane-3-carboxylic acid under Koch carboxylation conditions is that of fragmentation–recombination, including the intermediate formation of **21**.[59] This compound was synthesized and the protic acid catalyzed cyclization studied.[28] Since under all conditions studied only the unsubstituted methylene group is protonated and the 1,2,2-trimethyladamantane derivatives are formed,[28] along with results of deuteration experiments,[59] allows us to eliminate this possibility.

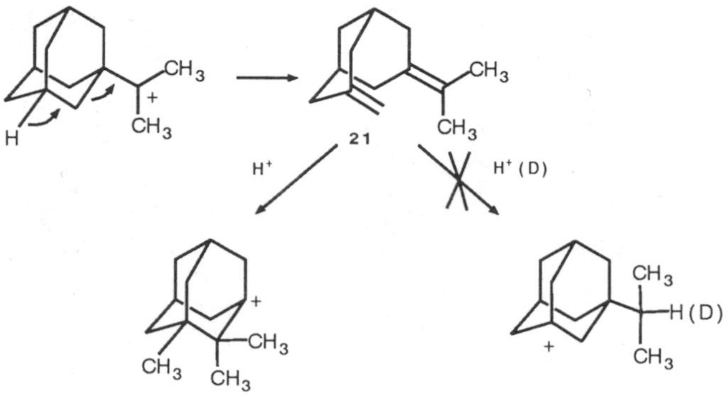

The regiospecific nature of this electrophilic addition to bicyclic allenes and cumulenes is noted above.

It should be emphasized that ionic transannular cyclization may occur only when the resulting cage molecule is not highly strained. All the cases studied showed that the noradamantane skeleton is not formed either by C-3–C-7 bond formation in the bicyclo[3.3.1]nonane system, or by introducing a methylene group between C-3 and C-7 positions in bicyclo[3.3.0]octane.

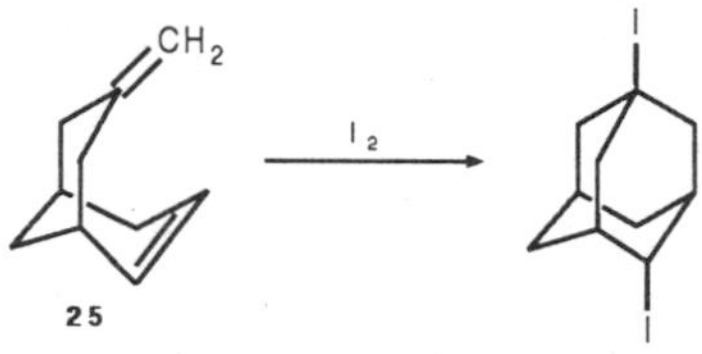

The basicity of the exo methylene group in **25** is higher than that of the C-6 position, and the carbonyl group in **28** is much more basic than the C-6 position. But just this position is attacked to form the reaction product.[33,41]

3,7-Dimethylenetricyclo[3.3.2.0]decane with hydrogen bromide or iodine undergoes only isomerization but not transannular cyclization.[60]

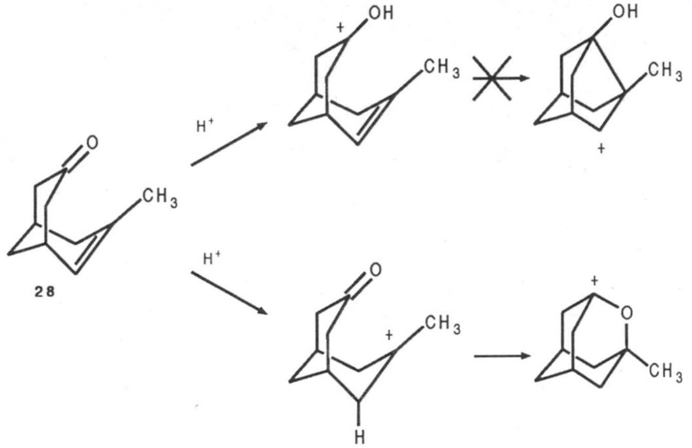

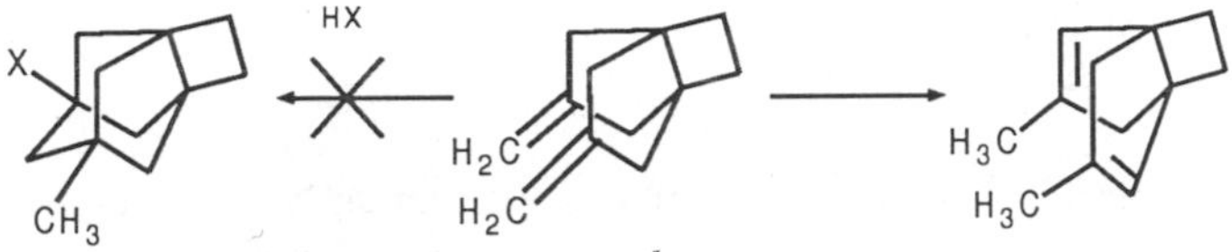

The introduction of a substituent into one of the exo methylene groups of compound **2** leads to loss of C_{2v} symmetry and gives rise to molecular chirality. A successful attempt of electrophilic cyclization of a chiral derivative of compound **1** was undertaken by McKervey,[61] although neither optical purity of the bicyclic compound nor the degree of retention of configuration in the course of cyclization were determined. This precedent gave us hope that we could prepare the chiral adamantanes from derivatives of compound **2**, provided successful separation of the racemates was achieved.

Cope's method[62] proved to be unsuccessful in the case of chiral bicyclo[3.3.1]nonane olefins. Platinum complexes of compounds **18** and **19** were found to be unstable and decomposed with formation of metallic platinum. By means of liquid chromatography on silica gel impregnated with silver *d*-camphorsulfonate we succeeded in separating several chiral bicyclic diolefins into fractions enriched by one enantiomer.[63,64] The compounds, their specific rotations, and enantiomeric purity are listed in Table 2. Determination of the optical purity of chiral olefins is a complicated problem in itself. The shortest way to solve this problem is by employing NMR studies of diastereomeric

TABLE 2. Optical Properties of Bicyclic Diolefins

Compound[a]	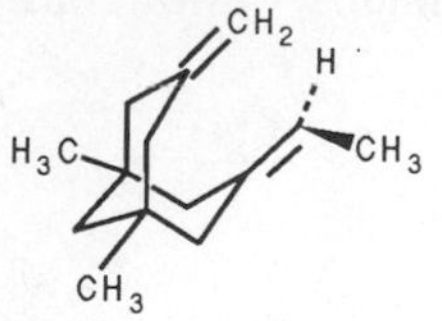 CH_2 H H_3C CH_3 CH_3		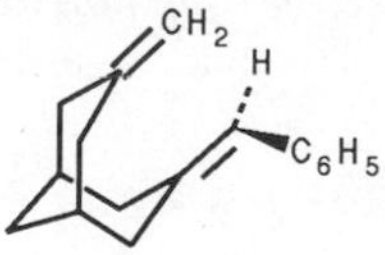 CH_2 H C_6H_5		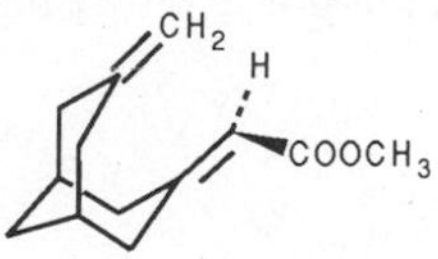 CH_2 H $COOCH_3$	
Isomer	(--)	(+)	(--)	(+)	(--)	(+)
$[\alpha]^{20}_{589}$ (deg, °)	-29.2	+29.1	-117	+119	-226.3	+226.3
Enantiomeric purity (%)[b]	29	23	60	64	62	33

[a](--) - isomers are depicted
[b]Enantiomeric purity listed was achieved in a single run through the chromatographic column.

complexes of the olefins with an optically active solvent of shift reagents. Success in distinguishing enantiomers can be expected in using the following combinations of reagents:

(a) Chiral olefin plus binuclear chiral silver–lantanide shift reagent;
(b) Chiral olefin plus π-complexing agent + chiral shift reagent;
(c) Chiral olefin plus chiral π-complexing agent plus shift reagent.

Case (a) was first proposed in 1981[65,66] and in many instances operates well, but is quite expensive. The two other cases have been used successfully for optical purity determinations.[63,67–69] In case (a) AgEu $(hfbc)_4$ was used as a binuclear reagent. In cases (b) and (c) silver nitrate was used as the π-complexing agent and silver 10-*d*-camphorsulfonate was used as the chiral π-complexing agent, and $Eu(fod)_3$ and $Eu(hfbc)_3$ were used as the shift reagent and chiral shift reagent, respectively. Proton chemical shifts in diastereomeric complexes of chiral diolefins with silver *d*-camphorsulfonate in the presence of $Eu(fod)_3$ differ markedly, thus the optical purity of samples can be determined by simple integration of the spectra.

Chiral derivatives of 3,7-dimethylenebicyclo[3.3.1]nonane (**2**) underwent electrophilic cyclization to form adamantane derivatives with peculiar "retention of configuration," though here we have the case of transformation of chirality (Scheme 5).

By means of a chiral shift reagent, $Eu(hfbc)_3$, alcohol **45**, produced by acid hydration of the enantiomeric diolefin **19**, was found to have the same optical purity as the starting substance.[69] Samples of the bromide (**44**) obtained either from diolefin (**19**) or from alcohol (**45**), had equal rotation.

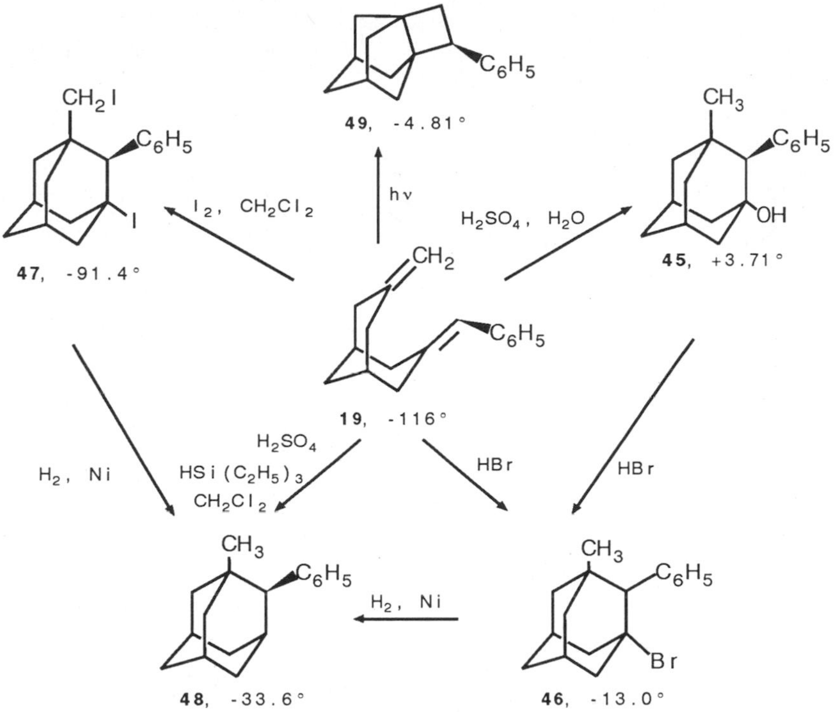

Scheme 5.

Samples of 1-methyl-2-phenyladamantane (**48**) were produced by three different routes:

1. By alcohol **45**, bromide **46**, and Raney Ni dehalogenation to **48**.
2. By diiodo derivative **47** and subsequent dehalogenation.
3. By the direct closure of hydrocarbon **19** in the presence of a hydride donor, that is, triethylsilane hydride.

These samples all have the same rotation. According to the principle of conservation of orbital symmetry, photocyclization of **19** also should retain optical purity. Unfortunately, there was no opportunity to check it for compound **49**. Chiral cage substances obtained in these studies are listed in the Table 3.

TABLE 3. Chiral Derivatives of Adamantane

Compound	$[\alpha]^{20}_{589}$ (deg,°)[a]	Optical Purity,(%)	Compound	$[\alpha]^{20}_{589}$ (deg,°)	Optical Purity, (%)
CH_2I, C_6H_5, I	–91.4 +92.2	60 64	CH_3, CH_3, CH_3, CH_3, OH	–2.11 +1.91	29 23
CH_3, C_6H_5, OH	–3.75 +3.71	70 80	CH_2Br, CH_3, CH_3, CH_3, Br	–1.29 +1.25	29 23
CH_3, C_6H_5, Br	–13.0 +12.7	80 70	Br, CH_3, CH_3, $CH(Br)-CH_3$	–3.52 +3.68	29 23
CH_3, C_6H_5, H	–33.5 +33.1	80 70	CH_3, $COOCH_3$, OH	–267 +271	33 62
C_6H_5	–4.81 +4.94	60 30			

[a]Maximum optical purity, achieved in different experiments.

The possibility of synthesizing different unsaturated derivatives of bicyclo[3.3.1]nonane from bridgehead substituted adamantanes, together with methods of separation of racemic chiral olefins into enantiomers, available optical purity control, and complete retention of configuration upon transannular cyclizations allow this sequence of reactions to be regarded as a suitable synthetic method for preparation of a wide variety of cage hydrocarbon derivatives including enantiomers of chiral compounds.

Proton Magnetic Resonance and the Thermochemical Study of π Complexes of Unsaturated Bicyclo[3.3.1]nonanes

The high stability of the π complexes of compound (**2**) and several of its derivatives was described in earlier synthetic papers,[18,27,28] but no details of these studies were reported. Evans, who first observed lanthanide induced chemical shifts (LIS) in the NMR spectra of olefins–π complexes,[70] encouraged us to study unsaturated bicyclo[3.3.1]nonane by means of the lanthanide shift reagents (LSR) method.[71,72] It was found that observed LIS in the diolefin–silver salt–LSR system becomes larger as the anion of the silver salt becomes more nucleophilic. This means that the anion serves as a coordination center for LSR. This observation considerably changes the concept of the mechanism of silver–LSR interaction[73] by Sievers (see Table 4).

Calculation of lanthanide and silver ion positions by means of correlation of the observed LIS and calculated according to the McConnell–Robertson equation[74] showed that part of the olefins formed chelate-type complexes. The

TABLE 4. Induced Chemical Shifts in the PMR Spectra of Complexes

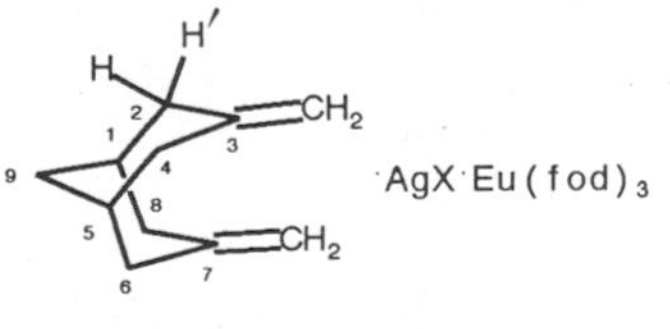

$AgX \cdot Eu(fod)_3$

	LIS (ppm)[a]				
	H-10,11	H-2,4,6,8	H-2′,4′,6′,8′	H-1,5	H-9
BF_4	0	0	0	0	0
ClO_4	0.29	0.20	0.20	0.06	0.06
CH_3CO_2	2.70	2.20	1.02	0.90	0.64
CH_3CO_2	3.24	2.60	1.14	0.96	0.78
fod	7.77	3.25	2.20	1.55	1.52

[a]Extrapolated to 1:1 complex/LSR ratio.

silver ion in such complexes forms coordination links with the two double bonds of the molecule. This was confirmed by X-ray analysis of the complex of compound **2** with $AgNO_3$ (see p. 168). Figure 3 summarizes the results. The dark circles represents the silver ion and its relative position in the studied diolefin π complexes.[58] As expected, the structure of the complexes is greatly affected by spatial hindrances of the double bonds (compare complexes of compounds **25** and **29**). The change from $AgNO_3$ to $AgEu(fod)_4$ as a complexing agent in some cases alters the mode of interaction. Compounds **19** and **24** form a chelate π complex with $AgNO_3$, though with $AgEu(fod)_4$, complexes were formed through unsubstituted exo-methylene groups as in compound **29**.

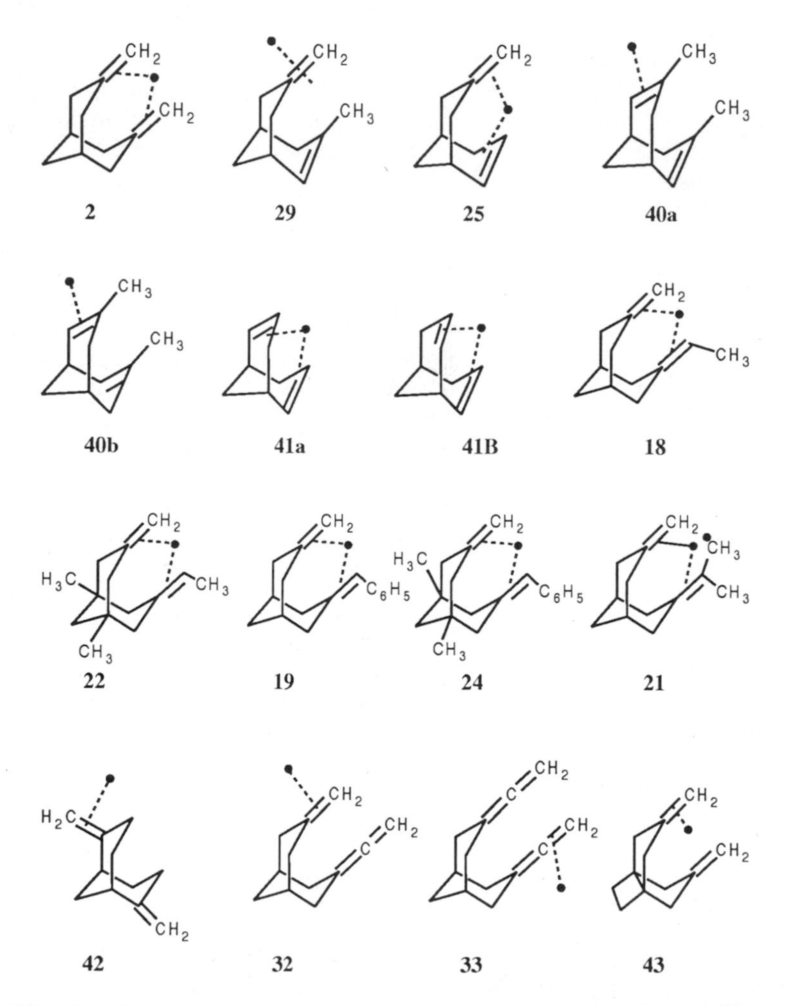

Fig. 3. Structure of π complexes of bicyclic olefins with $AgNO_3$.

TABLE 5. Complexing Parameters of Some Olefins with $AgNO_3$ in Ethanol at 25°C

Olefin	K(±10%)	- ΔH(kJ/mol)	- ΔS(J/mol•K)	Type of Complex[a]
2	430	55	130	Chelate
18	1450	40	70	Chelate
19	340	41	85	Chelate
21	2000	37	60	Chelate
22	1750	38	60	Chelate
25	190	30	55	Chelate
29	78	44	110	
40	78	33	75	
41	170	29	55	Chelate
42	80	26	50	
Cyclohexane	13	23	55	

[a]NMR with LSR estimation

The conclusions reached on the structure of the π complexes were supported by quantitative estimation of thermodynamic parameters of complexing. The complex formation was studied by calorimetric titration[75] and the data is collected in Table 5. Attempts to correlate ΔH and ΔS with spectral data or other parameters of olefins (e.g., ionization potentials) were unsuccessful. Equilibrium constants K seem more informative as they directly correlate with the structure of the complexes. As a rule, more stable complexes have a chelate structure, and K values less than 150 are evidence for the absence of chelation.

In general the study of complexing ability of bicyclic olefins is of significance for the understanding of electrophilic addition reactions in the studied systems.

Some Mechanistic Aspects of Transannular Addition to Unsaturated Bicyclo[3.3.1]nonanes

The mechanism of transannular electrophilic addition to 3,7-dimethylenebicyclo[3.3.1]nonane (**2**) and its derivatives can follow two different pathways, which should be different in their kinetic and thermodynamic parameters.

Route A corresponds to the classical stepwise carbenium ion mechanism[76] including consecutive participation of unsaturated systems in Ag_E reaction. Alternatively, route B is synchronous and includes intermediate formation of charge-transfer complexes (CTC) and simultaneous participation of double bonds in the transannular addition reaction. Until recently, route A was generally accepted for the mechanism of the transannular addition, without, however, sufficient evidence. Careful kinetic studies or different electrophilic additions to unsaturated bicyclo[3.3.1]nonanes show two different mechanisms that are operative depending on the nature of the reagent and the solvent.

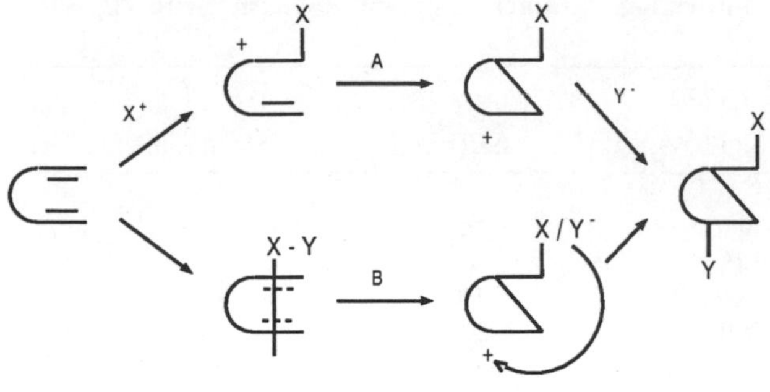

Compounds **2**, **29**, **32**, and **33** form corresponding closure products with 0.01–0.17 *M* H_2SO_4 in methanol. The rates W of these reactions follow second-order kinetics, the rate constants k and activation parameters being practically the same as those for acid catalyzed addition of methanol to simple olefins.[77]

$$W = k \cdot [\text{olefin}][H^+]$$

Obviously, the rate-determining step is protonation of one of the double bonds with carbenium ion formation, and the fast intramolecular cyclization makes the reaction irreversible. A byproduct, 1-methyladamantane, was detected upon the reaction of **2** under aforementioned conditions. This indirectly

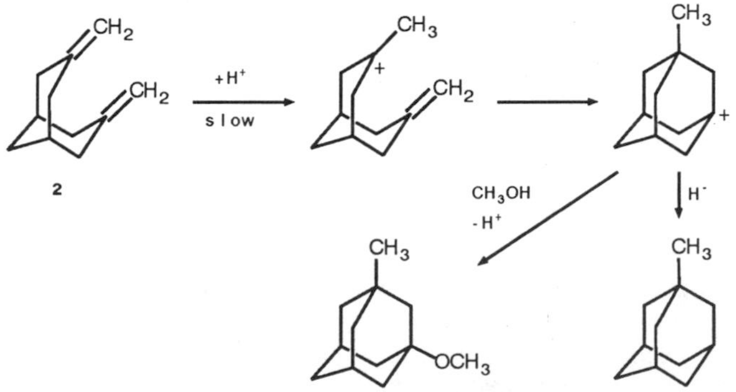

TABLE 6. Rate Constants and Activation Parameters of Additions in H_2SO_4–Methanol

Compound	$k_{298} \times 10^4$ (L/mol•k)	$\Delta H^‡$ (kJ/mol)	$-\Delta S$ (J/mol•k)
2	6.9±0.3	67±3	84±7
29	1.4±0.2	74±3	78±6
32	6.1±0.3	70±4	82±7
33	0.89±0.04	75±4	79±6

confirmed the intermediate formation of the 1-methyladamantyl-3-cation, which then underwent an intermolecular hydride shift. Activation parameters are listed in Table 6.

Another picture emerged when halogenation or hydrochlorination reactions were studied in nonpolar solvents (CCl_4 or hexane). Kinetics of iodination of several diolefins in CCl_4 showed general third-order reactions.

$$W = k \bullet [\text{olefin}][I_2]^2$$

The reaction has a negative temperature coefficient and proceeds via two charge-transfer complexes (CFC_1 and CTC_2).[78,79] The ΔS values are characteristic for molecular processes.[80] The same features were observed also for bromination in CCl_4 (Ref. 43) and addition of hydrogen chloride in hexane. Contrary

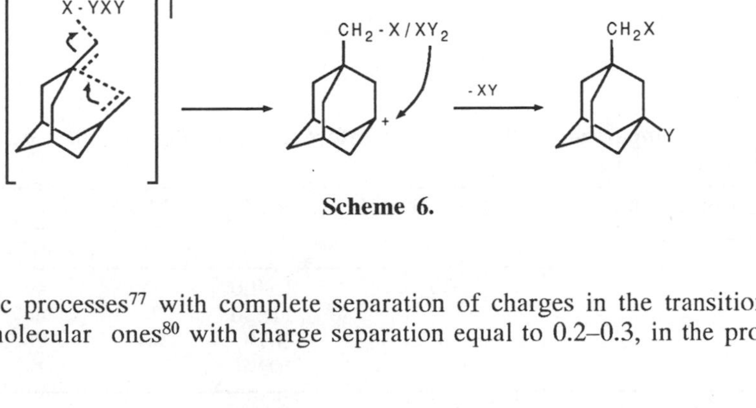

Scheme 6.

to ionic processes[77] with complete separation of charges in the transition state and molecular ones[80] with charge separation equal to 0.2–0.3, in the processes

TABLE 7. Rate Constants and Activation Parameters of Electrophilic Addition to Unsaturated Bicyclo[3.3.1]nonanes in a Nonpolar Media

Compound	Reagent and Solvent	k_{298} (l^2/mol^2•s)	ΔH (kJ/mol)	$-\Delta S$ (J/mol•K)
I	II	III	IV	V
1	I_2 in CCl_4	10.6	-10±2	255±5
28	I_2 in CCl_4	23.8	-18±2	280±5
21		29.9	-24±2	297±5
19		0.507	-14±2	299±5
24		18.5	-16±2	279±5
29		0.395	-17±2	310±10
25		0.70		
2	Br_2 in CCl_4	4.46×10^6	-32±3	185±10
29		1.49×10^5	-37±3	230±10
25		2.24×10^3	-49±3	310±10
Cyclohexene		8	-33.5	340
2	HCl in hexane	0.497	-10±2.5	285±8
32		0.55	-11±5	289±15
29		0.009	-23±6	361±20

studied charge separation in the transition state was ~ 0.6. As in the case of compound **2** and its derivatives, the molecular mechanism of addition was rejected for steric reasons and an ion–molecule mechanism of addition to unsaturated bicyclo[3.3.1]nonanes in nonpolar media was proposed. Obviously, the rate-determining step of the reaction is conversion of CTC_2 into a contact ion pair, which collapses into the reaction product. Scheme 6 depicts the generalized mechanism of electrophilic addition to unsaturated bicyclo[3.3.1]-nonanes in nonpolar solvents. Kinetic and thermodynamic parameters of the reactions studied are presented in Table 7.

Consequently, electrophilic addition to compound **2** and its structural relatives in nonpolar solvents proceeds by a synchronous molecule–ion process with considerable acceleration compared to a pure stepwise ionic pathway (see Table 8).

TABLE 8. Comparison Half-Life of Transformation for Compound 2(0.5 mol/L) in Different Systems

Starting Concentration of the Acid	HCl in Hexane $t_1^{1/2}$ (s)	H_2SO_4 in $t_2^{1/2}$ (s)	$\frac{t_2^{1/2}}{t_1^{1/2}}$
0.08	330	15400	47
0.1	29	11800	407
0.5	2.7	2070	767
1.0	0.7	1020	1457

FRAGMENTATION REACTIONS OF DIBROMOADAMANTANE DERIVATIVES

Preparative Aspects

Fragmentation of the adamantane nucleus was observed in series of reactions: solvolysis of 3-bromoadamantyl-1-amine,[81] methyl 3-bromoadamantyl-1-urethane,[82] 1,3-dibromoadamantane,[25] 4-bromoadamantanone,[83] 4-mesyloxy-adamantanone;[84] Beckmann rearrangement of adamantanone oxime;[85] treatment of 1-hydroxyadamantane with lead tetraacetate and iodine;[9] reaction of 4-mesyloxyadamantanone with $NaBH_4$ or Grignard reagent;[86] and the Schmidt reaction of adamantanone.[87]

The dibromo derivatives of adamantane seem to be attractive starting materials for preparation of diversely substituted unsaturated bicyclo[3.3.1]nonanes by zinc fragmentation.

Fragmentation reactions of the δ-dibromo derivatives of adamantane are carried out under rather severe conditions by refluxing the dibromide with zinc dust and sodium carbonate in dimethylformamide (DMF).[26–28] As a rule a mixture of substances is formed, but the desired diolefin prevails. As byproducts the reaction mixture contains compounds that are the result of complete and partial reduction and the hydrolysis of the dibromide, as well as isomerization of the resulting diolefins. The composition of the reaction mixture after the fragmentation of 1-bromo-3-bromomethyladamantane[88] is as follows:

CH_2Br, Br —Zn/DMF→ **2** (75%) + CH_3, Br (4%) + CH_3, OH (1%) + **44** (20%)

In some cases yields of diolefins were improved by using zinc–copper instead of zinc. Fast heating of the dibromide to the reaction temperature and removal of the diene from the reaction by distillation with DMF also improves the yields. Yields usually decrease when the dibromides become more complicated.

Dilution of the reaction mixture with DMF lowers the yield of the fragmentation products. As pointed out, earlier chromatographically pure bicyclic olefins from the reaction mixture were separated by conversion into π complexes with $AgNO_3$ and subsequent decomposition of the complexes with aqueous ammonia. The yield of the π complexes increased with the increase of the concentration of the diolefin solutions. The most complete extraction of the diolefin from the reaction mixture can be achieved by chromatographic analysis on silica gel or alumina impregnated with silver nitrate.

Stereochemistry and Some Aspects of the Fragmentation Mechanism

Though the fragmentation reaction was often used as a preparative method for alkene synthesis, the mechanism of the fragmentation of δ-dihalogeno compounds under treatment with metallic zinc was not yet studied. At the same time dibromoadamantane derivatives containing two bromide atoms at the δ-positions are the ideal rigid models for such study, because they possess a combination of stereoelectronic factors which favor synchronization of the multicenter fragmentation process.

The study of the fragmentation kinetics of δ-dibromides with zinc dust encounters difficulties because of the heterogeneous nature of the process. The nature of the reaction mechanism was deduced from a preparative fragmentation study of dibromides having a stereochemical marker and qualitative estimation of relative rates of the process.

There is no doubt that the fragmentation reaction of dibromoadamantane derivatives proceeds via formation of an intermediate organozinc compound.[26] In this case proper bond polarization in the system being fragmented can be achieved. Because all adamantane δ-dibromides hitherto studied contain chemically unequivalent bromine atoms, the possibility of the formation of two different intermediate organozinc compounds arises (α and δ route):

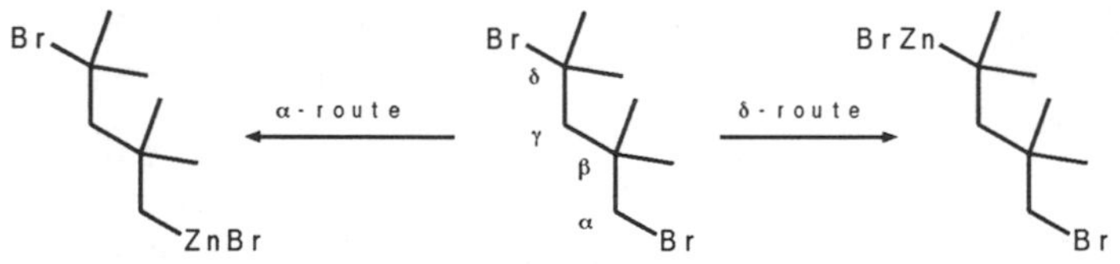

Br ZnBr ← α-route — Br δ γ β α Br — δ-route → BrZn Br

Organozinc compounds in polar DMF probably dissociate into a tight ion pair[89] and are capable of fragmentation by two routes, synchronous (S) and asynchronous (A):

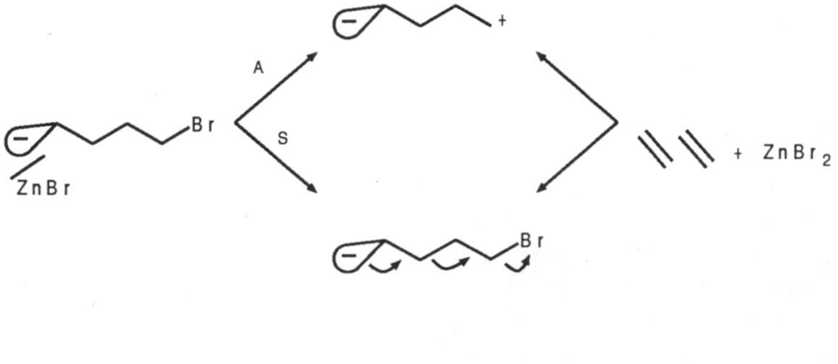

ZnBr Br A S + + ZnBr$_2$ Br

Consequently, there are four possible pathways of intermediate organozinc compound fragmentation: αA, αS, δA, and δS. Careful examination of the composition of the reaction mixture after treatment of *cis* and *trans*-1,4-dibromoadamantane with zinc dust in DMF did not reveal any differences, the qualitatively estimated rates of both reactions being almost the same.

This data did not contradict the stereoelectronic rules formulated by Grob[90] if organozinc compound formation at the bridge position of the cage is assumed. The tight ion pair thus formed undergoes rapid configurational isomerization, and the difference of reactivity between the two isomers disappears. In the case of formation of bridgehead organozinc compounds (δ route), *cis*-dibromides should react much slower than the trans isomers.

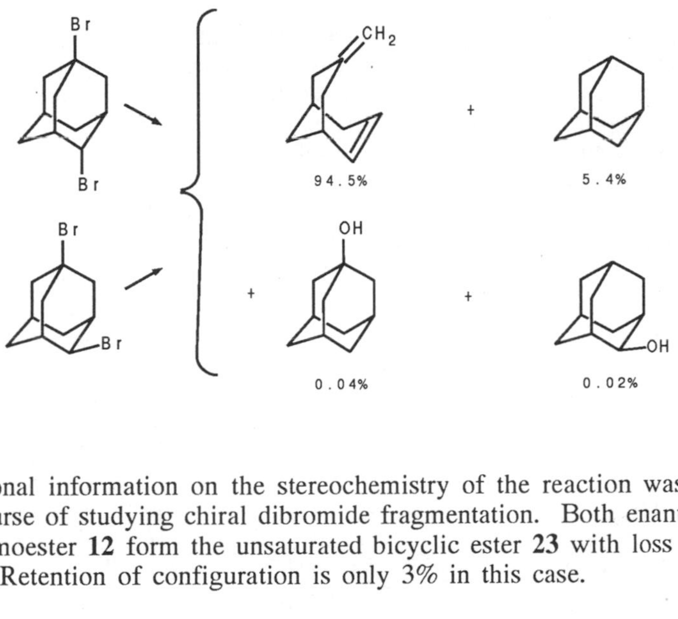

Additional information on the stereochemistry of the reaction was obtained in the course of studying chiral dibromide fragmentation. Both enantiomers of the dibromoester **12** form the unsaturated bicyclic ester **23** with loss of optical activity. Retention of configuration is only 3% in this case.

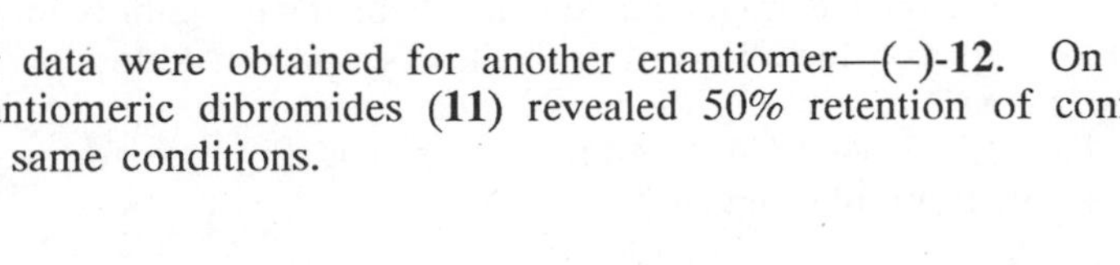

Similar data were obtained for another enantiomer—(–)-**12**. On the other hand, enantiomeric dibromides (**11**) revealed 50% retention of configuration under the same conditions.

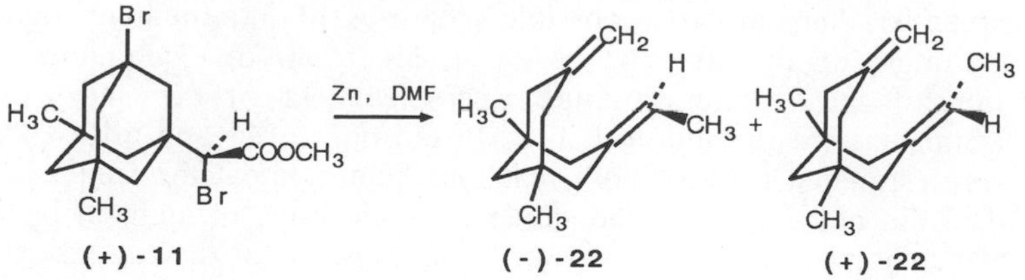

The stereochemical results of the fragmentation of chiral dibromides **11** and **12** offer several conclusions:

1. Intermediate organozinc compounds are formed because of the bromine atom in the side chain. Only in this case is racemization possible involving inversion of the carbanion at the reaction center and rotation of the C—C bond. Enolization of the dibromoester (**12**) makes the racemization easier, thus retention of configuration is only 3%. The alternative route to organozinc compound formation at the bridgehead of the adamantane nucleus eliminates the possibility of racemization if the stereoelectronic demand of Grob is operative.

2. Usually, organometallic compound formation from chiral α-halogen esters or α-halogen ketones results in complete loss of optical activity due to rapid enolization.[91] Since stereoselectivity is still observed in the case of the fragmentation of the dibromoester (**12**) supports the assumption that it proceeds via synchronous αS route. In this case the rate of fragmentation (k_f) is comparable with the enolization (k_e) and inversion (k_i) rates.

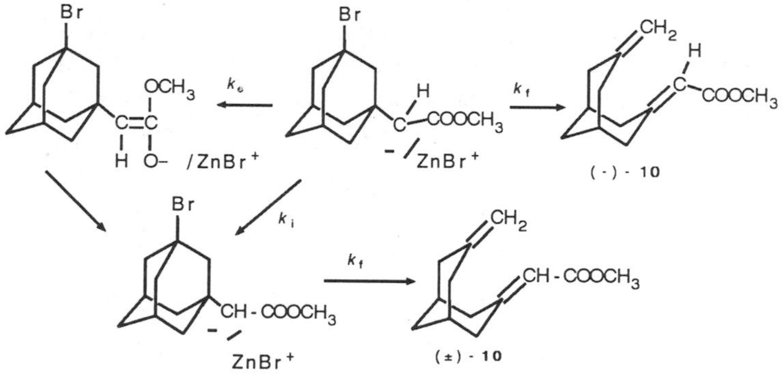

3. Experiments on the fragmentation of adamantane δ-dibromides indicate unfavorable energy considerations of bridgehead carbanion formation. This was also shown by self-consistent filled molecular orbital linear combination of atomic orbitals (SCF MOLCAO) estimation of enthalpy the formation of 1-adamantyl anion from adamantane.[92]

CONCLUSIONS

Thirty years of development in the field of adamantane and related cage hydrocarbon chemistry demonstrated the justified interest and promise of this area of organic chemistry due to the vide variety of possible chemical transformations and abundance synthetic opportunities. The accidental observations, which usually favors only the most attentive observant researchers, allowed Schleyer 30 years ago to create the necessary prerequisite for development of the field. He was also the first to explain some of the fascinating new effects found in cage compound chemistry. Among them are the cage effect hypothesis or carbocationic centers[93] and the 3-D aromaticity in the 1,3-dehydro-5,7-adamantane dication system,[94] besides his extensive synthetic and mechanistic work, including solvolytic reactions.

Wider practical application of adamantane derivatives seems to be only a question of time. Consequently, reviews on synthetic methods of cage compounds chemistry, similar to the present one, serve a useful purpose.

REFERENCES

1. P. v. R. Schleyer, *J. Am. Chem. Soc.* **1957**, *79*, 3292.
2. W. L. Davies, R. R. Grunert, R. F. Haat, J. Paulshock, J. C. Walts, T. R. Word, E. C. Hermann, and C. E. Hoffmann, *Science* **1964**, *144,* 862.
3. H. Meerwein, *J. Pract. Chem* **1922**, *104,* 179.
4. T. Mori, K. H. Yang, K. Kimoto, and H. Nazaki, *Tetrahedron Lett.* **1970**, 2419.
5. R. C. Fort, *Adamantane. The Chemistry of Diamond Molecules*, Marcel-Dekker, New York, 1976.
6. T. Sasaki, *Adv. Heterocycl. Chem.* **1982**, *30*, 79.
7. R. I. Yurchenko, E. E. Lavrova, and E. G. Martinuk, *J. Gen. Chem. USSR* **1984**, *54*, 1295; R. I. Yurchenko, E. E. Lavrova, S. M. Lukyanova, and N. S. Verpovsky, *J. Gen. Chem. USSR* **1983**, *53*, 242.
8. Yu. V. Migalina, V. I. Staninets, V. G. Lendel, A. S. Kozmin, and N. S. Zefirov, *Chem. Heterocycl. Comp. USSR* **1977**, N12, 1633.
9. R. M. Black and G. B. Gill, *Chem. Commun.* **1970**, 972; W. H. Lunn, *J. Chem. Soc. C*, **1970**, 2124.
10. J. H. Liu and P. Kovacic, *Chem. Commun.* **1972**, 564.
11. R. Bishop, W. Parker, and J. R. Stevenson, *J. Chem. Soc. Perkin Trans. 1* **1981**, 565.
12. B. R. Vogt, *Tetrahedron Lett.* **1968**, 1579.
13. R. Bishop and W. Parker, *Tetrahedron Lett.* **1973**, 2375.
14. N. V. Averina, G. V. Gleizene, N. S. Zefirov, and P. P. Kadziauskas, *J. Org. Chem. USSR* **1975,** *11*, N1, 77.

15. P. A. Knott and J. M. Mellor, *J. Chem. Soc. Perkin Trans. 1* **1972**, 1030.
16. I. A. McDonald, A. S. Dreiding, H. M. Hutmacher, and H. Musso, *Helv. Chim. Acta* **1973**, *56*, 1385.
17. H. Stetter, J. Gartner, and P. Tacke, *Berechte* **1965**, *98*, 3888.
18. H. Stetter and J. Gartner, *Chem. Ber.* **1966**, *99*, 925.
19. D. Faulkner, R. A. Glendinning, D. F. Johnston, and M. A. McKervey, *Tetrahedron Lett.* **1971**, 1961.
20. J. G. Korsloot and V. G. Keizer, *Tetrahedron Lett.* **1969**, 3517.
21. A. C. Udding, H. Wynberg, and J. Strating, *Tetrahedron Lett.* **1968**, 5719; M. A. Eakin, J. Martin and W. Parker, *Chem. Commun.* **1967**, 955.
22. H. Stetter, J. Gartner, and P. Tacke, *Chem. Ber.* **1966**, *99*, 1435.
23. S. J. Padegimas and P. Kovacic, *J. Org. Chem.* **1972**, *37*, 2672.
24. I. R. Likhotvoric, N. I. Dovgan, and G. I. Danilenko, *J. Org. Chem. USSR* **1977**, *13*, 897.
25. A. R. Gagneux and R. Meier, *Tetrahedron Lett.* **1969**, 1365.
26. F. N. Stepanov and W. D. Suchowerchov, *Angew. Chem.* **1967**, *79*, 860.
27. A. G. Yurchenko, Z. N. Moorsinova, and F. N. Stepanov, *J. Org. Chem. USSR*, **1972**, *8*, 2332.
28. A. G. Yurchenko, Z. N. Moorsinova, and S. D. Isaev, *J. Org. Chem. USSR*, **1975**, *11*, 1427.
29. P. A. Krasutsky, N. S. Chesskaja, V. N. Rodionov, O. P. Baula, and A. G. Yurchenko, *J. Org. Chem. USSR*, **1985**, *21*, 1677.
30. Z. N. Moorzinova, E. I. Dikolenko, and A. G. Yurchenko, Abstracts of the conference *Chemistry and Application Perspectives of Adamantane Hydrocarbons and Relative Compounds*, Kiev, 1974, p. 45.
31. P. A. Krasutsky, N. S. Chesskaja, V. N. Rodionov, and A. G. Yurchenko, *Theor. Exp. Chem. USSR*, **1985**, *21*, 620.
32. A. G. Yurchenko, P. A. Krasutsky, and N. A. Smirnova, *Theor. Exp. Chem. USSR* **1975** *11*, 552.
33. A. G. Yurchenko, Z. N. Moorzinova, and T. G. Fedorenko, *J. Org. Chem. USSR* **1974**, *10*, 1125.
34. J.-H. Lin, G. A. Gauger, and P. Kovacic, *J. Org. Chem.* **1973**, *38*, 543.
35. F. N. Stepanov, V. D. Suchoverhov, V. F. Baklan, and A. G. Yurchenko, *J. Org. Chem. USSR* **1970**, *6*, 884.
36. J. K. Williams and R. E. Benson, *J. Am. Chem. Soc.* **1962**, *84*, 1256.
37. R. B. Woodward and R. Hoffmann, *The Conservation of Orbital Symmetry*, Verlag Chemie, Academic Press, New York p. 107, 1971.
38. F. N. Stepanov, T. N. Utochka, and A. G. Yurchenko, *J. Org. Chem. USSR* **1972**, *8*, 1183.
39. F. N. Stepanov, R. A. Krasutsky, and A. G. Yurchenko, *J. Org. Chem. USSR* **1972**, *8*, 1179.
40. F. N. Stepanov, L. A. Zosim, E. N. Martinova, and A. G. Yurchenko, *J. Org. Chem. USSR* **1971**, *7*, 2533.
41. F. N. Stepanov, T. N. Utochka, A. G. Yurchenko, and S. D. Isaev, *J. Org. Chem. USSR* **1974**, *10*, 59.

42. F. N. Stepanov, T. N. Utochka, A. G. Yurchenko, and S. D. Isaev, *J. Org. Chem. USSR* **1974**, *10*, 1177.

43. P. A. Krasutsky, A. B. Hotkevich, Ju. A. Serguchev, and A. G. Yurchenko, *Theor. Exp. USSR* **1985**, *21*, 52.

44. P. A. Krasutsky, A. A. Fokin, E. D. Skoba, and A. G. Yurchenko, *J. Org. Chem. USSR* **1986**, *22*, 460.

45. P. A. Krasutsky, A. A. Fokin, and A. G. Yurchenko, *J. Org. Chem. USSR* **1985**, *21*, 2522.

46. P. A. Krasutsky, A. A. Fokin, N. I. Kulic, and A. G. Yurchenko, *J. Org. Chem.* **1985**, *21*, 2518.

47. P. A. Krasutsky, A. A. Fokin, and A. G. Yurchenko, *J. Org. Chem. USSR* **1986**, *22*, 459.

48. A. G. Yurchenko, L. A. Zosim, and N. L. Dovgan, *J. Org. Chem. USSR* **1974**, *10*, 1966.

49. A. G. Yurchenko, L. A. Zosim, N. L. Dovgan, and N. S. Verpovsky, *Tetrahedron Lett.* **1976**, 4843.

50. N. L. Dovgan, I. R. Likhotvoric, and A. G. Yurchenko, Abstracts of the conference *Chemistry of Organic Polyhedrones*, Volgograd, 1981, p. 44.

51. N. L. Dovgan, I. R. Likhotvoric, and A. G. Yurchenko, *Abstracts of XII Mendeleev Congress on Pure and Applied Chemistry*, N 2, Moscow, 1981, p. 71.

52. N. L. Dovgan, I. R. Likhotvoric, N. P. Danilenko, and A. G. Yurchenko, *J. Org. Chem. USSR* **1982**, *18*, 1774.

53. R. Bishop, W. Parker, and J. R. Stevenson, *J. Chem. Soc. Perkin Trans. 1* **1981**, 565.

54. A. G. Yurchenko, A. T. Voroshchenko, and F. N. Stepanov, *J. Org. Chem. USSR* **1970**, *6*, 189.

55. A. G. Yurchenko, P. A. Krasutsky, and N. A. Smirnova, *Theor. Exp. Chem.* **1982**, *18*, 189.

56. A. G. Yurchenko, P. A. Krasutsky, and N. I. Kulic, Abstracts of the conference *Chemistry of Organic Polyhedranes*, Volgograd, 1981, p. 18.

57. P. A. Krasutsky, A. G. Yurchenko, V. N. Rodionov, M. Yu. Antipin, and Yu. T. Struchov, *Theor. Exp. Chem. USSR* **1983**, *19*, 735.

58. P. A. Krasutsky, A.B. Hotkevich, Ju. A. Serguchev, and A. G. Yurchenko, *Theor. Exp. Chem. USSR* **1985**, *21*, 681.

59. D. J. Rober, R. C. Fort, E. Wiskott, C. W. Woodworth, P. v. R. Schleyer, J. Weber, and H. Stetter, *Tetrahedron* **1971**, *27*, 3.

60. A. G. Yurchenko and A. T. Voroshchenko, Abstracts of the conference *Chemistry and Application Perspectives of Adamantane Hydrocarbons and Relative Compounds*, Kiev, 1974, p. 44.

61. H. Hamill and M. A. McKervey, *J. Chem. Soc. Chem. Commun.* **1969**, 864.

62. A. C. Cope, W. R. Moore, R. D. Bach, and H. J. S. Winkler, *J. Am. Chem. Soc.* **1970**, *92*, 1243.

63. P. A. Krasutsky, V. N. Rodionov, V. P. Tikhonov, and A. G. Yurchenko, *Theor. Exp. Chem. USSR* **1984**, *20*, 58.

64. P. A. Krasutsky, N. S. Chesskaja, V. N. Rodionov, O. P. Baula, and A. G. Yurchenko, *J. Org. Chem.* **1985**, *21*, 1677.

65. W. Offerman and A. Mannschreck, *Tetrahedron Lett.* **1981**, *22*, 3227.
66. T. J. Wenzel and R. E. Sievers, *Anal. Chem.* **1981**, *53*, 393.
67. P. A. Krasutsky, A. G. Yurchenko, and V. N. Rodionov, *Tetrahedron Lett.* **1981**, *23*, 3719.
68. P. A. Krasutsky, V. N. Rodionov, and A. G. Yurchenko, *Theor. Exp. Chem.* **1983**, *19*, 126.
69. P. A. Krasutsky, V. N. Rodionov, N. S. Chesskaja, and A. G. Yurchenko, *J. Org. Chem. USSR* *1985*, 21, 1684.
70. D. F. Evans, J. N. Tucker, and G. C. Villardi, *J. Chem. Soc. Chem. Commun.* **1975**, 205.
71. A. G. Yurchenko, P. A. Krasutsky, M. Jones, M. Ju. Kornilov, and A. C. Degtyarev, *Spectroscopy of Coordination Compounds*, Krasnodar, 1980, p. 83.
72. P. A. Krasutsky, A. G. Yurchenko, and V. N. Rodionov, *Tetrahedron Lett.* **1982**, *23*, 3719.
73. K. A. Kime and R. E. Sievers, *Aldrichimica Acta* **1977**, *10*, 54.
74. H. M. McConnell and R. E. Robertson, *J. Chem. Phys.* **1958**, *29*, 1361.
75. M. A. Kovbuz, I. I. Artym, Ju. A. Serguchev, and A. B. Chotkevich, *Theor. Exp. Chem. USSR* **1984**, *20*, 631.
76. H. Brown and L. Kwang-Ting, *J. Am. Chem. Soc.* **1975**, *97*, 600.
77. *Rate and equilibrium constants tables of heterogeneous organic reactions*, V. A. Palm, Ed., Vol. 4/1, Moscow, 1977, p. 146.
78. P. A. Krasutsky, Ju. A. Serguchev, A. G. Yurchenko, and A. B. Chotkevich, *Theor. Exp. Chem.* **1983**, *19*, 229.
79. Ju. A. Serguchev, A. B. Chotkevich, V. B. Barabash, P. A. Krasutsky, and A. G. Yurchenko, *Theor. Exp. Chem.* **1984**, *20*, 732.
80. G. B. Sergeev, V. V. Smirnov, T. N. Rostovschikova, V. A. Poljakov, and O. S. Korinfskaja, *Kinet. Katal.* **1979**, *20*, 1466.
81. C. A. Grob and W. Schwarz, *Helv. Chim. Acta* **1964**, *47*, 1870.
82. H. Stetter and P. Tacke, *Chem. Ber.* **1963**, *96*, 694.
83. A. C. Udding, H. Wynberg, and J. Strating, *Tetrahedron Lett.* **1968**, 5719.
84. T. Sasaki, S. Eguchi, and T. Toru, *J. Org. Chem.* **1971**, *36*, 3460.
85. T. Sasaki, S. Eguchi, and M. Mizutani, *J. Org. Chem.* **1972**, *37*, 3961.
86. H. Stetter and V. Tillimans, *Chem. Ber.* **1972**, *37*, 3961.
87. T. Sasaki, S. Eguchi, and T. Toru, *J. Am. Chem. Soc.* **1969**, *91*, 3390.
88. P. A. Krasutsky and L. A. Zosim, *Vestn. Kiev. Polyteckh. Inst., Ser. Khim.* **1974**, *11*, 46.
89. S. Streitwieser and C. C. Chien, *Tetrahedron Lett.* **1979**, 327.
90. C. A. Grob, *Angew. Chem.* **1969**, *81*, 543.
91. R. Bates and C. Ogle, *Chemistry of Carbanions*, Springer-Verlag, 1983.
92. A. G. Yurchenko, V. V. Lobanov, and T. V. Fedorenko, Abstracts of the conference *Chemistry of Cage Compounds Development and Their Industrial Application*, Kiev, 1986, p. 57.
93. P. V. R. Schleyer, R. C. Fort, W. E. Watts, M. B. Comisarov, and G. A. Olah, *J. Am. Chem. Soc.* **1964**, *86*, 4195.
94. M. Bremer, P. v. R. Schleyer, K. Scholtz, M. Kausch, and M. Schindler, *Angew. Chem. Int. Ed. Engl.* **1987**, *26*, 761.

6 Bridgehead Reactivity in Solvolysis Reactions

PAUL MÜLLER and JIRI MAREDA

Department of Organic Chemistry
University of Geneva
Geneva, Switzerland

INTRODUCTION

Bridgehead molecules exhibit both remarkable and characteristic reactivities towards substitution reactions. Their studies have contributed much to our understanding of reaction mechanisms and of the preferred geometries of reactive intermediates. Bartlett and Knox[1] were the first to recognize the inertness of bridgehead positions in bicyclic molecules towards nucleophilic substitution. This observation laid the ground work for the now well-established two-step substitution mechanism via carbenium ion intermediates. Bartlett concluded that the lack of reactivity of triptycyl and other bicyclic derivatives towards nucleophiles was due to the impossibility, for geometric reasons, of backside attack and concomitant Walden inversion (S_N2 process) on one hand, and to the instability of the intermediate carbenium ion, constrained by the rigid bicyclic framework into an unfavorable nonplanar geometry on the other. Doering et al.[2] showed later that the reactivity of bridgehead derivatives in (S_N1) substitution reactions is very much dependent on the flexibility of the ring system. Bicyclic structures with large rings can almost accommodate a trigonal geometry at the bridgehead position and, unlike the more rigid structures studied by Bartlett, undergo nucleophilic substitution quite easily.

In contrast, the experimental evidence for bridgehead anions indicates that these species are often quite stable, and their stability is dictated by electronegativity and not by geometrical considerations. Bridgehead molecules undergoing radical reactions occupy a position intermediate between that observed for carbenium ions and carbanions. They react almost as readily as their acyclic counterparts and their range of reactivities is much smaller than that observed for reactions leading to carbenium ions in the same bridgehead series. This is in agreement with the picture that radicals, although preferring planar geometries, are relatively stable even when distorted.[3]

The topic of bridgehead reactivity in general has been reviewed splendidly and in full detail by Applequist and Roberts,[4] Schöllkopf,[5] Fort and Schleyer,[6] and Stirling.[7] Excellent reviews on bridgehead carbenium ions and radicals have been published by Fort[8] and Rüchardt,[9] respectively.

MOLECULAR MECHANICS TREATMENT OF BRIDGEHEAD SOLVOLYTIC REACTIVITY

Bridgehead molecules are particularly suitable substrates for investigations of solvolytic processes. Their range of reactivities spans ~ 22 orders of magnitude.[10] The rigid molecular skeleton precludes intervention of solvent participation (k_s processes), which can intervene in solvolysis of simple secondary and even tertiary systems.[11] In addition, possible competing E2 processes are unlikely owing to the high energy of the bridgehead (anti-Bredt) alkenes, which would be produced. For these reasons the bridgehead series provides a unique opportunity to investigate structure reactivity relationships. It may also be used as a mechanistic model, against which other, more complex compounds can be measured.

Starting in the early 1960s, Schleyer and his school[12–14] undertook a systematic investigation of solvolysis reactions of bridgehead derivatives. Because of the enormous variations in reactivities within the series, rate constants were determined with leaving groups of very different nucleofugalities, chlorides and bromides for the more reactive substrates, and *p*-toluenesulfonates and trifluoromethanesulfonates for the less reactive ones. For many compounds duplicate data for two leaving groups were determined and a reactivity scale for bridgehead derivatives was established. The data set was later expanded by inclusion of the rate constant of manxyl chloride (1-chlorobicyclo[3.3.3]-undecane, **11**), first synthesized by Parker et al.[15]

The rate variations could be rationalized in terms of changes in strain, as calculated by molecular mechanics. The molecular mechanics calculations are a computer application[16] of Westheimer's[17] classical treatment of steric effects. The steric energy of a molecule is in principle the sum of energy terms due to deviations of bond lengths and bond angles from their ideal values, and to torsional and nonbonded interactions. These energies may be converted to standard enthalpies of formation (ΔH_f°) by means of group or bond increments. Strain energies are calculated from enthalpies of formation relative to some arbitrary standard by using so-called strainless increments. The terminology used here is that of Allinger.[18] In early publications in this field the terms sometimes have different meanings, or still other terms are used.

Molecular mechanics calculations are particularly well suited for strain calculations of molecules in the ground state, because there is usually a substantial body of experimental data available, on which the parameters can be fitted. Reaction rates are related to the difference in free energy between reactants and transition states, and for the latter these calculations provide no information. However, it is possible to design a transition state model either by theoretical calculations[19] or by intuition, and parameterize it in such a way as to fit the experimental data. In the case of bridgehead solvolysis, Schleyer used the carbenium ion as a model for the transition state of the reaction on the assumption of a late transition state in solvolysis and therefore assuming energetic resemblance or proportionality between the transition state and the corresponding cationic intermediate.[20] Accordingly, a set of parameters for carbenium ions was designed to this effect and incorporated in the program. At the time when these calculations were performed, the method of molecular mechanics was at a stage of very early development, and the programs were not yet designed to treat functional groups. For this reason a simplification was introduced by replacing the leaving group of the substrate by hydrogen. Thus the strain effects operating in solvolysis were approximated by

$$R_3C-H \longrightarrow R_3C^{\oplus}$$

By using this approach Bingham and Schleyer rationalized the reactivity of bridgehead derivatives in terms of the strain changes between the bridgehead hydrocarbon and the corresponding bridgehead carbenium ion.[14,15] This pioneering study established molecular mechanics as a reliable method for the study of strain–reactivity relationships.

The investigation of Bingham and Schleyer[15] revealed two puzzling features: (1) It was found that separate correlation lines with significantly different slopes were obtained for chlorides (3.12), bromides (2.44), *p*-toluenesulfonates (1.11) and trifluoromethanesulfonates (0.94), and (2) one of the compounds, the perhydrotriquinacene-*p*-toluenesulfonate **10-OTs** reacted some 10^9 times slower than expected on the grounds of the strain calculations. The first phenomenon was attributed to the computational model. It was thought that the different slopes of the correlations were due to the use of hydrogen as a leaving group. Hydrogen is the smallest substituent possible, and for this reason it does not experience steric interactions (Front strain)[21] with the molecular skeleton, which the real leaving groups, particularly sulfonates, sometimes suffer. In agreement with this hypothesis, the rate differences between a given set of compounds tended to be larger for sulfonates than halides. The unexpectedly low reactivity of the perhydrotriquinacene derivative, on the other hand, was explained by the absence of the stabilizing C—C hyperconjugative interactions in the carbenium ion, which were thought to be present throughout the series of bridgehead derivatives, but impossible for reasons of molecular structure in the case of the perhydrotriquinacenyl cation **(10+)**.

FRONT STRAIN IN THE SOLVOLYSIS OF BRIDGEHEAD DERIVATIVES

The separation of strain into different components such as Front strain, Back strain, and Internal (I-) strain[22] was important for the development of the understanding of steric effects, but since the generalized use of ab initio, semiempirical, and molecular mechanics calculations in organic chemistry, this factorization of effects becomes unnecessary.[23] It is now recognized that strain can be spread over large areas rather than being localized in individual bonds or bond angles, the molecules adjusting their geometry in order to reach the conformation of minimal energy.[24] With respect to bridgehead reactivity, it must be recognized that the F-strain problem could only arise because a questionable leaving group model was used in the computations. If the strain calculations take the leaving group, from which the rate constants are determined, into consideration, then any strain due to the presence of this leaving group will naturally be included in the total strain of the reacting molecule.

Since the early work of Schleyer on bridgehead reactivities, molecular mechanics programs have been much improved, and it became possible to handle functional groups. In 1984 we decided to investigate the F-strain problem in bridgehead solvolysis by molecular mechanics calculations. The programs available to us at that time were BIGSTRN,[25] obtained by the courtesy of Schleyer, and Allinger's MM2, available from QCPE.[26] The BIGSTRN program was parameterized for carbenium ions, but not for functional groups, while MM2 contained parameters for several functional groups, although not for sulfonate esters, and not for carbenium ions. A force field for tertiary (UNICAT 1)[27,28] and secondary (UNICAT 2)[28] carbenium ions, similar to that used in BIGSTRN, was incorporated into MM2 and its performance tested by comparison of calculations with both programs. Acceptable correlations were obtained when steric energy differences ΔE_{st} (R_2CH^+–R_2CH_2) or ΔE_{st} (R_2CH^+–R_2CHCH_3) calculated by MM2 were plotted against values obtained with BIGSTRN for secondary and planar tertiary ions.[28] However, in the case of tertiary bridgehead cations, the structures obtained with both programs were quite different. The cations calculated with MM2 were pyramidalized to various degrees, and in this respect were similar to those reported by Bingham,[29] but the BIGSTRN structures were all practically planar at the bridgehead, while the rest of the molecular framework was heavily distorted. Since this anomaly was not specific for our (private) version of BIGSTRN, but also occurs in that obtained from QCPE, we abandoned this program and continued only with MM2.

Our initial objective was to simulate F-strain throughout the series of bridgehead substrates by using leaving group models of increasing bulk, namely, H, Cl, OH, CH_3, OCH_2CH_3, $OC(CH_3)_3$, and $C(CH_3)_3$. These 14 structures were selected from the work of Bingham and Schleyer[14,15] covering most of the experimental rate range, but with the exception of tricyclyl and nortricyclyl derivatives since MM2 was not parameterized for carbenium ions with cyclopropane rings in α, β, or γ position. The rate constants of interest were selected for chlorides and *p*-toluenesulfonates, and they were converted into free energies of activation ($\Delta G^{\ddagger}$) relative to that of the respective 1-adamantyl derivative **8**. By doing so we (inadvertently) avoided the ambiguities originating from the unexpected k_{Br}/k_{OTs} variations observed in the bridgehead series[14] (see below) in establishing a reactivity sequence for the entire rate range. The principal results of the calculations are summarized in Table 1. The strain energies of the carbenium ions are calculated from the steric energies with the general and strainless increments of MM2. For the cationic center the general increment for the total three 3 C^{+}—C bonds was 187.4[30] and the strainless increment 191.9 kcal/mol.[31] The values are selected such as to reproduce the experimental enthalpy of formation of the *t*-butyl cation of 166.2 kcal/mol[32] while its strain energy is arbitrarily set to zero. The definition of the front-strain parameters φ_F is given in Ref. 33 where most of the values used in Table 1 are also found; the missing ones were calculated by MM2.

The plot of $\Delta G^{\ddagger}$ versus ΔE_{st} (R^{+}–RH) for chloride solvolysis with H as the leaving group model has a slope of 0.69 (r = 0.99), while the slope of the plot for *p*-toluenesulfonates is 0.77 (r = 0.996)[27] with the same leaving group model. This surprising similarity suggests that the rate constants of both chlorides and *p*-toluenesulfonates can be correlated by one and the same equation. The correlation (Fig. 1) including all of the substrates is of the form:

$$\Delta G^{\ddagger} = 0.71\ \Delta E_{st}(R^{+}\text{–}RH) - 4.61 \qquad (1)$$

with r = 0.997 and a standard deviation σ of 0.54 in $\Delta G^{\ddagger}$. Correlations of similar, although slightly diminished quality, are produced with other leaving group models (Cl: r = 0.989, σ = 0.97; OCH_2CH_3: r = 0.989, σ = 1.25), but

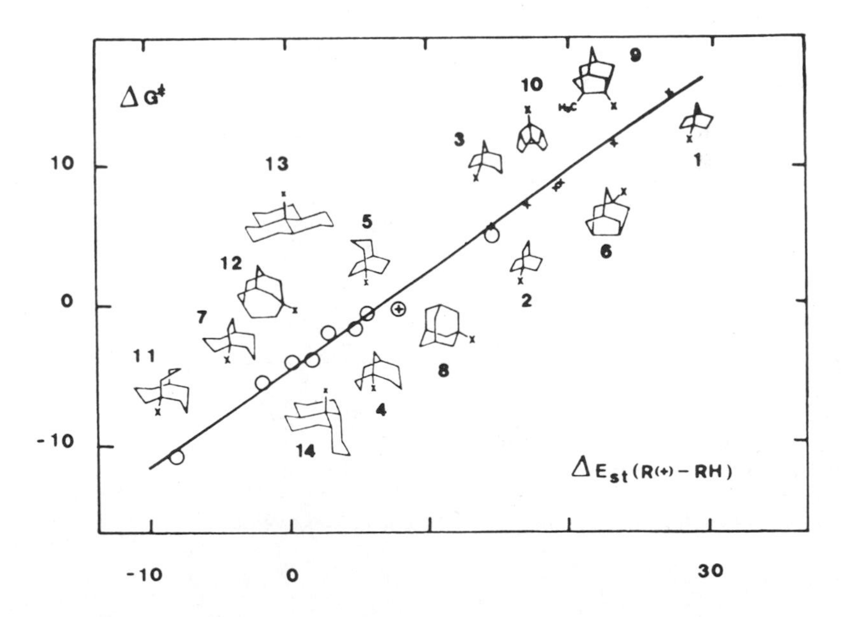

Fig. 1. Plot of $\Delta G^{\ddagger}$ versus ΔE_{st} (R^{+}–RH). Circles: Rates from chloride solvolysis; crosses: rates from *p*-toluenesulfonate solvolysis.

TABLE 1. Steric and Strain Energies of Bridgehead Carbenium Ions (kcal/mol) Calculated by MM2 (UNICAT 1) and Reactivities of Bridgehead Derivatives

No	Compound	E_{st} (R^+)	Strain (R^+)[a]	ΔE_{st} (R^+-RH)	ΔE_{st} (R^+-ROH)	Δ strain (R^+-RCl)	$\Delta G^{\dagger}_{-Cl}$	$\Delta G^{\dagger}_{-OTs}$	log k[d]	φ_f
1	1-Norbornyl	50.22	42.5	27.13	26.77[b]	23.8	-	15.62	-10.40	2.57
2	1-Bicyclo[2.2.2]octyl	33.88	25.5	14.24	13.72[b]	11.2	5.26	5.79	-4.00	3.49
3	1-Bicyclo[3.2.1]octyl	36.03	27.7	16.76	15.18	13.4	-	7.42	-5.17	3.39
4	1-Bicyclo[3.3.1]nonyl	22.76	13.7	4.50	3.01	1.1	-1.59	-	0.51	4.60
5	1-Bicyclo[3.2.2]nonyl	28.78	20.8	5.43	3.79	1.4	0.44	-	-0.13	4.33
6	1-Noradamantyl	46.95	37.9	18.93	17.12	16.0	-	8.66	-5.28	(2.64)
7	1-Bicyclo[3.3.2]decyl	28.00	18.3	-1.97	-3.93[b]	-7.0	-5.46	-	3.08	5.28
8	1-Adamantyl	24.80	15.1	7.71	7.17[b]	4.8	0.0	0.0	-0.40	4.08
9	7-Methyl-3-noradamantyl	51.15	41.8	23.35	22.45[b]	18.5	-	11.74	-7.96	(4.49)
10	10-Tricyclo-[5.2.1.0^{4,10}]decyl	45.94	36.2	19.39	18.12[b]	14.7	-	8.86	-6.16	(2.68)
11	1-Bicyclo[3.3.3]-undecyl	29.27	18.9	-8.12	-10.79	-14.4	-10.76	-	6.44	(4.58)
12	3-Homoadamantyl	28.19	17.8	0.15	-1.43	-4.1	-3.65	-	1.97	4.91
13	*t,t,t*-1-Tricyclo-[7.3.1.0^{5,13}]tridecanyl	18.18	6.1	2.67	-1.73	-4.1 (11.9)[c]	-1.96	-	0.8[e]	(12.69)
14	*c,c,t*-1-Tricyclo-[7.3.1.0^{5,13}]tridecanyl	20.68	8.6	1.48	-1.68	-4.5 (-4.1)[c]	-3.92	-	2.08[e]	(9.37)
15	1-Homoadamantyl	33.10[b]	22.7	-	4.09[b]	-	-	-	-0.20	7.70

[a]Relative to $(CH_3)_3C^+$.
[b]UNICAT 4.[34]
[c]For acetate leaving group (see text).
[d]Reactivity scale of *Bentley*.[10]
[e]For OPBN derivative, see Table 7.
[f]Reference 33, values in parentheses calculated by MM2.[37]

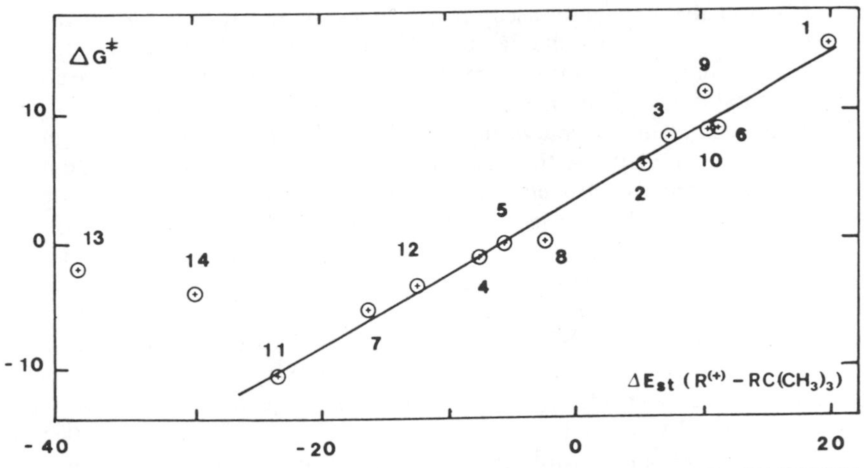

Fig. 2. Plot of $\Delta G^{\ddagger}$ for solvolysis of bridgehead derivatives versus ΔE_{st} (R^+-$RC(CH_3)_3$. Compounds **13** and **14** are excluded. Slope: 0.59; intercept: 3.10.

the best results for the plots of chlorides and *p*-toluenesulfonates, separately or combined, are obtained with H. Increase of the steric requirements of the leaving group results in deteriorations of the plot, but acceptable correlations are still produced with $OC(CH_3)_3$ (r = 0.99 with **13** excluded) and even for $C(CH_3)_3$ (r = 0.99 with **13** and **14** excluded) (Fig. 2).

This latter observation is particularly significant, since the $C(CH_3)_3$ group should be much more bulky than a *p*-toluenesulfonate and, therefore, should markedly overestimate F-strain in the substrate. We conclude therefore that, contrary to the hypothesis of Bingham and Schleyer, F-strain variations should be small throughout the series of bridgehead derivatives, and that they cannot explain the different slopes for different leaving groups in their structure–reactivity plots. These observations do not apply to **13** and **14**, which deviate from the general trend when bulky leaving group models are used. However, since in these two cases the rate constants used in the correlation refer to chloride solvolysis and not to that of *p*-toluenesulfonates, no F-strain arises. (The problem of F strain in the solvolysis of **14**–*p*-nitrobenzoate is discussed below.) The significance of F-strain in the series of bridgehead derivatives may be assessed by means of the φ_F parameters of Beckhaus[33] (Table 1), which express the difference in enthalpies of formation between R—$C(CH_3)_3$ and R—CH_3 with $\varphi_{F(R = CH_3)} = 0$. Values of φ_F increase in the series with the length of the bridge, from 2.57 (**1**) to 5.28 (**7**). This corresponds to an energy change of ~ 6.5 kcal/mol, a rather small variation in comparison to the 35 kcal/mol variation of ΔE_{st}. This 6.5 kcal/mol value should constitute an upper estimate for differential F-strain effects in solvolysis of bridgehead compounds, since it is based on the *t*-butyl group, which we believe to be too crowded for a realistic leaving group model. In addition, the compounds having longer bridges, and therefore higher φ_F values, are usually the more reactive ones and, for this reason, the chloro derivatives solvolyze in the experimentally convenient range. Sulfonate esters are used in the rate studies for the least reactive molecules, which are also the least sensitive towards F-strain, and for this reason, the eventual difficulties of having bulky leaving groups connected to molecules sensitive to F-strain were circumvented by coincidence.

While our calculations remove the causes for different slopes in Bingham's strain–reactivity correlations, they do not per se constitute proof that chlorides and *p*-toluenesulfonates must correlate with identical slopes. A priori straight line behavior over such a large rate range and for different leaving groups is rather unexpected, and there is no theoretical argument to support that this must occur. Conceivably, an inappropriate choice of the force-field parameters for the carbenium ions could produce straight line behavior as an artifact. This possibility was examined by strain–reactivity correlations obtained with different force fields.[34] Some 20 different sets of parameters were applied to a selection of bridgehead and nonbridgehead tertiary cations. The investigation produced a slightly modified force field (UNICAT 4), which is applicable to both types of ions; more importantly, with all sets of parameters tried, straight lines over the whole range were always obtained, and they only differed with respect to the quality of the fit.

At the same time, and independently from us, Bentley and Roberts[10] experimentally reinvestigated the solvolysis rates of several bridgehead derivatives. These authors were intrigued by the (predicted) abnormal k_{OTs}/k_{Br} rate ratio of less than unity for very unreactive substrates.[14] They attributed this and other anomalies to the fact that some of the rate data was obtained at rather high temperatures and then was extrapolated back over a considerable range, a procedure susceptible to substantial uncertainties when used for large temperature differences. Bentley and Roberts observed that relative rates of bridgehead compounds having the same leaving group are almost independent of the solvent. Making systematic corrections for changes in solvent and leaving groups, they established a reactivity scale for bridgehead solvolysis, based on *p*-toluenesulfonates (OTs) solvolyzing in 80% EtOH at 70°C. Bentley and Roberts found no variation in the relief of F-strain during ionization of typical tertiary *p*-toluenesulfonates, and they postulated one correlation line for all substrates in the bridgehead series.

The reactivity scale of Bentley and Roberts constitutes an ideal test for the molecular mechanics calculations. Indeed, if the ΔE_{st} (R^+–RH) values used in Figure 1 are correlated with the corresponding log k (Table 1, second-last entry), the straight line is defined by Eq. (2):

$$\log k = -0.45\ \Delta E_{st}\ (R^+\text{–RH}) + 2.49 \qquad r = 0.9964 \qquad (2)$$

The fit is about of the same quality as that for Eq. (1). Our hypothesis (based on calculations) of only one correlation for different leaving groups is therefore confirmed by (independent) experimental results.

The question then arises as to the causes of the differences between our calculations and those of Bingham.[14,29] It should be remembered that the program used by Bingham and Schleyer was one of the very first ones in use, and it was never meant to be the "final" version. At the time when it was applied to bridgehead reactivities, it was already out of date, and the authors recognized that it was "not as accurate as the best molecular mechanics programs available".[14] For this reason attention was focused on steric energy differences between R^+ and RH rather than on the absolute values of the energies themselves. It was thought that by doing so, deficiencies in the parameterization would be diminished or compensated, since the structures involved are very similar. With the observation of strain–reactivity correlations, demonstrated individually with chlorides, bromides, and sulfonates, this expectation was, at least in part, justified. More detailed inspection of the original data can be found in Bingham's thesis[29] which reveals important discrepancies about the steric energies of bridgehead hydrocarbons in comparison to those obtained with MM2 or BIGSTRN programs.[14,15] Discrepancies also occur with the cations, but they are less important. Figure 3 shows a comparison of the ΔE_{st} values calculated by Bingham with the corresponding MM2 data. The superposition of the discrepancies mentioned for hydrocarbons and carbenium ions produces two different correlations, one connecting the highly strained compounds (*p*-toluenesulfonates) and another the less strained ones (chlorides), similar to the correlation that is obtained, when Bingham's ΔE_{st} values are plotted versus log *k*. Figure 3 also shows that the unique position of **10** in Bingham's plot, which does not occur with MM2, must be attributed to differences in the calculations.

The absence of F-strain variation in the solvolysis of bridgehead compounds is the result of coincidence of two favorable factors, namely, the insensitivity of the substrates for which sterically demanding *p*-toluenesulfonates were used, and small chlorides used for the rate measurements of the compounds where F-strain might have occurred. This absence does in no way invalidate the

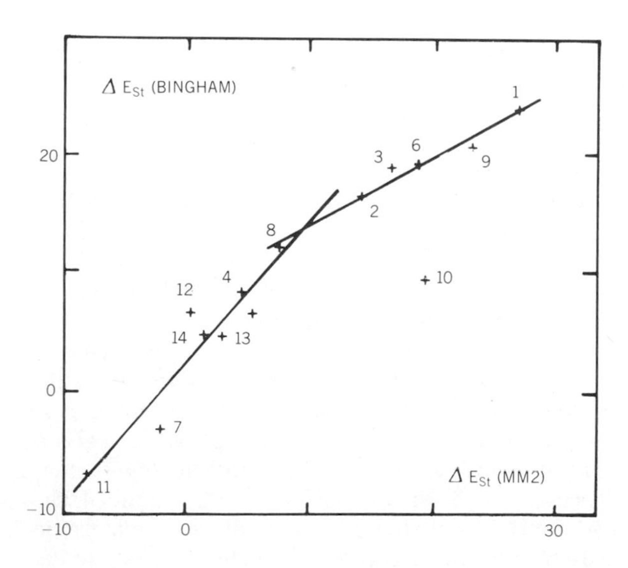

Fig. 3. Plot of ΔE_{st} (R^+–RH) calculated with the force field of Bingham versus ΔE_{st} (R^+–RH) calculated with MM2. For data see Refs. 27 and 29.

concept as such. Indeed, F-strain has been conclusively demonstrated in the case of the *trans,trans,trans*-perhydrophenalene derivative **13**.[35] When the chloro substituent is replaced by a *p*-nitrobenzoate (OPNB) the relative rate increases by ~ 10^4. This rate enhancement was attributed to the release of very strong steric interactions between the OPNB substituent and the hydrogens atoms. The interactions raise the energy of the substrate, but not that of the transition state, since they are released upon solvolysis. In contrast, the chloro substituent, which is sterically much less demanding, does not experience such interactions and therefore reacts normally in comparison to the isomeric **14**. In the case of **14** the interactions with the OPNB substituent do not occur, and therefore **14**-OPNB reacts normally.

That **13** and, to a lesser extent **14** are especially prone to F-strain interactions is qualitatively revealed by the φ_F parameters, which are 12.69 for **13**, and 9.37

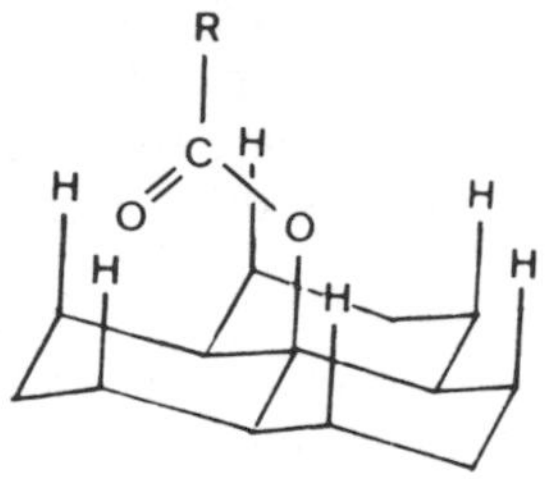

13

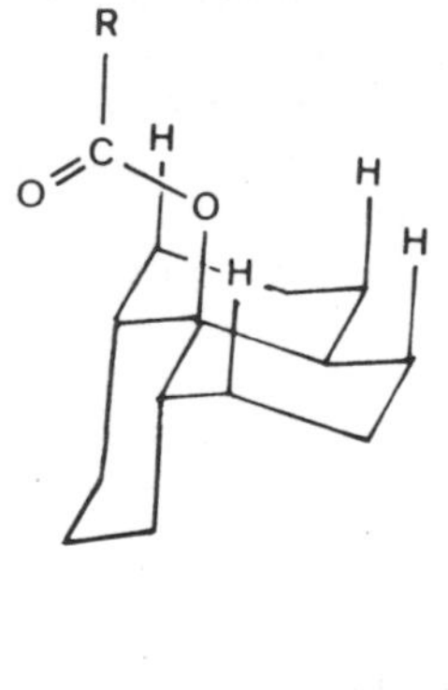

14

for **14**. The rate enhancement of **13**-OPNB can be reproduced by molecular mechanics calculations[36,37] using an acetate group to simulate the steric requirements of the OPNB. The most stable conformation of **13**-OAc, calculated by MM2 has the carbonyl group oriented towards the molecular skeleton, while the CH_3 substituent points away from it, in such a way that it experiences no steric interactions. This justifies use of the model, which, in addition, is easier to handle in the calculations than a benzoate group. Steric energies of chlorides are not directly comparable to those of acetates, and it is therefore indicated to work with strain energies. Strain energies are calculated from the steric energies and the bond increments of MM2. The rate constants for solvolysis of *p*-nitrobenzoates can be converted to the standard conditions of Bentley and Roberts (Table 7; see below).[10] As Figure 4 shows, the rate acceleration of **13**-OPNB is well reproduced with the acetate model, and both **13**-OPNB and **14**-OPNB are accommodated by the strain–reactivity plot including all of the bridgehead compounds.

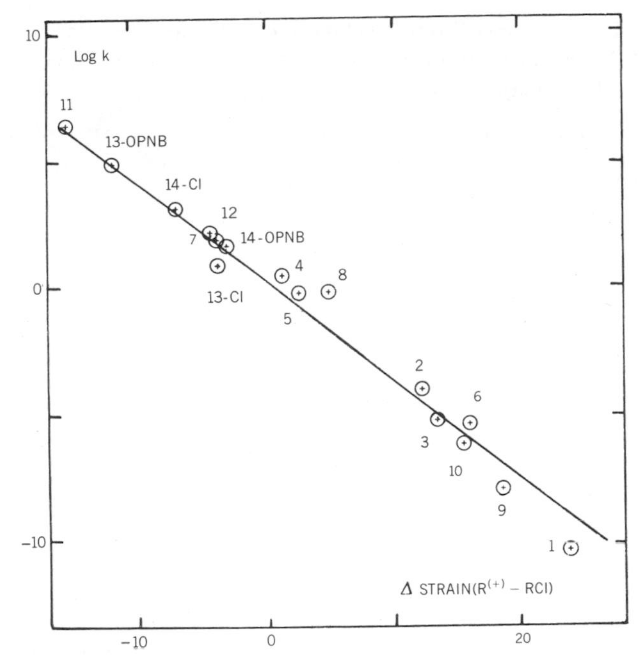

Fig. 4. Plot of log k for solvolysis of bridgehead chlorides versus Δ strain (R^+–RCl) or (R^+–ROAc) (for **13**-OPNB and **14**-OPNB), respectively. Slope: -0.41; intercept: 0.29; r = 0.991.

CARBENIUM ION STABILITY AND TRANSITION STATE FOR SOLVOLYSIS; THE CASE OF THE PERHYDROTRIQUINACYL CATION 10+

One of the principal hypotheses in the context of application of molecular mechanics calculations to bridgehead solvolysis is that the geometric and energetic properties of the transition state of the reaction should be reflected by those of the reaction intermediate, the carbenium ion. This hypothesis corresponds to the generally held view of the mechanism of solvolysis reactions.[38] Strain–reactivity correlations such as Eq. (1) constitute another a posteriori justification for it. The slope of the plot of $\Delta G^{\ddagger}$ versus ΔE_{st} indicates that 65–70% of the strain or steric energy difference between the bridgehead derivative and the bridgehead carbenium ion are expressed in the free energies of activation. However, this statement is only true if the calculations do reproduce the energies of the carbenium ions. Although initially the force field was parameterized for this, the subsequent adjustments were made in order to achieve the best possible correlation with the rates of solvolysis, while little attention was paid to the question of geometries and energies of the respective carbenium ions. Fortunately, it turns out that the calculated structures are quite realistic with respect to both. Recently, Laube[39] determined the X-ray structure

TABLE 2. Structural Data of 1-Adamantyl Cation 8+ (Bond Lengths in Å, Bond Angles in Degrees) compared with X-Ray Structure of the 3,5,7-Trimethyl-1-adamantyl Cation 8a+ Ref. 39

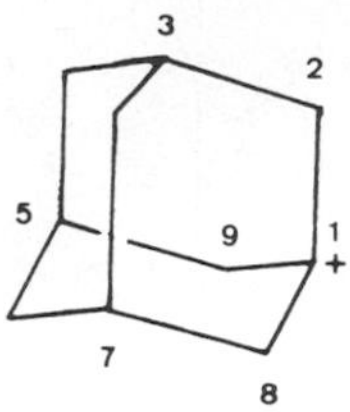

	MM2	MINDO/3	STO-3G	X-Ray
C1-C2	1.482	1.495	1.496	1.44
C2-C3	1.546	1.579	1.591	1.62
C2-C1-C8	117.3°	116.7°	117.4°	118.0°
D	0.243	0.277	0.243	0.21

of the 3,5,7-trimethyl-1-adamantyl cation (**8a+**). Some representative structural elements of this ion are compared in Table 2 with the structure of **8+**, calculated by MM2, MINDO/3, and STO-3G.[34] The MM2 structure is in reasonable agreement with the experimental one with respect to the valence bond angle at the cationic center and the pyramidalization of the bridgehead position (distance *D* from the plane defined by the carbon atoms adjacent to the bridgehead). The X-ray structure reveals stabilization of the positive charge by C—C -hyperconjugation,[40] which results in shortening of the C^+—C bonds, while the C_α—C_β bonds are weakened. This cannot be reproduced with the MM2 model, which at the present stage of refinement makes no provision for such effects. The trend does, however, appear in the ab initio structure, although for the latter only the minimal basis set was used (STO-3G).[34] Laube[41] also determined the X-ray structure of the 1,2,4,7-*anti*-tetramethyl-2-norbornyl cation (**28a+**), which is compared in Table 3 to the MM2 structure of the 1,2-dimethyl-2-norbornyl cation (**34+**).[43] The X-ray structure is unsymmetrically bridged, in agreement with high level ab initio calculations[42] for the 2-methyl-2-norbornyl cation (**28+**). Bridging results in a clearly shortened C-2—C-6 distance (2.09 Å). MM2, however, produces a classical structure with a C-2—C-6 distance of 2.41 Å. These discrepancies point to some of the limitations of the molecular mechanics approach which, by its nature, can only consider effects for which it is parameterized. The discrepancies for the 2-methyl-2-norbornyl cation (**28+**) are without consequences for the reactivity of bridgehead derivatives, but limit the applicability of the approach to classical structures. Some other data for MM2 calculated structures in comparison to MINDO/3 and ab initio results are shown in Table 4. In general the agreement between the various methods of calculation is reasonable. At least at this level of sophistication no evidence for unusual geometry of any ion shows up, including **10+**, which is discussed below in more detail. The 7-methyl-7-norbornyl cation (**32+**), however, seems to be

TABLE 3. Selected Structural Data for 1,2-Dimethyl-2-norbornyl Cation 33+

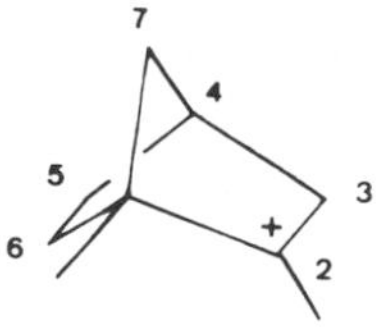

	MM2	X-Ray[a]
C1-C2	1.48	1.44
C1-C6	1.54	1.74
C2-C3	1.48	1.50
C2-C6	2.41	2.09
C1-C2-C3	114.8°	108.1°
C6-C1-C2	105.8°	81.7

[a]Structure of 1,2,4,7-*anti*-tetramethyl-2-norbornyl cation [41].

a notable exception. MM2 calculates a valence bond angle of 110.7°, while the corresponding bond angles obtained by MINDO/3 and STO-3G are 98.0° and 99.7°, respectively. This important deviation probably occurs because the force constant for valence bond angle deformation in MM2 is too high. The ion cannot adopt the ideal geometry required by the molecular structure because the energy needed to deform the angle at the cationic center is exaggerated. The problem appears to be general for ions suffering severe bond angle deformations and, at present, it is impossible to treat all of the ions with only one set of parameters.[30,34]

The performance of the force field with respect to energies is, in the context of bridgehead reactivity, more important than that with respect to structures. Only few experimental values for gas-phase enthalpies of formation of carbenium ions considered in this work are available from the literature, and they are compared with the results obtained from MM2 in Table 5. The steric energies calculated by MM2 are converted to enthalpies of formation using the general bond increments of MM2, including the translation–rotation term of 2.4 kcal/mol, a general increment of 187.4 kcal/mol[30] for the total of the three C^+—C bonds and corrections for inductive stabilization of 1.5 kcal/mol by each β-alkyl substituent.[43] This results in an arbitrarily selected reference value of 166.2 kcal/mol for the *t*-butyl cation.[32] Other reported experimental values for this ion are 169.1,[44] 169.2,[45] 166.5,[46] 165,[47] 164.8.[48] The inductive correction term is a rather crude approximation: It is now known from gas-phase studies,[52] NMR investigations under stable ion conditions,[54] the X-ray structure of **8a+**, and theoretical considerations[55] that the positive charge must be extensively delocalized even in aliphatic ions. The experimental and MM2 calculated ΔH_f values for the same structures cover roughly the same range (exp: 156.9–171;

TABLE 4. Calculated Structures of Carbenium Ions

	MM2	MINDO/3[a]	STO-3G[a]	4-31G[a]
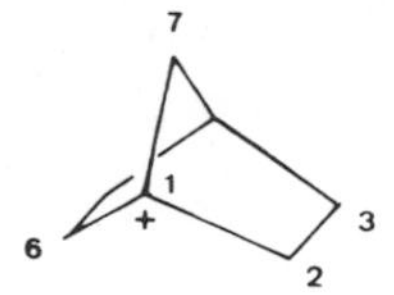 C1-C2	1.481	1.511	1.516	1.475
C1-C7	1.475	1.484	1.510	1.485
C2-C1-C6	116.8	121.5	120.1	120.7
D	0.354	0.319	0.388	0.349
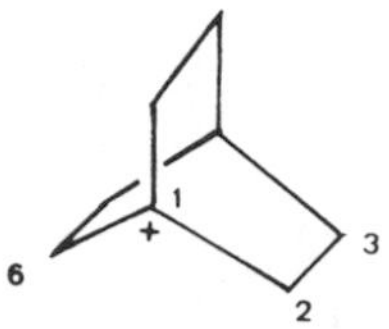 C1-C2	1.482	1.497	1.502	
C2-C1-C6	117.1	117.1	117.2	
D	0.255	0.254	0.252	
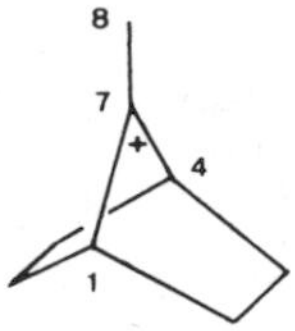 C1-C7	1.466	1.525	1.520	
C7-C8	1.481	1.455	1.493	
C1-C7-C4	110.7	98.0	99.5	
D	0.002	0.0	0.007	
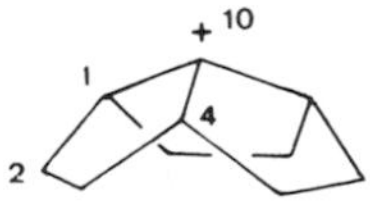 C1-C10	1.473	1.507	1.498	
C1-C2	1.540	1.548	1.568	
C1-C10-C4	118.6	117.0	117.4	
D	0.173	0.262	0.246	

[a]Reference 56; preliminary data was published.[34] Since then the structures have been reoptimized with an increased number of geometric parameters.

MM2 calc: 155.8–173.2 kcal/mol). For the 1-adamantyl cation (**8+**) MM2 gives 162.1, while 159[50] and 160.7[51] are measured experimentally. The MM2 value for the 2-methyl-2-norbornyl cation (**29+**) (173.2) corresponds to a classical structure; the calculated value is 2.2 kcal/mol higher than the experimental one, which refers to a (unsymmetrically) bridged ion.[41] Further comparisons are impossible owing to a lack of experimental energies for the ions of interest.

TABLE 5. Experimental and Calculated (MM2, MINDO/3) Energies of Tertiary Cations (ΔH_f, kcal/mol) and Propyl Cations Stabilization Energies (PCSE)[49]

Cpd	ΔH_f (R^+) exp	MM2		MINDO/3		
		ΔH_f (R^+)[a]	PCSE[b]	ΔH_f (R^+)	PCSE[b]	Charge on C^+
1	-	200.3	-3.4[k]	210.0[l]	-8.9[q]	-
2	-	177.6	-16.3[k]	191.2[m]	-9.2[q]	0.365
3	-	179.7	-	-	-	-
4	-	160.1	-26.0	170.1[n]	-16.0[m]	0.342
8	159[c]; 160.7[d]	162.2	-22.8	199.7[m]	-12.8[m]	0.330
10	-	178.9	-15.6	182.6[m]	-9.6[m]	0.338
11	-	153.7	-38.6	161.9[m]	-20.0[m]	0.375
17	166.2[e]	166.2	-18.2[k]	171.0[n]	-14.6[r]	-
21	167[f]; 169.4[g]	167.5	-23.2	164.0[o]	-18.1[o]	-
24	156.9[h]; 160.4[g]	155.8	-23.8[k]	154.6[p]	-21.9[q]	-
29	171.3[i]; (171)[i] 171[f]	173.2	-23.8	196.0[n]	-21.0[o]	-
32	-	188.9	-9.4	194.2[m]	-23.8[m]	0.371

[a]UNICAT 1 Ref. 27.
[b]See Eq. (3); ΔH_f ($CH_3CH_2CH_3$) = -24.77 (MM2) and -26.50 (MINDO/3),[q] ΔH_f ($CH_3CH^+{-}CH_3$) = 184.0 (MM2) and 191.8 (MINDO/3)[n].
[c]Ref. 50. [d]Ref. 51. [e]Ref. 31. [f]Ref. 44. [g]Ref. 45. [h]Ref. 47. [i]Ref. 53. [k]ΔH_f (R–H) from Ref. 54. [l]Ref. 55. [m]Ref. 56. [n]Ref. 57. [o]Ref. 49. [p]Ref. 58. [q]ΔH_f (RH) from Ref. 59. [r]ΔH_f from Ref. 60.

Accordingly, we have performed some semiempirical (MINDO/3) and ab initio (STO-3G) calculations for a limited number of structures, which are compared with the MM2 results in Tables 5 and 6. The ΔH_f values calculated by MINDO/3 have systematic uncertainties[49] and direct comparison with MM2 is not warranted. However, some of the systematic errors are eliminated by means of isodesmic relationships. Table 5 contains propyl cation stabilization energies (PCSE)[49] as defined by Eq. (3), and calculated by MM2 and MINDO/3.

$$R^+ + 2\text{-Pr—H} \rightarrow \text{R—H} + 2\text{-Pr}^+ \tag{3}$$

For simple ions the methods of calculation agree within ~5 kcal/mol, but in the bridgehead series the discrepancies are almost double this value. The most serious deviations occur with **11** (18.6 kcal/mol) and **32** (14.4 kcal/mol). The

TABLE 6. Isodesmic Relationships[49] for Tertiary Cations (STO-3G and MM2)

Cpd	STO-3G Energy (R^+)[a] (hartrees)	STO-3G Energy (RH) (hartrees)	STO-3G PCSE (kcal/mol)	PCSE MM2 (kcal/mol)
1	-268.03709	-268.89382[b]	-1.28	-3.42
2	-306.65516	-307.48656[a]	-17.17	-16.26
8	-382.69440[c]	-383.51550[d]	-23.64	-22.82
10	-382.65909	-383.49940[d]	-11.58	-15.61
17	-154.63918[e]	-155.46572[f]	-20.22	-18.20
32	-306.64438	-307.46083[a]	-26.56	-9.41
2-Propyl	-116.02765[g]	-116.88642[h]-	-	

[a]Ref. 56. [b]Ref. 61. [c]Higher energy structure, Ref. 40. [d]Ref. 62. [e]Ref. 63. [f]Ref. 64. [g]Ref. 65. [h]Ref. 66.

deviation with **32** is expected and probably due to the MM2 force field (see above). Even without **11** and **32** the PCSEs calculated by these methods correlate only very approximately.

The comparison of PCSEs obtained by MM2 and STO-3G (Table 6) is more satisfactory. The average discrepancy is only 2 kcal/mol, except for **32**. The significance of this agreement is, however, difficult to assess since the calculations refer to different standard states and must await the results of more sophisticated ab initio calculations which, at present, are beyond our possibilities.

In the context of bridgehead solvolysis the perhydrotriquinacene derivative **10** occupies a unique position. The compound deviates by many orders of magnitude from the original strain–reactivity correlations of Bingham and Schleyer.[14] At that time, this was attributed to an unusually low stability of the cation **10+** owing to the absence of stabilizing hyperconjugative C—C interactions. Such interactions are possible in bridgehead ions having C—C bonds parallel to the empty *p* orbital of the cationic center. For reasons of skeletal geometry, C—C hyperconjugation, as well as C—H hyperconjugation, are impossible in **10+**. However, in the calculations with MM2 **10** shows no anomaly and, therefore, there is no need for such an explanation, although the X-ray structure of the adamantyl cation (**8+**)[39] demonstrates clearly that the phenomenon of C—C hyperconjugation exists. Inspection of the series of cations **1+** to **14+** reveals that the geometric requirements for C—C hyperconjugation are ideally met in **2+**, **4+**, and **8+**, with three bonds parallel to the *p* orbital, while in other cases (**3+**, **11+**, **13+**) the situation is much less favorable and corresponds rather to that prevailing in **10+**. In spite of this, no systematic deviations occur in the plot of $\Delta G^{\ddagger}$ versus ΔE_{st} (Fig. 1). This could imply that the solvolysis transition state is not very sensitive towards electronic effects, an implication that is reinforced if the influence of β-alkyl substituents is considered. The enthalpies of formation of bridgehead carbenium ions are corrected for inductive stabilization owing to β branching. If the strain–reactivity plot is based on $\Delta\Delta H_f$ (R^+–RH) rather than on ΔE_{st}, then the structures **9**, **10**, **13**, and **14** deviate. The bridgehead ions generally have three β-alkyl substituents, while **9+** has five and **10+**, **13+**, and **14+** have six, but the additional stabilization of these ions is not reflected in the transition state for solvolysis. Similarly, we believe that C—C (and C—H) hyperconjugation, although significant in free ions,[67,68] is either compensated by inductive stabilization in the case of **10+** or, more generally, of little consequence to solvolytic reactivity. This is understandable since it was observed that the stabilities of carbenium ions in solution are compressed relative to their stabilities in the gas phase because of solvation.[69] In the transition state for solvolysis, the positive charge is further stabilized by the leaving group (counterion). As a result of these effects, the electronic demand of the cationic center decreases and so does the importance of stabilization by inductive effects and hyperconjugation, so that the reactivity is overridingly determined by strain effects.

The question concerning the absence of hyperconjugative stabilization in **10+** has recently been reopened. Prakash et al.[70] concluded from ^{13}C NMR and MNDO calculations that the cationic center of the dodecahydryl cation carries more positive charge than that of **8+**. The structural analogy with **10+** suggests that the same should hold for the perhydrotriquinacenyl cation (**10+**). However, the MINDO/3 calculated charge densities (Table 5) do not support this hypothesis.

SOLVOLYSIS OF TERTIARY NONBRIDGEHEAD DERIVATIVES

The mechanistic and structural uniformity of bridgehead compounds greatly facilitates studies of their reactivity. In the more general case of solvolysis of tertiary derivatives, the situation is much less favorable. The range of experimental reactivities is much smaller but the structural variation more important and, as a consequence, the relative computational error increases. Complications may arise because of secondary energy minima of flexible molecules. In addition, the mechanism of solvolysis may vary with structure and the reactivity does not necessarily depend on strain changes alone, but also on solvent participation,[11] anchimeric assistance,[71] leaving group hindrance,[72] or F-strain.[21,35] Some of these effects vary with the leaving group, others with the solvent, while still others are determined only by molecular structure. In this situation the strain calculations, treating only one aspect of the reactivity problem, do not necessarily reproduce the reactivity of the compounds under investigation. However, since they offer a good rationalization of the steric effects in the bridgehead series, the latter may be used as a mechanistic model, serving as reference points from which other compounds may deviate. Typically, in the past much debate was based upon comparison of exo/endo rate ratios, and depending on the viewpoint of the author, either of the isomers was considered to react at the "normal" rate, while the other was considered "accelerated" or "retarded." The correlation of bridgehead reactivities constitutes, in our view, a much more objective description of "normal" solvolysis.

A large number of rate constants for the solvolysis of tertiary molecules is available in the literature, mostly for chlorides and OPNBs. For the purpose of strain calculations it is desirable to establish a reactivity scale analogous to that of Bentley and Roberts[10] for bridgehead compounds and for the same reaction conditions (OTs leaving group, 80% EtOH, 70°C). This can be achieved by conversion of the original data by means of the Arrhenius equation or, when necessary, by extrapolation of relative rates from one system to another. Arbitrarily selected reference conditions are 80% EtOH, 70°C for chlorides and 80% acetone, 70°C for OPNBs. Details and references to the original literature are given in Ref. 36. Figure 5 is a plot of the rate constants for OPNB derivatives against those for chlorides of 3 bridgehead and 20 non-bridgehead derivatives (data from Ref. 36).

The two series of reactions are remarkably consistent; the principal deviation concerns **13**, and this has already been discussed and rationalized in terms of F-strain. Similarly, the other slight upwards deviation in the plot (**18**, **19**) should be due to the same cause. The structures believed to be retarded by leaving group hindrance, the 2-*endo*-norbornyl derivatives **29** and **34**,[72,73] deviate slightly downwards, as expected, but the effect is small and its significance therefore questionable. The remaining data are related by means of Eq. (4):

$$\log k(\mathrm{OPNB}) = 0.85 \log k(\mathrm{Cl}) - 4.89 \quad r = 0.984 \tag{4}$$

The chloride rates are converted to Bentley's conditions by multiplying log k(Cl) with the leaving group correction for *p*-toluenesulfonates of 1.6×10^5.[10] It is remarkable to find that structures as different as bridgehead-, 2-*exo*-norbornyl- , and *t*-butyl derivatives respond so uniformly to the leaving group and solvent change, although, admittedly, the solvent properties vary little between the systems used. It is clear that this uniformity must break down in more limiting solvents.[11,74] Table 7 contains the rate constants for chloride and OPNB solvolysis converted to standard conditions (log k_{calc}).

A similar parallel behavior results when the standardized rate constants for solvolysis are compared with those for dehydration of tertiary alcohols in acetic acid. This latter reaction, which has been investigated by Lomas et al.,[75] is an S_N1 process analogous to solvolysis. The comparison shown in Fig. 6 uses rate constants from chloride solvolysis whenever possible. The OPNB rates are included when data for chlorides are lacking, and in the case of **32** the comparison is based on OTs solvolysis.

As before, the reaction series are consistent, and only few serious deviations occur (**20**, **31**, **38**), which may be tentatively ascribed to F strain. Rather

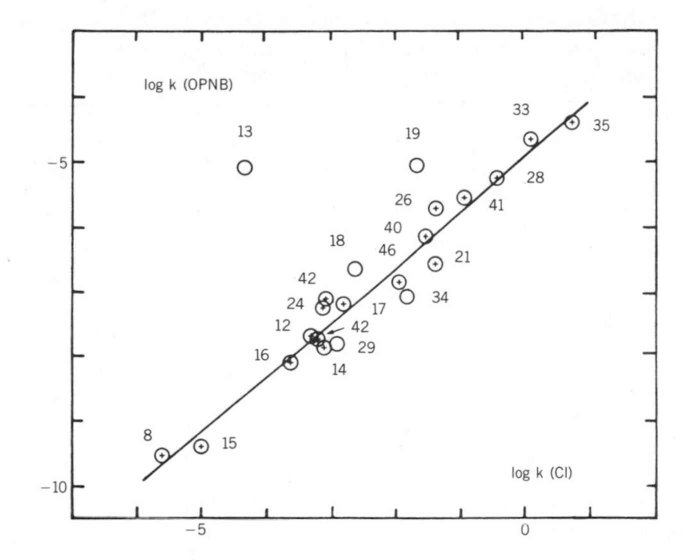

Fig. 5. Plot of log k for solvolysis of tertiary p-nitrobenzoates versus tertiary chlorides, extrapolated to standard conditions: OPNBs: 80% acetone, 70°C; chlorides: 80% EtOH, 70°C. For data see Ref. 36.

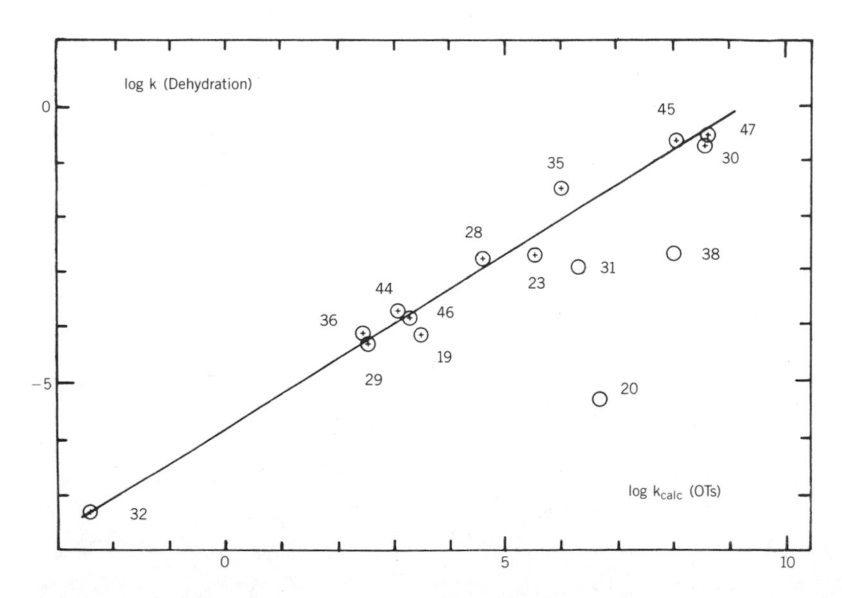

Fig. 6. Plot of log k for dehydration of tertiary alcohols (AcOH, 25°C Ref. 75) versus log k for chloride or p-nitrobenzoate solvolysis. Rate constants are converted to standard conditions via Eq. (4) and (5) and leaving group corrections of Bentley.[10] For data see Table 7.

TABLE 7. Rate Constants for Solvolysis of Tertiary Chlorides and *p*-Nitrobenzoates and for Alcohol Dehydration Converted to Standard Conditions.[a] Data from Refs. 30 and 76[b]

No	Substrate	$\log k_{calc}$ (Cl)	$\log k_{calc}$ (OPNB)	$\log k_{calc}$ (OH)	ΔE^{c}_{st} (R^+ - ROH)
8	1-Adamantyl	-0.41	-0.25	-	7.17
12	3-Homoadamantyl	1.91	1.87	-	-1.19
13	*t,t,t*-13-Tricyclo[7.3.1.0^{5,13}]tridecanyl	0.84	4.98	-	-1.73[d]
14	*c,c,t*-13-Tricyclo[7.3.1.0^{5,13}]tridecanyl	2.08	1.62	-	-1.68[d]
15	1-Homoadamantanyl	0.20	-0.08		4.09
16	*t,t,t*-1-Tricyclo[7.3.1.0^{5,13}]tridecanyl	1.49	1.46	-	
17	*t*-Butyl	2.38	2.46		1.93
18	*t*-Butyl(dimethyl)methyl	2.45	3.12		-0.87
19	Di-*t*-butyl(methyl)methyl	3.48	4.94	2.56	
20	Tri-*t*-butylmethyl		6.81	0.74	
21	1-Methylcyclopentyl	3.80	3.25		2.52
22	1-*t*-Butylcyclopentyl		5.96		-1.38
23	1,2,2,5,5-Pentamethylcyclopentyl	5.53	4.85		
24	1-Methylcyclohexyl	2.05	2.38		0.84
25	1-*t*-Butylcyclohexyl		4.65		-2.35
26	1-Methylcycloheptyl	3.77	4.24		-0.29
27	1-*t*-Butylcycloheptyl		6.80		-4.85
28	2-*exo*-2-Methylnorbornyl	4.58	4.57	5.04	2.56
29	2-*endo*-2-Methylnorbornyl	2.49	1.58	2.29	2.78
30	2-*exo*-2-*t*-Butylnorbornyl		8.69	8.10	
31	2-*endo*-2-*t*-Butylnorbornyl		6.40	4.49	-2.05
32	7-Methyl-7-methylnorbornyl	-1.66	-2.42	-2.50	19.64[d]
33	1,2-Dimethyl-2-*exo*-norbornyl	5.32	5.46		
34	1,2-Dimethyl-2-*endo*-norbornyl	3.37	2.64		1.70
35	2,3,3-Trimethyl-2-*exo*-norbornyl	5.98	5.72	6.77	
36	2,3,3-Trimethyl-2-*endo*-norbornyl		1.97	2.60	-0.99
37	2,7,7-Trimethyl-2-*endo*-norbornyl		4.27		-1.89
38	2-*t*-Butyl-3,3-dimethyl-2-*endo*-norbornyl		8.11	4.89	

TABLE 7. *(Continued)*

No	Substrate	log k_{calc} (Cl)	log k_{calc} (OPNB)	loc k_{calc} (OH)	ΔE_{st} (R^+ - ROH)
39	1-*trans*-bicyclo[3.3.0]octyl		5.88		0.30
40	1-*cis*-Hydrindanyl	3.38	3.71		-
41	1-*trans*-Hydrindanyl	4.26	4.41		-2.47
42	1-*cis*-Decalinyl	2.07	2.57		
43	1-*trans*-Decalinyl	1.96	1.83		
44	9-Methyl-9-bicyclo[3.3.1]nonyl		3.08	3.23	-2.32
45	9-*t*-Butyl-9-bicyclo[3.3.1]nonyl		8.21	8.18	-12.42
46	2-Methyl-2-adamantyl	3.21	2.88	3.10	-2.37
47	2-*t*-Butyl-2-adamantyl		8.21	8.40	-11.60

[a]Standard conditions: 80% EtOH, 70°C, OTs leaving group.
[b]Underlined values used for the correlations with ΔE_{st}.
[c]Calculated with UNICAT 4.
[d]Calculated with UNICAT 1.

unexpectedly, the very crowded *t*-butyl derivatives **45** and **47** are unaffected by this complication. The rate constants for dehydration (25°C) are transformed to standard conditions by Eq. (5):

$$\log k(\mathrm{OH}) = 0.625 \log k_{calc} - 5.77 \quad r = 0.9866 \qquad (5)$$

Table 7 reveals that, except for the deviations already mentioned, agreement between the extrapolated rate constants for a given compound is remarkable, and rarely exceeds 0.5 log units. This consistent behavior is a prerequisite for the application of strain calculations, and it helps to eliminate problematic cases, that is, compounds deviating from correlations (4) and (5).

A selected number of structures from Table 7 have been subjected to strain calculations[36,76] with the UNICAT 4[34,36] parameterization. For all compounds, OH is used as the leaving group model as a compromise between Cl, OH, and OPNB. For reasons of correspondence between the reaction and the leaving group model, the strain–reactivity correlation for nonbridgehead derivatives is based preferentially on the rate constants from alcohol dehydration. Rate constants from chloride solvolysis are selected in second preference and those from OPNBs in third. The values used in the correlation are underlined in Table 7. The examined structures cover the largest possible rate range and are normal according to Eqs. (4) and (5), although some examples are included without verification of "normal" behavior. For reasons mentioned earlier, the

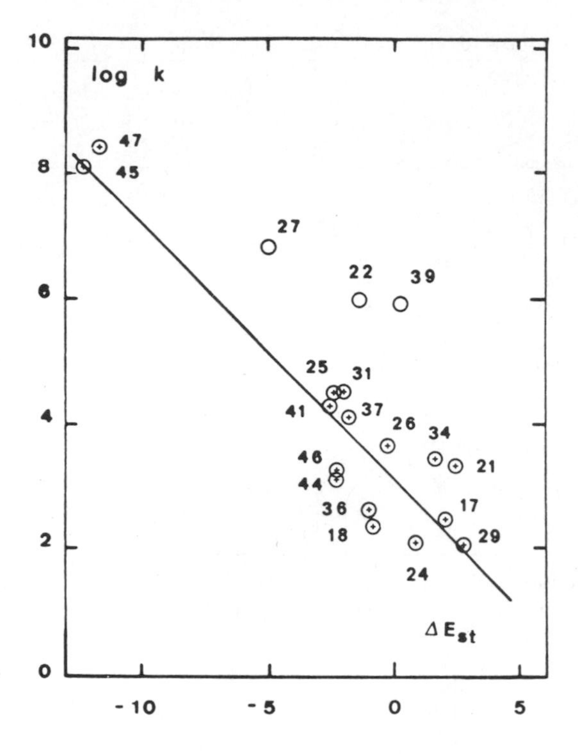

Fig. 7. Plot of log k for solvolysis of tertiary nonbridgehead derivatives versus ΔE_{st} (R^+–ROH). Slope: –0.397; intercept 3.17; r = 0.922. Compounds **22**, **27**, and **39** are excluded from correlation. For data see Table 7.

7-methyl-7-norbornyl derivative **32**, the least reactive compound of the series, is excluded from the strain–reactivity correlation. More importantly, all 2-*exo*-norbornyl derivatives are also excluded. There is now convincing evidence that not only secondary, but also tertiary, 2-*exo*-norbornyl derivatives solvolyze with anchimeric assistance.[71,77] Since the MM2 calculations refer to a "classical" structure for 2-norbornyl cations, they overestimate their energy and the calculated ΔE_{st} values for tertiary 2-*exo*-norbornyl derivatives come out too high. Typically, the 2-methyl-2-*exo*-norbornyl compound (**30**) deviates from the strain–reactivity correlation by ~ 2.5 log units,[34] while the reactivity of the endo isomer is correctly accounted for by the calculations. Figure 7 shows the strain-reactivity correlation for 16 nonbridgehead compounds. As predicted, the correlation is less satisfactory than that obtained in the bridgehead series, but the slope of the straight line of -0.397 compares well with that of -0.432 for the bridgehead compounds with OH as the leaving group.[76] The obtained intercept of 3.17 is somewhat higher than that of the bridgehead compounds, which is 1.69. The lower value of the correlation coefficient (r = 0.922 versus 0.996) can in part be ascribed to the smaller rate range; in addition, some of the scatter must be attributed to the extrapolations of the experimental data that

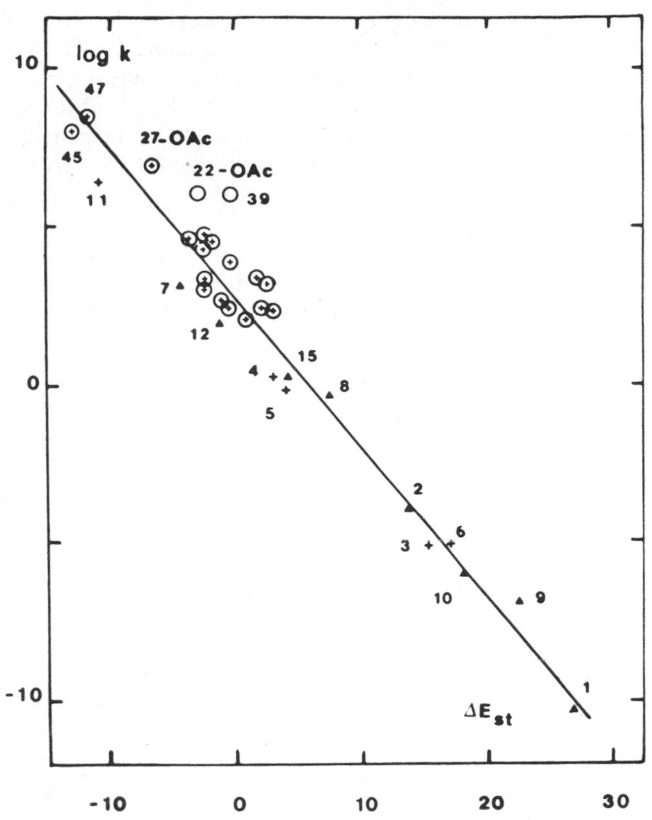

Fig. 8. Plot of log k for solvolysis of bridgehead and nonbridgehead derivatives versus ΔE_{st} (R^+–ROH). Crosses: bridgehead, calc. UNICAT 1; triangle: bridgehead, calc. UNICAT 4; circles w. cross: nonbridgehead, calc. UNICAT 4; hollow circles: excluded from correlation. For data see Tables 1 and 7.

are necessary for establishing the reactivity scale. Nevertheless, the correlation in Figure 7 is not very satisfactory. Most of the data points are clustered in a very restricted area, with the two compounds in the upper left dominating the slope of the straight line. Some improvement can be achieved if the BIGSTRN calculations of Lomas[75] are added to the MM2 results. Although based on a slightly different model (H as the leaving group) the calculations of Lomas are consistent with the MM2 results and can be extrapolated to the same scale.[76] Despite this, the significance of the correlation remains questionable as long as the nonbridgehead compounds are considered alone. The combination of the bridgehead with the nonbridgehead results is, however, very revealing. The combined plot (Fig. 8) contains 30 compounds that are correlated by Eq. (6):

$$\log k = -0.47 + 2.65\ \Delta E_{st}, \quad r = 0.983, \quad \sigma = 0.872 \tag{6}$$

In general the nonbridgehead compounds react at slightly accelerated rates in comparison to their bridgehead counterparts. This could be attributed to more efficient backside solvation of the developing charge; however, the acceleration is not dramatic in comparison with the scatter in the data, and it might not even be significant. As Figure 8 shows, according to the MM2 calculations, all 2-*endo*-norbornyl derivatives react normally in comparison with bridgehead compounds. This supports the assumption that leaving group hindrance is negligible in the compounds investigated.

The reasons why **22**, **27**, and **39** deviate from the correlations in Figures 7 and 8 are not yet clear. Since only rate constants for OPNB derivatives are available for them, the possibility of inconsistent solvolysis rates cannot be ruled out. The intervention of F-strain as a cause for the deviations can be examined by using the acetate as the leaving group. For **27**, replacing OH by $OCOCH_3$ changes ΔE_{st} to -6.48, and with this value, **27** falls within the correlation. The correction for **22** (ΔE_{st} = -2.61) is in the right direction but too small, while it brings practically no improvement in the case of **39**. It remains to be seen whether these deviations are chemically significant or whether they are rather due to inadequacies in the force-field calculations.

Despite these discrepancies, it is clear that the rates of solvolysis for tertiary derivatives of general structure are dominated by strain effects as are those of the bridgehead compounds, while differential k_s contributions and solvation play a minor role so long as solvolysis in the more traditional solvents (moderately nucleophilic and moderately limiting) are considered. The strain effects can be treated by the same molecular mechanics model. The latter is, however, limited and does not at present apply to compounds reacting with anchimeric assistance, or to those which, upon solvolysis, lead to carbenium ions suffering severe angle strain. The acetate model for simulation of F-strain requires further testing.

These principles apply also to solvolysis of secondary derivatives.[28,78] The correlation of $\Delta G^‡$ for acetolysis of 27 secondary *p*-toluenesulfonates, believed to react via the k_c mechanism with $\Delta E_{st}(R^+\text{–}ROH)$ has a slope of 0.67, almost identical to that found for tertiary bridgehead compounds, Eq. (1). It should be noted, however, that the slope of the correlation increases significantly if the solvolysis is carried out in more limiting solvents,[28,79] where more positive charge develops and the transition state occurs later on the reaction coordinate.

OTHER BRIDGEHEAD REACTIONS

It is beyond the scope of this chapter to discuss all other reactions occurring at bridgehead or adjacent to bridgehead positions in detail, but some recent work on radical forming processes must be mentioned. In view of the geometric similarities between carbenium ions and radicals, Bingham and

Schleyer[14] applied their strain calculations to the thermal decomposition of bridgehead peresters and diazoalkanes.[80] It was found that free radical reactivities parallel those for the corresponding carbenium ions. Both reactions were recently very carefully investigated, mainly by Rüchardt and his school. The rates of decomposition of bridgehead-*t*-butylperesters are parallel to those for solvolysis of bridgehead derivatives,[33b] but the rate range is about three times smaller on the logarithmic scale. Log k for perester decomposition can be correlated with ΔE_{st} (R^+–RH) with a slope of ~ 0.16. In contrast to this, when the rates of diazoalkane thermolysis are plotted against the rates for bridgehead solvolysis, a curve instead of the expected straight line is observed.[81] Accordingly, only a very approximate linear correlation exists between diazoalkane thermolysis and ΔE_{st} (R^+–RH). A tentative force field for radicals was developed by Beckhaus and incorporated into MM2,[82] and the homolytic bond dissociation energies of bridgehead hydrocarbons calculated with this force field correlate well with $\Delta G^{\ddagger}$ for thermolysis of bridgehead diazoalkanes.[82] The stabilities of bridgehead radicals were estimated by Lomas et al.[83] using a combination of rate measurements and molecular mechanics calculations. The strain energies are significantly higher than those calculated with the force fields of Beckhaus[84] or Allinger,[85] but they are probably still too low. Recent gas-phase studies attribute a strain energy of 3.7 kcal/mol to the 1-adamantyl radical (relative to *t*-butyl)[48] while the strain obtained by Lomas is 2.4 kcal/mol. The calculations with the parameters of Beckhaus and Allinger give 0.9 and 2.5 kcal/mol, respectively.

CONCLUSION

Bridgehead compounds have played a prominent role with respect to both the evolution of our understanding of solvolysis reactions and to the development of molecular mechanics calculations. It is amazing to see today, some 25 years after these calculations were first applied to solvolysis, that the main mechanistic concepts developed at that time still hold. The rates of solvolysis of bridgehead derivatives are dominated by steric effects, and the latter can be rationalized in terms of strain (or steric energy) changes between the starting compound and the corresponding carbenium ion, which serves as a transition state model. This applies also to the solvolysis of tertiary nonbridgehead (and secondary) derivatives, although there are exceptions, and the strain–reactivity correlations are less satisfactory than for the bridgehead series alone. The molecular mechanics model reproduces geometries and energies of carbenium ions which, because of their size, are still nearly inaccessible to high level ab initio calculations. Clearly, these calculations will never replace more advanced theoretical treatments or experiments, but their simplicity and speed make them very valuable for everyday use.

ACKNOWLEDGMENTS

This work was supported by the Swiss National Science Foundation (Project No. 2.602-0.87 and previous grants). The authors thank P. v. R. Schleyer, T. W. Bentley, J. S. Lomas, H. J. Schneider and (the late) W. Parker for numerous stimulating and clarifying discussions, M. Doyle for correcting the manuscript, and J. Blanc, who carried out the early calculations of bridgehead compounds.

REFERENCES

1. P. D. Bartlett and L. H. Knox, *J. Am. Chem. Soc.* **1939**, *61*, 3184.
2. W. v. E. Doering, M. Levitz, A. Sayigh, M. Sprecher, and W. P. Whelan, Jr., *J. Am. Chem. Soc.* **1953**, *75*, 1008.
3. C. Wentrup, *Reactive Molecules*, Wiley, New York, 1984; D.C. Nonhebel, J. M. Tedder, and J. C. Walton, *Radicals,* Cambridge University Press, 1979.
4. D. E. Applequist and J. D. Roberts, *Chem. Rev.* **1954**, *54*, 1065.
5. U. Schöllkopf, *Angew. Chem.* **1960**, *72*, 147.
6. R. C. Fort and P. v. R. Schleyer, *Adv. Alicyclic Chem.* **1961**, *1*, 283.
7. C. J. M. Stirling, *Tetrahedron*, **1985**, *41*, 1613.
8. R. C. Fort, in *Carbonium Ions*: G. A. Olah, P. v. R. Schleyer, Eds. Wiley, New York, 1973, Vol. IV, Chapter 32.
9. C. Rüchardt, *Angew. Chem. Int. Ed. Engl.* **1970**, *9*, 830.
10. T. W. Bentley and K. Roberts, *J. Org. Chem.* **1985**, *50*, 5852.
11. T. W. Bentley, C. T. Bowen, D. H. Morten, and P. v. R. Schleyer, *J. Am. Chem. Soc.* **1981**, *103*, 5466; T. W. Bentley and G. E. Carter, *J. Am. Chem Soc.* **1982**, *104*, 5741. See also: D. Farcasiu, J. Jähme, and C. Rüchardt, *J. Am. Chem. Soc.* **1985**, *107*, 5717, D. N. Kevill, S. W. Anderson, *J. Am. Chem. Soc.* **1986**, *108*, 1579.
12. P. v. R. Schleyer and R. D. Nicholas, *J. Am. Chem. Soc.* **1961**, *83*, 2700; J. L. Fry, C. J. Lancelot, L. K. M. Lam, J. M. Harris, R. C. Bingham, D. J. Raber, R. E. Hall, and P. v. R. Schleyer, *J. Am. Chem. Soc.* **1970**, *92*, 2538; J. Raber, R. C. Bingham, J. M. Harris, J. L. Fry, and P. v. R. Schleyer, *J. Am. Chem. Soc.* **1970**, *92*, 5977; D. N. Kevill, K. C. Kolwyck, and F. L. Weitl, *J. Am. Chem. Soc.* **1970**, *92*, 7300.
13. G. J. Gleicher and P. v. R. Schleyer, *J. Am. Chem Soc.* **1967**, *89*, 582. S. A. Sherrod, R. G. Bergman, G. J. Gleicher, and D. G. Morris, *J. Am. Chem. Soc.* **1972**, *94*, 4615.
14. R. C. Bingham and P. v. R. Schleyer, *J. Am. Chem. Soc.* **1971**, *93*, 3189.
15. W. Parker, R. L. Tranter, C. I. F. Watt, L. W. K. Chang, and P. v. R. Schleyer, *J. Am. Chem. Soc.* **1974**, *96*, 7121.
16. J. E. Williams, P. J. Stang, and P. v. R. Schleyer, *Ann. Rev. Phys. Chem.* **1968**, *19*, 531; P. v. R. Schleyer, J. E. Williams, and K. R. Blanchard, *J. Am. Chem.*

Soc. **1970**, *92*, 2377; K. B. Wiberg, *J. Am. Chem. Soc.* **1965**, *87*, 1070; H. A. Harris, Ph.D. thesis, Yale University, 1966.

17. F. H. Westheimer in *Steric Effects in Organic Chemistry*, M. S. Newman, Ed. Wiley, New York, 1956, Chapter 12.
18. U. Burkert and N. L. Allinger, *Molecular Mechanics*, ACS Monograph 177, American Chemical Society, Washington, D.C., 1982.
19. Y. D. Wu and K. N. Houk, *J. Am. Chem. Soc.* **1987**, *109*, 906; A. E. Dorigo and K. N. Houk, *J. Am. Chem. Soc.* **1987**, *109*, 3698; D. C. Spellmeyer and K. N. Houk, *J. Org. Chem.* **1987**, *52*, 959; A. E. Dorigo and K. N. Houk, *J. Am. Chem. Soc.* **1987**, *109*, 2195.
20. G. S. Hammond, *J. Am. Chem. Soc.* **1955**, *77*, 334.
21. P. D. Bartlett and T. T. Tidwell, *J. Am. Chem. Soc.* **1968**, *90*, 4421; T. T. Tidwell, *J. Org. Chem.* **1974**, *39*, 3533;
22. H. C. Brown, *Boranes in Organic Chemistry*, Cornell University Press, Ithaca, New York, 1972, Chapter VIII.
23. K. B. Wiberg, *Angew. Chem. Int. Ed.* **1986**, *25*, 312.
24. C. Rüchardt and H.-D. Beckhaus, *Angew. Chem. Int. Ed.* **1985**, *24*, 529.
25. E. M. Engler, J. D. Andose, and P. v. R. Schleyer, *J. Am. Chem. Soc.* **1973**, *95*, 8005; J. D. Andose, E. M. Engler, J. P. Hummel, K. Mislow, and P. v. R. Schleyer, *QCPE* **1979**, *11*, 348.
26. N. L. Allinger and D. Y. Chung, *J. Am. Chem. Soc.* **1976**, *98*, 6798; N. L. Allinger, *J. Am. Chem. Soc.* **1977**, *99*, 8127; N. L. Allinger, and Y. Yuh, *QCPE* **1979**, 395.
27. P. Müller, J. Blanc, and J. Mareda, *Chimia* **1984**, *38*, 389; *Helv. Chim. Acta* **1986**, *69*, 635.
28. P. Müller and J. Mareda, *Tetrahedron Lett.* **1984**, *25*, 1703; *Helv. Chim Acta* **1985**, *68*, 119.
29. R. C. Bingham, Ph.D. thesis, Princeton University, 1970.
30. P. Müller, J. Blanc, and J. Mareda, *Chimia*, **1985**, *39*, 234.
31. J. C. Traeger and R. G. McLoughlin, *J. Am. Chem. Soc.* **1981**, *103*, 3647.
32. P. Müller and J. Mareda, unpublished
33. (a) H. D. Beckhaus, *Angew. Chem. Int. Ed. Engl.* **1978,** *17*, 593; (b) C. Rüchardt, V. Golzke, and G. Range, *Chem. Ber.*, **1981**, *114*, 2769.
34. P. Müller and J. Mareda, *Helv. Chim. Acta*, **1987**, *70*, 1017.
35. J. Slutsky, R. C. Bingham, P. v. R. Schleyer, W. C. Dickason, and H. C. Brown, *J. Am. Chem. Soc.* **1974**, *96*, 1969.
36. P. Müller, J. Blanc, and J. Mareda, *Chimia*, **1987**, *41*, 399.
37. P. Müller, J. Blanc, and J. Mareda, unpublished, supplementary material to Ref. 27.
38. J. March, Advanced Organic Chemistry, 3rd ed. Wiley, New York, 1985.
39. T. Laube, *Angew. Chem. Int. Ed.* **1986**, *25*, 349.
40. D. E. Sunko, *Croat. Chem. Acta,* **1980,** 53, 525; D. E. Sunko, S. Hirsl-Starcevic, S. K. Pollack, and W. J. Hehre, *J. Am. Chem. Soc.* **1979,** *101*, 6163.
41. T. Laube, *Angew. Chem. Int. Ed. Engl.* **1987**, *26*, 560.

42. J. D. Goddard, Y. Osamura, and H. F. Schafer III, *J. Am. Chem. Soc.* **1982, 104,** 3258.

43. E. M. Engler, D. Farcasiu, A. Sevin, J. M. Cense, and P. v. R. Schleyer, *J. Am. Chem. Soc.* **1973,** *95*, 5769; T. M. Gund, P. v. R. Schleyer, G. D. Unruh, and G. J. Gleicher, *J. Org. Chem.* **1974,** *39*, 2995.

44. J. J. Solomon and F. H. Field, *J. Am. Chem. Soc.* **1976**, *98*, 1567.

45. R. H. Staley, R. D. Wieting, and J. L. Beauchamp, *J. Am. Chem. Soc.* **1977**, *99*, 5964; J. F. Wol, R. H. Staley, I. Koppel, M. Taagepera, R. T. McIver, Jr., J. L. Beauchamp, and R. W. Taft, *J. Am. Chem. Soc.* **1977**, *99*, 5417.

46. S. G. Lias, J. F. Liebman, and R. D. Levin, *J. Phys. Chem. Ref. Data*, **1984**, *13*, 695; T. Baer, *J. Am. Chem. Soc.* **1980**, *102*, 2482; H. M. Rosenstock, R. Buff, M. A. A. Ferreira, S. G. Lias, A. C. Parr, R. L. Stockbauer, J. L. Holmes, *J. Am. Chem. Soc.* **1982**, *104*, 2337.

47. D. H. Aue and M. T. Bower, in *Gas Phase Chemistry*, M. D. Owers, Ed. Academic Press, Vol. 2, Chap. 9, New York, 1979.

48. G. H. Kruppa and J. L. Beauchamp, *J. Am. Chem. Soc.* **1986**, *108*, 2162.

49. J. M. Harris, S. G. Shafer, and S. D. Worley, *J. Comput. Chem.* **1982**, *3*, 208.

50. P. Houriet and H. Schwarz, *Angew. Chem. Int. Ed. Engl.* **1979**, *18*, 951.

51. J. Allison and D. P. Ridge, *J. Am. Chem. Soc.* **1979**, *101*, 4998.

52. F. P. Lossing and J. L. Holmes, *J. Am. Chem. Soc.* **1984**, *106*, 6917.

53. R. B. Sharma, D. K. Sen Sharma, K. Hiraoka, and P. Kebarle, *J. Am. Chem. Soc.* **1985**, *107*, 3747.

54. N. L. Allinger, *J. Am. Chem. Soc.* **1977,** *99*, 8127.

55. G. Wenke and D. Lenoir, *Tetrahedron* **1979**, *35*, 489.

56. J. Mareda, unpublished results.

57. W. L. Jorgensen, *J. Am. Chem. Soc.* **1978**, *100*, 1049.

58. P. M. Viruela-Martin, I. Nabot-Gil, R. Viruela-Martin, and J. Planelles, *J. Chem. Soc. Perkin Trans. 2* **1987**, 307.

59. R. C. Bingham, M. J. S. Dewar, and D. H. Lo, *J. Am. Chem. Soc.* **1975**, *97*, 1294.

60. M. J. S. Dewar and H. Rzepa, *J. Am. Chem. Soc.* **1977,** *99*, 7432.

61. C. R. Castro, R. Dutler, A. Rouk, and H. Wieser, *J. Mol. Struct.* **1987**, *152*, 241.

62. J. M. Schulman and R. L. Disch, *Tetrahedron Lett.* **1985**, 5647.

63. W. J. Hehre, in *Methods in Electronic Structure Theory*, H.F. Schafer III, Ed., Plenum Press, New York, 1977, p. 277.

64. W. J. Hehre and J. A. Pople, *J. Am. Chem. Soc.* **1975**, *97*, 6941.

65. L. Radom, J. A. Pople, V. Buss, and P. v. R. Schleyer, *J. Am. Chem. Soc.* **1972**, *94*, 311.

66. L. Radom, W. A. Lathan, W. J. Hehre, and J. A. Pople, *J. Am. Chem. Soc.* **1971**, *93*, 5339.

67. G. A. Olah, G. K. Surya Prakash, J. G. Shih, V. V. Krishnamurthy, G. D. Matescu, G. Liang, G. Sipos, V. Buss, T. M. Gund, and P. v. R. Schleyer, *J. Am. Chem. Soc.* **1985**, *107*, 2764.

68. G. A. Olah, G. Liang, P. v. R. Schleyer, W. Parker, and C. I. F. Watt, *J. Am.*

Chem. Soc. **1977**, *99*, 966; G. A. Olah, G. K. S. Prakash, J. G. Shih, V. V. Krishnamurthy, G. D. Matescu, G. Liang, G. Sipos, V. Buss, T. M. Gund, and P. v. R. Schleyer, *J. Am. Chem. Soc.* **1985**, *107*, 2764.

69. E. M. Arnett, C. Petro, and P. v. R. Schleyer, *J. Am. Chem. Soc.* **1979**, *101*, 522; E. M. Arnett and N. J. Pienta, *J. Am. Chem. Soc.* **1980**, *102*, 3329.
70. G. A. Olah, G. K. Surya Prakash, W. D. Fessner, T. Kobayashi, and L. A. Paquette, *J. Am. Chem. Soc.* **1988**, *110*, 8599.
71. C. A. Grob, *Acc. Chem. Res.* **1983**, *16*, 426; G. A. Olah, G. K. Surya Prakash, and M. Saunders, *Acc. Chem. Res.* **1983**, *16*, 440; H. C. Brown, *The Non-classical Ion Problem*, Plenum Press, New York 1977.
72. S. Ikegami, D. L. Vander Jagt, and H. C. Brown, *J. Am. Chem. Soc.* **1968**, *90*, 7124.
73. H. C. Brown, *Acc. Chem. Res.* **1983**, *16*, 432.
74. T. W. Bentley, C. T. Bowen, W. Parker, and C. I. F. Watt, *J. Am. Chem. Soc.* **1979**, *101*, 2486.
75. J. S. Lomas, P. K. Luong, and J. E. Dubois, *J. Org. Chem.* **1979**, *44*, 1647.
76. P. Müller and J. Mareda, *J. Comput. Chem.* **1989**, *10*, 863.
77. D. Lenoir, Y. Apeloig, D. Arad, and P. v. R. Schleyer, *J. Org. Chem.* **1988**, *53*, 661; M. Saunders, L. A. Telkowski, and M. R. Kates, *J. Am. Chem. Soc.* **1977**, *99*, 8070.
78. M. R. Smith and J. M. Harris, *J. Org. Chem.* **1978**, *43*, 3588; D. Farcasiu, *J. Org. Chem.* **1978**, *43*, 3878; D. Lenoir and R. M. Frank, *Chem. Ber.* **1981**, *1145*, 3336.
79. H.-J. Schneider, N. Becker, G. Schmidt, and F. Thomas, *J. Org. Chem.* **1986**, *51*, 3602.
80. P. S. Engel, *Chem. Rev.* **1980**, *80*, 99.
81. V. Golzke, F. Groeger, and C. Rüchardt, *Nouv. J. Chim.* **1978**, *2*, 169.
82. H. D. Beckhaus, unpublished; see also A. Peyman, E. Hickl, and H. D. Beckhaus, *Chem. Ber.* **1987**, *120*, 713.
83. M. Schmittel and C. Rüchardt, *J. Am. Chem. Soc.* **1987**, *109*, 2750.
84. J. S. Lomas, *J. Org. Chem.* **1987**, *52*, 2627; **1985** *50*, 4291; J. S. Lomas and J. E. Dubois, *J. Org. Chem.* **1982**, *47*, 4505.
85. M. R. Imam and N. L. Allinger, *J. Mol. Struct.* **1985**, *126*, 345.

7 Stabilization of Cage Compounds through Steric Hindrance by *tert*-Butyl Groups

GÜNTHER MAIER, HARALD RANG, and DIETER BORN

Institute for Organic Chemistry
University of Giessen
Giessen, Federal Republic of Germany

INTRODUCTION

According to theory, tetrahedrane (**1**) is the most highly strained of the formally saturated hydrocarbons. In other words, the equating of the tetrahedron with the element fire[1] has a chemical parallel. If only a minimal lifetime is expected of the still unknown unsubstituted tetrahedrane (**1**), it is appropriate to replace the hydrogen atoms by *t*-butyl groups. Examples of this trick of attaining kinetic stabilization through steric hindrance by attack of external reaction partners are legion. Apart from the intermolecular effect of mutual shielding, the voluminous substituents in tetra-*t*-butyltetrahedrane (**2**) play a second role of decisive importance for the stability.

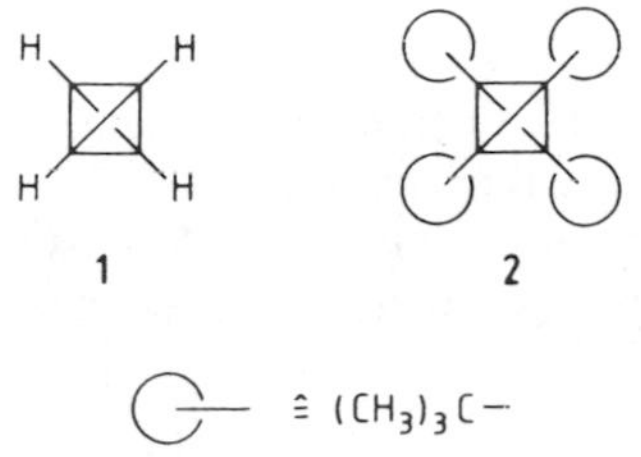

H H H H

1 2

≙ $(CH_3)_3C-$

The intramolecular repulsion of the four *t*-butyl groups is at a minimum when their mutual distance is at a maximum. This condition is fulfilled by the T_d symmetry of the tetrahedron. Any other imaginable arrangement causes the substituents to be closer to each other. The breaking of a tetrahedrane bond in **2** must equally lead to strong steric strain. In other words, the favored spherical arrangement of the four peripheral groups in **2** forces the tetrahedral geometry on the molecule. The two *t*-butyl groups, which would recede from each other on stretching a ring bond, are forced back again by the other two. This "corset effect" must stabilize **2** relative to **1**. If, however, only one of the substituents in **2** is replaced by a smaller group, the situation must change drastically. Since, in general, alkyl groups lower the thermal stability of a C—C bond, all tetrahedranes less highly substituted than **2** should undergo ring opening more easily than the parent molecule itself.

This concept has been the guideline for the synthesis of tetrahedrane (**2**), which we achieved 12 years ago.[2]

In this chapter we show how the idea arose that **2** might be a stable compound. We also include some additional examples in which steric repulsion plays a dominant role, and it can be seen as a kind of progress report completing our earlier summary.[3]

TETRA-*t*-BUTYLTETRAHEDRANE

History

The long route to synthesizing **2** began with two accidental observations. The first goes back to an experiment in which we tried to define the "baseline" for the ultraviolet (UV) absorption of certain cyclobutene derivatives. The model that we selected was tetramethylcyclobutenedicarboxylic anhydride (**3**).

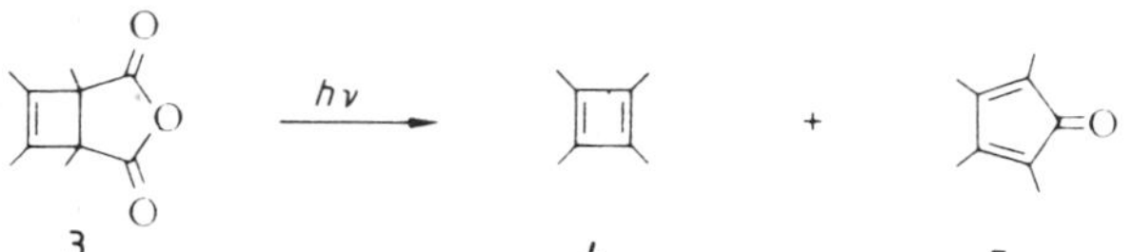

Surprisingly, this anhydride exhibits a pronounced maximum at 241 nm because of the homoconjugative interaction of the C=O group with the C=C double bond. The position of the maximum shifts towards longer wavelengths as the number of alkyl groups increases. It was obvious that due to this absorption the anhydride (**3**) should be an ideal candidate for a photofragmentation at low temperatures.

In the very first experiment we detected the formation of cyclobutadiene (**4**) and cyclopentadienone (**5**) when **3** was irradiated in an organic glass at –196°C.[4]

A second important experience was gained during the photochemical cleavage of *t*-butylated cyclobutenedicarboxylic anhydrides. The antiaromaticity of a cyclobutadiene like **4** demands a high reactivity, particularly with regard to a Diels–Alder-like dimerization. That is what really happens on thawing a matrix containing **4** and **5**.[4]

Introduction of *t*-butyl groups represents a method of preparing sterically hindered but electronically intact cyclobutadienes. Therefore, we attempted to apply the photofragmentation reaction to *t*-butylated anhydrides with the aim of preparing not only stable cyclopentadienones, but possibly even isolable cyclobutadienes.

The ideal starting material for the preparation of tri-*t*-butylcyclobutadiene, which, according to the above arguments, might be stable enough to be detectable as the monomer in solution, would appear to be the anhydride (**6**). However, its photocleavage leads exclusively to the two cyclopentadienones **7** and **8**. Neither cyclobutadiene (**9**) nor its dimer is found.

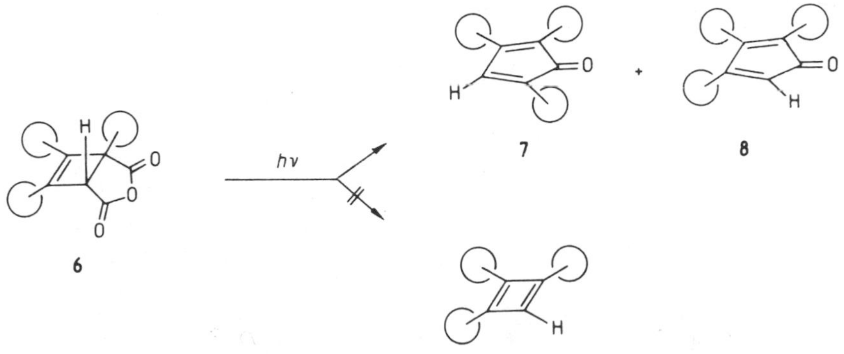

H O O O H O O H H

6 $h\nu$ 7 + 8

9

We explain this disappointing reaction course as being due to preferential formation of the former of the two conceivable diradical intermediates **10** and **11** because in this form the third *t*-butyl group possesses a staggered orientation. Elimination of CO from **10** would necessarily lead to an eclipsed arrangement of all *t*-butyl groups and is therefore precluded.

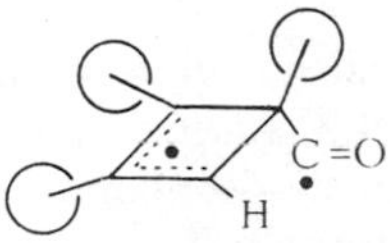

C=O H H C=O

10 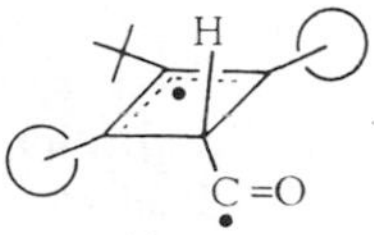 11

At this stage our efforts had come to an end. It was the perseverance of a highly talented student[5a] that was finally rewarded. He repeated the irradiation of **7** and **8** in spite of the fact that both dienones were reported in an earlier thesis to be photostable even at low temperatures. The new finding, which finally led to tetrahedrane **2** was the following: In contrast to the positional isomer **7**, **8** is photolabile at 77 K and gives the desired cyclobutadiene (**9**).[5b,6]

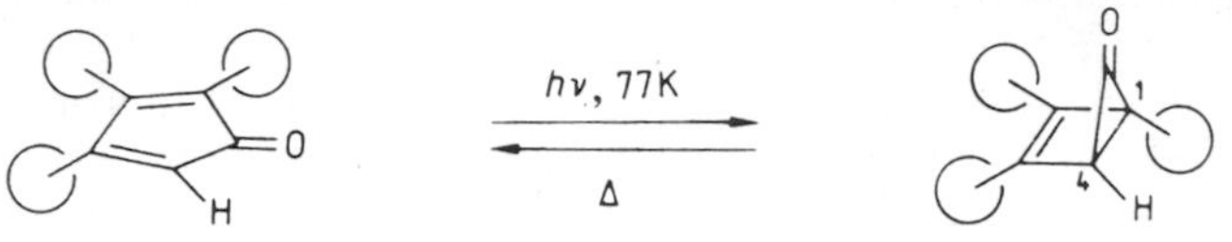

The mysterious origin of the strange temperature dependence of the photochemical behavior of **8** was solved:[7] Irradiation of **8** at 10 K gives the housenone (**12**), which subsequently is transformed into **9** via CO elimination. The thermal ring opening of **12** to **8** already takes place at –80°C.[8] Therefore, at room temperature the stationary concentration of **12** is so low that no formation of **9** is observable. The fact that the isomeric cyclopentadienone (**7**) does not undergo a photofragmentation at any temperature (room temperature, organic matrix at 77 K, or argon matrix at 10 K) demonstrates the importance of the mutual orientation of the *t*-butyl groups. Apparently, the corresponding housenone (**13**) is so strained that it is either not formed by photochemical excitation, or rapidly undergoes ring opening to **7** even at 10 K. This view is supported by theoretical studies.[9] Modified neglect of differential overlap (MNDO) calculations give a C1—C4 distance of 1.587 Å for **12**. The corresponding bond in **13** must be longer (1.647 Å) and hence weaker.

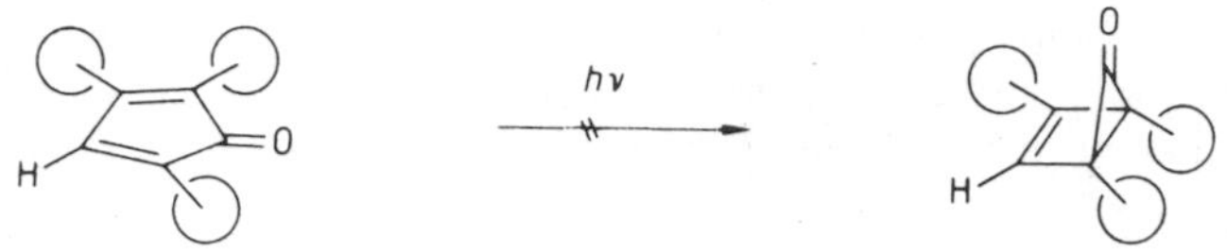

A peculiarity of cyclobutadiene (**9**) concerns the stereochemistry of its dimerization. In contrast to previous belief,[4] the dimer does not have the syn structure (**14**) expected from a reaction of the Diels–Alder type, but is in reality the anti-tricyclic compound **15**.[10]

It is again the influence of the *t*-butyl groups that governs the reaction path. Eventually a chain reaction catalyzed by the radical cation **16**[3]—as the "effectiveness" of the dimerization would suggest—is operating. In some way or other, traces of the radical cation (**16**) are formed by external effects. This reacts with an additional molecule of **9**—probably via a π-complex between

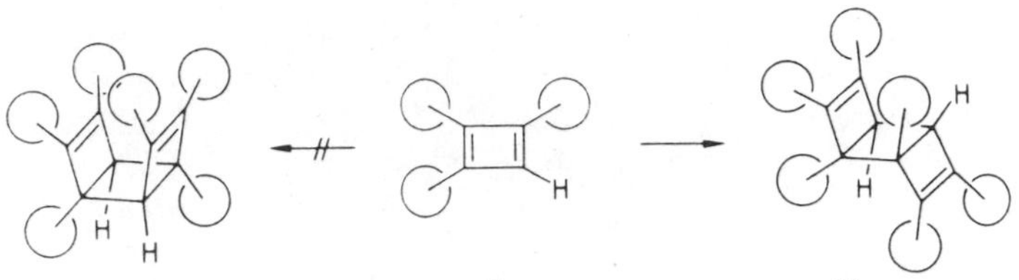

14 9 15

the two species—to give the coupling product **17**, which undergoes a redox process with **9** again, leading to renewed liberation of a radical cation **16**.

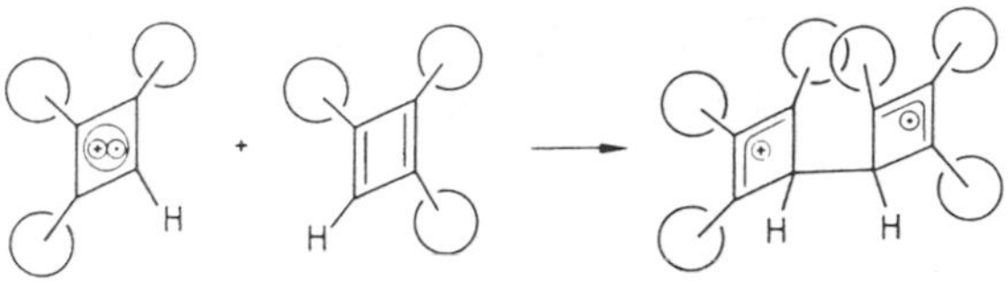

16 9 17

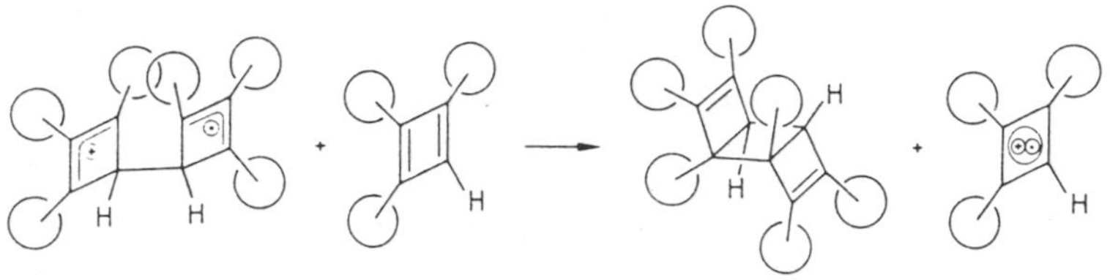

17 9 15 16

Synthesis

From the previous observations with the cyclopentadienones (**7** and **8**), it may be inferred that introduction of a fourth *t*-butyl group should drastically influence the ability to undergo photoisomerization. This expectation finally led to the successful synthesis of the tetrahedrane (**2**):[2] the photochemical behavior of tetra-*t*-butylcyclopentadienone (**18**) is entirely different from that of **8**. Excitation of the dienone (**18**) isolated in argon with 254-nm light results in exclusive crisscross addition to give the tricyclopentanone (**19**) as observed by infrared (IR) spectroscopy. Continued irradiation leads to carbon monoxide elimination. At the same time, the typical band of ketene (**20**) appears. This ketene is likewise photolabile and eliminates CO, albeit very slowly. Irradiation in an organic matrix at 77 K permits isolation of the hydrocarbons formed.

18 $\xrightarrow{h\nu}$ 19 $\xrightarrow{h\nu}$ 20

18 $\xrightarrow{h\nu}$ 21; 19 $\xrightarrow{h\nu}$ 2; 20 $\xrightarrow{h\nu}$ 22

21 $\xleftarrow{\Delta}$ 2; 21 $\xrightarrow{h\nu}$ 2; 2 $\xrightarrow{h\nu, \text{ sens}}$ 21; 2 $\xrightarrow{h\nu}$ 22

Apart from di-*t*-butylacetylene (**22**), which is the only end product of the room temperature irradiation, Pfriem[2] isolated the tetrahedrane (**2**) as stable, colorless crystals. On heating to 135°C, **2** is transformed into the corresponding cyclobutadiene (**21**). This valence isomerization is photochemically reversible. It may be assumed, therefore, that the bicyclobutanediyl diradical primarily formed on irradiation of **19** is converted into the cyclobutadiene (**21**) by cleavage of the backbone bond; the cyclobutadiene then furnishes the tetrahedrane derivative (**2**) under the conditions of irradiation.

If one tries to synthesize a sterically highly hindered compound one runs into a dilemma. On the one hand the bulky groups are expected to protect the target molecule from external attack, on the other hand one has to overcome the steric hindrance during its synthesis. Therefore it is very often necessary to use "unusual" procedures. One example is the preparation of cyclopentadienone **18**,[11] the starting material for tetrahedrane (**2**): Action of bromine on the tri-*t*-butyl derivative (**8**) led, in spite of steric hindrance, to a dibromide that readily eliminates hydrogen bromide on treatment with potassium hydroxide to form the bromodienone (**23**). Replacement of the halogen in **23** by the alkyl group of *t*-butyllithium is possible only under "forbidden" conditions, that is, in 1,2-dimethoxyethane at room temperature (–50°C should not really be exceeded if reaction of the alkyllithium with solvent is to be avoided).

The unusual reaction conditions for the steps **23** → **18** have their reason in the inability of the primary addition product (**25**) to eliminate a bromide ion. In a preceding step, it must undergo a 1,5 shift of a *t*-butyl group (**25** → **24**). This is favored by enolate formation in **24**. However, such reactions require higher temperatures than are normally used in reactions with *t*-butyllithium. In the last step **24** readily eliminates Br^- to give the product (**18**).

In this context it is important to realize that the steric hindrance may be carried too far. The following is another example in connection with the synthesis of **2**: The route to **2** should also be applicable to other highly strained tetrahedranes. Unfortunately, however, this is not so. An examination

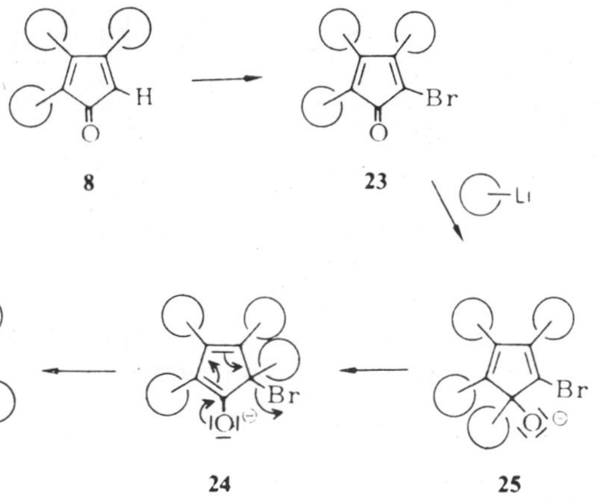

H Br O O
8 23 Li
Br Br O O
18 24 25

of the differently substituted cyclopentadienones (**26–36**)[12] showed that **26–31** are totally unchanged by irradiation (λ = 254 nm) in argon at 10 K. Only **32–34** undergo photoisomerization to the related tricyclopentanones of type **19** to an increasing extent. At least **34** undergoes the crisscross addition about as easily as the reference substance **18**. However, no tetrahedrane derivative is detectable on subsequent CO elimination. Although the extremely bulky adamantyl residue favors tricyclopentanone formation, it may already be too large to permit the planarization of the cyclobutadiene ring necessary for decarbonylation. Apparently, it is only in the tetra-*t*-butyl series, which also includes **35** and **36**, that everything fits nicely together.

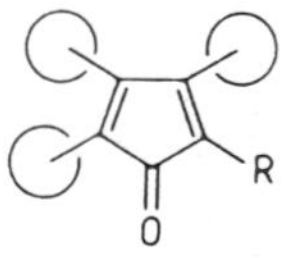

R O
26 — 36

Compound	R	Compound	R
26	Phenyl	**32**	Ethyl
27	Pentafluorophenyl	**33**	Isopropyl
28	*o*-Tolyl	**34**	1-Adamantyl
29	Benzyl	**35**	$C(CD_3)_3$
30	1-Phenylethyl	**36**	$^{13}C(CH_3)_3$
31	Trimethylsilyl		

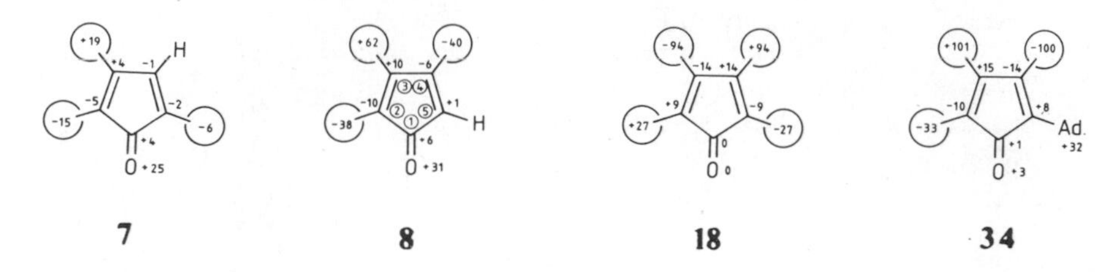

Fig. 1. Deformations (pm) of the cyclopentadienones **7**, **8**, **18**, and **34** relative to a "best" plane through the rings, determined by X-ray crystallography. +: above; -: below the plane. The distance to the "best" plane refers, in the case of the adamantyl moiety (Ad), to C1.

The very different photochemical behavior of the cyclopentadienones **7**, **8**, and **18** still requires an explanation. The X-ray structure analyses carried out by Boese provide an indication (Fig. 1).[13]

The structures demonstrate that the direction of the photochemically induced reaction is already given by the distortion of the rings in the ground state. The deviations are relatively small in the case of **7**. Therefore, no photoreaction takes place. In ketone **8** the three *t*-butyl groups and therefore also the neighboring ring carbon atoms are in a staggered arrangement. Here, less energy is required in order to achieve photoisomerization. From the structure, one could expect a crisscross reaction. Nevertheless, when a housenone is formed, this is because it is preferred over the tricyclopentanone on the grounds of ring strain. According to MNDO calculations,[14] the enthalpy of formation of **38** is 26.8 kcal/mol higher than that of **37**.

37 38 12 39

The introduction of three *t*-butyl groups reduces this energy difference between **39** and **12** to 4.8 kcal/mol. It is therefore no wonder that the tricycle (**39**) is known; however, the compound is completely photostable.[15] Following the trend, the tricyclopentanone (**19**) could already be more stable than the corresponding housenone in the tetra-*t*-butyl series. Furthermore, the structure of **18** (Fig. 1) implies that photoexcitation should lead to the tricycle (corset effect, see above). Since the deviations are larger still in the adamantyl derivative (**34**), the same reaction should be observed here, in agreement with experiment.

Another aspect has to be kept in mind: It is not only the size of a group that is important for the usage of the concepts discussed above. For instance, trimethylsilyl groups are bigger than *t*-butyl substituents. Protection from the attack of reagents should be better as long as the steric factor is dominating. But trimethylsilyl groups are sensitive against nucleophiles due to a S_N2-type substitution. Both effects can be used for the construction of polycyclic compounds.[16-20] As far as the "corset effect" is concerned, trimethylsilyl groups are less effective than *t*-butyl substituents. This difference results from the fact that the C—Si bond is longer than the C—C bond.

Last, but not least, one has to remember that trimethylsilyl groups have a high tendency for migration. For instance, tetrakis(trimethylsilyl)cyclopentadienone (**41**) behaves completely different compared to the analogous fourfold *t*-butyl substituted derivative **18** on photochemical excitation. The final product of the photolysis (Hg low-pressure lamp, rigisolve matrix, 77 K, 200 h) is not the tetrahedrane (**40**), but the tetrakis(trimethylsilyl)butatriene (**42**).[21]

hν
hν
C=C=C=C
40
41
42
Si— ≡ $(CH_3)_3Si$—

Additional Attempts Towards the Synthesis of Tetra-*t*-butyltetrahedrane

The long route to **2** described above is very tedious, and due to the low yields of some of the steps involved, the total amount of **2** available has to be small. Therefore, it would be nice to have a better procedure. One attractive

$BF_4^{\ominus}$
Li
CN_2
44
$C-N_2$
hν, Δ
43
45
22
2
46
21

way would be an entry via diazo compound **45**. The literature does not report any syntheses involving anions of alkylated diazomethanes. This is probably due to the belief that reagents of this type should be highly reactive. If this is true, anion **44** would be expected to be able to overwhelm the steric hindrance in the cyclopropenylium salt (**43**). Indeed, reaction between **43** and **44** results in the relatively stable diazo compound **45**.[12] Pyrolysis or photolysis of **45** yields only di-*t*-butylacetylene (**22**) and not even a trace of cyclobutadiene (**21**) or tetrahedrane (**2**).

Comparison with the reported[15] formation of cyclobutadienes **9** and **49** starting from **47** or **48** indicates that carbene **46** is "overloaded" with *t*-butyl groups. The carbenic center bearing a *t*-butyl group is no longer able to add to the double bond or to insert into the single bond of the three-membered ring.

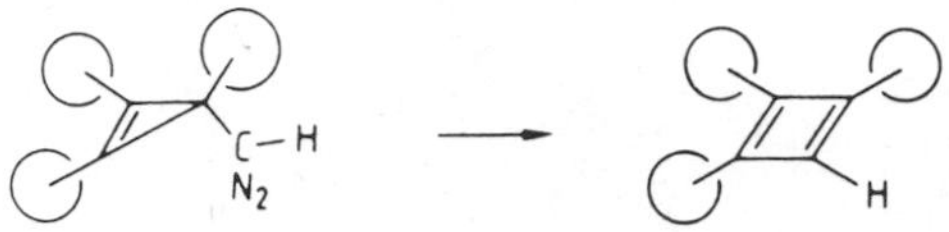

47 9

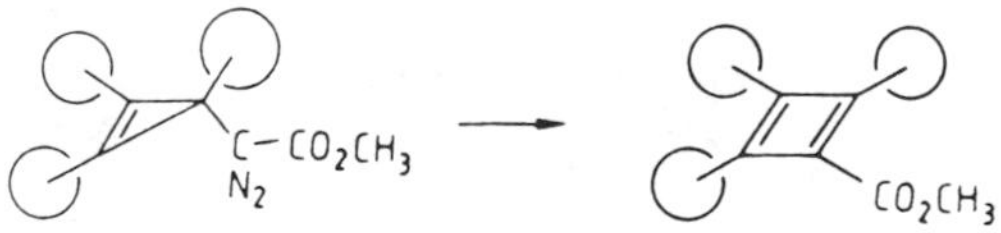

48 49

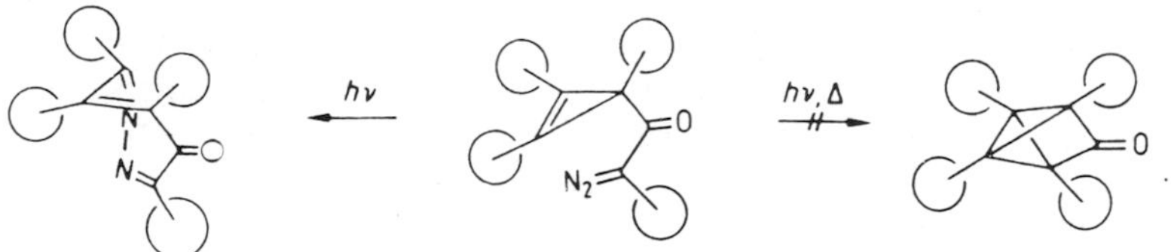

50 51 19

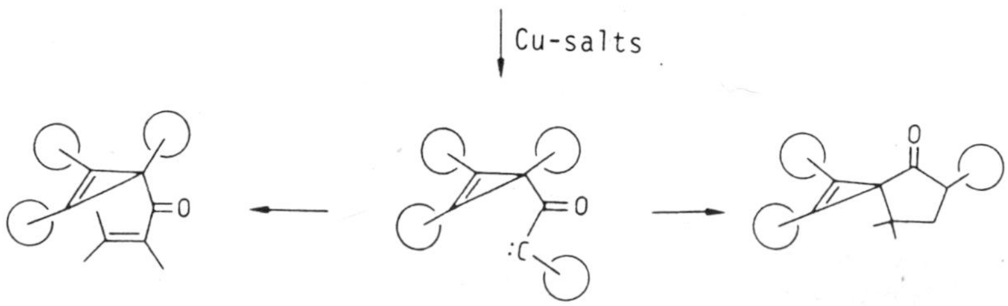

52 53 54

For the same reason diazoketone (**51**) does not yield tricyclopentanone (**19**) on photolysis or thermolysis. This reaction works nicely—as shown again by Masamune[15b,15c]—if the fourth *t*-butyl group on the diazo carbon is missing (formation of **39**). On the contrary, irradiation of **51** gives diazatropone (**50**). On heating in the presence of Cu(I) salts **51** splits off nitrogen, and two new compounds **52** and **54** can be isolated. Obviously in **53** the usual side reactions of a *t*-butyl substituted carbene, namely, insertion into a C—C (formation of **52**) or a C—H bond (formation of **54**), wins against an addition to the highly hindered double bond in the three membered ring. This borderline case illustrates an important principle: The best way to prepare highly substituted cage compounds is the elimination of small particles from precursors, which already contain all the bulky substituents. Addition or substitution reactions in the very last step do not work.

Additional Attempts towards the Synthesis of Tri-*t*-butyl(trimethylsilyl)tetrahedrane

As pointed out previously, tri-*t*-butyl(trimethylsilyl)cyclopentadienone (**31**) surprisingly failed as a starting material for the corresponding tetrahedrane. So far we were pleased when it turned out[22] that the lithium salt of trimethylsilyldiazomethane (**55**)[23] reacted easily with tri-*t*-butylcyclopropenylium tetrafluoroborate (**43**).

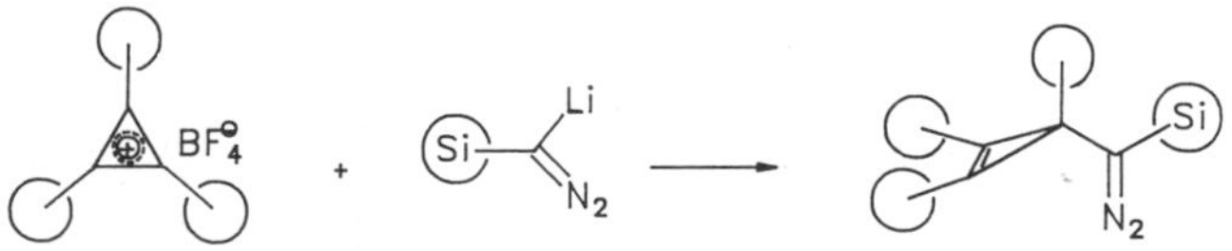

43 55 56

On irradiation, the product (**56**)[12a] behaves in a similar manner to the tetra-*t*-butyl derivative (**45**) and splits into **22** and **58** via the carbene (**57**). But, on flash thermolysis, the acetylenes are formed only in minor quantities, and a new reaction path seems to operate. Astonishingly (trimethylsilyl)cyanide (**65**) is one of the fragments. The other one has to be tri-*t*-butylazete (**64**)—a compound that has recently been described by Regitz[24]—as is shown by its adduct with dimethylacetylenedicarboxylate.[25] Eventually, Dewar pyridazine (**61**)[26a] functions as the precursor for **64** and **65**/**66** (thermal equilibrium).[27]

The situation is again different if the diazo compound (**56**) is heated in substance or in solution (e.g., benzene). Under these conditions **56** isomerizes exclusively and a green compound with the same composition as the starting material can be isolated. The spectroscopic data of this strange product are in accordance with structure **67**. This isomer is also formed via **68** if one adds nitrile **65**/isonitrile **66** to **64** independently (cyclobutadiene **49** adds isonitriles in the same way[26b]). In other words, these fragments also have to be the intermediates in the thermolysis of diazo compound **56**.

The reaction scheme given for the thermolysis of **56** is supported by the isolation of two iron carbonyl complexes **62** and **63** when **56** was heated in benzene together with diiron nonacarbonyl.

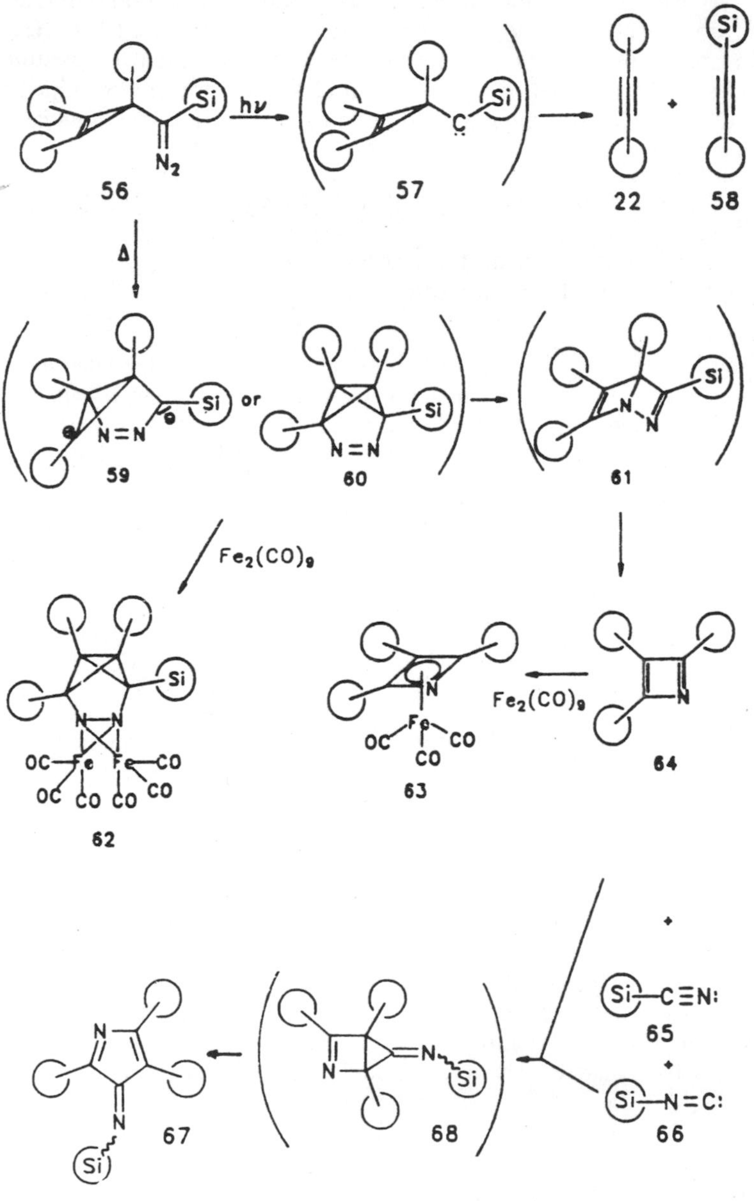

The addition of the diazo nitrogen atom to the double bond of **56** is analogous to the photochemical conversion of diazoketone **51** into diazatropone (**50**). The diazogroup is small enough to reach the π-system. The cheletropic addition of the carbenic center after nitrogen elimination is not possible for steric reasons. Under those circumstances the carbene undergoes "easier" reactions. This is another borderline case in the synthesis of sterically crowded compounds. Another disappointing finding should be added: All efforts to isolate cyclobutadiene (**69**) or tetrahedrane (**70**) starting from the benzvalene-type precursor **62**, which has been available only in small amounts, have failed.

If the thermal decomposition of **56** is catalyzed with CuCl, the situation changes dramatically. This is shown by the formation of adducts **71**–**73** when maleic anhydride, dimethyl acetylenedicarboxylate,[28a] or diiron nonacarbonyl[28b] are added to the reaction mixture after deazotation of **56**. The structure of complex **73** has been determined by X-ray diffraction.[13]

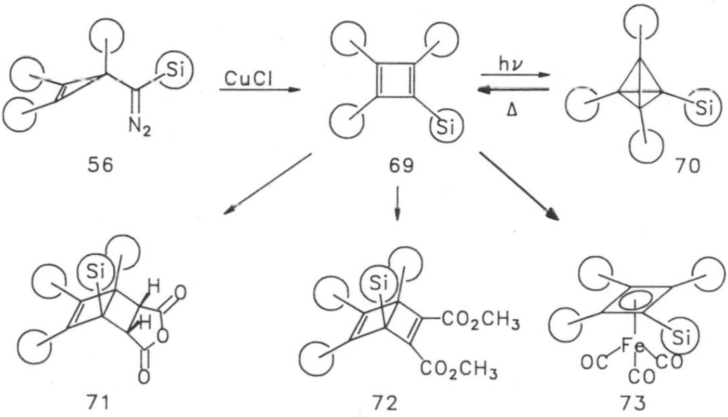

Our perseverance in looking for the free cyclobutadiene (**69**) was finally rewarded: Probably the cyclobutadiene present in the reaction mixture of the CuCl catalyzed decomposition is not completely free but still complexed with CuCl. In this situation it appeared attractive to add bis(diphenylphosphino)-ethane, a reagent that was used by Krebs[45,46] in order to release the first example of a kinetically stabilized cyclobutadiene from the corresponding $PdCl_2$ complex. Indeed, if the diphosphine is added to the crude reaction product, it is possible to isolate the target molecule **69** by sublimation. All spectroscopic data are in agreement with the proposed structure.

It is even possible to take the final step, namely, the photochemical crisscross photoisomerization of **69** to tetrahedrane (**70**). On irradiation ($\lambda \geq 300$ nm) **69** undergoes the valence isomerization to tetrahedrane (**70**) quantitatively. This second, spectroscopically identified tetrahedrane derivative is even more stable than tetra-*t*-butyltetrahedrane (**2**), in other words: The barrier for the thermal backreaction **70** $\rightarrow$ **69** is higher as in the case **2** $\rightarrow$ **21**. The new tetrahedrane melts at 179°C under the formation of cyclobutadiene **69**.[28b,c,d] Since the corset effect in **70** should not be stronger than in **2** (longer C—Si bond) it is obvious that a trimethylsilyl group is able to stabilize a tetrahedrane skeleton also electronically.

Structure and Spectroscopic Properties

Because of the erroneous announcements of successful synthesis of compounds with the tetrahedrane structure, we endeavored to gather as much data as possible and thus prove the structure of **2** unambiguously. The first indication was given by the ^{13}C NMR spectrum [in $CDCl_3$; three signals at δ 9.27 (ring C), 27.16, *C*$(CH_3)_3$, and 31.78 (CH_3)]. The observation that the quaternary carbon atoms of the *t*-butyl groups appeared at higher field than the primary ones was a first indication of the peculiarity of **2**. The bond between the *t*-butyl groups and the neighboring carbon atoms must possess a high degree of *s* character [as in di-*t*-butylacetylene (**22**)]. The high field position of the signal for the ring carbon atoms is even more revealing. It is in exact agreement with expectations.[29] Naturally, the ^{1}H NMR spectrum of **2** features only a single signal (even in a nematic phase)[30] at δ 1.18 $(CDCl_3)$. The mass spectrum (field desorption) gives the correct molecular mass (*m/z* 276.1). The PE (photoelectron) spectroscopic examination by Heilbronner et al.[31] played an important role in the structural analysis of **2** and **21**. As required, the lowest energy band of **2** (centered at **I** ≈ 7.9 eV) is a doublet due to Jahn–Teller splitting. Cyclobutadiene (**21**) shows a singlet at **I** = 6.35 eV.

In addition to the spectroscopic properties two investigations of isotopically labeled derivatives of **2** should be mentioned. Saunders, for example, queried whether in the case of **2** perhaps a dynamic system of rapidly equilibrating singlet diradicals exists rather than a static molecule. The tetrahedrane derivative of **2** possessing one perdeuterated *t*-butyl group is obtainable from the cyclopentadienone (**35**),[12b] and the isotopic perturbation observable by ^{13}C NMR spectroscopy was studied in this regard.[32] However, no effect over and above an "intrinsic shift" was detectable. Accordingly, **2** possesses a rigid, fixed structure. Another important point concerns the electron distribution in **2**. ^{13}C—^{13}C coupling constants can give information on this. The couplings are observable by labeling one of the *t*-butyl groups in **2** with ^{13}C at the quaternary center. The required substance can be obtained from the dienone **36**.[12b,33] The analysis of the ^{13}C NMR spectrum carried out by Lüttke et al.[34] allows the following conclusions to be drawn: The axes of the endocyclic hybrid orbitals ($sp^{5.20}$ hybrids) are bent outward by 28.4° with respect to the lines connecting the nuclei (record value for banana bonds). The bond connecting the tetrahedral corner with the *t*-butyl group has a high degree of *s* character ($sp^{1.04}$ hybridization; closely related to *t*-butylacetylene). These bond properties are in good agreement with predictions.[35–37]

The remaining spectroscopic information has become less significant as a consequence of a recent X-ray structure analysis of **2**. After many unsuccessful attempts with the room temperature modification of **2** (plastic, i.e., soft and ductile crystals) Irngartinger et al.[38] succeeded in growing a more highly ordered modification at low temperature. Recent, impressive results by Irngartinger et al.[39] show that this low-temperature form of **2** is "stabilized" by entrapped gases (N_2, Ar). This has allowed the solution of the structural problem. The result shown in Fig. 2[38] should remove any remaining doubt about the correctness of structure **2**. The bond lengths in the tetrahedron are relatively short (average

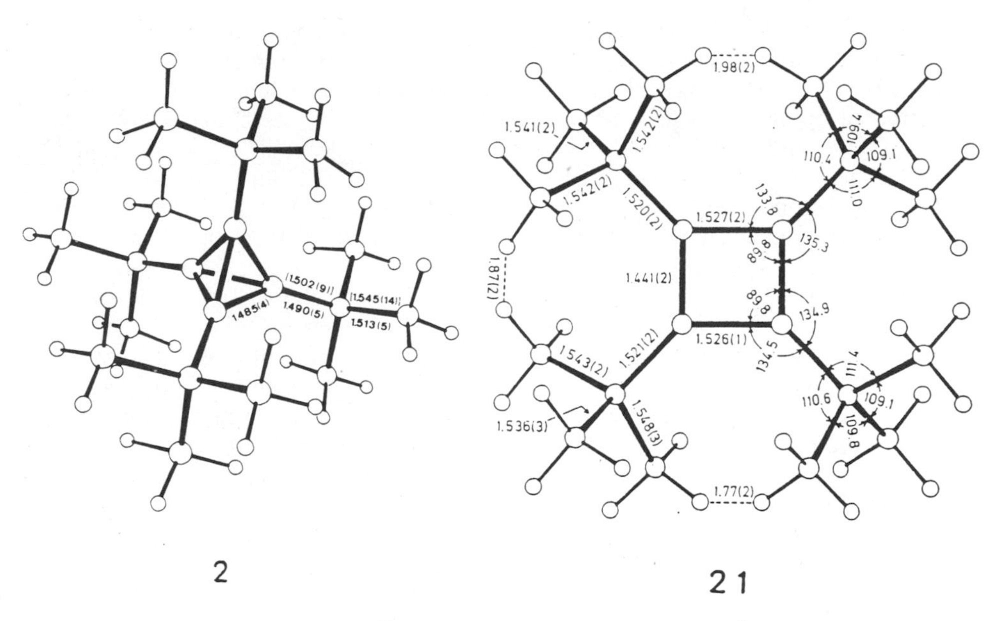

Fig. 2. Crystal structure of **2**[38] with averaged bond lengths ([Å]; deviations are in parentheses and corrected values are in square brackets); crystal structure of **21** at −150°C with bond lengths [Å] and angles [°].

1.485 Å[38]; 1.497 Å[39]) in comparison with normal C—C single bonds (1.54 Å) as a consequence of the bending of the bonds. The deformation densities determined for the Ar-clathrate of **2** demonstrate a bending of the ring bonds by 26°.[39] The agreement with the value mentioned above (28.4°) is remarkable. Incidentally, the quantum chemically calculated bond lengths vary between 1.470 and 1.520 Å.[35,37,40–42] Of these, the distance originally computed by Schulman and Venanzi[35] using an ab initio method comes closest to reality. The appearance of relatively large vibrational ellipsoids of the methyl carbon atoms could indicate that enantiomeric molecules of chiral T symmetry are disorderly arranged about the mirror plane in the crystal. Mislow[43] predicted such deviation from the maximal T_d symmetry to the lower T symmetry on the basis of force-field calculations for the free molecule **2**. The MNDO calculations have confirmed this.[41]

In this context, it is worth mentioning the curious history of the structural analysis of tetra-*t*-butylcyclobutadiene (**21**). At room temperature, slight but nevertheless significant alternation is found (1.482 and 1.464 Å) for the ring bonds. The quaternary carbon atoms of the *t*-butyl groups deviate from the ring plane by 0.37 Å above or below in an alternating fashion. Also, the four-membered ring itself is not planar but folded with a dihedral angle of 170°.[44] Whereas in other stable cyclobutadienes prepared by Krebs[45,46] and Masamune[47] the single bond is stretched, thereby causing a clear alternation (up to 0.26 Å)[48,49], the double bond in **21** is elongated, and therefore the difference in the bond lengths reduced. It has been argued that the stretching of the double bonds (caused by the amassing of four *t*-butyl groups) should lead to a shortening of the single bonds and hence a nearly square structure.[50] Doubt[51] about this X-ray structure of **21** with "almost equal bond lengths in the four-membered ring" motivated Irngartinger to repeat the structure analysis at –150°C.[52a] Under these conditions, the constitution given in Figure 2 results. The folding of the four-membered ring remains at -150°C also. However, the bond alternation is now found to be noticeably larger (0.086 Å). The strange "temperature dependence of the structure" can be explained by disordering phenomena. The experimentally determined bonding electron density distribution in **21**[49, 52a] shows bent bonds in the four-membered ring. The single bond having a lower electron density is more bent than the double bond. A part of the above-mentioned disorder is possibly still present even at –150°C. A detailed analysis of the anisotropic displacement parameter of ring carbon atoms of **21** shows that the measured data can be reconciled with an averaging, in the sense of a superposition of two rectangular rings rotated through 90° with the bond alternations (1.34 and 1.60 Å) found for the other cyclobutadienes mentioned above.[52b]

Physical Properties

A ball-shaped molecule like **2** should also show peculiar physical properties. Indeed, tetra-*t*-butyltetrahedrane forms colorless crystals, stable in air and melting at 135°C with red coloration; that is, at this temperature **2** is transformed into the valence isomeric cyclobutadiene (**21**). This extreme thermal stability appeared to make a tetrahedrane structure very unlikely. On the other hand, the high melting point is an expression of the high symmetry of cage compound **2**. A model based on the van der Waals radii leads to a sphere whose surface is covered with hydrogen atoms. The volatility is correspondingly high: **2** sublimes very easily in a water aspirator vacuum at room temperature.

Of course the ring strain of **2**, which must be an absolute record, is of most interest. There is no other molecule that could surpass tetrahedrane in terms of the strain theory postulated by Baeyer[53] just over 100 years ago.

What does the experiment tell? The enthalpy of combustion of **2** determined by Månsson[54] (ΔH°_c = 3099.8 ± 1.8 kcal/mol) leads to an enthalpy of formation in the condensed phase of ΔH°_f(c) = –10.9 ± 1.9 kcal/mol. The enthalpy of sublimation measured by Rüchardt et al.[55] (ΔH_{sub} = 17.1 ± 0.8 kcal/mol) gives an enthalpy of formation of **2** in the gas phase of ΔH°_f (g) = +6.2 ± 2.1 kcal/mol. Schleyer's increments[56] sum up to a calculated enthalpy of formation for the strain free molecule ΔH°_f (g) = –123.0 kcal/mol. Thus, the strain enthalpy of **2** is calculated as ΔH_s = 129.2 ± 2.1 kcal/mol (541 kJ/mol). Since the *t*-butyl groups contribute very little to the strain of **2**,[57] each ring bond in the tetrahedrane skeleton of **2** can be assigned a strain enthalpy of 21.5 kcal/mol. This is the highest value ever found.[58,59]

According to force-field calculations by Mislow,[60] the substitution by four *t*-butyl groups in the system **2/21** causes a strong preference for the tetrahedrane form (by 43.8 kcal/mol). The MNDO calculations[41] give the same result (preference by 39.2 kcal/mol; without *t*-butyl groups ΔH° = 45.9 kcal/mol, that is, **2** is only 6.7 kcal/mol higher in energy than **21**).

These numbers illustrate nicely the corset effect: The big energy difference between tetrahedrane (**2**) and cyclobutadiene (**21**) shrinks to a small value because of a strong destabilization of **21** (39.3 kcal/mol) and a stabilization of **2** (–4.5 kcal/mol, Ref. 60).

The isolation of **2** also allowed us to determine the activation barrier for the symmetry forbidden process **2** → **21**. An activation enthalpy of 20 kcal/mol was predicted for the process **2** → **21**.[41] It was possible to monitor the kinetics of isomerization of the tetrahedrane (**2**) to cyclobutadiene (**21**). The activation parameters thus derived were $\Delta G^{\ddagger} = 28.6 \pm 2.9$ kcal/mol, $\Delta H^{\ddagger} = 25.5 \pm 2.3$ kcal/mol, and $\Delta S^{\ddagger} = -10.3 \pm 5.8$ cal/mol·K.[30]

As described in more detail in the next section, the most important chemical property of **2** is its ease of oxidation.[61] This is to be expected from the exceptionally low ionization potential of **2**[31] and is confirmed by results from chemical ionization.[62] These show that **2** is not protonated by the plasma ion, but oxidized. It was now enticing to attempt the deliberate oxidation of **2**. Several aspects were of interest: (a) Is the same radical cation **74** formed from both **2** and **21**? (b) How high is the barrier for a possible isomerization of the radical cation **75** to **74**? (c) Is **74** a π or a σ radical? (d) What is the structure of the cyclobutadiene radical cation?

Bock et al.[4,61] succeeded in finding answers to all these questions.

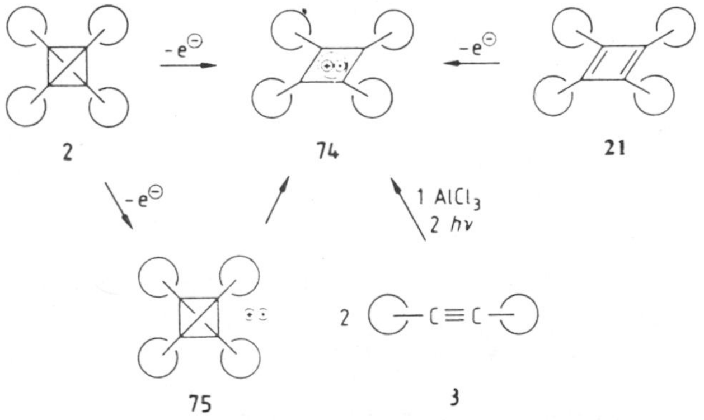

-e⊖ -e⊖ 2 74 21 -e⊖ 1 AlCl₃ 2 hν 75 2 —C≡C— 3

The compounds **2** and **21** afford the same radical cation (**30**) on one-electron oxidation with aluminum chloride in dichloromethane at –70°C. The radical cation (**75**) initially formed from **2** spontaneously isomerizes to **74**. As indicated by calculations, no activation barrier is expected for the conversion of **75** into **74** (**75** is a C_4 cluster with the "wrong" number of electrons). Thus, the radical cation of **2** behaves in a completely different manner from the neutral precursor (see above). The ESR spectrum of **74** features 36 equivalent, coupled protons and 4 equivalent ring carbon atoms. The coupling constants argue for a π radical. Moreover, the values measured [$a(H_\gamma)$ = 0.28 G; $a(^{13}C_\alpha)$ = 10.0 G] can be taken as indications of rooflike bending. This interpretation is supported by subsequently published work. Hogeveen et al.[63] prepared the radical cation of tetramethylcyclobutadiene by treatment of 2-butyne with aluminum chloride followed by irradiation of the σ complex so formed. This method can be applied to other acetylenes.[63,64] According to Davies et al.[64] it is even possible to generate the radical cation **74** in sufficient quantity for ESR spectroscopic detection, starting from di-*t*-butylacetylene (**22**). These studies also reveal that the cyclobutadiene radical cations are to be regarded as planar π radicals. The folding, in the case of **74**, is an exception, although it is in agreement with the structure of the electroneutral cyclobutadiene **21**. Furthermore, the bending becomes still more noticeable in the radical cation bearing four 1-adamantyl groups.[65,66]

The picture is rounded-off by experiments on electrochemical oxidation of **2** and **21** by Fox and Hünig et al.[67] The oxidation potentials determined by cyclic voltammetry are exceptionally low and again constitute record values for saturated and unsaturated hydrocarbons, respectively. At E_{pa} = 0.50 ± 0.10 V the tetrahedrane (**2**) gives a radical cation (**75**), which is transformed into **74** in an unmeasurably fast reaction.[61] In turn, **74** is reversibly formed from cyclobutadiene (**21**) at $E_{1/2}$ = 0.15 ± 0.10 V. Knowing this data, we examined[68] whether **2** might be convertible into cyclobutadiene (**21**) in a photochemically induced electron transfer reaction.[69] In fact, **2** isomerizes smoothly and quantitatively to the cyclobutadiene (**21**) on irradiation in tetrahydrofuran in the presence of catalytic amounts of 9,10-dicyanoanthracene, and even in the absence of the acceptor by irradiation in benzene solution. Thus, the equilibrium **2** → **21** can be entirely displaced in one or the other direction depending on the conditions of irradiation (see the earlier section on the synthesis of **2**); this is not an everyday occurrence.

Chemical Properties

In contrast to the valence isomeric cyclobutadiene (**21**), tetra-*t*-butyltetrahedrane (**2**) is insensitive to air and humidity under ordinary conditions. It is now known that only two reaction channels are open to **2**, namely, protonation and oxidation.[68,70,71] A proton can just fit in between the *t*-butyl groups. Apart from this, all reactions start with an oxidation step, and the chemistry of **2** is characterized by the primary formation of the radical cation **74**.

Steric shielding by the four *t*-butyl groups hinders the immediate conversion of **74**. In this regard **74** is an ideal molecule for the study of the chemical properties of radical cations.[72] It is also understandable why **2** is a potential precursor for **74**: Ordinary reactants cannot approach the ring skeleton of **2** sufficiently. This is less serious in the oxidation. The electron can overcome a distance of up to 15 Å[73] in SET (single electron transfer) reactions.

A first indication of how to "activate" the unreactive tetrahedrane derivative **2** was obtained in the attempt to catalyze the valence isomerization **2** → **21**, which requires heating to 135°C in the absence of a catalyst. The use of Ag^+ ions as a catalyst[74] did not result in the expected reaction; instead **2** was oxidized. The nature of the isolable endproducts, cyclobutenol (**79**) or ketone (**80**), depends on the reaction conditions. It has taken some effort to unravel the ensuing intricacies.[75] Silver tetrafluoroborate, silver hexafluoroantimonate, and silver perchlorate cause oxidation of **2** and **21** to the radical cation (**74**) (both valence isomers **2** and **21** were always employed in parallel in these experiments).

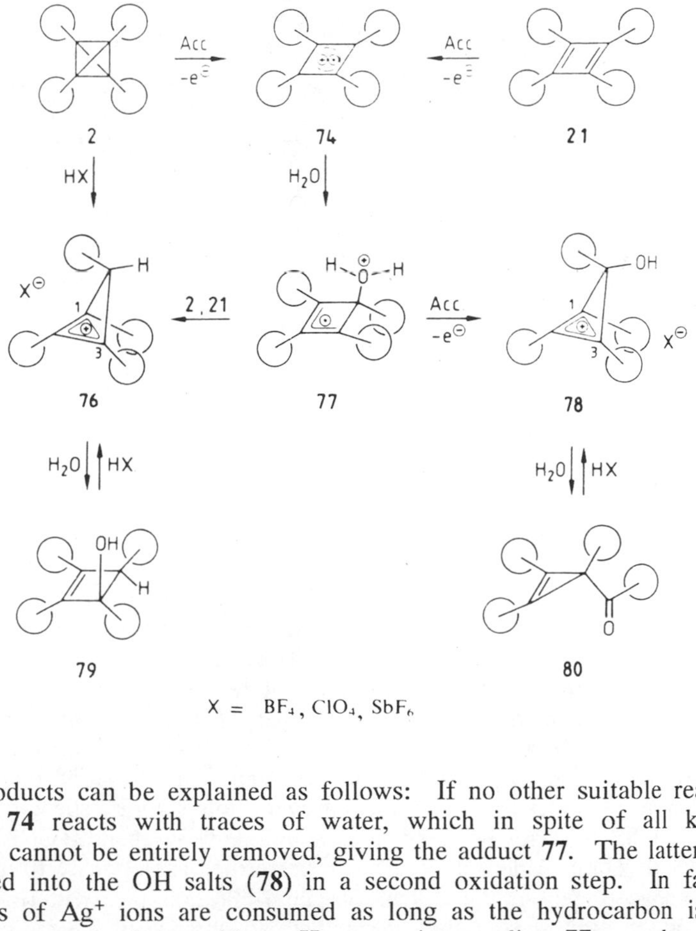

X = BF_4, ClO_4, SbF_6

The products can be explained as follows: If no other suitable reagent is available, **74** reacts with traces of water, which in spite of all kinds of precaution cannot be entirely removed, giving the adduct **77**. The latter can be transformed into the OH salts (**78**) in a second oxidation step. In fact, two equivalents of Ag^+ ions are consumed as long as the hydrocarbon is added slowly to an excess of the oxidant. However, intermediate **77** can also transfer a proton to **2** or **21**. This latter reaction occurs on inverse addition. Here, radical cation **77** acts as an acid toward **2** or **21** still present.[76] In this manner the salts **76** are formed. Hydrolysis of the OH salts **78** leads to ketone **80**, the salts (**76**) give cyclobutenol (**79**).

Also halogenations of **2** may be looked upon as redox reactions. The first step is again oxidation to the radical cation **74**. The subsequent course is determined by the nature of the halogen. Xenon difluoride, (dichloroiodo) benzene (or chlorine), and pyridinium perbromide (or bromine) react with **2** to give dihalocyclobutenes **82** via **74** → **81**. The stability of the dihalocyclobutenes decreases strongly from the difluoro to the dichloro and dibromo compound. The iodide ion is too large for an addition to **74**, and hence with iodine the OH salt (**78**) is obtained exclusively. Depending on the number of equivalents of iodine the mono- , tri- or heptaiodide is formed.

81 **a**: X = F
b: X = Cl
c: X = Br

The above mentioned properties engendered the hope that examples of Diels–Alder-type reactions with radical ion pair intermediates could be found for **2** or **21**. Although the two reactions described below do not constitute a proof, they provide strong indication of such a course of reaction.[77] Both **2** and **21** give rise to the 1:2 adduct **85** with tetracyanoethene (TCNE). This substance is a rarity with regard to structure and reactivity.[78] It may be assumed that the radical ion pair **74·83** is formed in the first step. Due to steric hindrance, the radical anion **83** attacks **74** with the nitrogen atom as the "spear head." The end product **85** is formed via **84** after addition of a second TCNE molecule. The "normal" Diels–Alder adduct **88** can be formed analogously from **2** and **21** and dicyanoacetylene via the radical ion pair **74·86** and the zwitterion **87**.[77,79,80]

What is the evidence for the intermediate appearance of the radical ion pair **74·86**? It is simply the fact that while cyclobutadiene (**21**) gives a Diels–Alder adduct with dimethylacetylenedicarboxylate (see the section on valence isomers of benzene) tetrahedrane (**2**) remains unchanged under the same conditions. If, however, dicyanoacetylene is used, both **2** and **21** give adduct **88**. The explanation is obvious: The oxidation potential of dicyanoacetylene, but not that of acetylene dicarboxylic acid esters, suffices in order to generate the radical cation **74** from **2** (and **21**).

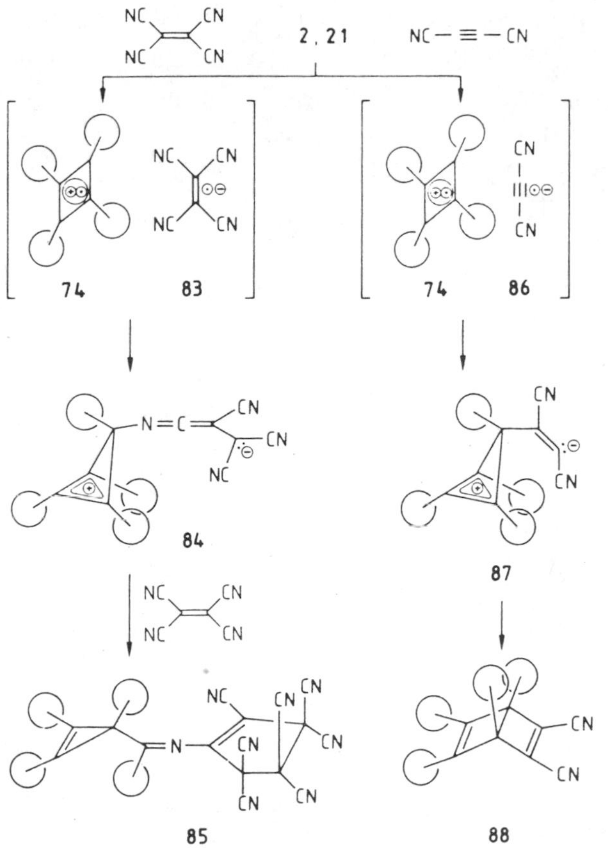

Autoxidation is a subject where the importance of tetrahedrane **2** for cognition in apparently very remote areas becomes particularly clear. Tetrahedrane (**2**) and cyclobutadiene (**21**) behave very differently in the presence of oxygen. Furthermore, the product formation is also very sensitive to reaction conditions: Tetra-*t*-butyltetrahedrane (**2**) is completely stable in air. The corresponding cyclobutadiene (**21**) reacts smoothly with oxygen to give the dioxetane (**90**). This picture changes completely in the presence of an acceptor. Addition of 0.1–0.3 equivalents of silver ions or iodine to an O_2-saturated solution of **2** results in the formation of at least four products, of which the oxete (**91**) is the major component.

The other products are the diketones **92** and **93**, which incidentally are photochemically interconvertible, and the cyclopropenyl ketone (**80**). The latter is formed also when **2** is allowed to stand for several days in O_2-saturated chloroform, and by oxidation of cyclobutadiene (**21**) in the presence of 0.1–0.5 equivalents of the acceptor. The use of an excess of oxygen in diiodomethane, or the addition of a trace of iodine in chloroform or carbon tetrachloride results in the exclusive formation of the cyclic peroxide **89**. This is the only product from **2** in which only one bond has been broken.

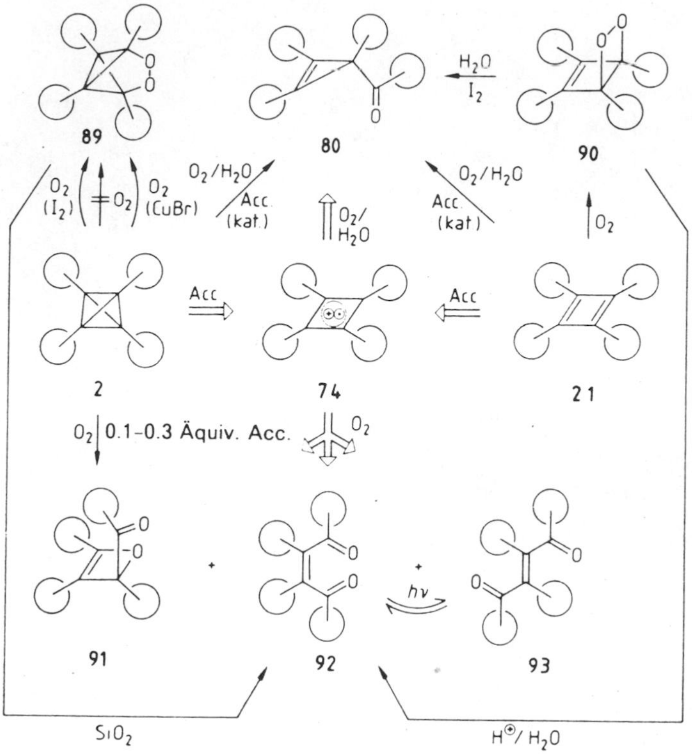

Apart from **89**, all the products of the reaction of **2** with oxygen are likely formed via the radical cation **74**. It is difficult to provide an overall picture of the branching observed in the autoxidation of **2** and **21**. Nevertheless, it will be attempted: The normal potential of oxygen [$E_{1/2}$ ($O_2/O_2^{\bullet-}$) = –0.57,[81a] –0.8,[82] –0.83,[83], –0.94 V[81b]] is too low to oxidize **2** (E_{pa} = 0.5 V, see above). In order to make **2** reactive enough, a sufficiently strong acceptor is needed (Ag^+ ions, iodine, TCNE, and similar acceptors). The crucial intermediate **74** is formed via **75**. At this stage oxygen is consumed. A chain process (SET oxidation; for a detailed discussion see Ref. 3) follows. Such a reaction can explain the rapid oxidation of **2**. The formation of ketone **80** is more difficult to interpret. As described above **80** can be formed from **74** by addition of water and repeated oxidation. But another route is also open.[3] Finally, the question of how the peroxide **89** is formed from **2** is still unanswered. Several interpretations are possible[3]: Perhaps the attacking oxygen is activated rather than **2**. This phenomenon is not altogether unknown.[84] This would imply that some iodine–oxygen complex is the attacking agent. This type of oxygen activation with iodine has not to our knowledge been reported in the literature.[85] There are, however, accounts of a corresponding function of transition metal–oxygen complexes.[86] Accordingly, we also treated **2** with copper(I)

bromide in acetonitrile and oxygen.[87] Indeed, peroxide **89** is the exclusive product of this reaction.[68] Since **2** is inert towards copper bromide alone (the normal potential of copper(I) is strongly reduced under these conditions[83]), oxygen activation cannot be dismissed as the mechanism of the reaction **2** → **89**.

As mentioned previously, oxete (**91**) is believed to be the main product in the oxidation of **2**. At least the spectra support this unusual structure. The last doubt has not been removed by these means, however, for in contrast to known oxetes,[88] **91** is stable up to 200°C. Perhaps the accumulation of four sterically demanding groups is once again responsible. In order to get a definite answer additional experiments have been carried out with **91**. Irradiation yields the isomeric diketones **92** and **93**. The final elucidation of the structure of **91** is based on a recent X-ray structure analysis.[89] In contrast to oxete (**91**) or its reduction product (**94**) the esters **95** and **96** form decent crystals.

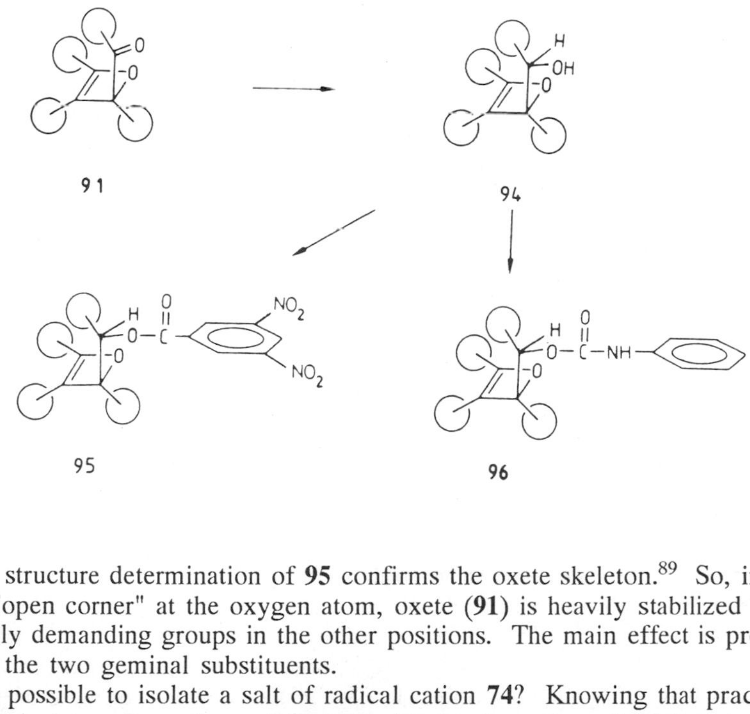

91 94 95 96

The structure determination of **95** confirms the oxete skeleton.[89] So, in spite of an "open corner" at the oxygen atom, oxete (**91**) is heavily stabilized by the sterically demanding groups in the other positions. The main effect is probably due to the two geminal substituents.

Is it possible to isolate a salt of radical cation **74**? Knowing that practically all chemical transformations of **2** start with the formation of **74** it is tempting to isolate such a species. The goal is to have the X-ray structure of such a salt.

Even if one takes into account what is known about radical cation **74** (see above), there remains one problem: Inspired by the preparation of **74**, Borden et al.[90] carried out an ab initio calculation of the cyclobutadiene radical cation. As expected, this species is subject to Jahn–Teller distortion. The rectangular geometry represents the minimum (the same is true of the radical cation of tetramethylcyclobutadiene[61,91]). The two rectangular valence isomers can be transformed into each other via two 6 kcal/mol higher energy rhombohedral forms. In this regard the situation differs from that of cyclobutadiene itself, in which the transition state for valence isomerization is described by a square geometry. Accordingly, one may very well conclude that the experimental ESR spectrum of **74**[61] is the result of exchange between two folded structures with alternating bond lengths, occurring rapidly on the ESR time scale. In order to have experimental proof of this one needs a crystalline salt of **74**. This is only possible if the counterion is not nucleophilic and has no propensity to transfer an electron to the radical cation. Unfortunately even 7,7,8,8-tetracyano-*p*-quinodimethane[71, 119a] or 2,3-dichloro-5,6-dicyano-p-benzoquinone fail to give a stable radical cation salt. If one uses nitrosyl salts (**97**) as oxidating agents the primarily formed radical cation is again attacked either by unavoidable traces of water (formation of **78**) or by the nitrogen monoxide resulting from the oxidation step (nitrile **100** as the main product via **74** → **98** → **99** → **100**).

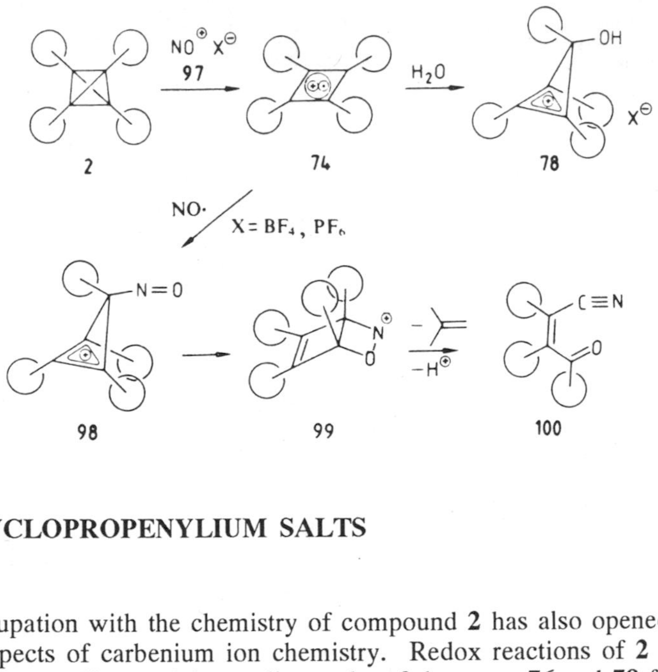

HOMOCYCLOPROPENYLIUM SALTS

Our occupation with the chemistry of compound **2** has also opened our eyes to some aspects of carbenium ion chemistry. Redox reactions of **2** permit the isolation of crystalline cyclobutenylium salts of the types **76** and **78** for the first time, as well as the X-ray determination of their crystal and molecular structures (Fig. 3).[75] This is of importance in connection with Winstein's[92] postulate that these ions should have homoaromatic character. Numerous investigations of this point have been carried out—all were in solution.[93] The structural data now available for **76** and **78**[75] (folded rings; relatively short C1—C3 distances) justify the description of these compounds as homocyclopropenylium salts in spite of the skepticism of theoreticians.[94]

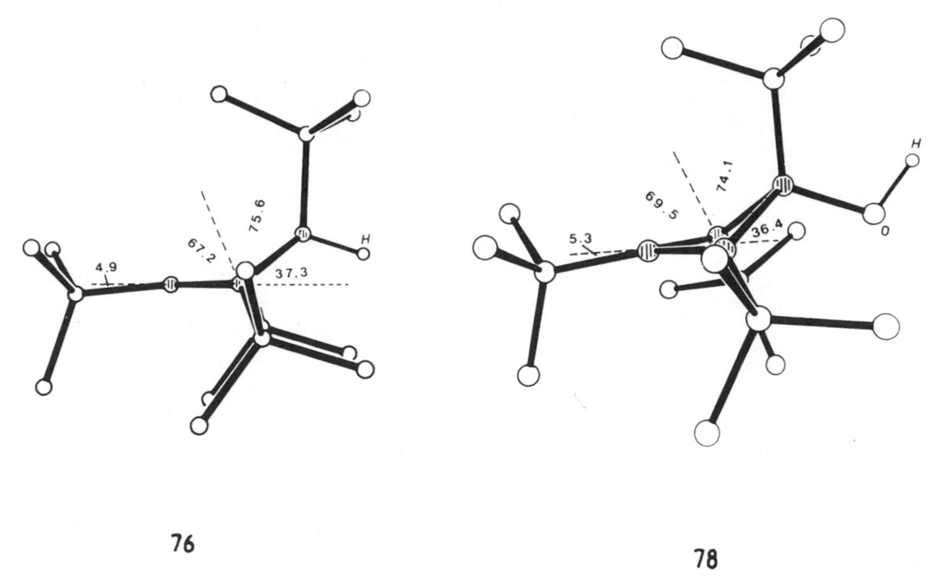

Fig. 3. Crystal structure [folding angle (°)] of the cyclobutenylium ions of **76** (X = I_3) (C1—C3 = 1.806 Å) and of **78** (X = SbF_6) (C1—C3 = 1.833 Å). C1 and C3 are the atoms at the site of folding.

Ion **78** offers a further surprise. Its NMR spectrum is temperature dependent as a result of a carousel rearrangement **78** ⇌ **101** ⇌ **78′**.[95]

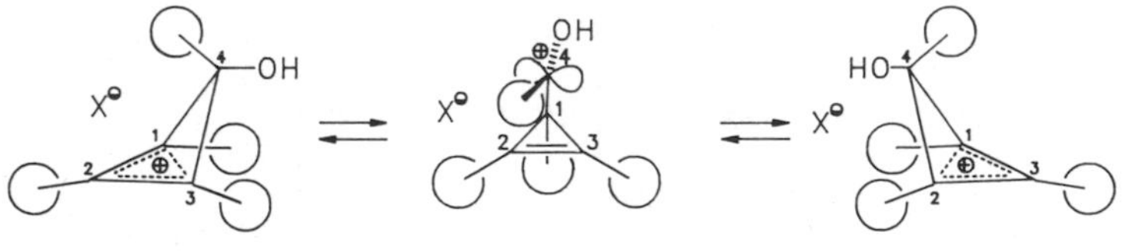

$X^{\ominus}$ 78 ⇌ $X^{\ominus}$ 101 ⇌ $X^{\ominus}$ 78'

It is of particular interest that, in contrast to the predictions of the Woodward–Hoffmann rules, this reaction proceeds with inversion at the migrating carbon atom, that is, the bisected transition state (**101**) is preferred—as predicted.[96]

Incidentally, if the *t*-butyl group at C1 is missing in cation **101**, the cyclopropenylmethylium system becomes an energy minimum relative to the corresponding homocyclopropenylium system; accordingly, direct observation is possible.[97] The interplay between the cyclobutenylium and the cyclopropenylmethylium systems culminates in the dynamic behavior of cation **102**.[98] The latter is obtainable from the diazo compound (**56**) via an unconventional route. Compound **102** can undergo a tandem walk rearrangement, that is, a consecutive migration of C4 around the segment C1/C2/C3 accompanied by a mutual exchange of ring moieties in the sense **102** ⇌ **102′**. As a consequence, the carbon atom carrying the hydroxyl group migrates around six equivalent ring positions.

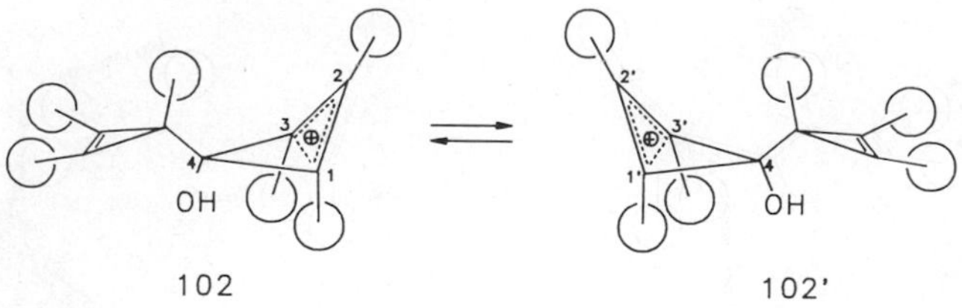

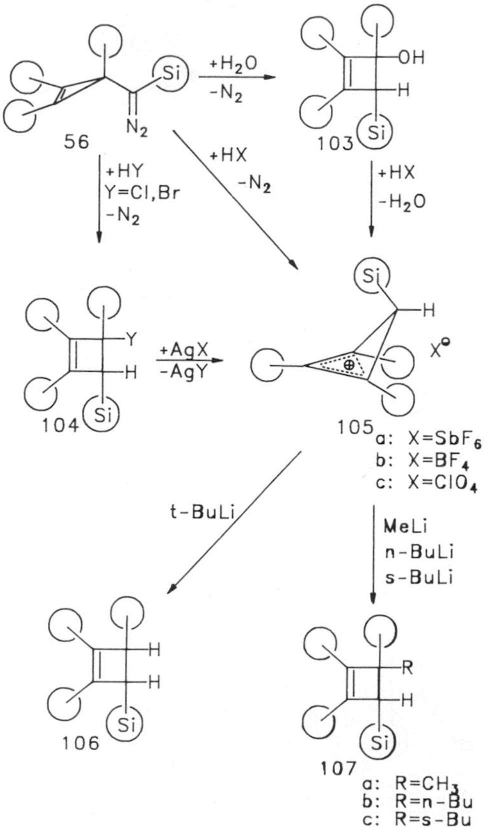

Another series of homocyclopropenylium salts can be derived from the silylated diazo compound **56**. The H-salts (**105**) are crystalline stable compounds that are directly or indirectly (via alcohol **103** or adducts **104**) accessible by treatment of **56** with acids. There was some hope that the proton in the α position to the trimethylsilyl group might be acidic enough to be removed with bases like organolithium compounds. This would be an effective way of preparing cyclobutadiene (**69**). However, the reaction follows another route. Whereas metal amides cause unspecific decompositions, alkyl lithium reagents prefer an addition as compared with deprotonation. *t*-Butyllithium yields cyclobutene (**106**), less hindered lithium compounds form the adducts **107a–c**. In conclusion, one can argue that the stabilization of the homoaromatic cation (**105**) prevents the molecule from being transformed into the antiaromatic cyclobutadiene (**69**).

Astonishingly, the presence of a trimethylsilyl substituent can have a remarkable effect on the chemical behavior of homocyclopropenylium salts. For instance, if diazo compound **45** is treated with chlorine, the primary product is dihalide **108**.[68] In methylene chloride **108** isomerizes to **82b** and it is reasonable to assume that homocyclopropenylium salt (**81b**)—which cannot be detected directly—is an intermediate in this reaction.

On the contrary, the same process does not lead to a cyclobutene in the case of the diazo precursor **56**. Again one isolates geminal dihalides (**109**) and finds an isomerization in methylene chloride.[28] But in this case the ring-opened products **110** are formed.

If the homocyclopropenylium ion (**112**) is an intermediate, it is hard to understand how the trimethylsilyl group in position 4 can change the reaction pathway completely. The two other ions **111** and **113**, which may play a role in the ring opening, also cannot explain the experimental result. Only one point seems to be clear, independent from the starting material (**108** or **109**): Halogen atoms in the 4 position reduce the stability of homocyclopropenylium ions.

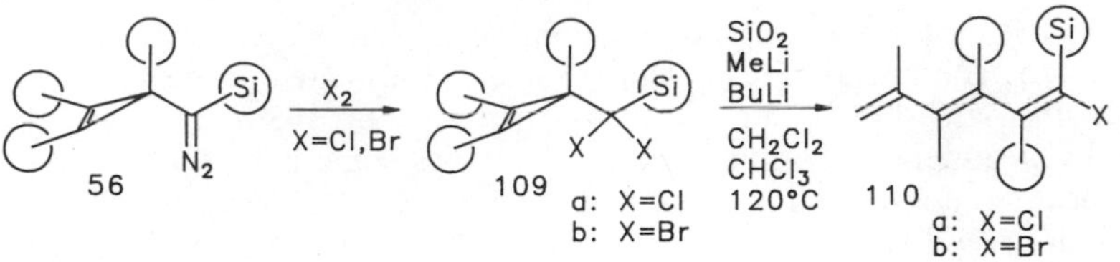

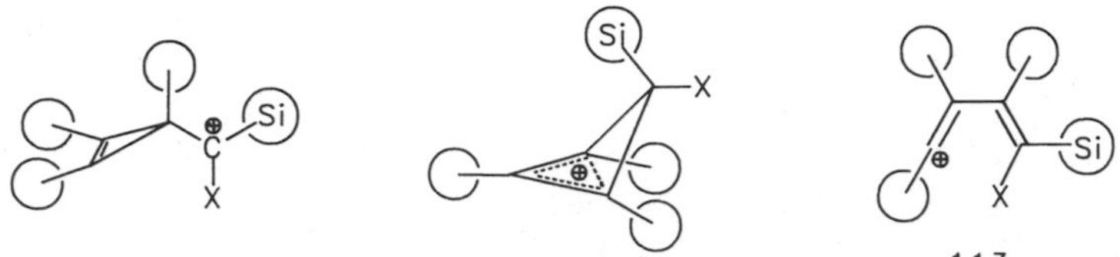

VALENCE ISOMERS OF BENZENE

The Diels–Alder adduct (**114**) between cyclobutadiene (**21**) and dimethyl acetylenedicarboxylate also has no shortage of surprising properties.[99] The genetic connections between the valence isomers **114–116** are shown below. The crucial point is the thermal isomerization of the arene (**115**) to the thermodynamically more stable Dewar benzene (**116**).

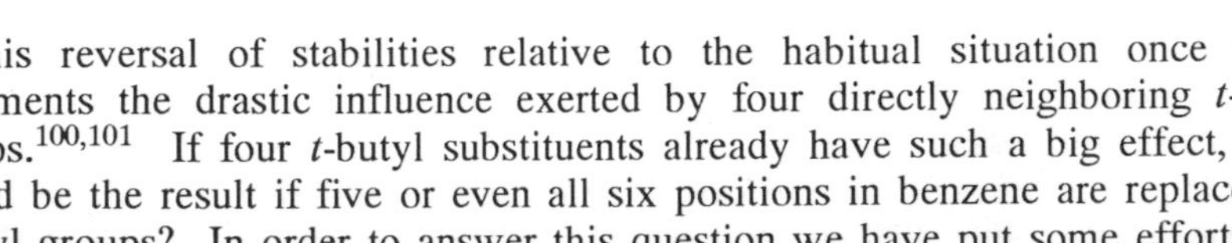

This reversal of stabilities relative to the habitual situation once again documents the drastic influence exerted by four directly neighboring *t*-butyl groups.[100,101] If four *t*-butyl substituents already have such a big effect, what would be the result if five or even all six positions in benzene are replaced by *t*-butyl groups? In order to answer this question we have put some efforts into attempts to synthesize penta- and hexa-*t*-butylbenzene or its valence isomers. Since persubstitution by *t*-butyl groups favors spherical structures (corset effect) we hoped that in the hexa-*t*-butyl substituted series the prismane might—in absolute contrast to the parent compounds—be the most stable valence isomer.

One way to enter this field would be the transformation of the ester groups in **115** or its Dewar isomers into *t*-butyl substituents. It is not a big surprise that this procedure does not work.[28] It makes more sense to use the high reactivity of tetra-*t*-butylcyclobutadiene (**21**) and to test its tendency to undergo cycloaddition reactions with dienophiles that already contain additional *t*-butyl groups. The reactions carried out so far give an indication about the resistance of **21** to become overloaded with bulky substituents. The molecule either does not react at all or it finds a route to a less hindered species than the expected adduct would be.

Not unexpected, di-*t*-butylacetylene (**22**) does not react with **21**. The same is true for di-*t*-butylcyclopropenone (**121**). Even di-*t*-butylmaleicanhydride (**126**),[102] which can be prepared[68] in an effective manner from 3,4-di-*t*-butylcyclopentadienone[103] with singlet oxygen, gives no addition to **21**.

no addition

22 Δ

118 117 Δ

Δ, p 119 CO_2CH_3 120

21

Δ no addition 121

122

126 Δ no addition

123

Δ

124 + 125

Obviously 1,2-substitution with *t*-butyl groups prevents a π system from becoming bound to the hindered diene (**21**). A better chance exists if the dienophile bears only one *t*-butyl substituent and is also activated by an electron withdrawing group. Nitrile **117** reacts with **21** quite easily, but it is not the acetylenic triple bond that adds. In a mechanism similar to the one discussed in the reaction of **2** or **21** with tetracyanoethene, pyridine (**118**) is formed instead.[100,104] Ester **119** needs some pressure (10 kbar)[105] in order to induce a reaction. Again the isolated product is not the expected persubstituted Dewar valence isomer of methyl benzoate. In this case the steric hindrance can only be overcome by a radical-type addition with **120** as the final product.

Another interesting example is the combination **21**/**122**. The corresponding Diels–Alder adduct would be an ideal precursor for penta-*t*-butyl benzene and its isomers. Indeed, if one mixes the two components a deep red color is observed. This phenomenon is strongly suggestive of the matrix-isolated charge-transfer complexes **127** and **128**.[106,107]

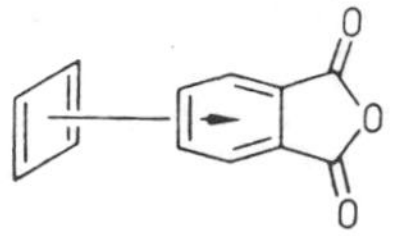

127

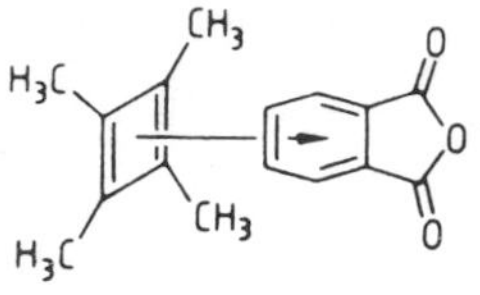

128

Complex **123** is stable even at room temperature due to the crowding of *t*-butyl groups. On heating the red color slowly disappears and two compounds (**124** and **125**) can be isolated. The structures illustrate once more the difficulties in trying to synthesize a penta-*t*-butylated Dewar benzene derivative. On the other hand, **124** is the first example for the formation of a molecule with a bicyclobutane skeleton[108,109] starting from cyclobutadiene (**21**).

In the synthesis of sterically overcrowded molecules it is a better strategy to start from a precursor that already contains all the building blocks and to

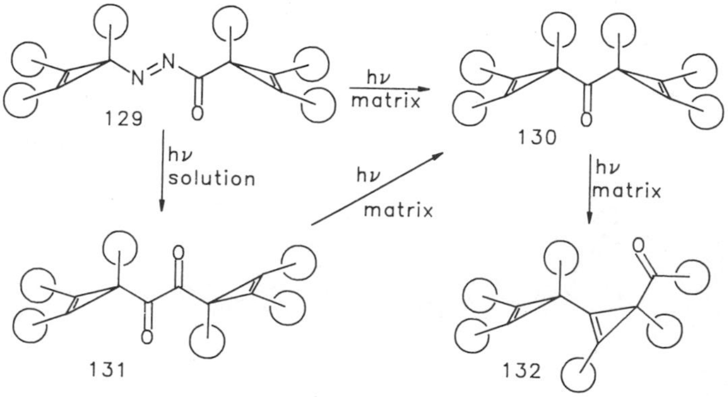

129 hν matrix 130

hν solution hν matrix hν matrix

131 132

eliminate a decent leaving group in the very last step. In this context diazoacyl imide (**129**), which can also be obtained from diazocompound (**56**),[98] appears as an ideal candidate for the preparation of per-*t*-butylated benzene and/or the corresponding valence isomers. Unfortunately the photochemical behavior of **129** is quite unusual: Irradiation at room temperature (254 nm; benzene, cyclohexane, or THF) results in gas evolution. However, instead of **130** the α-diketone **131** is isolated.

Compound **131** is formed by dimerization of the acyl radical, which is less hindered than the cyclopropenyl radical.[110] Upon irradiation of **129** in an organic matrix at –196°C, the acyl radical formed has no other choice than to combine with the cyclopropenyl radical formed in the same solvent cage. Indeed, under these conditions, ketone **130** is obtained.

If **130** is irradiated in an organic matrix continuously a surprising isomerization to ketone **132** occurs. This observation indicates that the acyl fragment can also undergo a *t*-butyl migration followed by the formation of isomeric ketone **132** (structure proven by X-ray diffraction[13]).

PYRAMIDAL CATIONS

Masamune et al.[111] proposed that a nonclassical cation with pyramidal structure[112] **134** is formed on treatment of **133**, **135**, or **136** with superacids. Due to the stabilization of spherical geometries by *t*-butyl groups the cation, which can be formally derived from cyclopentanol (**138**), eventually should be rather stable even at room temperature.

HO H H_3C CH_3 133 → CH_3 + CH_3 134 ← HO H CH_3 CH_3 135

CH_3 OCH_3 CH_3 136 → 134

This could offer an opportunity to carry out an X-ray structure analysis of such a pyramidal species.

Reduction of tricyclopentanone **19** with an excess of lithium aluminum hydride gives hydrocarbon **137**. Only if one equivalent of the reagent is used the alcohol (**138**) can be isolated, which isomerizes to **139** on standing in aprotic solvents.[113]

LAH

19

137

LAH

HO H

rt

HO

138

139

Both alcohols **138** and **139** give the same cation "X," which is stable up to –20°C when an excess of trifluoroacetic acid or antimony pentafluoride is added to a methylene chloride solution of either precursor. Hydrolysis of the cation with sodium hydroxide yields two products, alcohol **139** and cyclobutene **141**.

HO H

138

$-OH^{\ominus}$

HO

$-OH^{\ominus}$

$+OH^{\ominus}$

X

$+OCH_3^{\ominus}$

$-OCH_3^{\ominus}$

H OCH_3

139

140

$+OH^{\ominus}$

O

141

Reaction with sodium methoxide in methanol gives exclusively methoxy derivative **140**, which also forms cation "X" when trifluoroacetic acid or antimony pentafluoride is added.

The structure of cation "X" is very unusual. Whereas the highly symmetrical pyramidal assembly **142** is strictly excluded by NMR spectra, a decision between the two spherical arrangements **143** and **144** is difficult to make.

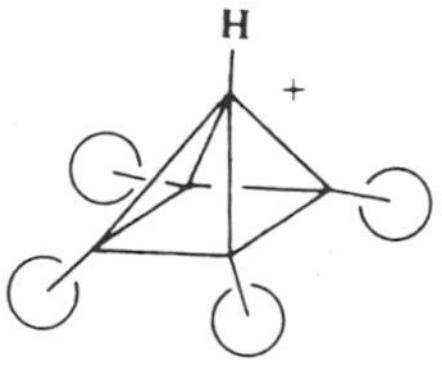

142

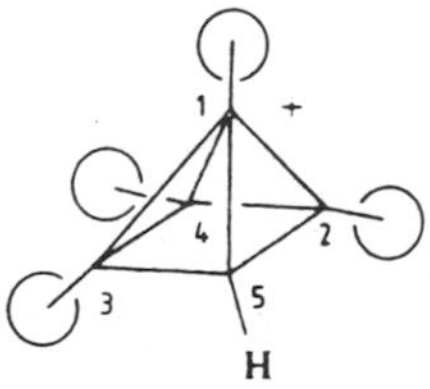

143

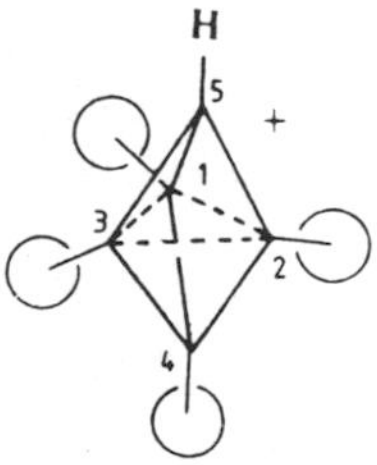

144

At –100°C the ^{13}C NMR spectrum shows all the signals expected for cation **143** in comparison to Masamune's data for **134**.[111] After warm up to –60°C the signals that can be assigned to the apical C1 (δ=–6.8) and the two identical carbon atoms C2 and C3 disappeared. It is reasonable to assume that this behavior originates from a dynamic process. Such a coalescence phenomenon should create an averaged signal. But even at –20°C such an absorption cannot be detected. In addition, the pyramidal cation (**143**) would not be in accordance with such a fluctional property.

A dynamic process would better fit a cation of type **144** with the geometry of a trigonal bipyramid. This structure may represent the transition state in the equilibration of three cations of lower symmetry.[114] Also, the analogy of cation "X" with cyclopropylcarbinyl cation **145**[115] or 1-methylcyclopropylcarbinyl cation **146**[116] cannot be overseen. In both cases an equilibrium between nonclassical cations exists, as can be demonstrated by the study of deuterated samples.[117] In the case of the methyl derivative (**146**) the dynamic interconversion can be frozen out.[118]

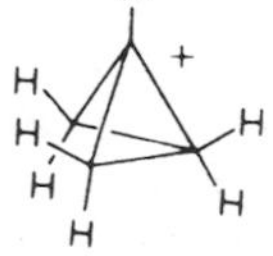

145

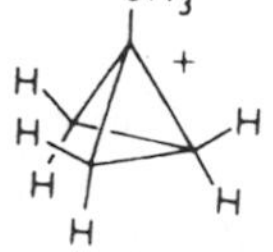

146

The same type of process may also occur in the series of cations with a $(CH)_5$ ring skeleton. If this is true, "X" has to be taken as an equilibrating system in which carbonium ions **147**, **147**′, and **147**″ are transferred into each other.

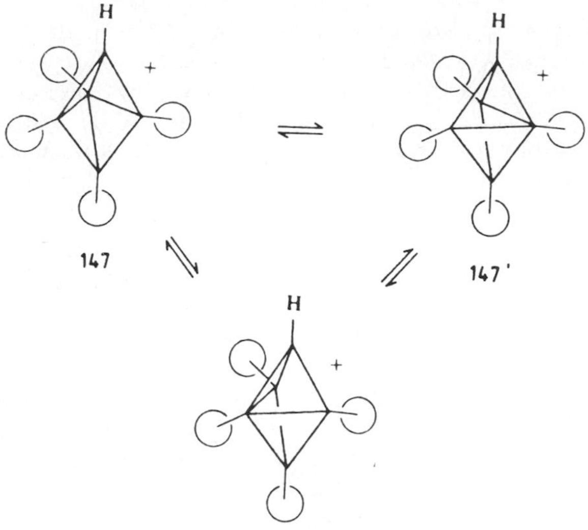

If this reaction is sufficiently rapid the observed NMR spectrum should have the appearance that one would expect for a symmetry shown in structure **144**. At –100°C the interconversion should be slow enough for us to observe directly the static carbonium ion (**147**). The final proof that this interpretation is correct follows from the temperature dependence of the signals originating from the *t*-butyl groups in positions 1–3, which show the expected coalescence phenomenon in the temperature range from –82 to –116°C.[119]

CONCLUSION

The study of tetra-*t*-butyltetrahedrane has (a) sharpened our feeling for the subtle influence of sterically demanding groups (favoring of otherwise impracticable synthetic routes; reversal of thermodynamic stabilities), (b) led to broadly applicable concepts (corset principle, formation of radical ion pairs), (c) opened up access to several unknown or at least difficultly accessible classes of substances, (d) allowed the isolation of crystalline derivatives of highly reactive systems, (e) caused X-ray investigations with surprising results (extremely bent bonds, dependency of alternation on degree of substitution, gas inclusion compounds, and structural indicators of "homoaromatic" character), (f) allowed detailed insight into the behavior of radical cations, and (g) made the discovery of new reaction mechanisms possible (in addition reactions, autoxidation, and cationic rearrangements). The molecules **2** and **21** have delivered nonstop record values of physical data (NMR, ionization potentials, redox potentials, etc.).

Having successfully prepared the highly strained tetrahedrane 2, any shyness towards other sterically loaded molecules disappears. The dramatic stabilizing effect of *t*-butyl groups on the spherical structures demonstrated in this chapter for several classes of compounds can be a guide to the synthesis of many other types of cage molecules.

REFERENCES

1. Plato: *Timaeus*
2. (a) G. Maier and S. Pfriem, *Angew. Chem.* **1978**, *90*, 551; *Angew. Chem. Int. Ed. Engl.* **1978**, *17*, 519. (b) G. Maier, S. Pfriem, U. Schäfer, and R. Matusch, *Angew. Chem.* **1978**, *90*, 552; *Angew. Chem. Int. Ed. Engl.* **1978**, *17*, 520. (c) G. Maier, S. Pfriem, U. Schäfer, K.-D. Malsch, and R. Matusch, *Chem. Ber.* **1981**, *114,* 3965.
3. Summary: G. Maier, *Angew. Chem.* **1988**, *100*, 317; *Angew. Chem. Int. Ed. Engl.* **1988**, *27*, 309.
4. Summary: G. Maier, *Angew. Chem.* **1974**, *86*, 491; *Angew. Chem. Int. Ed. Engl.* **1974**, *13*, 425.
5. (a) A. Alzérreca, Ph.D. thesis, Universität Marburg, 1974. (b) G. Maier and A. Alzérreca, *Angew. Chem.* **1973**, *85*, 1056; *Angew. Chem. Int. Ed. Engl.* **1973**, *12*, 1015.
6. Compound **10** was obtained independently and simultaneously by Masamune et al. using another route: S. Masamune, N. Nakamura, M. Suda, and H. Ona, *J. Am. Chem. Soc.* **1973**, *95*, 8481.
7. G. Maier, U. Schäfer, W. Sauer, H. Hartan, R. Matusch, and J. F. M. Oth, *Tetrahedron Lett.* **1978**, 1837.
8. W. Sauer, Ph.D. thesis, Universität Marburg, 1977.
9. A. Schweig and W. Thiel, *J. Comput. Chem.* **1980**, *1*, 129.
10. G. Maier, K. Euler, H. Irngartinger, and M. Nixdorf, *Chem. Ber.* **1985**, *118,* 409.
11. S. Pfriem, Ph.D. thesis, Universität Marburg, 1978.
12. (a) G. Maier, K. A. Reuter, L. Franz, and H. P. Reisenauer, *Tetrahedron Lett.* **1985**, *26*, 1845. (b) L. Franz, Ph.D. thesis, Universität Giessen, 1982.
13. R. Boese, unpublished. We are very grateful to Dr. Boese, Universität Essen, for the structural investigations.
14. W. Thiel, unpublished. We thank Professor Thiel, Universität Wuppertal, for these data (private communication).
15. (a) S Masamune, N. Nakamura, M. Suda, and H. Ona, *J. Am. Chem. Soc.* **1973**, *95*, 8481. (b) T. Bally and S. Masamune, *Tetrahedron* **1980**, *36*, 343; in Scheme 10, the arrows should start from **22a** instead of **22**. (c) M. Suda, Ph.D. thesis, University of Alberta, 1975.
16. H. W. Lage, Ph.D. thesis, Universität Giessen, 1982.
17. M. Hoppe, Ph.D. thesis, Universität Giessen, 1984.

18. B. Wolf, Ph.D. thesis, Universität Giessen, 1985.

19. D. Volz, Ph.D. thesis, Universität Giessen, 1987.

20. G. Maier and B. Wolf, *Synthesis* **1985**, 871.

21. G. Maier, H. W. Lage, and H. P. Reisenauer, *Angew. Chem.* **1981**, *93*, 1010; *Angew. Chem. Int. Ed. Engl.* **1981**, *20*, 976.

22. K. A. Reuter, Ph.D. thesis, Universität Giessen, 1985.

23. (a) D. Seyfarth and T. C. Flood, *J. Organomet. Chem.* **1971**, *29*, C25. (b) U. Schöllkopf, H.-U. Scholz, *Synthesis* **1976**, 271.

24. U.-J. Vogelbacher, M. Regitz, and R. Minott, *Angew. Chem.* **1986**, *98*, 835; *Angew. Chem. Int. Ed. Engl.* **1986**, *25*, 842.

25. (a) U.-J. Vogelbacher, M. Ledermann, T. Schach, G. Michels, U. Hees, and M. Regitz, *Angew. Chem.* **1988**, *100*, 304; *Angew. Chem. Int. Ed. Engl.* **1988**, *27*, 272. (b) U. Hees, U.-J. Vogelbacher, G. Michels, and M. Regitz, *Tetrahedron*, in print.

26. (a) P. Eisenbarth and M. Regitz, *Chem. Ber.* **1984**, *117*, 445. (b) J. Fink and M. Regitz, *Chem. Ber.* **1986**, *119*, 2159.

27. (a) M. R. Booth and S. G. Frankiss, *J. Chem. Soc. Chem. Commun.* **1968**, 1347. (b) *Spectrochimi. Acta* **1970**, *26A*, 895. (c) J. A. Seckar and J. S. Thayer, *Inorgan. Chem.* **1976**, *15*, 501.

28. (a) I. Bauer, Ph.D. thesis, Universität Giessen, 1986. (b) D. Born, Ph.D. thesis, Universität Giessen. (c) G. Maier and D. Born, *Angew. Chem.*, **1989**, *101*, 1085; *Angew. Chem. Int. Ed. Engl.*, **1989**, *28*, 1050. (d) R. Wolf, diploma work, Universität Giessen, 1990.

29. A chemical shift of $\delta = -15.9$ is calculated for the unsubstituted tetrahedrane: M. Schindler and W. Kutzelnigg, *J. Am. Chem. Soc.* **1983**, *105*, 1360; application of a low-field shift of 25 ppm due to the *t*-butyl groups gives a value of $\delta \approx 9$ for **2**. For the calculation of shielding tensors, see also: J. C. Facelli, A. M. Orendt, M. S. Solum, G. Depke, D. M. Grant, and J. Michl, *J. Am. Chem. Soc.* **1986**, *108*, 4268.

30. Summary of the spectroscopic properties of **2**: G. Maier, S. Pfriem, K.-D. Malsch, H.-O. Kalinowski, and K. Dehnicke, *Chem. Ber.* **1981**, *114*, 3988; see also Ref. 3.

31. E. Heilbronner, T.-B. Jones, A. Krebs, G. Maier, K.-D. Malsch, J. Pocklington, and A. Schmelzer, J. Am. Chem. Soc. **1980**, *102*, 564.

32. M. Saunders and G. E. Walter, unpublished; G. E. Walter, Ph.D. thesis, Yale University, 1983.

33. T. Loerzer, Ph.D. thesis, Universität Gottingen, 1983.

34. T. Loerzer, R. Machinek, W Lüttke, L. H. Franz, K.-D. Malsch, and G. Maier, *Angew. Chem.* **1983**, *95*, 914; *Angew. Chem. Int. Ed. Engl.* **1983**, *22*, 78.

35. J. M. Schulman and T. J. Venanzi, *J. Am. Chem. Soc.* **1974**, *96*, 4739.

36. The NMR coupling constants of the unsubstituted tetrahedrane ($^1J_{C—C} = 7.1$, $^1J_{C—H} = 240$ Hz) should be $^1J_{C—C} = 14.5$, $^1J_{C—H} = 273$ Hz, according to another calculation (coupled Hartree–Fock perturbation theory; J. M. Schulman, private information, 1978).

37. K. Kovacevic and Z. B. Maksic, *J. Org. Chem.* **1974**, *39*, 539.

38. H. Irngartinger, A. Goldmann, R. Jahn, M. Nixdorf, H. Rodewald, G. Maier, K.-D. Malsch, and R. Emrich, *Angew. Chem.* **1984**, *96*, 967; *Angew. Chem. Int. Ed. Engl.* **1984**, *23*, 993.
39. H. Irngartinger, R. Jahn, G. Maier, and R. Emrich, *Angew. Chem.* **1987**, *99*, 356; *Angew. Chem. Int. Ed. Engl.* **1987**, *26*, 356.
40. W. J. Hehre and J. A. Pople, J. Am. Chem. Soc. **1975**, *97*, 6941.
41. A. Schweig and W. Thiel, *J. Am. Chem. Soc.* **1979**, *101*, 4742.
42. H. Kollmar, *J. Am. Chem. Soc.* **1980**, *102*, 2617.
43. W. D. Hounshell and K. Mislow, *Tetrahedron Lett.* **1979**, 1205.
44. H. Irngartinger, N. Riegler, K.-D. Malsch, K.-A. Schneider, and G. Maier, *Angew. Chem.* **1980**, *92*, 214; *Angew. Chem. Int. Ed. Engl.* **1980**, *19*, 211.
45. A. Krebs, H. Kimling, and R. Kemper, *Liebigs Ann. Chem.* **1978**, 431.
46. H. Kimling and A. Krebs, *Angew. Chem.* **1972**, *84*, 952; *Angew. Chem. Int. Ed. Engl.* **1972**, *11*, 932; see also Ref. 41.
47. L. T. J. Delbaere, M. N. G. James, N. Nakamura, and S. Masamune, *J. Am. Chem. Soc.* **1975**, *97*, 1973.
48. H. Irngartinger and H. Rodewald, *Angew. Chem.* **1974**, *86*, 783; *Angew. Chem. Int. Ed. Engl.* **1974**, *13*, 740; see also Lit. 28.
49. H. Irngartinger, M. Nixdorf, N. H. Riegler, A. Krebs, H. Kimling, J. Pocklington, G. Maier, K.-D. Malsch, and K. A. Schneider, *Chem. Ber.* **1988**, *121*, 673.
50. W. T. Borden and E. R. Davidson, *J. Am. Chem. Soc.* **1980**, *102*, 7958.
51. O. Ermer and E. Heilbronner, *Angew. Chem.* **1983**, *95*, 414; *Angew. Chem. Int. Ed. Engl.* **1983**, *22*, 402.
52. (a) H. Irngartinger and M. Nixdorf, *Angew. Chem.* **1983**, *95*, 415; *Angew. Chem. Int. Ed. Engl.* **1983**, *22*, 403. (b) J. D. Dunitz, C. Krüger, H. Irngartinger, E. F. Maverick, Y. Wang, and M. Nixdorf, *Angew. Chem.* **1988**, *100*, 415; *Angew. Chem. Int. Ed. Engl.* **1988**, *27*, 387.
53. (a) A. Baeyer, *Ber. Dtsch. Chem. Ges.* **1885**, *18*, 2269; summaries: (b) R. Huisgen, *Angew. Chem.* **1986**, *98*, 297; *Angew. Chem. Int. Ed. Engl.* **1986**, *25*, 297. (c) K. B. Wiberg, *Angew. Chem.* **1986**, *98*, 312; *Angew. Chem. Int. Ed. Engl.*. **1986**, *25*, 312; see also: (d) A. Greenberg and J. F. Liebman, *Strained Organic Molecules,* Academic Press, New York, 1978.
54. M. Månsson, unpublished. We thank Dr. Månsson, Universität Lund, Sweden, for carrying out the calorimetric measurements.
55. C. Rüchardt, H.-D. Beckhaus, and B. Dogan, unpublished. We also thank the group in Freiburg for the determination of the heat of sublimation.
56. P. v. R. Schleyer, J. E. Williams, R. R. Blanchard, *J. Am. Chem. Soc.* **1970**, *92*, 2377.
57. According to force field calculations the strain energy of **2** is only 33 kJ/mol (7.9 kcal/mol) higher than that of the unsubstituted tetrahedrane (C. Rüchardt and H.-D. Beckhaus, private communication).
58. The strain is also greater than in [1.1.1]propellane, in which it is 14 kcal/mol per ring bond (Ref. 53c).
59. M. D. Newton and J. M. Schulman, *J. Am. Chem. Soc.* **1972**, *94*, 773.
60. K. Mislow and W. D. Hounshell, private communication, March 1978.

61. (a) H. Bock, B. Roth, and G. Maier, *Angew. Chem.* **1980**, *92*, 213; *Angew. Chem. Int. Ed. Engl.* **1980**, *19*, 209. (b) *Chem. Ber.* **1984**, *117*, 172.

62. Experiments by H. Schwarz, Technische Universität Berlin; already cited in Ref. 30. We thank Professor Schwarz for these measurements.

63. (a) Q. B. Broxterman, H. Hogeveen, and P. M. Kok, *Tetrahedron Lett.* **1981**, *22*, 173. (b) Q. B. Broxterman and H. Hogeveen, *Tetrahedron Lett.* **1983**, *24*, 639.

64. (a) J. L. Courtneidge, A. G. Davies, and J. Lusztyk, *J. Chem. Soc. Chem. Commun.* **1983**, 893. (b) J. L. Courtneidge, A. G. Davies, and J. E. Parkin, *J. Chem. Soc. Chem. Commun.* **1983**, 1262. (c) J. L. Courtneidge, A. G. Davies, E. Lusztyk, and J. Lusztyk, *J. Chem. Soc. Perkin Trans. 2* **1984**, 155. (d) J. L. Courtneidge and A. G. Davies, *Acc. Chem. Res.* **1987**, *20*, 90.

65. Q. B. Broxterman, H. Hogeveen, and R. F. Kingma, *Tetrahedron Lett.* **1984**, *25*, 2043.

66. M. Chan, J. L. Courtneidge, A. G. Davies, J. C. Evans, A. G. Neville, and C. C. Rowlands, *Tetrahedron Lett.* **1985**, *26*, 4121.

67. M. A. Fox, R. A. Campbell, S. Hünig, H. Berneth, G. Maier, K.-A. Schneider, and K.-D. Malsch, *J. Org. Chem.* **1982**, *47*, 3408.

68. R. Emrich, Ph.D. thesis, Universität Giessen, 1986.

69. Summaries: (a) G. J. Kavarnos and N. J. Turro, *Chem. Rev.* **1986**, *86*, 401. (b) M. A. Fox, *Adv. Photochem.* **1986**, *13*, 237; cf. also (c) P. G. Gassman and B. A. Hay, *J. Am. Chem. Soc.* **1986**, *108*, 4227. (d) J. Mattay, *Angew. Chem.* **1987**, *99*, 849; *Angew. Chem. Int. Ed. Engl.* **1987**, *26*, 825.

70. K.-D. Malsch, Ph.D. thesis, Universität Giessen, 1982.

71. K.-A. Schneider, Ph.D. thesis, Universität Giessen, 1982.

72. Reviews on the generation and reactions of radicalions: (a) L. Eberson, *Adv. Phys. Org. Chem.* **1982**, *18*, 79. (b) O. Hammerich and V. D. Parker, *Adv. Phys. Org. Chem.* **1984**, *20*, 55. (c) A. Pross, *Acc. Chem. Res.* **1985**, *18*, 212. (d) L. Eberson: *Electron Transfer Reactions in Organic Chemistry*, Springer, Berlin, 1987.

73. (a) J. R. Miller, J. A. Peeples, M. J. Schmitt, and G. L. Closs, *J. Am. Chem. Soc.* **1982**, *104*, 6488. (b) L. T. Calcaterra, G. L. Closs, and J. R. Miller, *J. Am. Chem. Soc.* **1983**, *105*, 670.

74. Reviews on Ag catalysis: (a) L. A. Paquette, *Synthesis* **1975**, 347. (b) R. C. Bishop III, *Chem. Rev.* **1976**, *76*, 461.

75. G. Maier, R. Emrich, K.-D. Malsch, K.-A. Schneider, M. Nixdorf, and H. Irngartinger, *Chem. Ber.* **1985**, *118*, 2798.

76. F. G. Bordwell and J. Bausch, *J. Am. Chem. Soc.* **1986**, *108*, 2473.

77. This extreme case of a Diels–Alder reaction has barely been discussed previously, also not in: (a) J. Sauer and R. Sustmann, *Angew. Chem.* **1980**, *92*, 773; *Angew. Chem. Int. Ed. Engl* **1980**, *19*, 779. (b) Meanwhile R. Sustmann et al. have reported convincing evidence for a radical ion pair intermediate in another system: M. Dern, H.-G. Korth, G. Kopp, and R. Sustmann, *Angew. Chem.* **1985**, *97*, 324; *Angew. Chem. Int. Ed. Engl.* **1985**, *24*, 337.

78. G. Maier, K.-A. Schneider, K.-D. Malsch, H. Irngartinger, and A. Lenz,

Angew. Chem. **1982**, *94*, 446; *Angew. Chem. Int. Ed. Engl.* **1982**, *21*, 437; *Angew. Chem. Suppl.* **1982**, 1072.

79. Similar zwitterions have been discussed in the context of addition of tetracyanoethylene to olefins: (a) R. Huisgen and J. P. Ortega, *Tetrahedron Lett.* **1978**, 3975; relevant summaries: (b) R. Gompper, *Angew. Chem.* **1969**, *81*, 348; *Angew. Chem. Int. Ed. Engl.* **1969**, *8*, 312. (c) R. Huisgen, *Acc. Chem. Res.* **1977**, *10*, 117.

80. A detailed discussion of SET activation of Diels–Alder reactions between tetracyanoethylene and anthracene has appeared meanwhile: S. Fukuzumi and J. K. Kochi, *Tetrahedron* **1982**, *38*, 1035.

81. (a) J. Eriksen, C. S. Foote, and T. L. Parker, *J. Am. Chem. Soc.* **1977**, *99*, 6455. (b) J. Eriksen and C. S. Foote, *J. Am. Chem. Soc.* **1980**, *102*, 6083.

82. K. Okada, K. Hisamitsu, Y. Takahashi, T. Hanaoka, T. Miyashi, and T. Mukai, *Tetrahedron Lett.* **1984**, *25*, 5311.

83. H. Bock and D. Jaculi, *Angew. Chem.* **1984**, *96*, 298; *Angew. Chem. Int. Ed.* Engl. **1974**, *23*, 305.

84. G. Henrici-Olivé, *Angew. Chem.* **1974**, *86*, 1; *Angew. Chem. Int. Ed. Engl.* **1974**, *13*, 29.

85. See, however: (a) $Cl_2 \cdot O_2$: G. Porter and I. J. Wright, Discuss. Faraday Soc. **1953**, *14*, 23. (b) $ICl \cdot O_2$: M. Hawkins, L. Andrews, A. J. Downs, and D. J. Drury, *J. Am. Chem. Soc.* **1984**, *106*, 3076.

86. Reviews: (a) S. Fallab, *Angew. Chem.* **1967**, *79*, 500. *Angew. Chem. Int. Ed. Engl.* **1967**, *6*, 496. (b) L. Vaska, *Acc. Chem. Res.* **1976**, *9*, 175. (c) A. B. P. Lever, H. B. Gray, *Acc. Chem. Res.* **1978**, *11*, 348. (d) D. Cremer in S. Patai, Ed., *The Chemistry of Functional Groups, Peroxides,* Wiley, London, 1983, pp. 1–84.

87. R. D. Gray, *J. Am. Chem. Soc.* **1969**, *91*, 56.

88. (a) J. Hollander and C. Woolf, Belgium Patent 671439, (1966); *Chem. Abstr.* **1966**, *65*, 8875a. (b) J. M. Holovka, P. D. Gardner, C. B. Strow, M. L. Hill, and T. V. van Auken, *J. Am. Chem. Soc.* **1968**, *90*, 5041. (c) L. E. Friedrich and G. B. Schuster, *J. Am. Chem. Soc.* **1969**, *91*, 7204. (d) L. E. Schuster and G. B. Schuster, *J. Am. Chem. Soc.* **1971**, *93*, 4602. (e) L. E. Friedrich and J. D. Bower, *J. Am. Chem. Soc.* **1973**, *95*, 6869. (f) J. Y. Koo and G. B. Schuster, *J. Am. Chem. Soc.* **1977**, *99*, 5403. (g) Y. Kobayashi, Y. Hanzawa, W. Miyashita, T. Kashiwagi, T. Nakano, and I. Kumadaki, *J. Am. Chem. Soc.* **1979**, *101*, 6445. (h) P. C. Martino and P. B. Shevlin, *J. Am. Chem. Soc.* **1980**, *102*, 5429.

89. We are grateful to Professor Irngartinger, Universität Heidelberg, for carrying out the X-ray structure analysis.

90. W. T. Borden, E. R. Davidson, and D. Feller, *J. Am. Chem. Soc.* **1981**, *103*, 5725.

91. M. Shiotani, K. Ohta, Y. Nagata, and J. Sohma, *J. Am. Chem. Soc.* **1985**, *107*, 2562.

92. Summaries: (a) S. Winstein, *Q. Rev. Chem. Soc.* **1969**, *23*, 141. (b) P. Warner, *Top. Nonbenzenoid Aromat. Chem.* **1977**, *2*, 283. (c) L. A. Paquette, *Angew. Chem.* **1978**, *90*, 114; *Angew. Chem. Int. Ed. Engl.* **1978**, *17*, 106.

93. (a) G. A. Olah, J. S. Staral, and G. Liang, *J. Am. Chem. Soc.* **1974**, *96*, 6233; G. A. Olah, J. S. Staral, R. J. Spear, and G. Liang, *J. Am. Chem. Soc.* **1975**,

97, 5489; however see also: W. J. Hehre and A. J Devaquet, *J. Am. Chem. Soc.* **1976**, *98*, 4370. (b) A. E. Lodder, H. M. Buck, and L. J. Oosterhoff, *Recl. Trav. Chim. Pays-Bas* **1970**, *89*, 1229. (c) G. A. Olah and R. J. Spear, *J. Am. Chem. Soc.* **1975**, *97*, 1845.

94. More recent theoretical studies: (a) R. C. Haddon and K. Raghavachari, *J. Am. Chem. Soc.* **1983**, *105*, 118. (b) D. Cremer, E. Kraka, T. S. Slee, R. F. W. Bader, C. D. H. Lau, T. T. Nguyen-Dang, and P. J. MacDougall, *J. Am. Chem. Soc.* **1983**, *105*, 5069. (c) D. Cremer, J. Gauss, P. v. R. Schleyer, and P. H. M. Budzelaar, *Angew. Chem.* **1984**, *96*, 370; *Angew. Chem. Int. Ed. Engl.* **1984**, *23,* 370.

95. G. Maier, R. Emrich, and H.-O. Kalinowski, *Angew. Chem.* **1985**, *97*, 427; *Angew. Chem. Int. Ed. Engl.* **1985**, *24*, 429.

96. (a) Ab initio: W. J. Hehre and A. J. P. Devaquet, *J. Am. Chem. Soc.* **1976**, *98*, 4370; **1974**, *96*, 3644. (b) MINDO/2: K. Morio and S. Masamune, *Chem. Lett.* **1974**, 1251.

97. G. Maier, K. Euler, and R. Emrich, *Tetrahedron Lett.* **1986**, *27*, 3607.

98. G. Maier, I. Bauer, D. Born, and H.-O. Kalinowski, *Angew. Chem.* **1986**, *98*, 1132; *Angew. Chem. Int. Ed. Engl.* **1986**, *25*, 1093.

99. G. Maier and K.-A. Schneider, *Angew. Chem.* **1980**, *92*, 1056; *Angew. Chem. Int. Ed. Engl.* **1980**, *19*, 1022.

100. 2,3,4,5-Tetra-*t*-butylbiphenyl shows the same phenomenon (A. Krebs, private communication). Compare also: A. Krebs, E. Franken, and S. Müller, *Tetrahedron Lett.* **1981**, *22*, 1675.

101. (a) The Dewar-benzene analogue of hexakis(pentafluoroethyl)benzene becomes energetically favored above 278°C for entropic reasons: E. D. Clifton, W. T. Flowers, and R. N. Hazeldine, J. Chem. Soc. Chem. Commun. **1969**, 1216; A.-M. Dabbagh, W. T. Flowers, R. N. Hazeldine, and P. J. Robinson, *J. Chem. Soc. Chem. Commun.* **1975**, 323. (b) for a thermal valence isomerization between the dianions of a naphthalene and a hemi Dewar-naphthalene see: I. B. Goldberg, H. R. Crowe, and R. W. Franck, *J. Am. Chem. Soc.* **1976**, *98*, 7641.

102. V. Jäger and H. G. Viehe, *Angew. Chem.* **1970**, *82*, 836; *Angew. Chem. Int. Ed. Engl.* **1970**, *9,* 795.

103. G. Maier, G. Fritschi, and B. Hoppe, *Angew. Chem.* **1970**, *82*, 551; *Angew. Chem. Int. Ed. Engl.* **1970**, *9,* 529.

104. J. Fink and M. Regitz, *Bull. Soc. Chim. Fr.* **1986**, 239.

105. We are grateful to Professor Klärner, Universität Bochum, for carrying out the high pressure investigations.

106. G. Maier and H. P. Reisenauer, *Tetrahedron Lett.* **1976**, 3591.

107. G. Maier, H. P. Reisenauer, and H.-A. Freitag, *Tetrahedron Lett.* **1978**, 121.

108. P. Reeves, J. Henery, and R. Pettit, *J. Am. Chem. Soc.* **1969**, *91*, 5888.

109. G. Michels, M. Hermersdorf, J. Schneider, and M. Regitz, *Chem. Ber.* **1988**, *121*, 1775.

110. (a) A. Berndt and K. Schreiner, *Angew. Chem.* **1976**, *88*, 764; *Angew. Chem. Int. Ed. Engl.* **1976**, *15*, 1625. (b) H. M. Walborsky, *Tetrahedron* **1981**, *37*, 1625.

111. (a) S. Masamune, M. Sakai, H. Ona, and A. J. Jones, *J. Am. Chem. Soc.* **1972**, *94*, 8956. (b) S. Masamune, *Pure Appl. Chem.* **1975**, *44*, 861.

112. Concerning the synthesis of trimethyl substituted pyramidal cations see: V. I. Minkin, N. S. Zefirov, M. S. Korobov, A. V. Averina, A. M. Boganov, and L. E. Nivorozhkin, *Zh. Org. Khim.* **1981**, *17*, 2616.

113. Compare: S. Masamune, M. Sakai, H. Ona, and A. J. Jones, *J. Am. Chem. Soc.* **1972**, *94*, 8955.

114. See also: W. D. Stohrer and R. Hoffmann, *J. Am. Chem. Soc.* **1972**, *94*, 1661.

115. G. A. Olah, D. P. Kelly, C. L. Jeuell, and R. D. Porter, *J. Am. Chem. Soc.* **1970**, *92*, 2544.

116. (a) M. Saunders and J. Rosenfeld, *J. Am. Chem. Soc.* **1970**, *92*, 2548. (b) G. A. Olah, C. L. Jeuell, D. P. Kelly, and R. D. Porter, *J. Am. Chem. Soc.* **1972**, *94*, 146.

117. (a) M. Saunders and H.-U. Siehl, *J. Am. Chem. Soc.* **1980**, *102*, 6869. (b) W. J. Brittain, M. E. Squillacote, and J. D. Roberts, *J. Am. Chem. Soc.* **1984**, *106*, 7280. (c) H.-U. Siehl, *J. Am. Chem. Soc.* **1985**, *107*, 3390. (d) G. K. S. Prakash, M. Arvanaghi, and G. A. Olah, *J. Am. Chem. Soc.* **1985**, *107*, 6017. (e) H.-U. Siehl and E.-W. Koch, *J. Chem. Soc. Chem. Commun.* **1985**, 496. (f) H.-U. Siehl, *Adv. Phys. Org. Chem.* **1987**, *23*, 63.

118. R. P. Kirchen and T. S. Sorensen, *J. Am. Chem. Soc.* **1977**, *99*, 6687.

119. (a) H. Rang, Ph.D. thesis, Universität Giessen, 1989. (b) G. Maier, H. Rang, and H.-O. Kalinowski, *Angew. Chem.*, **1989**, *101*, 1293; *Angew. Chem. Int. Ed. Engl.* **1989**, *28*, 1232.

8 Homologues of Barrelene, Bullvalene, and Benzene: Concepts, Questions, and Results

ARMIN DE MEIJERE
Institute for Organic Chemistry
University of Göttingen
Göttingen, Federal Republic of Germany

INTRODUCTION

The cage hydrocarbons barrelene (**1**, bicyclo[2.2.2]octatriene)[1] and bullvalene (**2**, tricyclo[3.3.2.0^{4,6}]deca-2,7,9-triene),[2] as well as the planar benzene (**3**) have one common feature, though at first sight only a formal one: All three of them contain three C=C double bonds in cyclic arrays, yet under different geometric constraints (see Fig. 1).

Conceived to have a barrel-shaped π-electron distribution—thus its name—the D_{3h} symmetric barrelene (**1**) has been the subject of many theoretical[3] and experimental[1,4] investigations. In contrast to the cyclic 6π-electron ribbon of benzene, which is a Hückel type, that of barrelene has a node in the ground state and thus is a Möbius type.[1b,5] As was recently verified by ESR spectroscopy in a Freon matrix, the radical cation ($\mathbf{1}^{\bullet+}$) has a nondegenerate A'_2 ground state[6] in full accordance with theoretical predictions.

Bullvalene (**2**) is best known for its unique feature of being a fluctuating structure. Predicted by Doering[7] and first synthesized by Schröder,[2] bullvalene with its interconnected *cis*-1,2-divinylcyclopropane subunits at room temperature (25°C) undergoes ~4800 Cope rearrangements per second,[2b] and these average out all differences between the 10 hydrogen atoms and the 10 carbon atoms in over 1.2 million different connectivities.

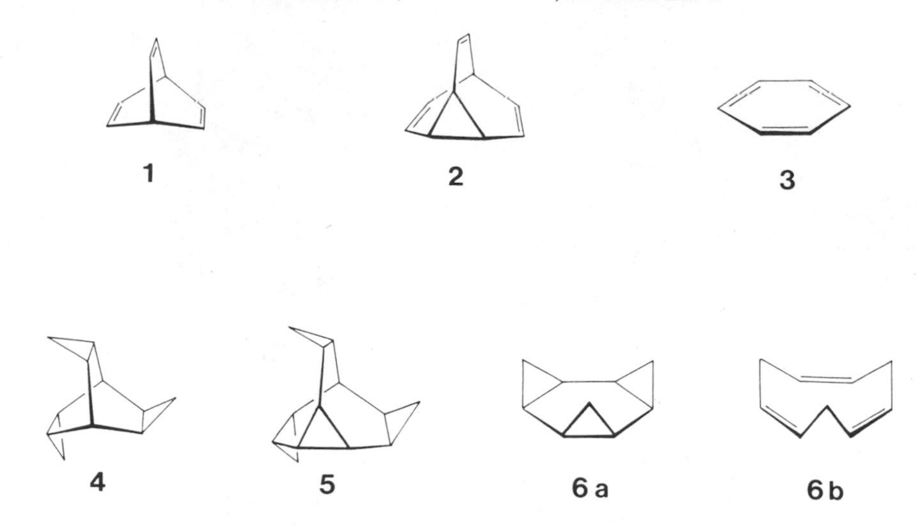

Fig. 1. Barrelene (**1**), bullvalene (**2**), benzene (**3**), and their trishomologues **4**, **5**, **6a**/**6b**.

Interesting homologues of **1**, **2**, and **3** arise by replacing one, two, or all three double bonds with three-membered rings. The resulting trishomologues are **4**, **5**, and **6a** (see Fig. 1). A different type of homologue is created by insertion of CH_2 groups into a C—C single bond of any such skeleton. As cyclonona-1,4,7-triene (**6b**) had been termed trishomobenzene earlier,[8] it has been suggested to differentiate between the two as tris-σ-homobenzene (**6a**) and tris-π-homobenzene (**6b**).[9] This account will primarily deal with the σ homologues **4**, **5**, and **6a**, although, as we will see, **6a** and **6b** are interdependent. In fact, this chapter will concentrate on cage versions of **6a**, which result by attaching additional bridges onto the corners of all three cyclopropyl groups. The hexacyclic cages **7** and **8** (see Fig. 2) are derivatives of *cis*-tris-σ-homobenzene (**6a**), the half-cage hydrocarbon **10** is a bridged version of *trans*-tris-σ-homobenzene **9**.

All the reasons for seriously dealing with such σ homologues of **1**, **2**, and **3** are derived from the structural and electronic peculiarities of the three-membered ring. As the well-known molecular orbital (MO) description of cyclopropane by Walsh illustrates,[10] the three-membered ring is not only formally, but electronically a homoethylene: Its three occupied MOs, one σ- and two degenerate π-type MOs, closely resemble the familiar bonding picture of ethylene, the σ,π description by Mulliken. As a consequence of this electronic relationship cyclopropane derivatives in many chemical reactions behave like olefins.[11,12] In one of its properties, though, the cyclopropyl group is outstanding: In its ability to act as an electron donor for electron deficient centers attached to it, the cyclopropyl group even surpasses a vinyl and a phenyl group.

The geometric requirements for the stabilization of, for example, a carbenium ion center by an adjacent vinyl or phenyl group on one side and a cyclopropyl

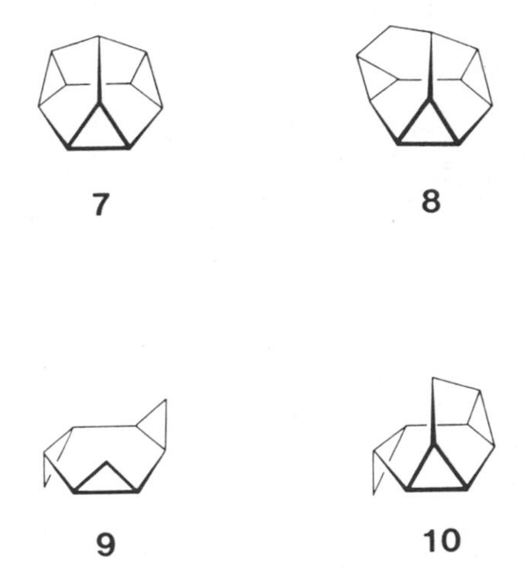

Fig. 2. Diademane (**7**) (hexacyclo[$4.4.0.0^{2,10}.0^{3,5}.0^{4,8}.0^{7,9}$]decane), 1,6-homodiademane (**8**) (hexacyclo[$5.4.0.0^{2,11}.0^{3,5}.0^{4,9}.0^{6,8}$]undecane), *trans*-tris-σ-homobenzene (**9**), and pentacyclo[$5.3.0.0^{2,10}.0^{3,5}.0^{6,8}$]decane (**10**).

group on the other side are, however, quite different in one respect. In both cases maximum stabilization is attained with parallel orientation of the empty *p* orbital and the π or quasi-π orbitals of the donor group, that is, in a coplanar arrangement of an allyl and a bisected orientation of a cyclopropylcarbinyl cation [see Fig. 3, (A)/(B) and (D)/(E)]. In a perpendicular orientation a vinyl and a cyclopropyl group (C) and (F) exert a σ-electron withdrawing effect and thus destabilize an adjacent carbenium ion center (see below). The important geometric difference, which can play a role especially in conformationally rigid systems, is the following: Whereas the four substituents attached to the two carbon atoms of a double bond are located in one plane (*S*) including the carbon atoms, and this plane is oriented perpendicular to the plane (*O*) of the π-orbital axes (interplanar angle $\alpha = 90°$ in Fig. 3), four substituents attached to two carbon atoms of a cyclopropane ring define, together with the carbon atoms, two distinct planes (S^1 and S^2) enclosing an angle of ~ 120° and forming angles of ~ 60° with the plane of the three-membered ring, and thus also with that of the *p*-orbital axes (*O*) ($\alpha = 60°$ in Fig. 3).

As a consequence of the geometric requirements for allyl- and cyclopropylcarbinyl cation stabilization the bridgehead cations of barrelene (**11**) and bullvalene (**14**) would be destabilized rather than stabilized with respect to the 1-bicyclo[2.2.2]octyl (**12**) and 1-tricyclo[$3.3.2.0^{4,6}$]decyl cation (**15**) because the α-vinyl-groups on C-1 in **11** and **14** are rigidly held orthogonal in relation to the empty bridgehead orbitals (see Fig. 4). Aside from the fact that it would be difficult to obtain a pure bridgehead derivative of **2** due to its rapid valence isomerization,[13] bridgehead derivatives of **1** and **2** would solvolyze extremely

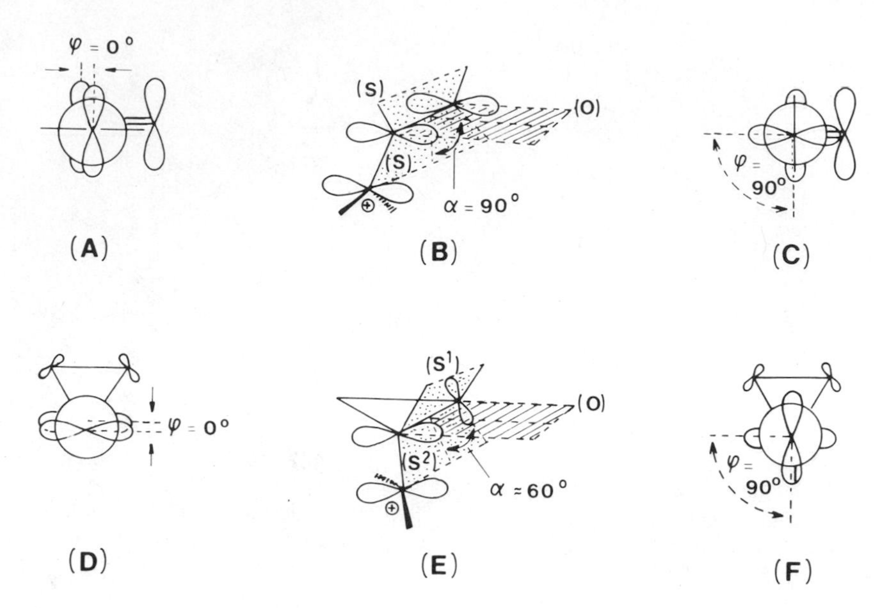

Fig. 3. Newman and sawhorse projections of coplanar [(A) and (B), respectively] and perpendicular (C) allyl cation as well as cyclopropylcarbinyl cation [(D) and (E), respectively] and (F).

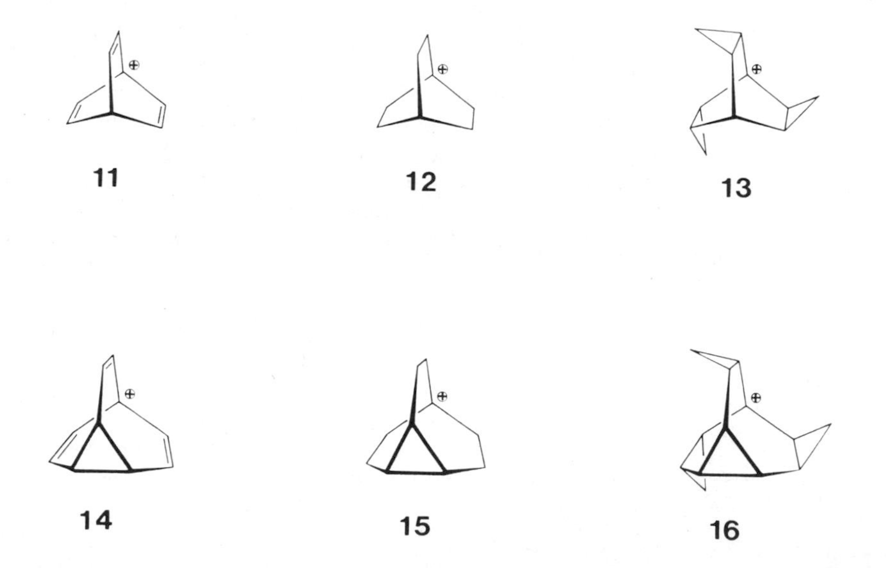

Fig. 4. Bicyclo[2.2.2]octyl and tricyclo[3.3.2.$0^{4,6}$]decyl (hexahydrobullvalenyl) derived bridgehead carbenium ions.

slowly, much like 1-halotriptycene.[14] In contrast, the cyclopropyl homologues of **11** and **14**, that is, the trishomobarrelenyl (**13**) and trishomobullvalenyl cation (**16**) ought to be stabilized with respect to the saturated parent systems **12** and **15**, because the interorbital angle between the empty bridgehead orbital and that of the adjacent cyclopropyl carbon will always be definitely smaller than 90°—somewhere ~ 60°—thus permitting substantial orbital overlap. In fact, there ought to be a whole series of cyclopropyl stabilized bridgehead carbenium ions derived from the bicyclo[2.2.2]octyl and tricyclo[3.3.2.$0^{4,6}$]decyl cages by attachment of one, two, or three cyclopropyl groups onto the two carbon bridges α to C-1. All these bridgehead cyclopropylcarbinyl cations are rigidly held at conformational angles between the two extremes 90° and 0°. Such bridgehead derivatives are thus ideal models to experimentally test the angular dependence of cyclopropylcarbinyl cation stabilization and complement the two rigid adamantane derived systems, **17** with $\varphi = 90°$ (Ref. 15) and **19** with $\varphi = 0°$ (see Fig. 5).[16] The total range of stabilization as defined by the two end points **17** and **19** is 16 kcal/mol corresponding to an observed total rate ratio of 5×10^{11} after normalization.

HOMOBARRELENES, HOMOBULLVALENES, AND SOME OF THEIR BRIDGEHEAD DERIVATIVES

Although barrelene (**1**) would be a logical precursor to any of its homologues, the fact that it cannot easily be prepared in large quantities[1] is somewhat restricting. Since monohomobarrelene (**21**) is accessible in one step from the Diels–Alder adduct of cycloheptatriene and maleic anhydride,[17] it was first cyclopropanated by the cuprous chloride catalyzed addition of diazomethane and under Simmons–Smith conditions to yield both stereoisomeric bishomobarrelenes (**22**/**23**) and trishomobarrelene **4** (Fig. 6).[18] As the *exo,exo*-bishomobarrelene (**23**), which cannot be methylenated any further, predominates in both cases, a stereoselective approach to *endo,exo*-bishomobarrelene (**22**) and thus to **4** was developed, starting from the ethylene acetal of tricyclo-[3.2.2.$0^{2,4}$]non-6-en-8-one (**24**). In a sequence of Simmons–Smith cyclopropanation, deprotection of the carbonyl group and Shapiro reaction **22** could be obtained in reasonable quantities and further methylenated with diiodomethane/diethylzinc to give **4** in good yield (69%; total yield based on tricyclononenone **24**: 24%).[18]

The corresponding homologues **27**–**29** and **5** of bullvalene (**2**) were obtained by methylenation of this easily accessible hydrocarbon (Fig. 6). With the modified Simmons–Smith reaction of Furukawa and Kuwabata the cyclopropanation proceeded with a diastereoselectivity of ~ 2:1 in favor of the *endo,exo* isomer **29** and thus trishomobullvalene (**5**), when an excess of reagent was employed. The total isolated yield of **28** and **5** was 86%.[18]

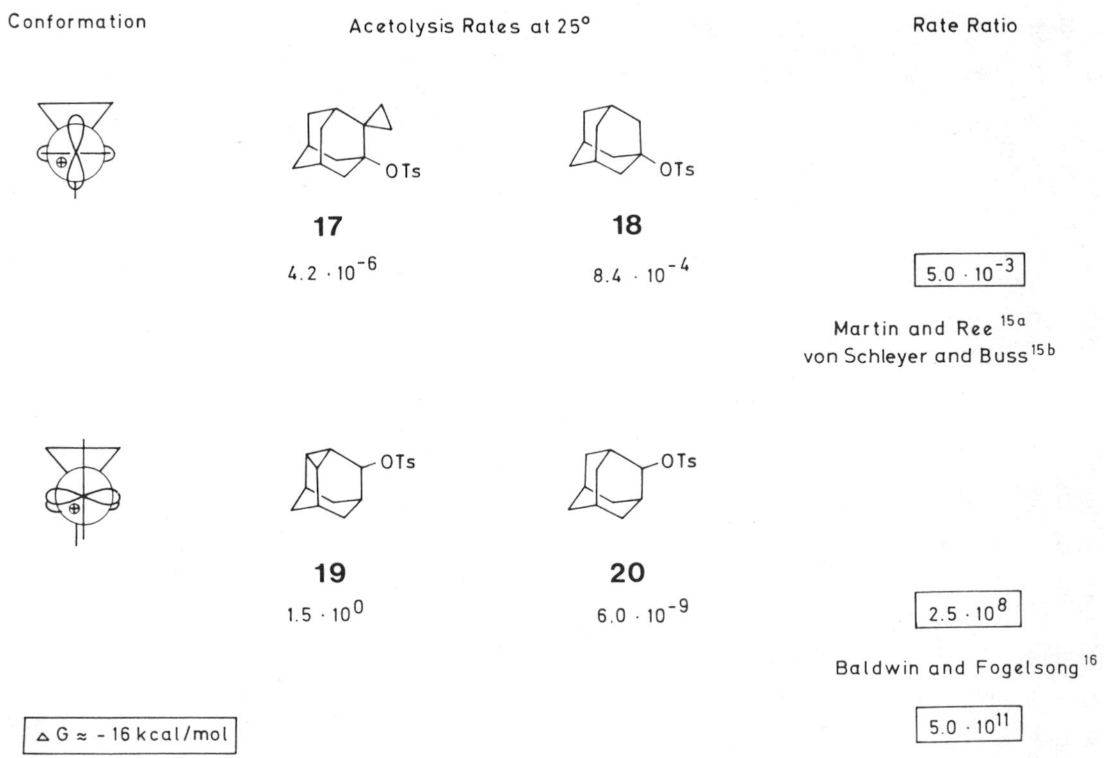

Conformation

Acetolysis Rates at 25°

Rate Ratio

OTs

17

$4.2 \cdot 10^{-6}$

OTs

18

$8.4 \cdot 10^{-4}$

$5.0 \cdot 10^{-3}$

Martin and Ree [15a]
von Schleyer and Buss [15b]

OTs

19

$1.5 \cdot 10^{0}$

OTs

20

$6.0 \cdot 10^{-9}$

$2.5 \cdot 10^{8}$

Baldwin and Fogelsong [16]

$\Delta G \approx -16$ kcal/mol

$5.0 \cdot 10^{11}$

Fig. 5. Rates of acetolysis of rigid perpendicular (**17**) and bisected cyclopropylcarbinyl (**19**) tosylates compared to 1- and 2-adamantyl tosylates (**18** and **20**), respectively.

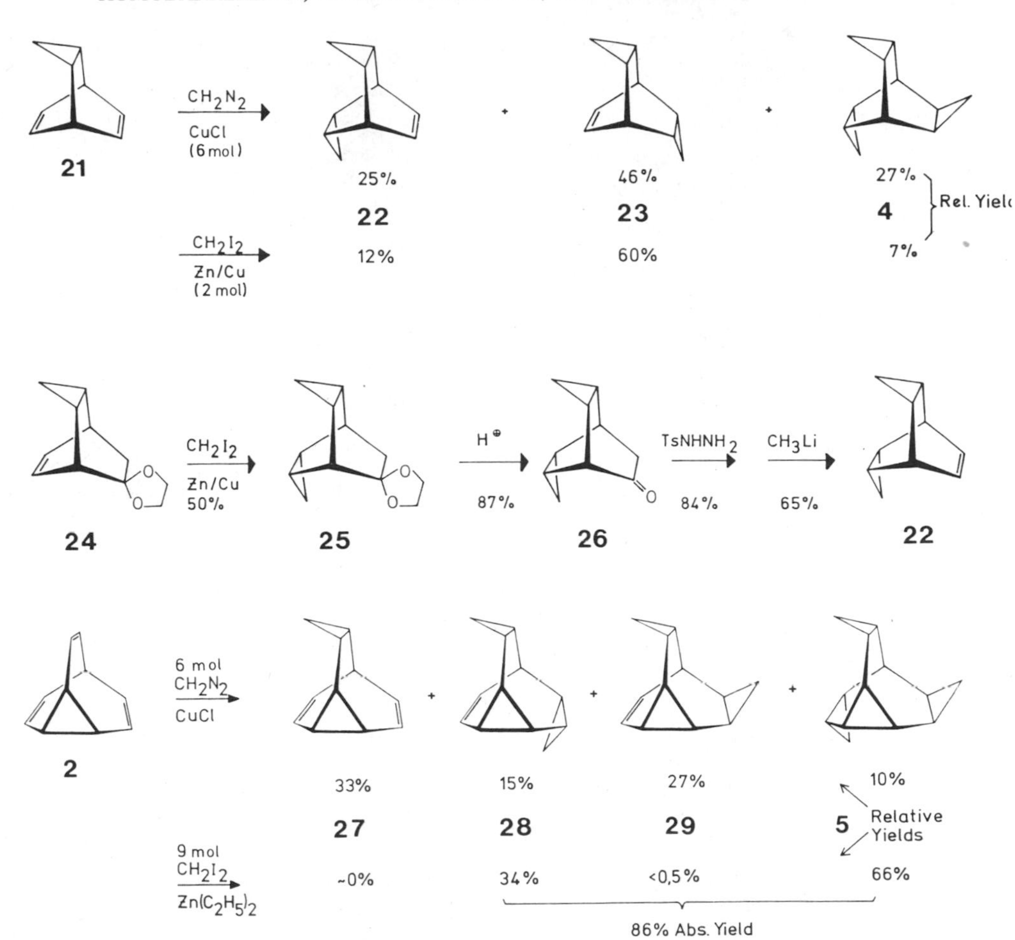

Fig. 6. Homologues of **1** and **2** by cyclopropanation of homobarrelene (**21**) and bullvalene (**2**).

The hydrocarbons **4** and **5** can selectively be functionalized at their bridgehead positions by low-temperature photochlorination with *t*-butyl hypochlorite.[19,20] 1-Chloro-trishomobarrelene (**30-Cl**) and 1-chlorotrishomobullvalene (**33-Cl**) readily solvolyze in aqueous dioxane or ethanol to the bridgehead alcohols **30-OH** and **33-OH**, respectively, without any skeletal rearrangement (Fig. 7). Compound **33-Cl** is indeed so reactive that it hydrolyzes from the moisture in air to give **33-OH** simply upon standing.[19] The alcohols **30-OH** and **33-OH** can be obtained directly from the hydrocarbons in high yields (72 and 60%) with ozone adsorbed on silica gel (so-called dry ozonation),[20,21] the latter one also by autoxidation.[19]

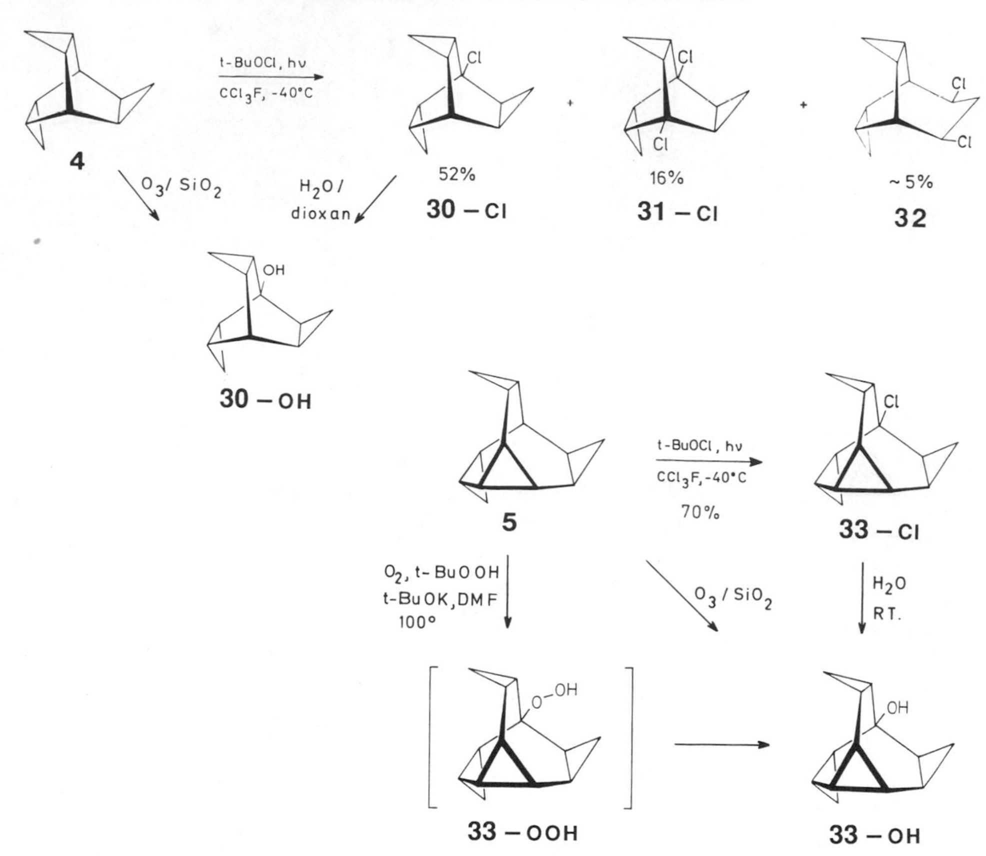

Fig. 7. Bridgehead functionalization of trishomobarrelene (**4**) and trishomobullvalene (**5**).

The complete selectivity for bridgehead attack in **4** and **5** in radical substitution reactions is not only due to the weak, yet definite radical stabilization by α-cyclopropyl groups,[11] but also to the so-called polar effect in carbon–hydrogen abstractions by electronegative radicals[22] causing the development of a partial positive charge at the corresponding carbon atom. Such a positive charge is apparently well stabilized by the three cyclopropyl groups adjacent to C-1 in **4** and **5**, less so, of course, by only one (as in **36**) or two α-cyclopropyl groups (as in **37–39**). Consequently the degree of preferred bridgehead hydrogen abstraction decreases with a decreasing number of cyclopropyl groups annelated at the two carbon bridges of a bicyclo[2.2.2]octyl skeleton (see Table 1) and even more with a spirocyclopropyl group attached as in **35**, which corresponds to a perpendicular cyclopropylcarbinyl conformation as in **17**.

The decomposition of *t*-butyl peresters also proceeds with a late transition state involving a certain degree of charge separation.[25] In accord with this mechanistic implication, the bridgehead *t*-butyl peroxycarboxylates **41** and **42**

TABLE 1. Hydrogen Abstraction by *t*-Butoxy Radicals from Different Hydrocarbons with Bicyclo[2.2.2]octyl Skeletons.[23,24]

	Tert. position (bridgehead-CH)	Sec. position (bridge - CH_2)	Statistical weight tert. : sec	Ratio tert./sec. statist. corr.
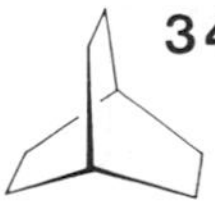 34	25 %	75 %	1 : 6	2.0
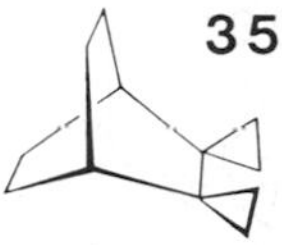 35	23 %	77 %	1 : 4	1.2
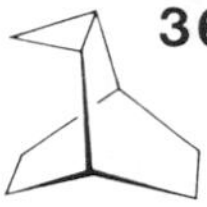 36	57 %	43 %	1 : 4	5.3
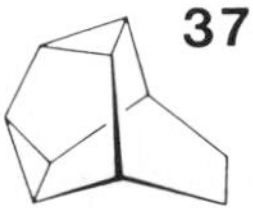 37	65 %	35 %	1 : 2	3.8
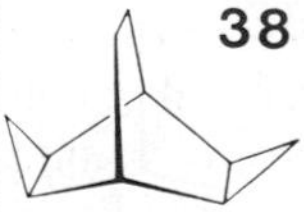 38	86 %	14 %	1 : 2	12.3
 39	89 %	11 %	1 : 2	16.2

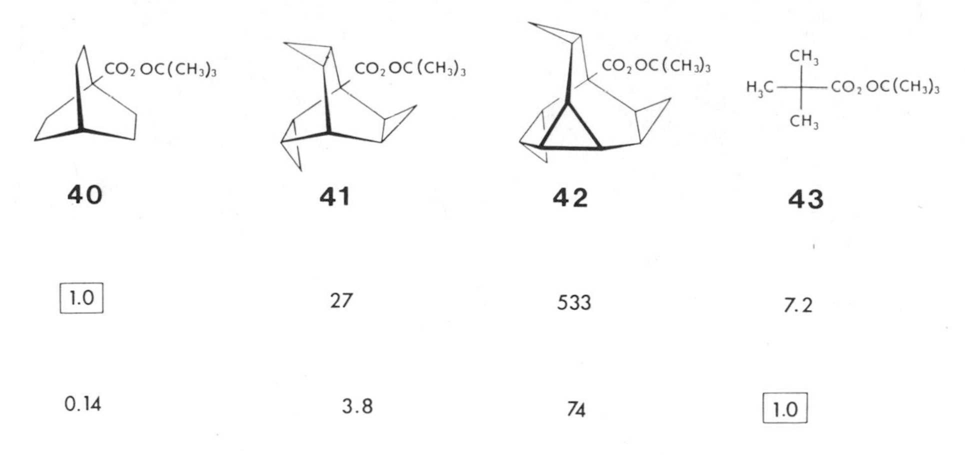

40	41	42	43
1.0	27	533	7.2
0.14	3.8	74	1.0

Fig. 8. Relative thermolysis rates of some bridgehead *t*-butyl peroxycarboxylates.[24,26]

fragment 27 and 533 times more rapidly than the bicyclooctyl compound (**40**) and even faster than *t*-butyl peroxypivaloate (**43**) (Fig. 8).[24,26]

BRIDGEHEAD REACTIVITIES OF HOMOBARRELENES AND HOMOBULLVALENES

Since bridgehead derivatives can only solvolyze by the S_N1 mechanism, their solvolysis rates correlate particularly well with relative stabilities of the corresponding intermediate carbenium ions. Solvolysis rates were determined for 1-chlorotrishomobarrelene (**30-Cl**),[19,23] 1-chlorotrishomobullvalene (**33-Cl**),[19,27] and 1-chlorohexahydrobullvalene (**45**)[19,27] in aqueous dioxane and/or aqueous ethanol and recalculated for 80% aqueous ethanol according to the Grunwald–Winstein relationship.[28] Comparing these values with the known rate for *t*-butyl chloride (**46**) and the estimated one for 1-chlorobicyclo[2.2.2]octane (**44a**)[29] (see Fig. 9) shows rate enhancements by factors of 2.7×10^8 and 8.6×10^{11} for **30-Cl** and **33-Cl**, respectively, over that of **44a**. Both chlorides **30-Cl** and **33-Cl** are more reactive than *t*-butyl chloride (**46**), the latter one by a factor of 1.7×10^5. With a half-life of only 2 s at room temperature **33-Cl** solvolyzes so rapidly that the stopped flow method had to be employed for the kinetic measurements.[19,27] In fact, **33-Cl** exhibits the greatest bridgehead reactivity found so far for any system, reacting about 10 times faster than 1-chlorobicyclo[3.3.3]undecane (**49**) (manxyl chloride).[30] While **30-Cl** has virtually the same bridgehead geometry as the bicyclo[2.2.2]octyl system (**44**), the trishomobullvalene skeleton in **33-Cl** has not. The reactivity of **33-Cl** should therefore be compared to that of 1-chlorohexahydrobullvalene (**45**), which has a similar bridgehead geometry with larger C—C—C bond angles. Thus, only the fraction 3.9×10^6 of the total rate ratio 8.6×10^{11} is due to the effect

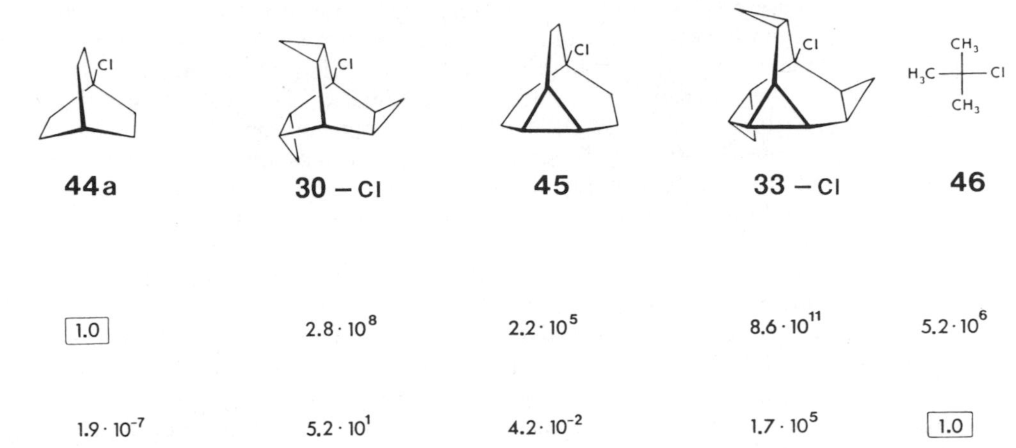

Fig. 9. Relative rates of solvolysis (in 80% aqueous ethanol) for bridgehead chlorides compared to *t*-butyl chloride.[19,23,27]

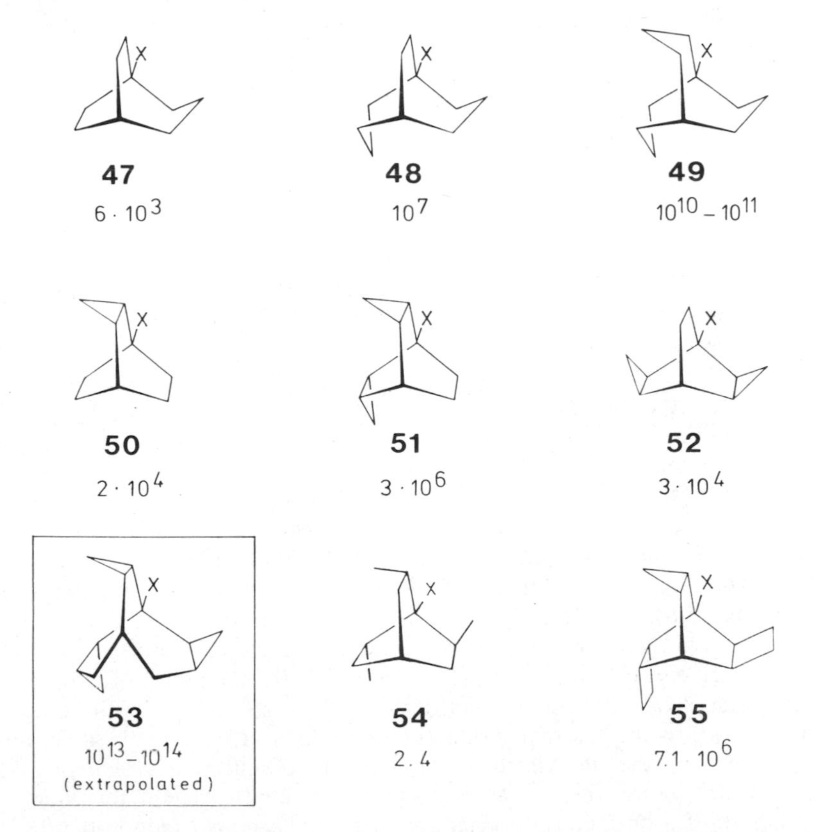

Fig. 10. The effect of ring size and α-cyclopropyl groups on bridgehead reactivities of selected systems (relative rates based on 1-chlorobicyclo[2.2.2]octane k_{rel} = 1.0).

of three α-cyclopropyl groups, the remaining part stems from the fact that the bridgehead carbon in skeletons with larger bridges can more easily be planarized on going to the more stable sp^2-hybridized carbenium ion. In accordance with this general phenomenon for bridgehead carbenium ion stabilities,[31] the solvolysis rates for the series **47–49** increase by about the same factors or even more than those for **50**, **51**, **30-Cl** (Fig. 10). But in the first series the enhanced carbenium ion stability is due to a smaller strain increase on planarizing one bridgehead carbon, while in the second series it is caused by the electron donating effect of the α-cyclopropyl groups. In combining the two effects, that is, by attaching three cyclopropyl groups onto for example, a bicyclo[3.3.2]decyl system, one ought to obtain a bridgehead derivative **53**, which would surpass the most reactive systems by a factor of 10^2–10^3, only a 1-chlorobicyclo[4.4.4]tetradecyl derivative might be even more reactive.[32]

As the bicyclo[3.3.2]deca-2,7,9-triene can be prepared by alkali metal reduction of bullvalene (**2**), the hydrocarbon precursor to **53** can be prepared just like trishomobullvalene by exhaustive cyclopropanation of the triene.[21]

In order to clearly separate geometry and electron donor effects on the bridgehead reactivities in the two series **47–49** and **50**, **51**, and **30-Cl**, the solvolysis rates of 1-chloro-2-*endo*,6-*exo*,7-*syn*-trimethylbicyclo[2.2.2]octane (**54-Cl**) and 1-chloro-*exo*,*exo*-dihydrobishomobarrelene (**52-Cl**) were also determined. Compound **54-Cl** reacts only 2.4 times as fast as the parent compound, which clearly demonstrates the efficiency of the α-cyclopropyl groups as electron donors in these cage systems.[33] The donor ability of a cyclobutyl neighboring group as in **55** must be somewhere intermediate between that of an α-methyl and an α-cyclopropyl neighbor, since the solvolysis rate of **55-Cl** is increased by a factor of 15.6 for each cyclobutyl group over that of **50-Cl**.[34]

The solvolysis rate of the bridgehead chloride **52-Cl** in comparison to that of its endo,exo-isomer **51-Cl** provided the clue to a detailed analysis with respect to the real geometries of the intermediate bridgehead carbenium ions in the whole series.[23] The compounds **51-Cl** and **52-Cl** both have two α-cyclopropyl groups and virtually the same bridgehead geometries, yet their solvolysis rates differ by a factor of 175, corresponding to a free energy difference of ~ 3 kcal/mol for the intermediate carbenium ions **56** and **57**, respectively. This additional stabilization of **56** can only stem from an increased overlap between the empty bridgehead orbital and the adjacent Walsh orbitals due to a twisting of the skeleton (see Fig. 11). As **57** cannot gain any stabilization from such a twist in either direction, the geometry of its precursor chloride **52-Cl** resembles that of the transition state leading to **57**. With **52-Cl** as a model for untwisted carbenium ions in this series, the complete analysis of solvolysis rates for **50**, **51**, and **30-Cl** (see Fig. 10) reveals a decreasing degree of twisting upon attaching one, two, and eventually three cyclopropyl groups onto the bridges of a bicyclo[2.2.2]octyl cation (**12**).[23] The same holds for the series of cyclopropyl stabilized bridgehead carbenium ions derived from **15**, and in both series the trishomologues are the least distorted (twist angle in **13** indirectly determined as 3.6°) next to the untwisted exo,exo-bishomologues. Consequently, the tricyclo[$3.2.2.0^{2,4}$]nonyl (monohomotetrahydrobarrelenyl) system, for which the solvolysis rate of a bridgehead tosylate had been determined independently,[35] is the least best model to derive an accurate quantitative relationship between

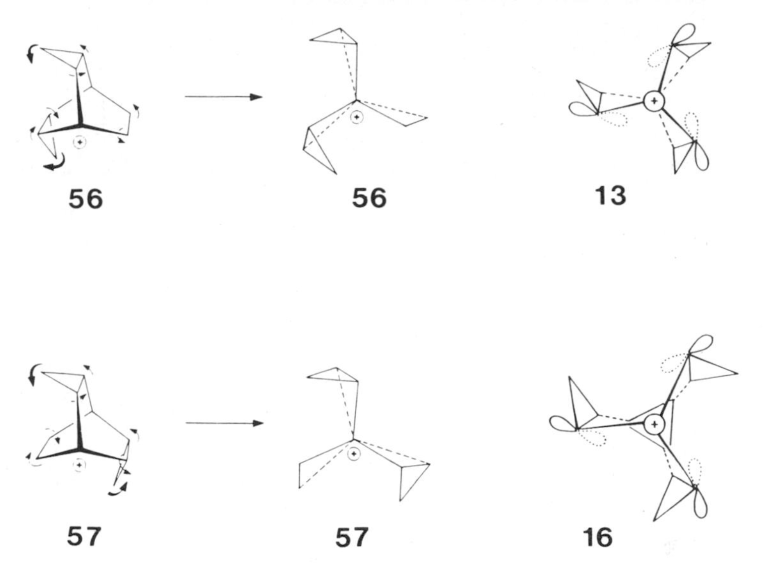

Fig. 11. Twisted geometries of *endo,exo*- (**56**) and *exo,exo*-dihydrobishomobarrelene (**57**) as well as trishomobarrelene (**13**) and trishomobullvalene (**16**) bridgehead carbeniums ions.

conformation and stability of a cyclopropylmethyl cation, because the transition structure leading to the tricyclo[3.2.2.0^{2,4}]nonyl bridgehead cation is twisted by ~ 12°.[23]

Paying due attention to all relevant structural factors, the relative stabilization energy E_φ of a cyclopropylmethyl cation was found to depend upon the torsional angle φ following the $\cos^2_\varphi$ function

$$E_\varphi = E_{\varphi=0} \cdot \cos^2\varphi$$

where $E_{\varphi=0}$ = 16.2 kcal/mol.[23] Apart from this value for maximum stabilization the function is identical to that calculated by the complete neglect of differential overlap (CNDO) method.[36] The trishomobarrelenyl cation (**13**) corresponds to a point at φ = 49.1°, the untwisted cation from **57** has φ = 52.7°as taken from the trishomobarrelene skeleton.[27]

An analogous detailed analysis for a complete family of bridgehead derivatives with homobullvalene skeletons has not yet been carried out, though it may well be deserving. Experimental difficulties prevented the determination of solvolysis rates for the highly reactive bridgehead chlorides (**58-Cl**, **59-Cl**, and **60-Cl**). While **58-Cl** and **59-Cl** could be prepared from the corresponding alcohols with thionyl chloride in the presence of diethylaminomethyl-polystyrene,[21] the dihydro-*exo,exo*-bishomobullvalenylalcohol (**60-OH**) under these conditions only gave the rearranged product 10-*exo*-chlorobicyclo[4.4.2]-dodeca-1,4,7-triene (**62**) (see Fig. 12). Whether **62** is formed from **60-Cl** by consecutive cyclopropylcarbinyl to homoallyl rearrangements or via the interesting ethanobridged trishomocycloheptatrienyl cation (**61**), could not be

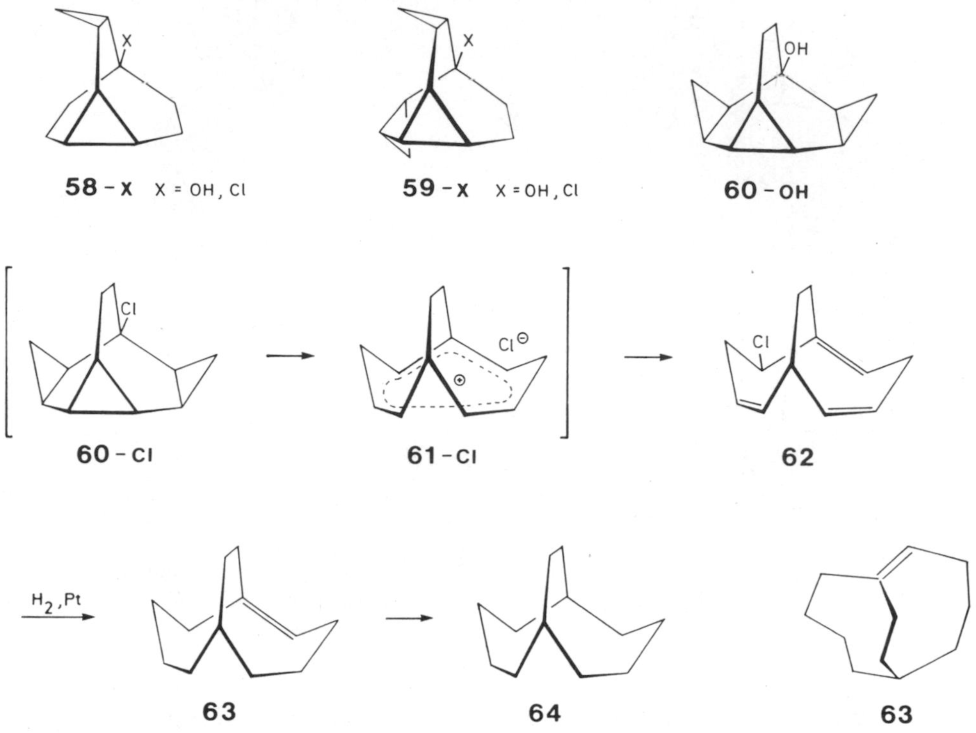

Fig. 12. Selected bridgehead derivatives **57–59** with a common hexahydrobullvalene skeleton and preparation of the "hyperstable" bridgehead olefin **63**.

verified experimentally.[37] But **62** itself is an interesting compound in that it belongs to the rare class of so-called hyperstable bridgehead olefins.[38] This is evidenced by the fact that catalytic hydrogenation of **62** leads predominantly to bicyclo[4.4.2]dodec-1-ene (**63**), the strain of which has been calculated to be 13 kcal/mol smaller than that of the corresponding saturated bicyclododecane (**64**).[38] Thus, **63** apparently contains an almost undistorted bridgehead double bond (see Fig. 12), whose catalytic hydrogenation leading to the saturated **64** proceeds extremely slowly.

FREE BRIDGEHEAD CARBENIUM IONS WITH HOMOBARRELENE AND HOMOBULLVALENE SKELETONS

Although the trishomobarrelenyl bridgehead cation **13** experiences only 43% of the maximum stabilizing power of its α-cyclopropyl groups—and in the trishomobullvalenyl cation (**16**) the effective fraction is even lower

(30–35%)—the cumulative effect of three adjacent cyclopropyl groups stabilizes **13** and **16** by 11.5 and 9.0 kcal/mol, respectively, in relation to the bicyclo[2.2.2]octyl (**12**) and hexahydrobullvalenyl cation (**15**).[27] Consequently, the "free" bridgehead cations **13** and **16** can be readily generated from the corresponding chlorides **30-Cl** and **33-Cl**[23,39] under long lived ion conditions as developed by Olah et al. (Fig. 13).[40] A comparison of their ^1H NMR chemical

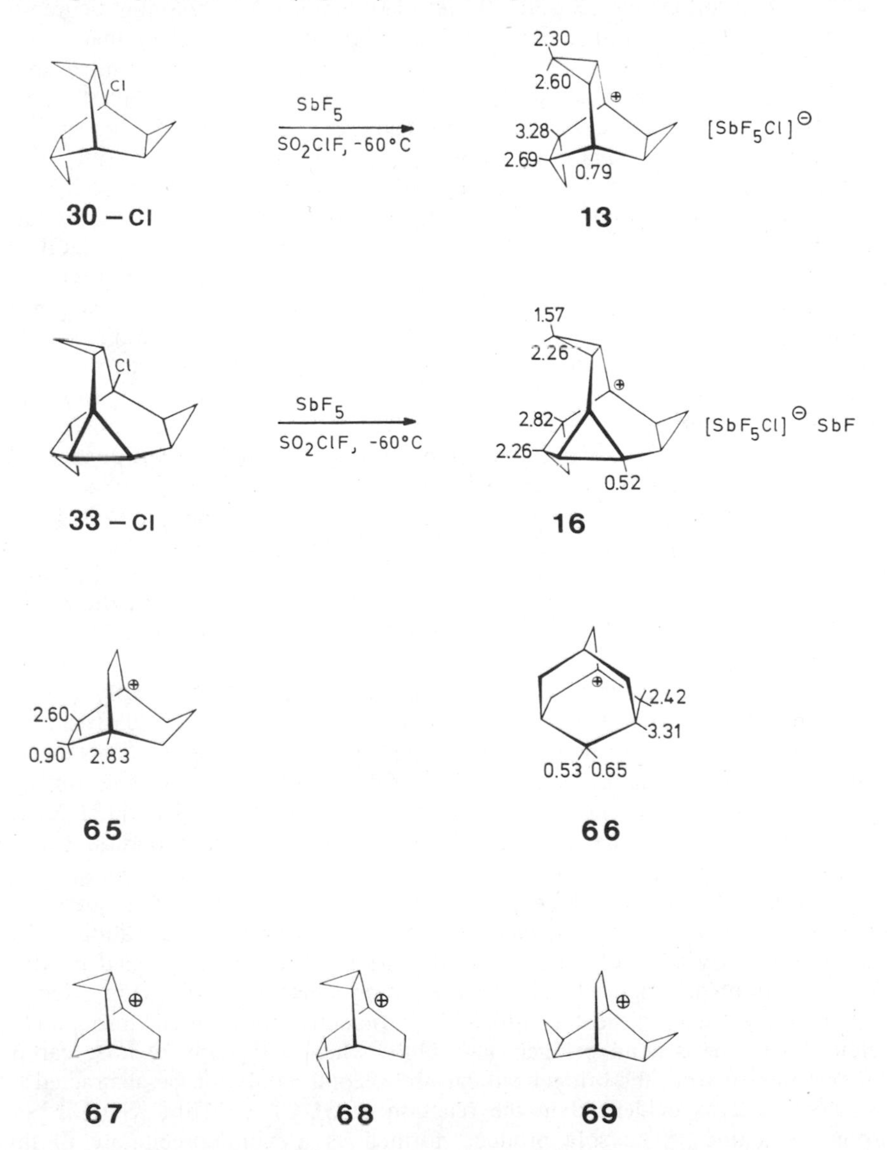

Fig. 13. Bridgehead carbenium ions with homobarrelene and homobullvalene skeletons. ^1H NMR chemical shift (δ_{TMS}) for **13**, and **16** in comparison to 1-bicyclo[3.2.2]nonyl (**65**) and 1-adamantyl cation (**66**).

shifts with those of 1-bicyclo[3.2.2]nonyl (**65**) and 1-adamantyl cation (**66**) suggests considerable delocalization of the positive charge into the cyclopropyl groups, as has been demonstrated for cyclopropyl methyl cations with other conformations.[41] In spite of the documented limitations to a quantitative correlation between ^{13}C chemical shifts and charge densities,[42] the variation in C-1 chemical shifts in the series of cations **67** [δ(C-1) = 306.6 ppm], **68** (300.7), **69** (306.1), and **13** (287.7) definitely indicates a varying degree of charge density at the bridgehead position, which is smaller than that on the central carbon in *t*-butyl cation (δ(C-2) = 330 ppm]. The ^{13}C chemical shifts of the cyclopropyl carbon atoms in **67–69** and **13** vary accordingly.[23,43]

All these bridgehead carbenium ions can be quenched with nucleophiles yielding bridgehead derivatives without any skeletal rearrangement. The outstanding stability of the trishomobullvalenyl cation (**16**) is evidenced in its reaction with carbon monoxide. While **13** readily adds to CO giving predominantly the carboxylic acid **30-CO$_2$H** after treatment with NaOMe/MeOH at –78°C and H_2O at room temperature, **16** yields mainly the methyl ether **33-OMe** under the same conditions (Fig. 14). Apparently, the bridgehead carbenium ion **16** is more stable than the corresponding carboxonium ion (**16**-CO$^+$ and predominates in the equilibrium **16** $\rightleftharpoons$ **16**-CO$^+$. In fact, **16** is so stable that it can be generated directly by hydride abstraction from the hydrocarbon **5** with either isopropyl, *t*-butyl or even the trishomobarrelenyl cation **13** (Fig. 15).[24] The tropylium ion, however, is more stable than **16** as cycloheptatriene converts **16** back to **5**. The solution of **16** can be treated with nucleophilic reagents such as potassium cyanide in methylene chloride in the presence of 18-crown-6 to give the bridgehead nitrile **33-CN** with up to 80% yield on a small scale; for larger scale preparations of **33-CN**, the ideal precursor to the carboxylic acid **33-CO$_2$H** (see below), treatment of the methyl ether **33-OMe** (or the alcohol **33-OH**) with KCN in trifluoroacetic acid is preferable.[24]

As much as the trishomobarrelenyl monocation **13** is energetically favored over the hypothetical 1-bicyclo[2.2.2]octyl cation **12**, the 1,5-trishomobarrelenediyl dication **71** should be stabilized over the bicyclo[2.2.2]octanediyl dication (**70**), at least as predicted by MINDO/3 calculations.[43] In spite of this, the bridgehead cations generated from 1,5-dihalotrishomobarrelenes (**31-X**) (X = F, Cl, Br, and I) and the diol (**31-OH**) under long lived ion conditions were only the 5-substituted monocations (**72-X**) (X= F, Cl, Br, I, and OH), unequivocally identified by their ^1H and ^{13}C NMR spectra as well as quenching products (Fig. 16).[43] Although there is substantial charge delocalization in **72-X**, as revealed by the ^{13}C chemical shifts, the bridgehead–bridgehead dication (**71**) is not formed, supposedly due to the strong electron withdrawing effect of the positively charged center which is efficiently transmitted through the skeleton onto the second bridgehead. Only when the charge is one carbon further removed from the bridgehead can the second halide ion be abstracted by the Lewis acid as evidenced in the reaction of **31-Cl** with SbF_5/SO_2ClF and carbon monoxide. The sole product, formed as a white precipitate in this reaction, was the biscarboxonium salt of **31-CO$^+$**, as characterized by its ^{13}C

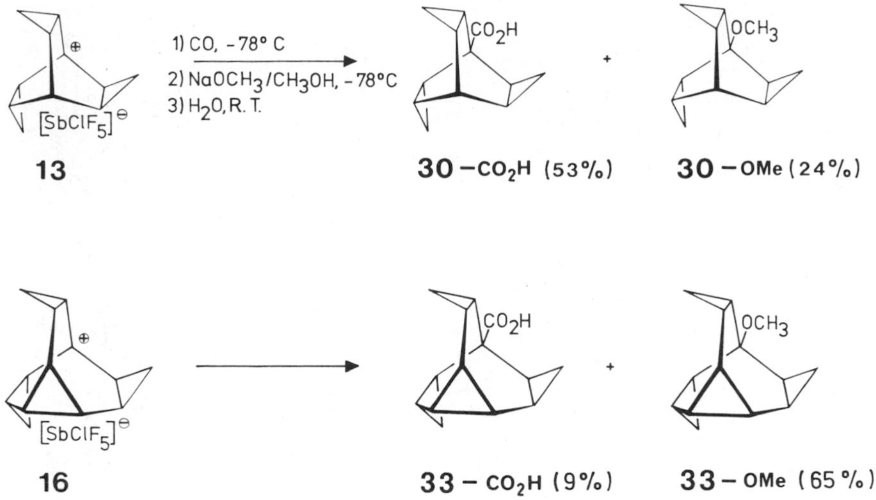

[SbClF$_5$]$^{\ominus}$ 1) CO, −78° C 2) NaOCH$_3$/CH$_3$OH, −78°C 3) H$_2$O, R. T. CO$_2$H + OCH$_3$

13 **30**−CO$_2$H (53%) **30**−OMe (24%)

[SbClF$_5$]$^{\ominus}$ CO$_2$H + OCH$_3$

16 **33**−CO$_2$H (9%) **33**−OMe (65%)

Fig. 14. Carbonylation reactions of trishomobarrelenyl (**13**) and trishomobullvalenyl cations (**16**).

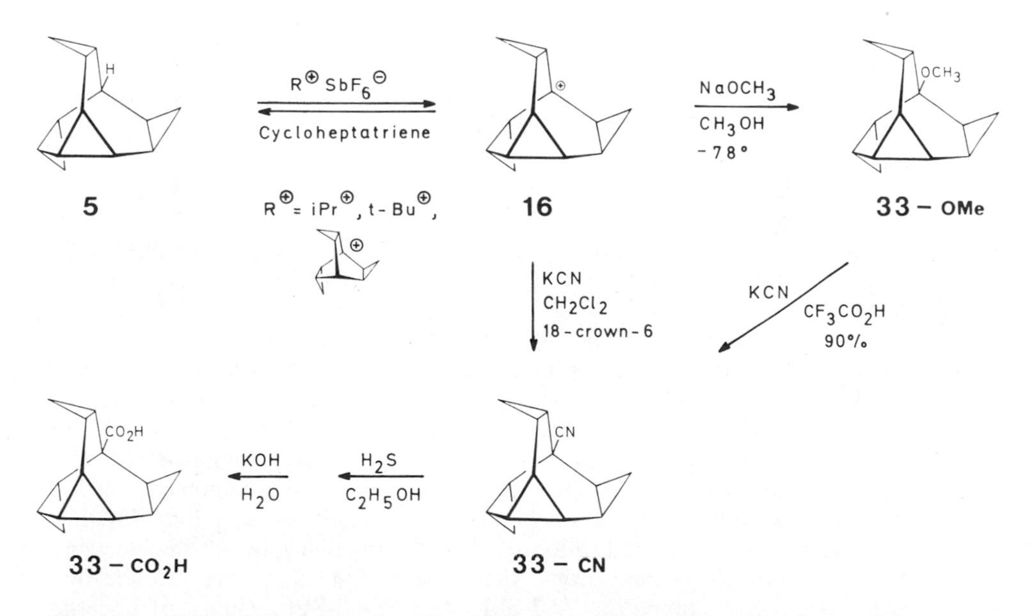

Fig. 15. Trishomobullvalenyl cation (**16**) by hydride abstraction from trishomobullvalene (**5**), and formation of substitution products (**33-X**).

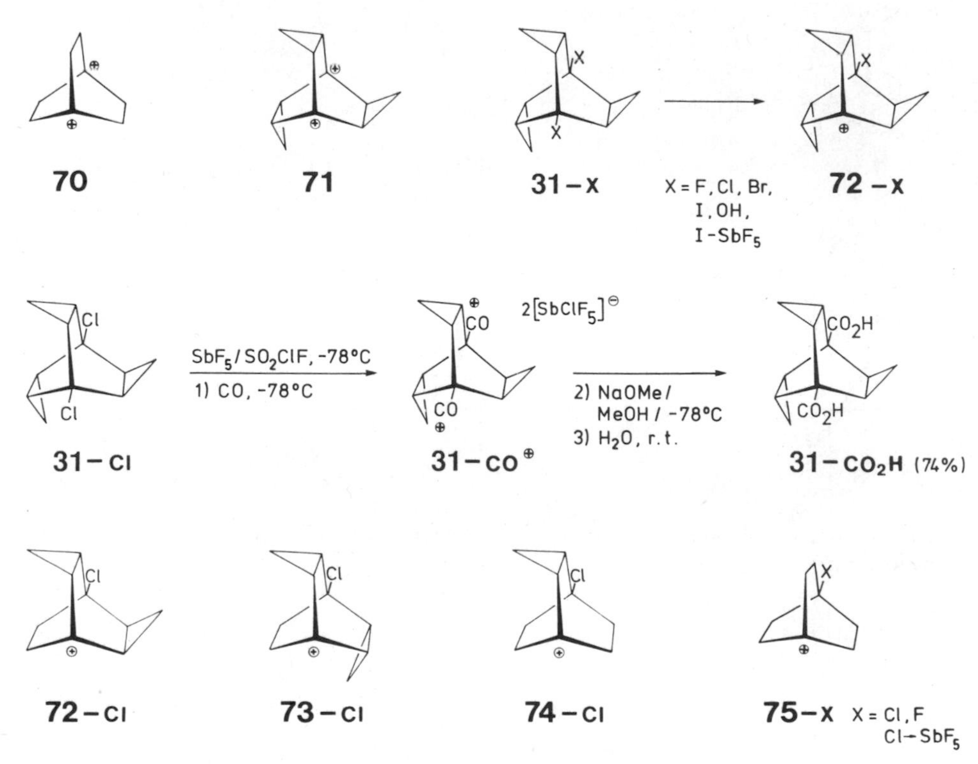

Fig. 16. Five-substituted monocations from bridgehead dihalides with homobarrelene skeletons and 4-halobicyclo[2.2.2]octyl cation (**75-X**).

NMR spectrum in SO_2 solution. Upon treatment with sodium methoxide in methanol and subsequently with water a high yield of the 1,5-dicarboxylic acid (**31-CO_2H**) was obtained.[43]

Even the 1,5-dichlorotetracyclo[3.3.2.0^{2,4}.0^{6,8}]decanes and 1,5-dichlorotricyclo[3.2.2.0^{2,4}]nonane derived from **51-Cl**, **52-Cl**, and **50-Cl** as well as the parent 1,4-dihalobicyclo[2.2.2]octane gave only the substituted bridgehead monocations **72-Cl**, **73-Cl**, **74-Cl**, and **75-X** (X = F, Cl),[43,44] although the transmitting ability for electronic effects of their C—C σ framework is less pronounced than that of **72-X** (see below).[24] The 1,5-trishomobarrelenediyl dication (**71**) as well as the parent bicyclo[2.2.2]octanediyl dication (**70**) thus remain elusive. The only indication of their intermediacy in yet undetectably small concentrations stems from the observation that the monocation monodonor–acceptor complexes **72-I-SbF_5** and **75-Cl-SbF_5** (Fig. 16) undergo halide exchange with external halide.[43]

BRIDGEHEAD DERIVATIVES OF TRISHOMOBARRELENE AND TRISHOMOBULLVALENE BY S_N1 DISPLACEMENTS AND OTHER TRANSFORMATIONS

Due to the high reactivities of 1-chlorotrishomobarrelene (**30-Cl**) and 1-chlorotrishomobullvalene (**33-Cl**) and the remarkable stability of the corresponding bridgehead cations **13** and **16**, almost any desirable bridgehead derivative can be prepared by nucleophilic displacement with suitable reagents under a variety of conditions (see Figs. 17, 18, and Table 2). In no case was skeletal rearrangement observed. It is noteworthy that the bridgehead azide, amine, and imidazole derivatives are formed in high yield. Hydroxy **30-OH** or the methoxy derivative **30-OMe**, obtained in high yield upon solvolysis of the chloride (**30-Cl**), can be converted to any of the halides (**30-X**) (X = Cl, Br, and I) with HX or **30-OH** to **30-F** with 2-chloro-1-diethylamino-1,1,2-trifluoroethane. The alkyl and aryl derivatives can be obtained from **30-Cl** with the corresponding Grignard or alkyllithium reagents and from the alcohol (**30-OH**) with trialkylaluminium. The bridgehead Grignard derivative **30-MgCl** can

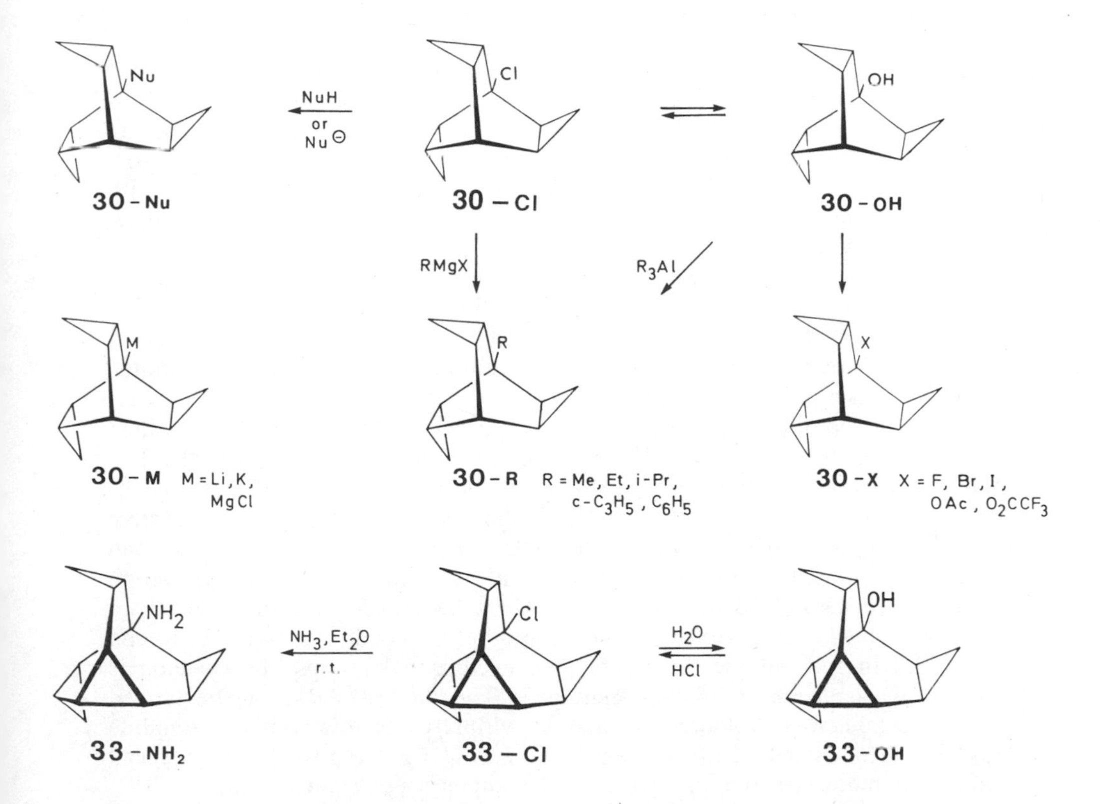

Fig. 17. Bridgehead derivatives of trishomobarrelene and trishonobullvalene from the chlorides or the alcohols (**30-OH/33-OH**) (for expermental conditions see Table 2).

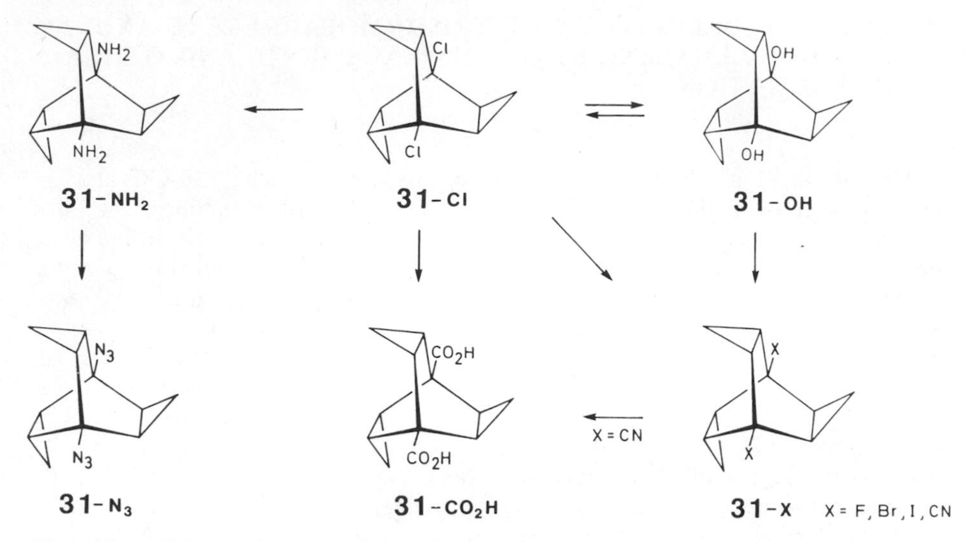

Fig. 18. 1,5-Disubstituted trishomobarrelene derivatives (**31-X**) from 1,5-dichloro-trishomobarrelene (for experimental conditions see Table 3).

be prepared in poor yield from **30-Cl** and active magnesium powder,[48] its carboxylation leads to **30-CO_2H**. Better yields of the carboxylic acid are obtained via the potassium or lithium derivative **30-K** or **30-Li** generated from **30-Cl** with Na/K alloy or from **30-I** with *t*-butyllithium.[46] The best route to **30-CO_2H**, however, is via the nitrile (**30-CN**) (from **30-Cl** with KCN/CuI in *N*-methylpyrrolidone at 200°C).

A similar set of transformations can be performed with 1-chlorotrishomo-bullvalene (**33-Cl**) (from **5** by photochlorination) and 1-hydroxytrishomobull-valene (**33-OH**) (from **5** by dry ozonation, see Fig. 7). The best preparation of analytically pure **33-Cl** is by evaporating a solution of **33-OH** and concentrated hydrochloric acid in acetone to dryness.[24]

Unlike trishomobullvalene (**5**), trishomobarrelene (**4**) has two equivalent reactive bridgehead positions. With an excess of *t*-butyl hypochlorite the 1,5-dichlorotrishomobarrelene (**31-Cl**) is therefore formed either photochemically[18,24] or thermally.[46] The dichloride (**31-Cl**) is, however, significantly less reactive than the monochloride (**30-Cl**). This is an apparent effect of the second chlorine substituent, which is rather efficiently transmitted through the carbon skeleton (see below). Nonetheless, **31-Cl** can be transformed to a variety of 1,5-disubstituted derivatives (**31-X**) under appropriate conditions (see Fig. 18 and Table 3). In spite of the three annelated cyclopropyl groups, the skeleton is remarkably stable under acidic conditions. The diol (**31-OH**) can be treated with concentrated hydroiodic acid to give virtually quantitatively the diiodide (**31-I**). The isolated yield is lower than for **31-Br**, because the bridgehead diiodide is more sensitive and partially decomposes upon purification.

TABLE 2. Bridgehead Derivatives of Trishomobarrelene (4) and Trishomobullvalene (5) (See Figs. 17 and 15)

Starting material	Reagent/ Method	Product	Yield [%]	Ref.
Trishomobarrelene (4) bridgehead substituent X:				
H	$CHCl_3$/NaOH/TEBACl	$CHCl_2$	65	45
H	1) i-$C_3H_7^+$ SbF_6^- 2) NaOMe/MeOH	OCH_3	11	24
H	t-BuOCl/hν	Cl	52	18
H	O_3/SiO_2	OH	73	46
Cl	H_2O/dioxane	OH	91	20, 45
Cl	CH_3OH/CH_3ONa	OCH_3	95	46
Cl	NH_3/Et_2O	NH_2	90	45
Cl	$NaN_3/ZnCl_2/CS_2$	N_3	95	46
Cl	KCN/CuI/NMP [a], 200°C	CN	69	46
Cl	Ag_2CO_3/H_2O	[c]	92	47
Cl	i-C_3H_7MgBr	i-C_3H_7	54	46
Cl	*cyclo*-C_3H_5MgBr	*cyclo*-C_3H_5	53	46
Cl	C_6H_5MgBr	C_6H_5	72	46
Cl	Imidazole (Im-H)	Im	91	46
Cl	CuCN/NMP [a], 70°C	NC	70	45
Cl	DMF, 130°C, 12h	$N(CH_3)_2$	>80	51
Cl	C_2H_5OH/K_2CO_3, Δ	OC_2H_5	>95	24
Cl	1) $(Mg)_x$ [d]/THF, r.t. 2) CO_2 3) H_2O	CO_2H	10-15	46
Cl	1) Na-K/pentane, 0°C 2) CO_2 3) H_2O	CO_2H	50-60	46
Cl	n-BuMgBr/n-hexane, r.t.	n-C_4H_9	62	46
Cl	$Zn(C_2H_5)_2/C_6H_6/CH_2Cl_2$, r.t.	C_2H_5	63	46
Cl	n-C_8H_{17}MgCl/Et_2O, Δ	n-C_8H_{17}	47	46

TABLE 2. (*Continued*)

Starting material	Reagent/ Method	Product	Yield [%]	Ref.
Cl	$MgBr_2$/THF, 130°C	Br	71	46
I	1) t-BuLi/Et_2O, -110°C 2) $(CH_3)_2CO$ 3) H_2O	$(CH_3)_2CHOH$	52	46
I	1) t-BuLi, 2) CO_2	CO_2H	60	46
OH	$SOCl_2$/polym. base [b]	Cl	95	46
OH	HBr/H_2O	Br	81	46
OH	HI/H_2O	I	92	46
OH	$ClFCH\text{-}CF_2NEt_2$	F	40	46
OH	$Al(CH_3)_3$	CH_3	88	46
OCH_3	HCl/Me_2CO/H_2O	Cl	81	46
NH_2	CF_3NO/CH_3OH,-78°C	$N{=}N\text{-}CF_3$	80	49, 50
N_3	$LiAlH_4$/Et_2O	NH_2	80	46
$N{=}N\text{-}CF_3$	hν, t-BuOH	CF_3	65	49, 50
Trishomobullvalene (**5**) bridgehead substituent X:				
H	t-BuOCl/hν	Cl	30	18, 45
H	O_3/SiO_2	OH	60	22
H	1) $i\text{-}C_3H_7^+$ SbF_6^-/CH_2Cl_2 2) KCN/CH_2Cl_2/18-C-6	CN	80	24
H	$CHCl_3$/NaOH/TEBACl	$CHCl_2$	60	45
Cl	NH_3/Et_2O	NH_2	82	45
Cl	CuCN/DMF, 100°C	CN	33	45
Cl	NaOH/H_2O/dioxane	OH	85	18
OH	HCl/H_2O/acetone	Cl	100	24
OH	KCN/CF_3CO_2H	CN	90	24
NH_2	CF_3NO/CH_3OH,-78°C	$N{=}N\text{-}CF_3$	50	49, 50
$N{=}N\text{-}CF_3$	hν, t-BuOH	CF_3	40	49, 50

[a] NMP= N-methylpyrrolidone. [b] (Dimethylaminomethyl)polystyrene. [c] Bistrishomobarrelenylether. [d] Reactive magnesium prepared according to R.D. Rieke et al..[48]

TABLE 3. 1,5-Disubstituted Trishomobarrelene Derivatives (31-X) (see Fig. 18)

Starting material (bridgehead substituents)	Reagent/ conditions	Product (bridgehead substituents)	Yield [%]	Ref.
H	3 t-BuOCl/hν	Cl	80	18, 24
H	t-BuOCl/CCl_4/AIBN, Δ	Cl	63	46
Cl	Dioxane/H_2O (3:2), 120°C, 24 h	OH	81	18
Cl	MeOH/NaOMe, 120°C, 65 h	OCH_3	81	45, 46
Cl	NH_3/Et_2O, 160°C	NH_2	67	46
Cl	CuCN/NMP [a], 200°C, 3 h	CN	51	51
Cl	$AgNO_3$/CH_3CN, 60°C, 12 h	ONO_2	90	51
Cl	C_2H_5OH/H_2O (4:1)	OC_2H_5	25	24
Cl	DMF, 130°C, 12 h	$N(CH_3)_2$	72	51
Cl	CuCN/NMP [a], 100°C, 12 h	NC	60	51
Cl	MeMgI/Et_2O, 100°C, 1 h	CH_3	65	46
Cl	1) SbF_5/SO_2ClF, 2) CO 3) NaOMe/MeOH 4) H_2O	CO_2H	74	43
OH	ClFCH-CF_2NEt_2, 70°C, 2 h	F	39	43, 45
OH	NaF/HF/pyridine, r.t., 2 h	F	52	46
OH	HBr/Me_2CO/H_2O (15:1), 50°C, 1 h	Br	91	43, 46
OH	HI/Me_2CO/H_2O, 40°C	I	62	43, 46
NH_2	NaH/TsN_3/THF, r.t., 3 d	N_3	47	46
NH_2	CF_3NO/CH_3OH, -78°C	N=N-CF_3	65	49, 50
N=N-CF_3	hν, t-BuOH	CF_3	32	49, 50

[a] NMP = N-methylpyrrolidone.

TRANSMISSION OF SUBSTITUENT EFFECTS AND BRIDGEHEAD—BRIDGEHEAD ELECTRONIC INTERACTIONS IN THE TRISHOMOBARRELENE CAGE

Whereas the hydrolysis of 1-chlorotrishomobarrelene (**30-Cl**) in 60% aqueous dioxane at 80°C is complete within 10 min, that of the dichloride requires 24 h at 120°C in the same solvent. Another observation indicating the efficient transmission of a substituent effect is the fact that **31-X** could not be ionized to the dication **71** even with strong Lewis acids. The relative transmitting ability of the trishomobarrelene skeleton was quantified by determining the solvolysis rates of a series of 5-substituted 1-chlorotrishomobarrelenes (**76-X**) (see Table 4).[24]

The logarithms of the relative rates log k_X/k_H correlate well with the inductive substituent constants σ_q^* of the Taft equation[52] (see Fig. 19), the parent system (X = H) falls off the line as does the parent in the series of 4-substituted 1-bicyclo[2.2.2]octyl *p*-bromobenzenesulfonates (**77-X**).[53] The correlation is significantly better with the σ_q^I constants introduced by Grob.[54] The slopes of the regression lines, that is, the reaction constants $\rho^*_{CH_2} = -3.44$

TABLE 4. Relative Solvolysis Rates (k_X/k_H) for 5-Substituted 1-Chlorotrishomobarrelenes (76-X) (in 80% Aqueous Dioxane at 25°C), $\sigma^*_{CH_2}$ (Ref. 52) and σ_q^I Constants[54] of the Substituents X

X	k_X/k_H	$-\log k_X/k_H$	$\sigma^*_{CH_2}$	σ_q^I
H	1	0	0.0	0.0
CH_3	$1.15 \cdot 10^{-1}$	0.939	-0.100	0.11
C_2H_5	$8.18 \cdot 10^{-2}$	1.087	-0.115	0.03
i-C_3H_7	$9.04 \cdot 10^{-2}$	1.044	-0.125	-0.08
C_6H_5	$1.35 \cdot 10^{-2}$	1.870	0.215	0.94
CO_2CH_3	$1.35 \cdot 10^{-3}$	2.869	0.71	1.70
OCH_3	$6.06 \cdot 10^{-4}$	3.218	0.52	1.81
OC_2H_5	$6.98 \cdot 10^{-4}$	3.156	0.52	1.81
CF_3	$2.65 \cdot 10^{-5}$	4.577	0.92	-
Cl	$6.61 \cdot 10^{-5}$	4.180	1.050	2.51
F	$2.29 \cdot 10^{-5}$	4.640	1.10	2.57

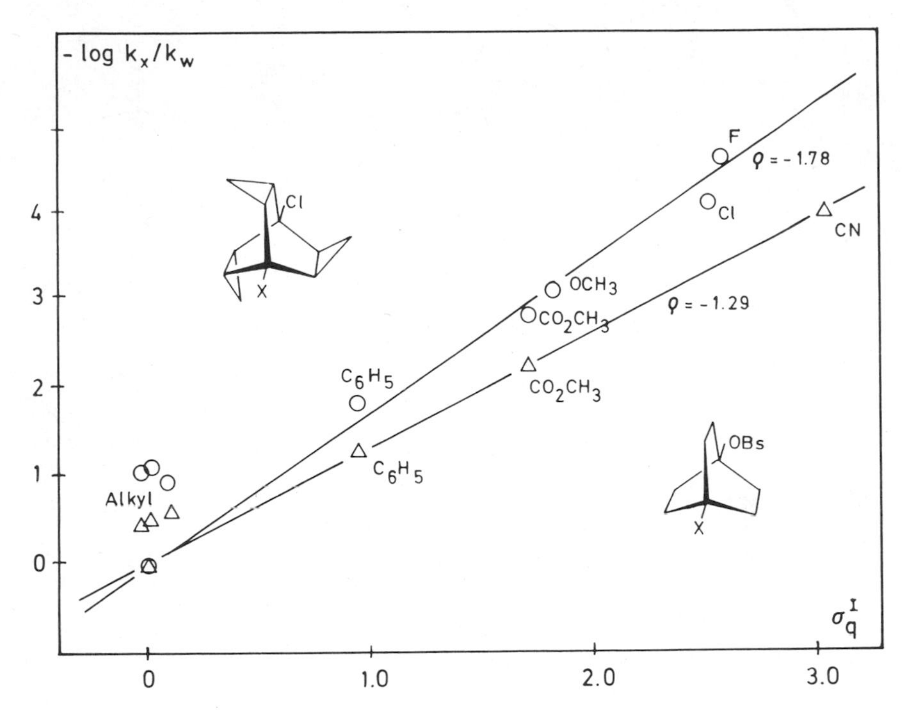

Fig. 19. Correlation between log k_X/k_H for 5-substituted 1-chlorotrishomobarrelenes (**76-X**) (o) and 4-substituted 1-bicyclo[2.2.2]octyl *p*-bromobenzenesulfonates (Δ) with Grob σ_q^I (Ref. 54) constants (alkyl substituted systems are not included).

(at 25°C) and $\rho_q^I = -1.78$ for **76-X** are 46 and 37% greater[55] than those for **77-X** ($\rho^*_{CH_2} = -2.36$) (see Fig. 19) and for 4-substituted 1-bicyclo[2.2.2]octyl *p*-bromobenzenesulfonates ($\rho_q^I = -1.30$),[56] respectively. Hence the ability to transmit inductive substituent effects must be considerably greater for cyclopropane C—C σ bonds than for normal C—C σ bonds. This property is a result of the greater *p* character of cyclopropane σ bonds and therefore does not depend as much on the conformational orientation of the cyclopropyl group relative to the carbon-substituent bond as does the conjugative effect (see above).[55]

With its two bridgehead bonds colinear, the trishomobarrelene skeleton also offered itself for an investigation of direct bridgehead interactions as should be observed in the elusive dication (**71**) as well as in the hypothetical radical cation (**78**), diradical (**81**), and dianion (**83**) (see Fig. 20 and 21).

The radical cation (**78**) was approached in two steps from the 1,5-disubstituted derivative **76-N_2CF_3** [prepared from the amine (**30-NH_2**) by

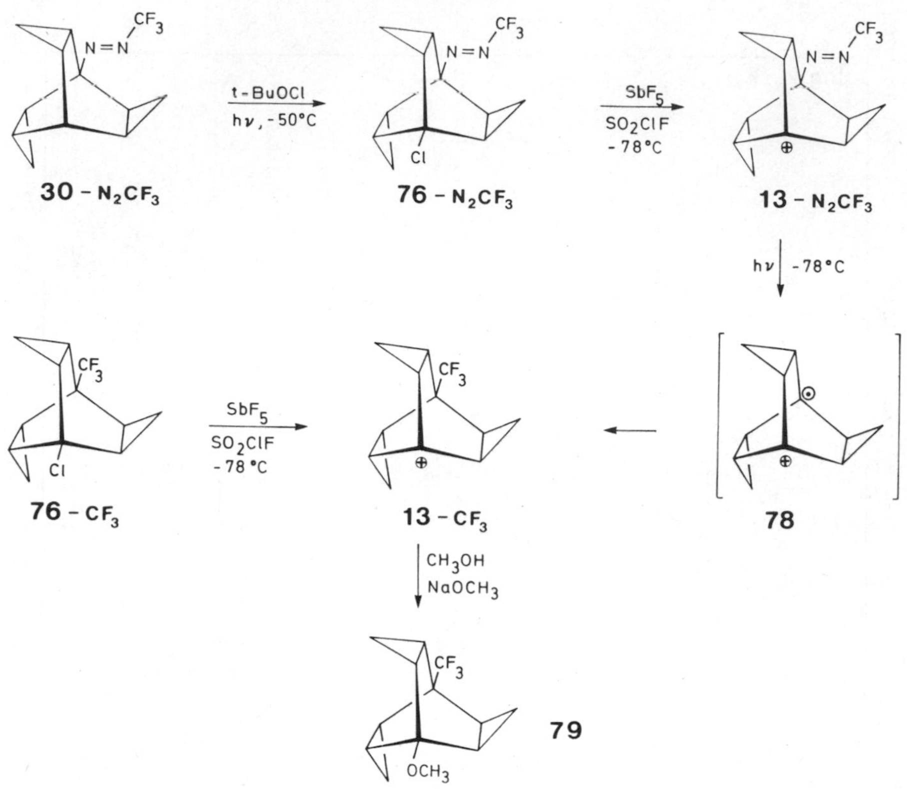

CF_3 N=N
30 - N_2CF_3
t-BuOCl
hν, -50°C
CF_3 N=N Cl
76 - N_2CF_3
SbF_5
SO_2ClF
- 78°C
CF_3 N=N ⊕
13 - N_2CF_3
hν -78°C
⊙ ⊕
78
CF_3 Cl
76 - CF_3
SbF_5
SO_2ClF
- 78 °C
CF_3 ⊕
13 - CF_3
CH_3OH
$NaOCH_3$
CF_3 OCH_3
79

Fig. 20. Generation of the trishomobarrelene bridgehead radical bridgehead cation (**78**) as an intermediate.

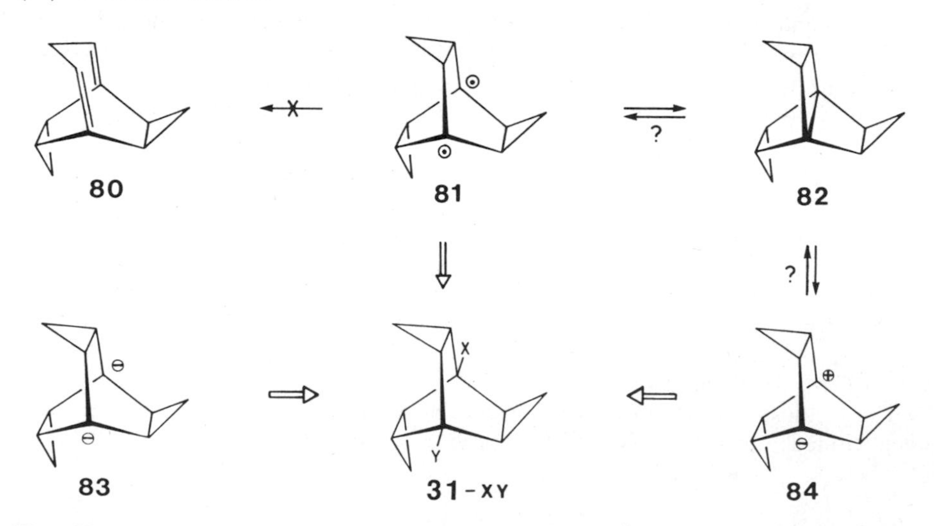

Fig. 21. Species with bridgehead bridgehead interaction in a trishomobarrelene skeleton.

reaction with CF_3NO and subsequent chlorination of **30-N$_2$CF$_3$**]; as the trifluormethyldiazenyl substituent is stable under strongly Lewis acidic conditions, the bridgehead cation (**13-N$_2$CF$_3$**) could readily be generated under long lived ion conditions (Fig.20). When **13-N$_2$CF$_3$** was irradiated at –78°C, its disappearance and the quantitative formation of the new bridgehead cation (**13-CF$_3$**) could be monitored by ^{1}H and ^{13}C NMR spectroscopy.[57] Compound **13-CF$_3$** was independently generated from **76-CF$_3$** and trapped with methanol to give the methyl ether (**79**). There is no doubt that the radical cation (**78**) must have been formed upon going from **13-N$_2$CF$_3$** to **13-CF$_3$**, whatever its lifetime may be. It is an open question whether the substantial cation stabilization by the three α-cyclopropyl groups and the through-space and through-bond interaction with the unpaired electron make **78** sufficiently long lived for a direct spectroscopic detection. In essence, **78** is the radical cation of the elusive propellane (**82**); because a one-electron single bond is longer than a two electron bond, **78** would at least be less strained than **82**. More information about the electronic symmetry and the lifetime of **78** under the conditions employed, should be available upon using the precursor **76-N$_2$CF$_3$** in an optically active form (see below).

The elusive propellane (**82**) appears to approach the limits of what is conceivable on the basis of current bonding theory. Its three bicyclo[2.1.0]-pentane units, which all share the same central bond, constitute a total strain energy of at least 160 kcal/mol;[58] the fraction of ~ 78 kcal/mol, which would be released upon opening the central bond, very closely resembles the energy of a normal C—C single bond. Yet, the two radical centers in a diradical (**81**) cannot avoid "seeing" each other through space and through the cyclopropyl σ bonds, and in addition there ought to be a certain degree of radical stabilization by the adjacent cyclopropyl groups.[11,59] Consequently, when **81** is generated from an appropriate precursor, it would probably prefer to exist with some sort of a central bond, especially since it does not have the option of the parent 1,4-bicyclo[2.2.2]octanediyl, which readily opens up to 1,4-dimethylenecyclohexane.[60] The same type of ring opening would transform **81** to the extremely strained bridgehead–bridgehead diene (**80**) containing a *trans,trans*-homotropylidene subunit.

The generation of **82** was attempted via a synthetic equivalent of the hypothetical 1,5-zwitterionic intermediate **84**. Towards this end the 1,5-diiodide was treated with 1–9 equivalents of *t*-butyllithium at –110°C. Upon aqueous work-up, 1-*t*-butyltrishomobarrelene (**30-*t*-Bu**) was isolated in yields ranging from 3 to 25%.[50] No **30-*t*-Bu** was formed from the monoiodide (**30-I**) under the same conditions. Since a direct *t*-butylation of **31-I** can thus be excluded, it is most probable that **30-*t*-Bu** is in fact formed via **85** and the propellane (**82**) (Fig. 22). It is reasonable to assume that the highly strained **82**, just like [2.2.1]propellane,[60] adds the strongly basic *t*-butyllithium to give **87**, which upon protonation yields **30-*t*-Bu**. It is far less probable that **87** is formed by direct *t*-butylation of **85** with *t*-butyllithium. The intermediacy of **87** was proved by carboxylation at –110°C and isolation of the *t*-butylsubstituted carboxylic acid (**88**). When the solution of metallated **31-I** was quenched with

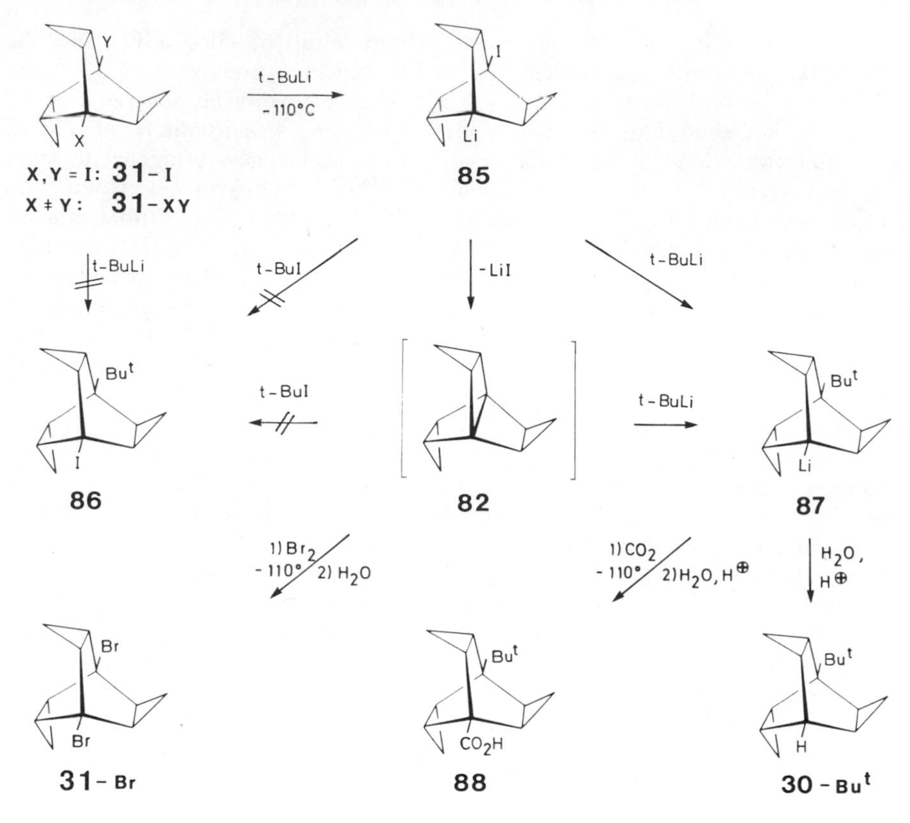

Fig. 22. Metallation of 1,5-diiodotrishomobarrelene (**31-I**) and quenching of the reactive intermediates.

bromine at –110°C, an 85% yield of the dibromide (**31-Br**) was isolated. This proof of the intermediacy of the propellane (**82**) is in line with the published one for the transient formation of [2.2.2]propellane,[60] but it is not totally conclusive, as a control experiment with just **31-I** and elemental bromine also gave **31-Br**. But it is highly unlikely that all the **31-Br** in the quenching experiment could have arisen from unreacted **31-I**.

Final proof for the formation of a C_{3h} symmetrical intermediate like **82** [or the corresponding diradical (**81**)] in this type of reaction would be possible with the use of a 1,5-disubstituted trishomobarrelene derivative (**31-XY**) with two different leaving groups in optically active form (see below). So far, experimental difficulties in preparing such optically active derivatives with two good enough leaving groups have prevented this test. Currently, a direct spectroscopical identification of **82** in an argon matrix after gas-phase reaction of **31-I** with alkali metal vapor is being developed in collaboration with Michl.[61]

OPTICALLY ACTIVE DERIVATIVES OF TRISHOMOBARRELENE AND TRISHOMOBULLVALENE

Trishomobarrelene derivatives with one (**30-X**) or with two different bridgehead substituents (**31-XY**), as well as trishomobullvalene derivatives (**33-X**) and even trishomobullvalene (**5**) itself, are C_3-symmetrical molecules and thereby chiral (see Fig. 23). Optically active compounds **30-X** and **33-X** are accessible by resolution of the racemic bridgehead carboxylic acids (**30-CO_2H**, **33-CO_2H**) by way of diastereomeric salts with quinine or α-phenylethylamine, respectively.[62] Reaction of the optically active acids (+)-**30-CO_2H** and (+)-**33-CO_2H** with lead tetraacetate in the presence of *N*-chlorosuccinimide gave (+)-**30-Cl** and (+)-**33-Cl** with the acetates (+)-**30-OAc** and (+)-**33-OAc** as byproducts. The latter could be transformed to the chlorides by treatment with concentrated hydrochloric acid. Hydrolysis of the chlorides lead to the alcohols (+)-**30-OH** and (+)-**33-OH**. In none of these transformations, even when performed via the "free" bridgehead carbenium ions **13** (= **30-+**) and **16** (= **33-+**) (see Figs. 24 and 25), did any loss of optical activity occur. It is interesting to note that dextrorotatory chlorides (+)-**30-Cl** and (+)-**33-Cl** gave levorotatory carbenium ions (-)-**13** and (-)-**16**, respectively, with remarkably high rotatory power. This is an expression of their UV absorption maxima, which are shifted to significantly longer wavelengths than those of the precursor chlorides.

The maximum rotations of **30-CO_2H** and **33-CO_2H** were established for the [carbonyl-^{14}C]-carboxylic acids by the isotopic dilution method, and all the other $[\alpha_{max}]_D$ [°] of optically active trishomobarrelene (**30-X**) and trishomobullvalene derivatives (**33-X**). All trishomobullvalene derivatives (**33-X**) show higher

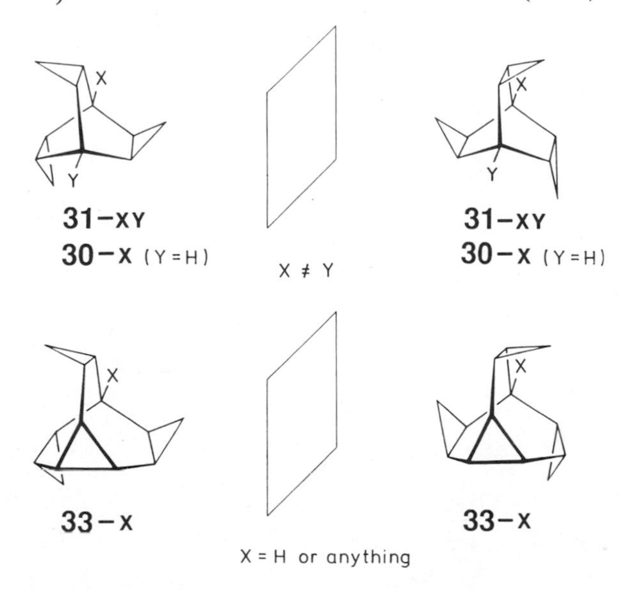

Fig. 23. The propeller type chirality of trishomobarrelene and trishomobullvalene derivatives.

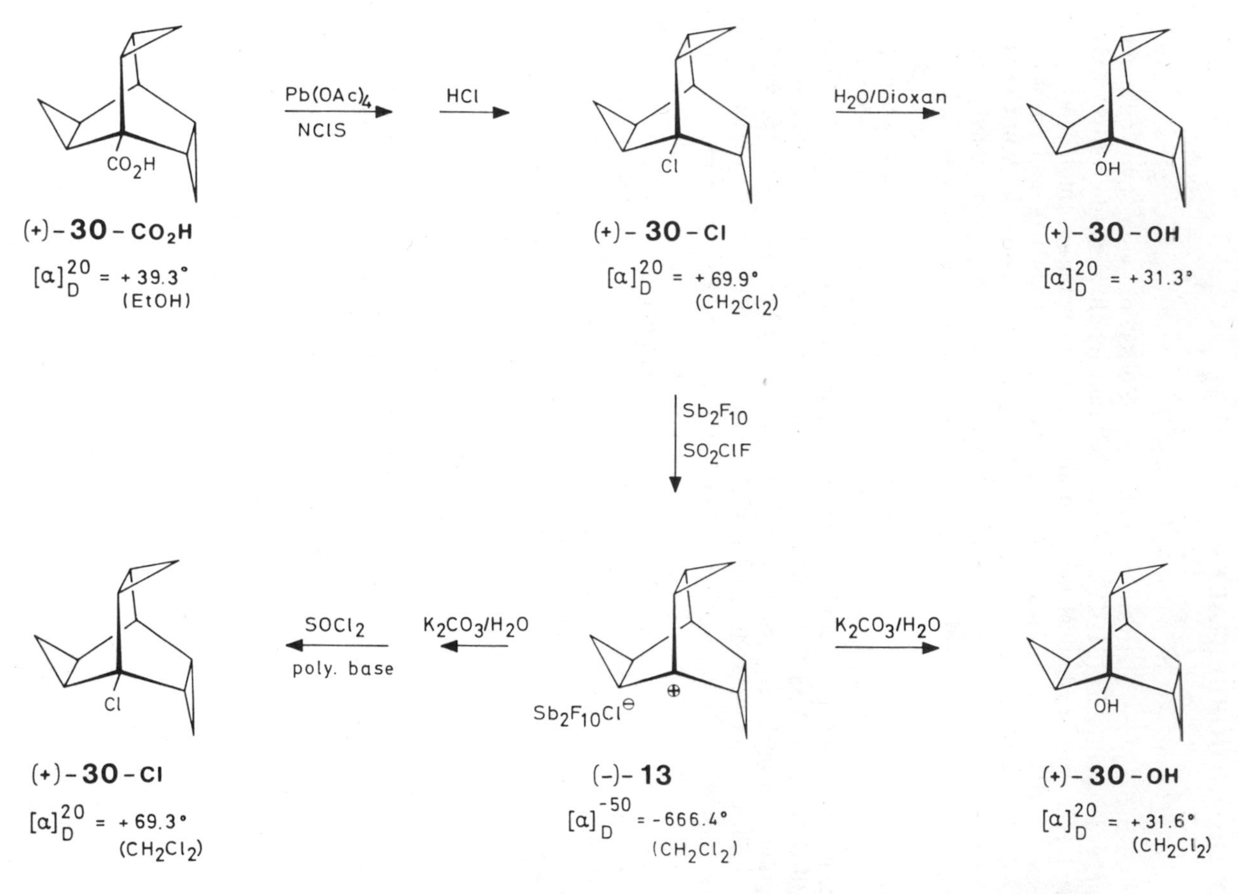

Fig. 24. Chemical transformations of optically active trishomobarrelene derivatives (**30-X**).

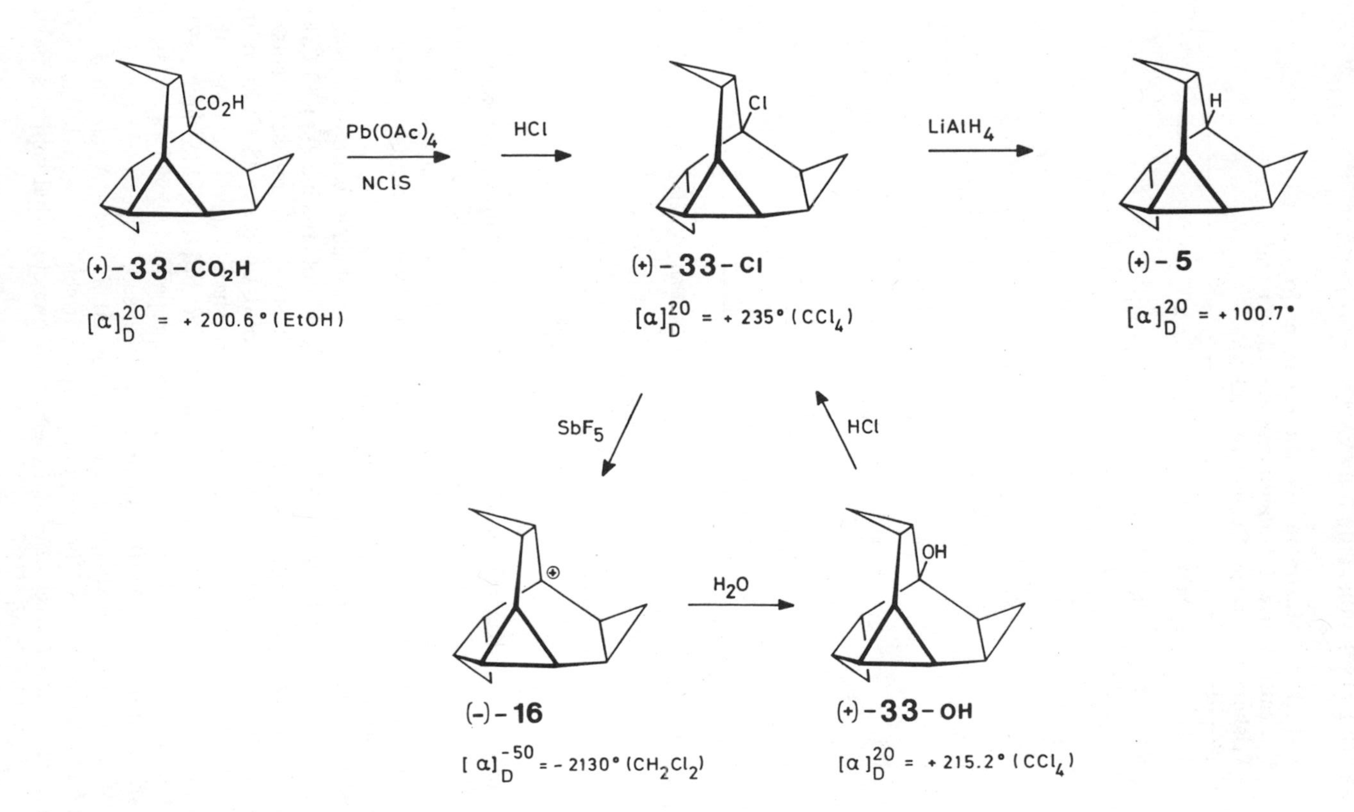

Fig. 25. Chemical transformations of optically active trishomobullvalene derivatives.

TABLE 5. Maximum Specific Rotations $[\alpha_{max}]^{20}$ [°] of Optically Active Trishomobarrelene (30-X) and Trishomobullvalene Derivatives (33-X).

X	30-X $[\alpha_{max}]^{20}$ [a]	33-X $[\alpha_{max}]^{20}$ [a]
H	0	+ 117 ± 3 (CCl_4)
CO_2H	+ 125 ± 3 (EtOH)	+ 233 ± 4 (EtOH)
Cl	+ 159 ± 4 (CH_2Cl_2)	+ 273 ± 5 (CCl_4)
OH	+ 70 ± 2 (CH_2Cl_2)	+ 250 ± 5 (CCl_4)
OAc	+ 103 ± 4 (CH_2Cl_2)	--
D	- 1.2 (CCl_4)	--
+	-1516 ± 40 (CH_2Cl_2)[b]	-2473 ± 53 (CH_2Cl_2)[b]

[a] Relative direction of rotation based on the (+)-carboxylic acid: average values of several measurements at concentrations c = 0.3-0.7 g/100 ml.

[b] Measured at -50°C.

specific rotations than their trishomobarrelene counterparts (**30-X**) and the misfit between calculated and experimental molar rotations for **33-X** is also much larger (see Table 5).[62] This may be taken to indicate that the chiral propeller structure especially of the trishomobullvalene carbon skeleton makes a significant contribution to the overall rotation.

The absolute configurations of all these propeller molecules were established by an X-ray structure analysis of (+)-1-chloro-trishomobarrelene [(+)-**30-Cl**][63] and correlations on the basis of circular dichroism (CD) measurements.[64] According to these, all derivatives related to (+)-**30-Cl** and (+)-**33-Cl** have an (all-S)-configuration except for the hydrocarbon trishomobullvalene (+)-**5**, the configuration of which has to be designated (2*S*, 4*R*, 8*R*, 10*S*, 11*S*, and 13*R*). But all of these molecules resemble dextroverse propellers upon looking from the side of the bridgehead or the substituted bridgehead, respectively (see Fig. 26). The CD of trishomobullvalene (**5**) is of special interest. Although this hydrocarbon contains no real chromophoric group, it shows a well-defined CD at 210 nm, which must arise from its three chiral bicyclopropyl subunits fixed in a synclinal conformation.[64]

Since it had been established that the more stable trishomobullvalenyl cation (**16**) can be generated by hydride abstraction with the less stable trishomobarrelenyl cation (**13**) from the hydrocarbon (**5**) (see above), optically active (+)-**13**, generated from (-)-**30-Cl**, was treated with two equivalents of racemic **5**.

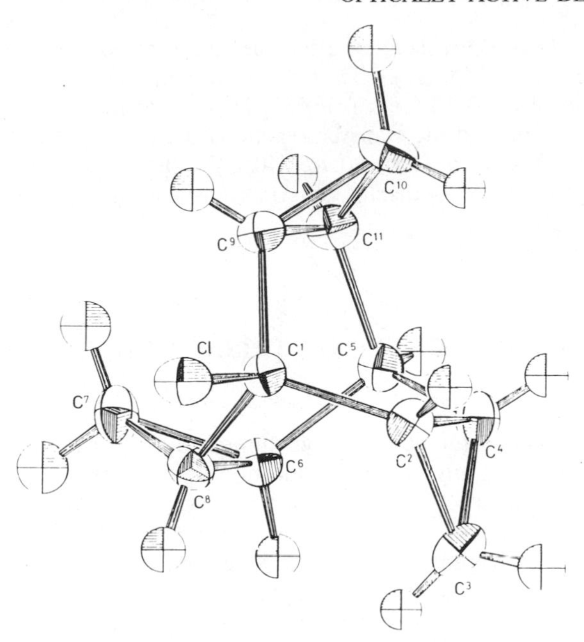

Fig. 26. The dextroverse propeller of (2*S*, 4*S*, 6*S*, 8*S*, 9*S*, 11*S*)-(+)-1-chlorotrishomobarrelene [(+)-**30-Cl**] according to a X-ray structure analysis.[63]

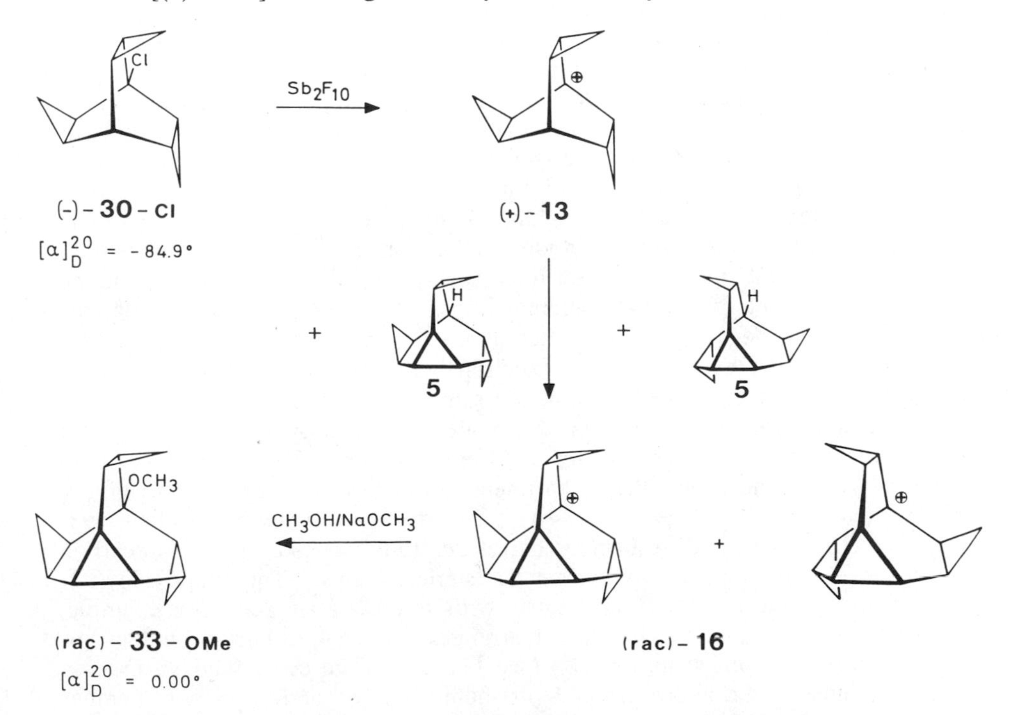

Fig. 27. The question of asymmetrically induced hydride abstraction.

The methyl ether (**33-OMe**) isolated after quenching the cation solution showed no optical activity (see Fig. 27). Thus no asymmetric induction occurs in the hydride abstraction by (+)-**13** from (rac)-**5**. This is not too surprising, since the activation energies for hydride transfer reactions between two tertiary positions are very small (3–4 kcal/mol at –80 to –40°C)[65] and consequently the energy differences between the two diastereomeric transition states leading to (+)-**16** should be extremely small.

HETEROCYCLIC ANALOGUES OF TRISHOMOBARRELENE

The rigid trishomobarrelene skeleton turned out to be ideally suited to ultimately answer the long standing question about the relative donor (or acceptor) ability of heterocyclic analogues of cyclopropyl groups.[11,66] In all previous kinetic and mechanistic investigations on aziridylmethyl and oxirylmethyl derivatives,[67–69] ring opening of the three-membered ring and neighboring group participation of the heteroatom played an important role. In contrast to this, the 1-chloro-3-methyl-3-azatrishomobarrelene (**90-Cl**) and the 1-chloro-3-oxatrishomobarrelene (**91-Cl**) solvolyze exclusively to the corresponding unrearranged bridgehead derivatives (**90-OH/90-OEt** and **91-OH/91-OEt** in 80% aqueous ethanol). Their solvolysis rates therefore provide reliable information pertaining to the question of how an intact three-membered heterocycle influences the relative stability of an adjacent carbenium ion center (see Fig. 28). Taking into account all relevant structural features, the detailed analysis of the rate data (see Table 6) shows that in these systems with conformations rigidly held at dihedral angles ~ 50–60°C, the electron donating ability of an unsubstituted aziridyl group is ~ 13 times weaker than that of a cyclopropyl group, and the oxiryl group even exhibits a strong electron withdrawing effect leading to a destabilization of the bridgehead cation from **91-Cl** by 1.9 kcal/mol with respect to that from the correct reference system. The decreasing donating ability of the hetero three-membered rings correlates very well with the increasing electronegativity of the heteroatoms,[66] which lowers the energy of its highest occupied Walsh MO.[71] Electron withdrawing substituents at the cyclopropyl group as in **89-Cl** act in the same way (see Table 6).

Trisheterotrishomobarrelenes **92** or **93** are consequently not of interest for their bridgehead reactivities, but for a different reason. With an excess of *m*-chloroperbenzoic acid (*m*-CPBA), barrelene (**1**) is readily epoxidized to give a 15:85 mixture of the diepoxide (**94-O$_2$**) and the trioxatrishomobarrelene (**92-O$_3$**).[72] The *exo,exo*-dioxabishomobarrelene (**94-O$_2$**) like its hydrocarbon analogue **23** cannot react any further for steric reasons. The triepoxide (**92-O$_3$**), when heated to 200°C or treated with traces of an acid, for example, $BF_3 \cdot OEt_2$ at –25°C cleanly rearranges to 4,7,11-trioxapentacyclo-[6.3.0.0^{2,6}.0^{3,10}.0^{5,9}]undecane (**95-O$_3$**) (see Fig. 29). The compound **95-O$_3$** is a trioxaanalogue of the hydrocarbon D_3-trishomocubane[73] and as such a member

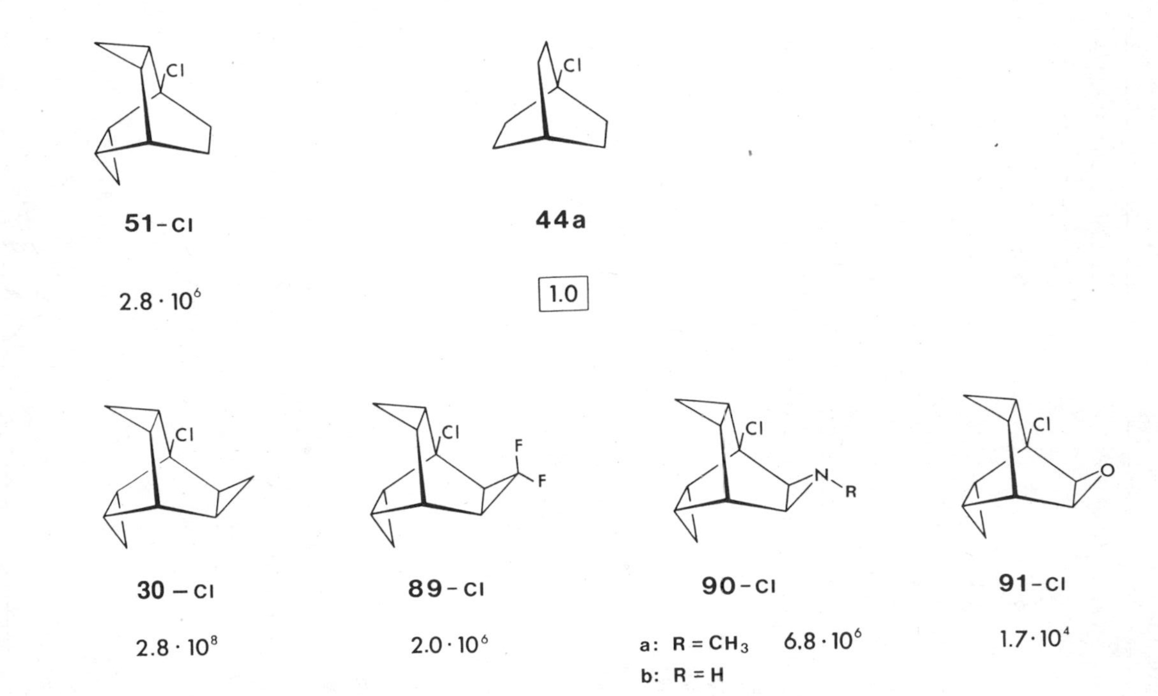

Fig. 28. Relative solvolysis rates (at 25°C in 80% aqueous ethanol) of heterologous 1-chlorotrishomobarrelenes and reference compounds.

TABLE 6. Rate Constants and Free Energies of Activation of Heterologous 1-Chlorotrishomobarrelenes and Reference Compounds[66]

	k_{25}[sec^{-1}]	k_{rel}	$\Delta G^{\#}_{25}$[kcal/mol]	$\Delta\Delta G^{\#}_{25}$[kcal/mol]
51-Cl	$5.8 \cdot 10^{-6}$	1.0	24.6	0.0
30-Cl	$5.8 \cdot 10^{-4}$	$1.0 \cdot 10^{2}$	21.9	-2.7
89-Cl	$4.2 \cdot 10^{-6}$	$7.2 \cdot 10^{-1}$	24.8	+0.2
90b-Cl	$1.4 \cdot 10^{-6}$ [a]	$2.4 \cdot 10^{-1}$	25.4	+0.8
91-Cl	$3.6 \cdot 10^{-8}$	$6.2 \cdot 10^{-3}$	27.6	+3.0

[a] Calculated from the rate constant for **90a-Cl** and a factor of 10^{-1} for the methyl group.[70]

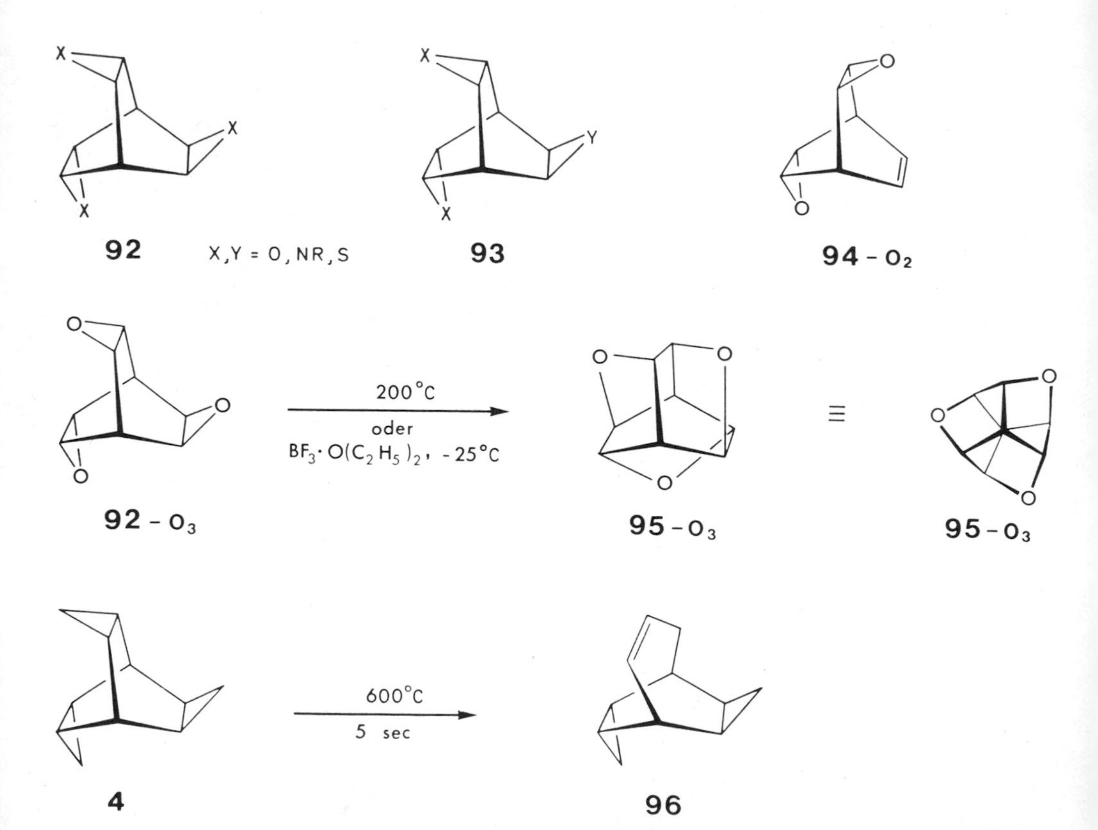

Fig. 29. Trisheterotrishomobarrelenes. Formation and rearrangement of barrelene tri-epoxide.

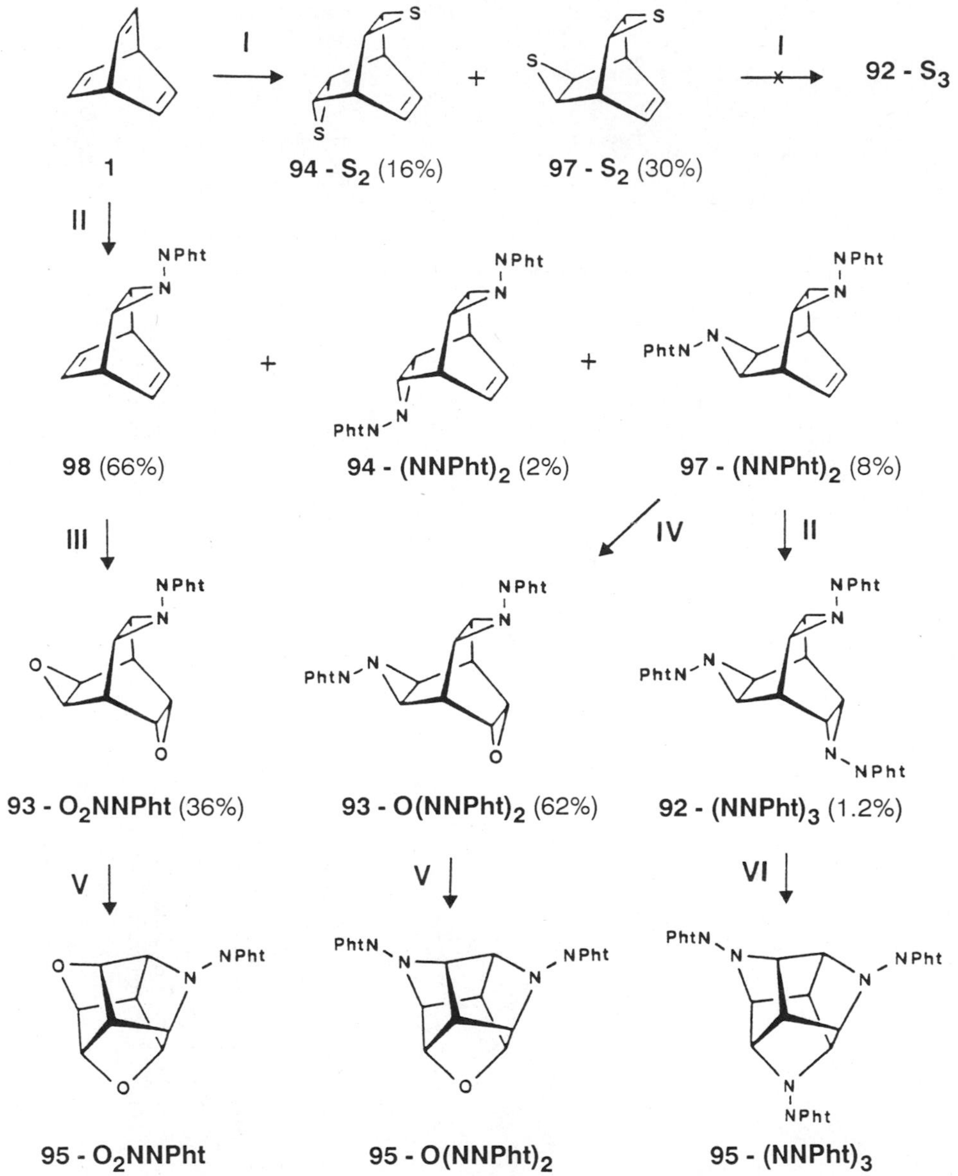

1 —I→ **94 - S_2** (16%) + **97 - S_2** (30%) —I (✗)→ **92 - S_3**

1 —II→ **98** (66%) + **94 - $(NNPht)_2$** (2%) + **97 - $(NNPht)_2$** (8%)

98 —III→ **93 - O_2NNPht** (36%)

97 - $(NNPht)_2$ —IV→ **93 - $O(NNPht)_2$** (62%)

97 - $(NNPht)_2$ —II→ **92 - $(NNPht)_3$** (1.2%)

93 - O_2NNPht —V→ **95 - O_2NNPht**

93 - $O(NNPht)_2$ —V→ **95 - $O(NNPht)_2$**

92 - $(NNPht)_3$ —VI→ **95 - $(NNPht)_3$**

I: 1)PhtNSCl, CH_2Cl_2, R.T., 7d; 2)$LiAlH_4$, THF, -78° to R.T., 3h. II: $PhtNNH_2$, $Pb(OAc)_4$, CH_2Cl_2, -10° to R.T., 2h. III: mClPBA, CH_2Cl_2, R.T., 2d. IV: mClPBA, CH_2Cl_2, R.T., 15h. V: Amberlyst 15, $CHCl_3$, R.T., 15h. VI: Amberlyst 15, $CHCl_3$, 60°, 12h. Pht = phthalimido.

Fig. 30. Formation and transformations of heterohomobarrelenes.

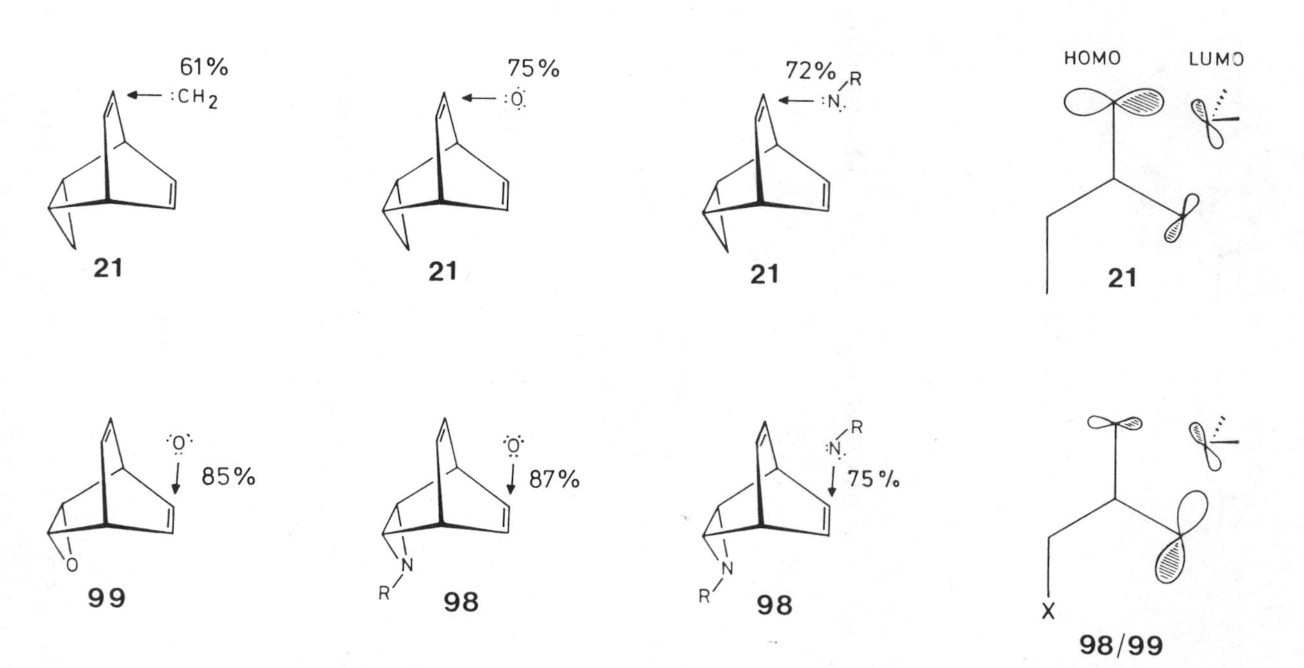

Fig. 31. Preferred modes of electrophilic cheletropic additions onto homobarrelene (**21**), oxahomobarrelene (**99**) and *N*-phthalimido-azahomobarrelene (**98**).[77,79]

of the interesting class of gyrochiral D_3-symmetrical molecules[74] containing a D_3-*twist*-bicyclo[2.2.2]octane core skeleton.[75]

This rearrangement is typical for trisheterotrishomobarrelenes (**92/93**), but the hydrocarbon (**4**) is quite stable towards acid and thermally rearranges by a totally different mode to give **96** as the primary product (Fig. 29).[76]

Upon treatment of barrelene (**1**) with an excess of *N*-(chlorothio)phthaloylimid only two equivalents of reagent were consumed, and the subsequent reaction of the adducts with $LiAlH_4$ lead to a mixture of only the diastereomeric dithiabis-homobarrelenes **94-S$_2$** (16% isolated yield) and **97-S$_2$** (30%).[77] The trithiatrishomologue could not be obtained in this way. But the *endo,exo*-bisadduct [**97-(NNPht)$_2$**] of phthalimidonitrene onto **1**, isolated from the mixture with the monoadduct **98** and the exo,exo-isomer [**94-(NNPht)$_2$**], could be reacted with excess phthalimidonitrene to give the trisphthalimido-3,7,10-triazatrishomo-barrelene [**92-(NNPht)$_3$**], albeit in extremely poor yield (1.2 %). The mixed oxaazatrishomologues **93-O$_2$NNPht** and **93-O(NNPht)$_2$**, however, were readily formed in 36 and 62% yield, respectively, upon epoxidation of **98** and **94-(NNPht)$_2$** with *m*-CPBA (Fig. 30). In the presence of a strongly acidic ion exchange resin (Amberlyst 15) in chloroform solution, **93-O$_2$NNPht** and **93-O(NNPht)$_2$** underwent quantitative rearrangement to the corresponding trishetero-trishomocubanes **95-O$_2$NNPht** and **95-O(NNPht)$_2$** within 15 h at room temperature, the triazatrishomobarrelene derivative [**95-(NNPht)$_3$**] required 12 h at 60°C. This retarded reactivity of **95-(NNPht)$_3$** is probably mainly caused by the low basicity of the nitrogen atoms due to the strongly electron withdrawing phthaloylimido substituents.[77]

It is particularly noteworthy that the epoxidation of barrelene (**1**) and the addition of phthalimidonitrene both predominantly yield the endo,exo-isomeric bishomologues, whereas the cyclopropanation of homobarrelene (**21**) gives mainly *exo,exo*-bishomobarrelene (**23**) (see above). This is a general phenomenon for electrophilic cheletropic additions to homobarrelene (**21**) in contrast to its oxa- (**99**) and aza analogue (**98**) and an example of efficiently frontier orbital controlled regioselectivity (see Fig. 31).[78] The heteroatom in **98/99** causes a switch of the larger coefficients in the highest occupied molecular orbital (HOMO) of the dienes **21** and **98/99**, respectively. The cycloaddend prefers that double bond with which its lowest unoccupied molecular orbital (LUMO) can achieve the larger overlap, and the preferred anti-attack in **21** (leading to *exo,exo*-bishomobarrelenes) is a result of a secondary interaction between the frontier orbitals of the electrophile and the two different π systems of the diene.[78]

PREPARATION AND PROPERTIES OF BRIDGED TRISHOMOBENZENES[80]

According to the concept of homoconjugation and homoaromaticity,[81] compounds like all-*cis*- (**6b**) or *cis,trans,trans*-1,4,7-cyclononatriene (**108**) as well as their valence isomers *exo,exo*- (**6a**) and *endo,exo*-tetracyclo-

[6.1.0.0^{2,4}.0^{5,7}]nonane (**9**) (*cis*-or *trans*-tris-π-homobenzene **6b/108**, *cis*-and *trans*-tris-σ-homobenzene **6a/9**, see Figs. 1, 2, and 35) ought to exhibit physical and chemical peculiarities, which would justify their classification as trishomoaromatic compounds. It has indeed been proved by PE spectroscopy that the electronic interaction between the formally nonconjugated double bonds in the all-*cis*-triene (**6b**) leads to a small, but significant "resonance" effect and that there is significant electronic interaction in the cyclic array of three cis-oriented adjacent cyclopropyl groups in the bridged *cis*-tris-σ-homobenzene (**7**).[82] While **6a** is still unknown, hexacyclo[4.4.0.0^{2,10}.0^{3,5}.0^{4,8}.0^{7,9}]decane (**7**) (diademane)[83] and its homologue hexacyclo[5.4.0.0^{2,11}.0^{3,5}.0^{4,9}.0^{6,8}]undecane (**8**) (1,6-homodiademane) remain the only hydrocarbons with the characteristic feature of *cis*-tris-σ-homobenzene (**6a**).[84] The $(CH)_{10}$ hydrocarbon diademane (**7**) is readily obtained by photoisomerization of pentacyclo[4.4.0.0^{2,4}.0^{3,8}.0^{5,7}]dec-9-ene (**100**) (snoutene),[85] a hydrocarbon corresponding to a bridged *endo,endo*-bishomobarrelene. The most striking property of diademane (**7**) is its facile thermal rearrangement, with a half-life of 186 min at 80°C, to triquinacene (**101**), yet another $(CH)_{10}$ hydrocarbon. The remarkably low Arrhenius activation energy of 31.6 ± 0.7 kcal/mol (log A = 14.64 ± 0.3) for this rearrangement is an indication of a concerted process, in which all three cyclopropyl σ bonds open and all three π bonds form simultaneously. According to MINDO/3 model calculations,[86] the transition state of this rearrangement has the same C_{3v} symmetry as the starting material **7** and the product **101**, as expected for a $[_{\sigma}2_s + _{\sigma}2_s + _{\sigma}2_s]$ cycloreversion.

The symmetry of the molecule and the transition structure is not distorted by a substituent at C-1, the central bridgehead. 1-Substituted diademane derivatives (**7-R**) are in fact formed with a high degree of regioselectivity (> 85:15 for R = CH_3 and > 95:5 for R = $SiMe_3$) in the photochemical rearrangements of 4-substituted snoutene derivatives.[87] As one would expect from simple frontier orbital energy considerations, the electron donating methyl group in **7-CH$_3$** facilitates the cycloreversion to **101-CH$_3$**, while the net electron accepting trimethylsilyl group retards it (see Fig. 32). Although diademane (**7**) and 1-methyldiademane (**7-CH$_3$**) crystallize very well, several attempts to determine their structure by X-ray crystallography have failed, because the crystals were plastic and completely disordered. This phenomenon is frequently encountered with spherically shaped hydrocarbon molecules.

The rearrangement of **7** to **101** is catalyzed by silver ions and copper(I) complexes, while rhodium(I), gold(I), gold(0), and palladium(0) cause diademane (**7**) to rearrange to snoutene (**100**), that is, they exactly reverse the photochemical formation of **7**.[88] Apparently, the same type of absorption on the metal surface occurs upon hydrogenation on a palladium catalyst, leading predominantly to secosnoutane and further to protoadamantane, but not at all to adamantane, the most stable of all possible $C_{10}H_{16}$ hydrocarbons.[88]

Whereas the 1-substituted derivatives **7-R** are only altered electronically, compound **8** with its CH_2 bridge between C-1 and C-6 of diademane and therefore called 1,6-homodiademane, is distorted both electronically and geometrically. Compound **8** is formed analogously to **7** by an intramolecular photochemical [2 + 2] cycloaddition in 4,5-homosnoutene (**102**), prepared from the [2 + 2 + 2] cycloadduct of maleic anhydride to barbaralane.[84] The thermal

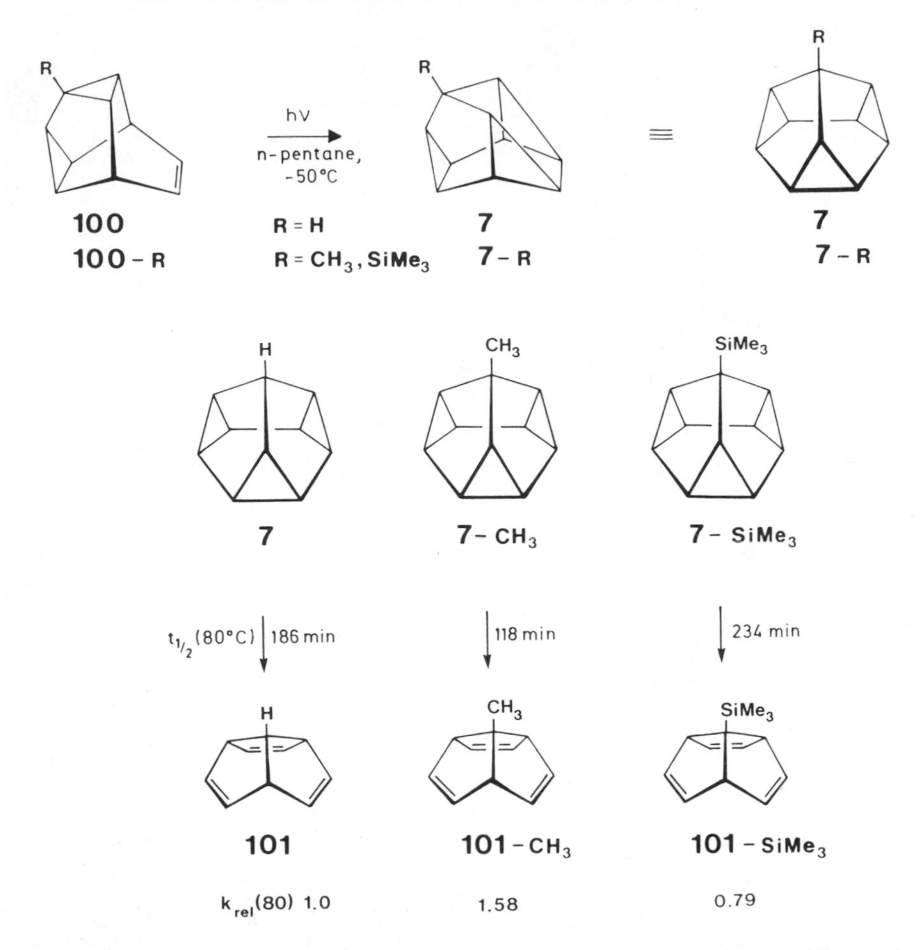

Fig. 32. Photochemical formation and thermal rearrangement of diademane (**7**) and 1-substituted diademanes (**7-R**).

rearrangement of **8** to tricyclo[5.2.2.$0^{4,8}$]undeca-2,5,10-triene (**103**) (1,10-homotriquinacene) has an activation energy of only 28.3 ± 1.0 kcal/mol (log A = 14.0 ± 0.5), significantly lower than that of **7**. This difference may be attributed mainly to the increased overlap of cyclopropyl Walsh orbitals on the side of the six-membered ring in **8** caused by the downward bending of one cyclopropyl group (increased angle α with respect to α in **7**, see Fig. 33). In essence, one of the cyclopropyl groups in **8** approaches an orientation, which all three cyclopropyl groups must adopt in the unbridged *cis*-tris-σ-homobenzene (**6a**). If this were the main factor determining the stability of such systems, the difference between **7** and **8** leads to a realistic estimate for the activation energy of the rearrangement **6a** → **6b** to be ~ 22 kcal/mol. Consequently, any attempted synthesis of the elusive **6a** would have to concentrate on methods, which permit its formation and isolation at temperatures below 20°C.[84]

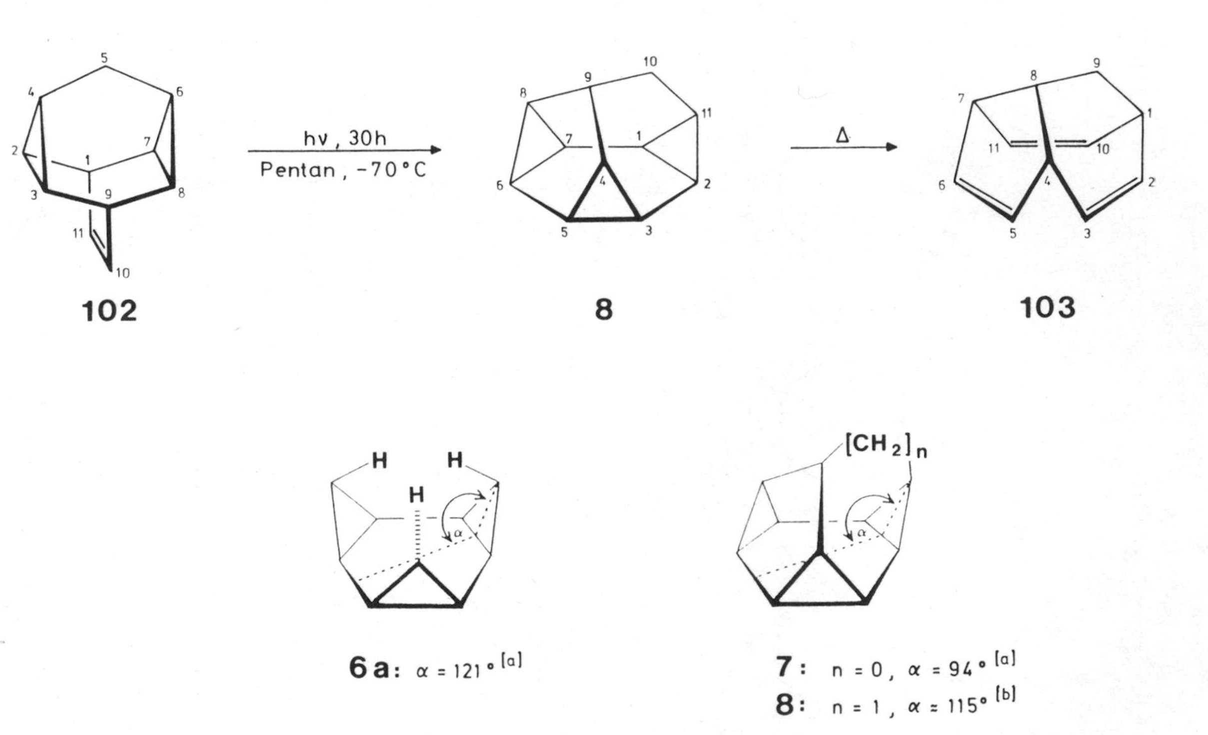

Fig. 33. Photochemical formation, thermal rearrangement, and structural features of 1,6-homodiademane (**8**) in relation to *cis*-tris-σ-homobenzene (**6a**). (*a*) According to MINDO/3 calculations.[86] - (*b*) Estimated from a molecular model.

Of course, substituents and heteroatoms can considerably stabilize the *cis*-tris-σ-homobenzene system, the first such derivatives have been reported by Prinzbach and Stusche in 1970.[80] A great number of heterocyclic analogues of **6a** with and without additional substituents have since been prepared and investigated.[89–91] Two interesting representatives of *cis*-trioxatris-σ-homobenzenes, which combine the features of a trisheterotrishomobenzene with that of a cage hydrocarbon are the [2.2]paracyclophane derivatives **104** and **105** (Fig. 34).[92] Surprisingly, the thermal rearrangements of **104** and **105** to their tris-π-homobenzene counterparts **106** and **107**, require about the same drastic conditions as the parent trioxa-*cis*-tris-σ-homobenzene in spite of the structural deformation caused by the phane bridging.

Trans-tris-σ-homobenzene (**9**) has been prepared along three different routes.[93–95] Although the overlap between neighboring *p* orbitals in two out of three bicyclopropyl groups is very small, **9** rearranges formally according to the same scheme as all the *cis*-tris-σ-homobenzene derivatives; however, reaction of **9** requires an activation energy of 42 kcal/mol.[84,95] Thus, under the conditions employed, the initial product, namely, the highly strained *cis,trans,trans*-1,4,7-cyclononatriene (**108**), could not be isolated but detected by the isotopic labeling pattern in the final product **109b** from **9b** (see Fig. 35).

Only when an additional bridge is present, as in **10**, does the system avoid the excessive strain energy in the bridged intermediate of type **108** and instead

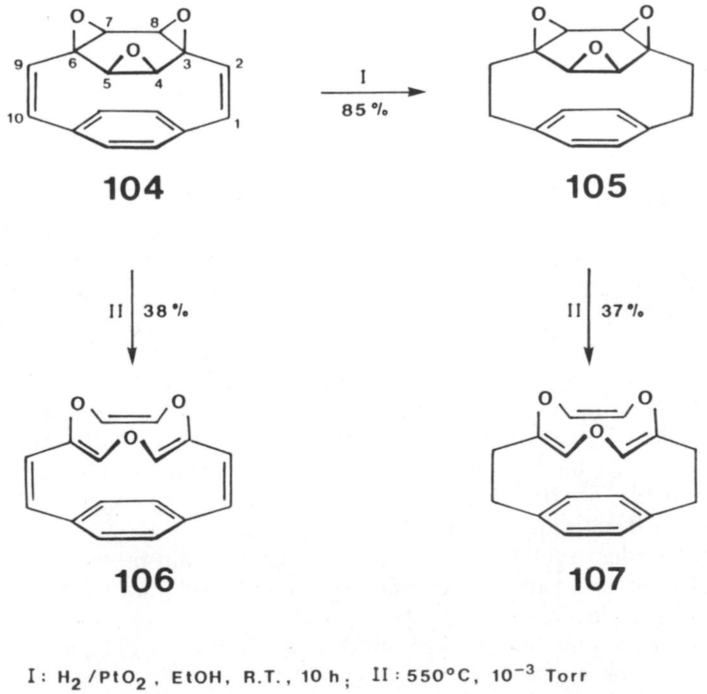

104 → 105 (I, 85 %)

104 → 106 (II, 38 %)

105 → 107 (II, 37 %)

I: H_2/PtO_2, EtOH, R.T., 10 h; II: 550°C, 10^{-3} Torr

Fig. 34. Trioxatrishomo[2.2]paracyclophanes, cage molecules with two faces.

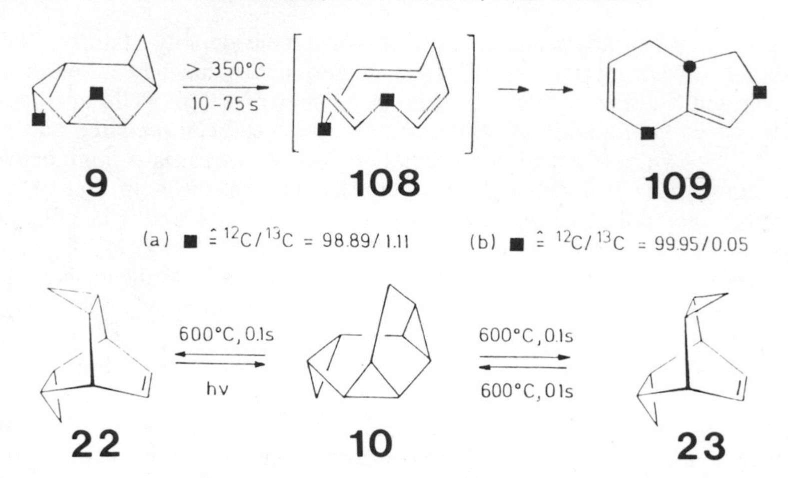

Fig. 35. Thermal rearrangements of *trans*-tris-σ-homobenzene (**9**) and its bridged analogue **10**.

undergoes a $[_{\sigma}2 + _{\pi}2]$ cycloreversion probably via diradical intermediates at higher temperatures giving the isomeric bishomobarrelenes **22** and **23**, from which it was in turn generated photochemically[96] or thermally.[84,97]

AN OUTLOOK

While diademane (**7**) thermally rearranges to triquinacene (**101**), the latter cannot photochemically be transformed back to **7**. Low temperature photolysis of **101** in solution predominantly yields two new $(CH)_{10}$ isomers, pentacyclo-$[4.4.0.0^{2,4}.0^{3,10}.0^{5,9}]$dec-7-ene (**101**) and hexacyclo$[4.4.0.0^{2,4}.0^{3,10}.0^{5,8}.0^{7,9}]$decane (**111**), along with five other C_{10} hydrocarbons.[98] The saturated compound **111**, nicknamed "barettane",[83] is a particularly interesting isomer of **7** in several respects. With its C_2 symmetry it is a chiral hydrocarbon consisting of three-, four-, five-, and six-membered rings. Its two bicyclo[2.1.0]pentane subunits, which are linked head to tail, each contribute 53.6 kcal/mol of strain energy. But in spite of its total strain in excess of 107 kcal/mol, **111** is thermally completely stable at 200°C. Yet it can easily be hydrogenated to yield tetracyclo$[5.2.1.0^{2,6}.0^{4,8}]$decane (**113**) (see Fig. 36).

In an independent synthesis, **111** is formed as the main product by a double carbene insertion upon alkaline thermolysis of tetracyclo$[5.2.1.0^{2,6}.0^{4,8}]$decane-5,10-dione-bis(tosylhydrazone) (**112**).[98] It would certainly be interesting to further investigate possible reaction modes of barettane (**111**), determine its molecular structure, and its chiroptical properties.

The long sought for $(CH)_{12}$ hydrocarbon truncahedrane (**115**) (heptacyclo$[5.5.0.0^{2,12}.0^{3,5}.0^{4,10}.0^{6,8}.0^{9,11}]$dodecane)[99,100] is the most interesting

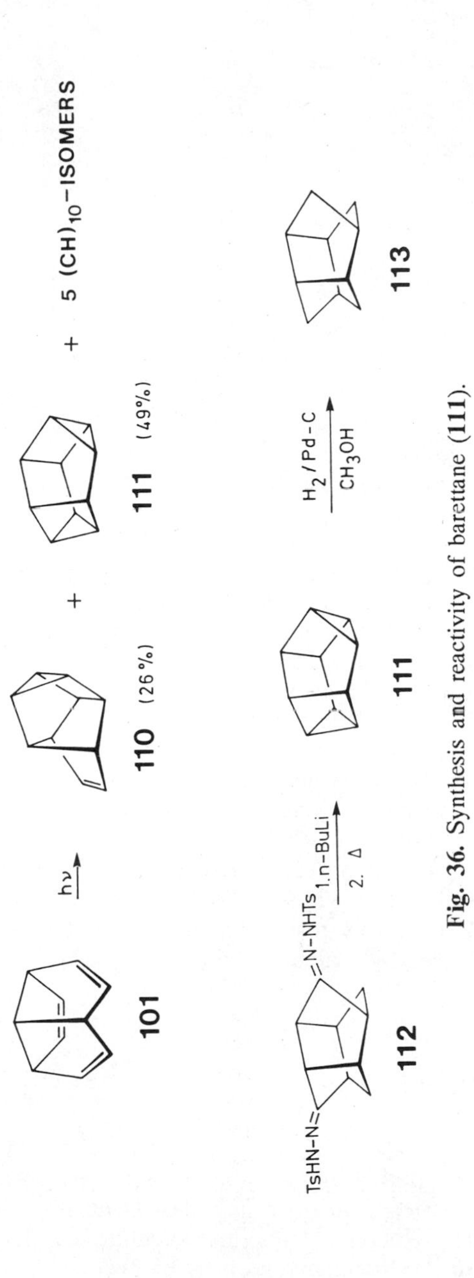

Fig. 36. Synthesis and reactivity of barettane (**111**).

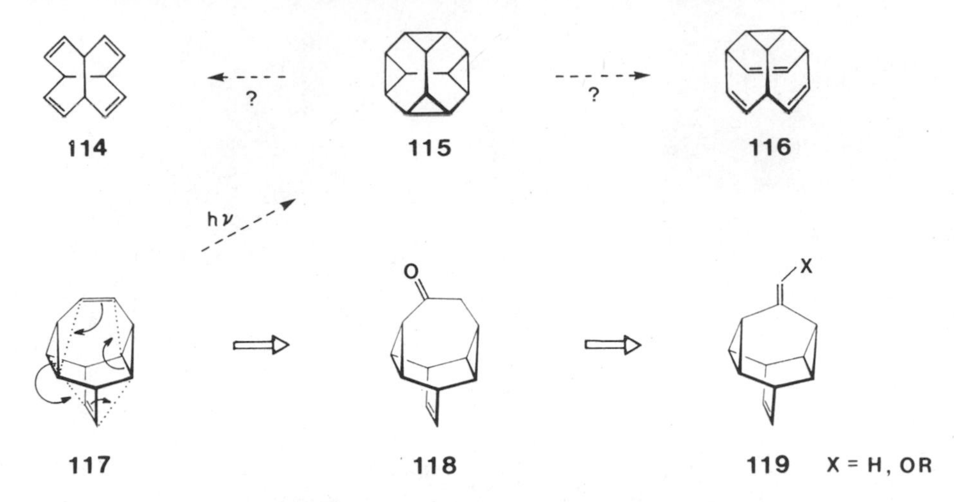

Fig. 37. A concept for the synthesis of truncahedrane (**115**).[101]

relative of *cis*-tris-σ-homobenzene (**6a**). All around, it consists of four such units. Although one conceivable mode of rearrangement would transform **115** to the tetraene **114**, which had been conceived as a potential precursor to **115**, ab initio calculations predict that **114** is less stable than **115** by ~ 24 kcal/mol, whereas the triene **116** should be more stable than **115** by ~ 16 kcal/mol.[99] Consequently, **115** would probably rearrange by the same $[_{\sigma}2_s + _{\sigma}2_s + _{\sigma}2_s]$ mode as diademane (**7**). A possible route to **115** that we have embarked on is directed towards the $(CH)_{12}$ analogue of snoutene, the pentacyclic diene (**117**).[101] So far, all attempts to extend the one carbon bridge in methylenehomosnoutene **119-H** and its derivatives **119-OR** to give the two carbon bridged bishomosnoutene skeleton as in **118**, have failed. Once the diene could be made, a photochemical $[_{\sigma}2_a + _{\pi}2_a + _{\sigma}2_a + _{\pi}2_a]$ process could very well lead to truncahedrane (**115**), as this rearrangement would be exothermal by ~ 25 kcal/mol according to MNDO calculations (see Fig. 37).[101]

ACKNOWLEDGMENTS

This research was supported by the Deutsche Forschungsgemeinschaft, Fonds der Chemischen Industrie, Studienstiftung des Deutschen Volkes, Herrmann Schlosser Stiftung and Norges Almenvitenskapelige Forskningsrad. We are indebted to the BASF AG for generous gifts of cyclooctatetraene and to Shell Chemie GmbH for cycloheptatriene. The work described here was carried out at the Universities of Göttingen and Hamburg by a group of highly talented and motivated co-workers to whom I am especially grateful. In chronological order they were: Christian Weitemeyer, Otto Schallner, Dieter Kaufmann, Dieter

Bosse, Hans-Heinrich Fick, Bernhard Schrader, Lüder-Ulrich Meyer, Ehrhardt Proksch, Ihsan Erden, Werner Spielmann, Peter Gölitz, Thomas Preuss, Bengt Bengtson, Michael Stöbbe, Hans-Peter Kukuk, Stephan Kirchmeyer, Jürgen Höfer and Thomas Lendvai.

REFERENCES

1. (a) H. E. Zimmerman and R. M. Paufler, *J. Am. Chem. Soc.* **1960**, *82*, 1514. (b) H. E. Zimmerman, G. L. Grunewald, R. M. Paufler, and M. A. Sherwin, *J. Am. Chem. Soc.* **1969**, *91*, 2330. (c) C. Weitemeyer, T. Preuss, and A. de Meijere, *Chem. Ber.* **1985**, *118*, 3993.
2. (a) G. Schröder, *Angew. Chem.* **1963**, *75*, 722; *Angew. Chem. Int. Ed. Engl.* **1963**, *2*, 707; *Chem. Ber.* **1964**, *97*, 3140. (b) R. Merényi, J. F. M. Oth, and G. Schröder, *Chem. Ber.* **1964**, *97*, 3150. (c) cf. the review: G. Schröder and J. F. M. Oth, *Angew. Chem.* **1967**, *79*, 458; *Angew. Chem. Int. Ed. Engl.* **1967**, *6*, 414 and references cited therein.
3. G. Giacometti and G. Rigatti, *Ric. Sci.* **1960**, *30*, 1061; J. Paldus and J. Koutecky, *Collect. Czech. Chem. Commun.* **1962**, *27*, 2139; J. Koutecky and J. Paldus, *Tetrahedron* **1963**, *19*, 201; F. A. Van-Catledge, *J. Am. Chem. Soc.* **1971**, *93*, 4365; M. H. Perrin and M. Gouterman, *J. Chem. Phys.* **1967**, *46*, 1019; R. Hoffmann, E. Heilbronner, and R. Gleiter, *J. Am. Chem. Soc.* **1970**, *92*, 706.
4. E. Haselbach, E. Heilbronner, and G. Schröder, *Helv. Chim. Acta* **1971**, *54*, 153; F. A. Van-Catledge and C. E. McBride, Jr., *J. Am. Chem. Soc.* **1976**, *98*, 304; F. A. Van-Catledge and C. E. McBride, Jr., *J. Phys. Chem.* **1976**, *80*, 2987; N. L. Allinger and J. C. Tai, *J. Am. Chem. Soc.* **1965**, *87*, 2081; A. Gedanken and A. de Meijere, *J. Chem. Phys.* **1988**, *88*, 4153.
5. E. Haselbach, L. Neuhaus, R. P. Johnson, K. N. Houk, and M. N. Paddon-Row, *Helv. Chim. Acta* **1982**, *65*, 1743.
6. F. Gerson, A. de Meijere and X.-Z. Quin, *J. Chem. Soc. Chem. Commun.* **1989**, 1077.
7. W. von E. Doering and W. R. Roth, *Angew. Chem.* **1963**, *75*, 27; *Angew. Chem. Int. Ed. Engl.* **1963**, *2*, 24; *Tetrahedron* **1963**, *19*, 715.
8. S. Winstein, *J. Am. Chem. Soc.* **1959**, *81*, 6524; *Quart. Rev. Chem. Soc.* **1969**, *23*, 141.
9. H. W. Whitlock and P. F. Schatz, *J. Am. Chem. Soc.* **1971**, *93*, 3837.
10. (a) A. D. Walsh, *Trans. Faraday Soc.* **1949**, *45*, 179. (b) cf. also: E. Heilbronner, *Isr. J. Chem.* **1972**, *10*, 143.
11. Compare the review by A. de Meijere, *Angew. Chem.* **1979**, *91*, 867; *Angew. Chem. Int. Ed. Engl.* **1979**, *18*, 809 and references cited therein.
12. Compare Z. Rappoport Ed., *The Chemistry of the Cyclopropyl Group*, Wiley, New York, 1988, Vols. 1 and 2.
13. (a) J. F. M. Oth, R. Merényi, F. Nielsen, and G. Schröder, *Chem. Ber.* **1965**, *98*, 3385. (b) H. Röttele, P. Nikoloff, J. F. M. Oth, and G. Schröder, *Chem. Ber.* **1969**, *102*, 3367.

14. P. D. Bartlett and F. D. Greene, *J. Am. Chem. Soc.* **1954**, *76*, 1088.
15. (a) B. R. Ree and J. C. Martin, *J. Am. Chem. Soc.* **1970**, *92*, 1660. (b) V. Buss, R. Gleiter, and P. von R. Schleyer, *J. Am. Chem. Soc.* **1971**, *93*, 3927.
16. J. E. Baldwin and W. D. Fogelsong, *J. Am. Chem. Soc.* **1968**, *90*, 4303, 4311.
17. H. H. Westberg and H. J. Dauben, Jr., *Tetrahedron Lett.* **1968,** 5123; P. Radlick, R. Klem, S. Spurlock, J. J. Sims, E. E. van Tamelen, and T. Whitesides, *Tetrahedron Lett.* **1968**, 5117.
18. A. de Meijere, C. Weitemeyer, and O. Schallner, *Chem. Ber.* **1977**, *110*, 1504.
19. A. de Meijere, O. Schallner, and C. Weitemeyer, *Angew. Chem.* **1972**, *84*, 63; *Angew. Chem. Int Ed. Engl.* **1972**, *11*, 56.
20. E. Proksch and A. de Meijere, *Angew. Chem.* **1976**, *88*, 802; *Angew. Chem. Int. Ed. Engl.* **1976**, *15*, 761.
21. E. Proksch, Ph.D. thesis, Universität Göttingen, 1977.
22. Compare W. A. Pryor, *Free Radicals*, McGraw-Hill, New York 1966, pp. 170ff.
23. A. de Meijere, O. Schallner, C. Weitemeyer, and W. Spielmann, *Chem. Ber.* **1979**, *112*, 908.
24. W. Spielmann, Ph.D. thesis, Universität Göttingen, 1976.
25. Compare C. Rüchardt, *Fortschr. Chem. Forsch.* **1966**, *6*, 251; *Angew. Chem.* **1970**, *82*, 845; *Angew. Chem. Int. Ed. Engl.* **1970**, *9*, 830.
26. Ch. Rüchardt, H. Golzke, W. Spielmann, and A. de Meijere, unpublished results.
27. B. Andersen, O. Schallner, and A. de Meijere, *J. Am. Chem. Soc.* **1975**, *97*, 3521.
28. E. Grunwald and S. Winstein, *J. Am. Chem. Soc.* **1948**, *70*, 846; S. Winstein, E. Grunwald, and H. W. Jones, *J. Am. Chem. Soc.* **1951**, *73*, 2700; A. H. Fainberg and S. Winstein, *J. Am. Chem. Soc.* **1956**, *78*, 2770.
29. Calculated from the rate for 1-bromobicyclo[2.2.2]octane (C. A. Grob, K. Kostka, and F. Kuhnen, *Helv. Chim. Acta* **1970**, *53*, 608 and the known ratio between *t*-butyl chloride and *t*-butyl bromide.[27]
30. W. Parker, R. L. Tranter, C. I. F. Watt, L. W. K. Chang, and P. von R. Schleyer, *J. Am. Chem. Soc.* **1974**, *96*, 7121.
31. G. J. Gleicher and P. von R. Schleyer, *J. Am. Chem. Soc.* **1967**, *89*, 582.
32. Compare J. E. McMurry and C. N. Hodge, *J. Am. Chem. Soc.* **1984**, *106*, 6450.
33. A. de Meijere and B. Schrader, unpublished results. Compare loc. cit.[34]
34. I. Erden and A. de Meijere, *Tetrahedron Lett.* **1980**, *21*, 3179.
35. Y. E. Rhodes and V. G. DiFate, *J. Am. Chem. Soc.* **1972**, *94*, 7582.
36. K. B. Wiberg and J. G. Pfeiffer, *J. Am. Chem. Soc.* **1970**, *92*, 553.
37. H.-P. Kukuk, E. Proksch, and A. de Meijere, *Angew. Chem.* **1982**, *94*, 304; *Angew. Chem. Int. Ed. Engl.* **1982**, *21*, 306; *Angew. Chem. Suppl.* **1982,** 696.
38. W. F. Maier and P. von R. Schleyer, *J. Am. Chem. Soc.* **1981**, *103*, 1891.
39. A. de Meijere and O. Schallner, *Angew. Chem.* **1973**, *85*, 400; *Angew. Chem. Int. Ed. Engl.* **1973**, *12*, 399.

40. Compare G. A. Olah, G. Liang, J. R. Wiseman, and J. A. Chong, *J. Am. Chem. Soc.* **1972**, *94*, 4927.

41. Compare G. A. Olah, G. Liang, K. A. Babiak, T. K. Morgan, Jr., and R. K. Murray, Jr., *J. Am. Chem. Soc.* **1976**, *98*, 576 and earlier work cited therein.

42. J. F. Wolf, P. G. Harch, R. W. Taft, and W. J. Hehre, *J. Am. Chem. Soc.* **1975**, *97*, 2902 and references cited therein.

43. A. de Meijere, O. Schallner, G. D. Mateescu, P. Gölitz, and P. Bischof, *Helv. Chim. Acta* **1985**, *68*, 1114.

44. A. de Meijere, O. Schallner, P. Gölitz, W. Weber, P. von R. Schleyer, G. K. S. Prakash, and G. A. Olah, *J. Org. Chem.* **1985**, *50*, 5255.

45. O. Schallner, Ph.D. thesis, Universität Göttingen, 1974.

46. B. Schrader, Ph.D. thesis, Universität Göttingen, 1977.

47. A. de Meijere, O. Schallner, B. Schrader, W. Spielmann, and P. Gölitz, unpublished results.

48. R. D. Rieke and P. M. Hudnall, *J. Am. Chem. Soc.* **1972**, 94, 7178; R. D. Rieke and S. E. Bales, *J. Chem. Soc. Chem. Commun.* **1973**, 879; *J. Am. Chem. Soc.* **1974**, *96*, 1775.

49. P. Gölitz and A. de Meijere, *Angew. Chem.* **1977**, *89*, 892; *Angew. Chem. Int. Ed. Engl.* **1977**, *16*, 854.

50. P. Gölitz, Ph.D. thesis, Universität Göttingen, 1978.

51. P. Gölitz and A. de Meijere, unpublished results; P. Gölitz, *Diplomarbeit*, Universität Göttingen, 1975.

52. R. W. Taft, Jr., in *Steric Effects in Organic Chemistry,* M. S. Newman, Ed. Wiley, New York, 1956.

53. P. von R. Schleyer and C. W. Woodworth, *J. Am. Chem. Soc.* **1968**, *90*, 6528.

54. C. A. Grob, *Angew. Chem.* **1976**, *88*, 621; *Angew. Chem. Int. Ed. Engl.* **1976**, *15*, 569.

55. A. de Meijere and W. Spielmann, unpublished results.

56. C. A. Grob and R. Rich, *Tetrahedron Lett.* **1978**, 663.

57. P. Gölitz and A. de Meijere, *Angew. Chem.* **1981**, *93*, 302; *Angew. Chem. Int. Ed. Engl.* **1981**, *20*, 298.

58. Estimated assuming additivity of strain contributions from all subunits. Compare L. N. Ferguson, *Highlights of Alicyclic Chemistry,* Franklin Publishing Co., Palisade, NJ, 1973.

59. Compare D. Kaufmann and A. de Meijere, *Chem. Ber.* **1984**, *117*, 3134 and references cited therein.

60. Compare K. B. Wiberg, W. E. Pratt, and W. F. Bailey, *J. Am. Chem. Soc.* **1977**, *99*, 2297.

61. J. Michl, T. Lendvai, and A. de Meijere, unpublished results.

62. W. Spielmann and A. de Meijere, *Angew. Chem.* **1976**, *88*, 446; *Angew. Chem. Int. Ed. Engl.* **1976**, *15*, 429.

63. D. A. Adamiak, W. Saenger, W. Spielmann, and A. de Meijere, *J. Chem. Res. (S)* **1977**, 120; (M) **1977**, 1507.

64. W. Spielmann, C. Weitemeyer, T.-N. Huang, A. de Meijere, F. Snatzke, and G. Snatzke, *Isr. J. Chem.* **1977**, *15*, 99.

65. Compare S. Brownstein and J. Bornais, *Can. J. Chem.* **1971**, *49*, 7.

66. C. Weitemeyer, Ph.D. thesis, Universität Göttingen, 1976.

67. J. A. Deyrup and C. L. Moyer, *Tetrahedron Lett.* **1968**, 6179; J. A. Deyrup, C. L. Moyer, and P. S. Dreifus, *J. Org. Chem.* **1970**, *35* 3428; V. R. Gaertner, *Tetrahedron Lett.* **1968**, 5919; *J. Org. Chem.* **1970**, *35*, 3952.

68. (a) G. Szeimies, *Chem. Ber.* **1973**, *106*, 3695. (b) G. Szeimies, K. Mannhardt, and M. Junius, *Chem. Ber.* **1977**, *110*, 1792.

69. Compare (a) D. L. Whalen, *J. Am. Chem. Soc.* **1970**, *92*, 7619. (b) W. C. Danen, *J. Am. Chem. Soc.* **1972**, *94*, 4835, and references cited therein.

70. Compare P. von R. Schleyer and G. W. Van Dine, *J. Am. Chem. Soc.* **1966**, *88*, 2321; R. S. Brown and T. G. Traylor, *J. Am. Chem. Soc.* **1973**, *95*, 8025.

71. Compare H. Basch, M. B. Robin, N. A. Kuebler, C. Baker, and D. W. Turner, *J. Chem. Phys.* **1969**, *51*, 52; M.-M. Rohmer and B. Roos, *J. Am. Chem. Soc.* **1975**, *97*, 2025.

72. C. Weitemeyer, T. Preuss, and A. de Meijere, *Chem. Ber.* **1985**, *118*, 3993.

73. D_3-Trishomocubane (a) and D_3-trishomocubanetrione (b) have been prepared in optically active form. (a) G. Helmchen and G. Staiger, *Angew. Chem.* **1977**, *89*, 119; *Angew. Chem. Int. Ed. Engl.* **1977**, *16,* 116. (a) and (b) W.-D. Fessner and H. Prinzbach, *Tetrahedron*, **1986**, *42*, 1797.

74. Compare M. Nakazaki, *Top. Stereochem.* **1984**, *15*, 199 and references cited therein.

75. Recently, several new examples of such molecules have been reported. H. Müller, J.-P. Melder, W.-D. Fessner, D. Hunkler, H. Fritz, and H. Prinzbach, *Angew. Chem.* **1988**, *100*, 1140; *Angew. Chem. Int. Ed. Engl.* **1988**, *27*, 1103.

76. A. de Meijere and D. Kaufmann, unpublished results.

77. T. Preuss, Ph.D. thesis, Universität Hamburg, 1983.

78. R. Gleiter, M. C. Böhm, A. de Meijere, and T. Preuss, *J. Org. Chem.* **1983**, *48*, 796.

79. The previously reported[78] percentages of predominating products in cheletropic additions to **21**, **98** and **99** have been slightly corrected.[77]

80. H. Prinzbach and D. Stusche, *Angew. Chem.* **1970**, *82*, 836; *Angew. chem. Int. Ed. Engl.* **1970**,*9*, 799. An extensive review on this class of compounds will appear elsewhere. H. Prinzbach and C. Rücker, *Chem. Rev.*, to be published.

81. S. Winstein, *J. Am. Chem. Soc.* **1959**, *81*, 6524; *Quart. Rev., Chem. Soc.* **1969**, *23*, 141.

82. E. Heilbronner, R. Gleiter, T. Hoshi, and A. de Meijere, *Helv. Chim. Acta* **1973**, *56*, 1594.

83. Compare A. Nickon and E. F. Silversmith, *Organic Chemistry: The Name Game,* Pergamon Press, New York, 1987, pp. 20 and 201.

84. D. Kaufmann, H.-H. Fick, O. Schallner, W. Spielmann, L.-U. Meyer, P. Gölitz, and A. de Meijere, *Chem. Ber.* **1983**, *116*, 587.

85. (a) W. G. Dauben, C. H. Schallhorn, and D. L. Whalen, *J. Am. Chem. Soc.* **1971**, *93*, 1446. (b) L. A. Paquette and J. C. Stowell, *J. Am. Chem. Soc.* **1971**, *93*, 2459. (c) I. Erden, A. de Meijere, *Tetrahedron Lett.* **1980**, *21*, 1837.

86. J. Spanget-Larsen and R. Gleiter, *Angew. Chem.* **1978**, *90*, 471; *Angew. Chem. Int. Ed. Engl.* **1978**, *17*, 441.

87. B. Bengtson, Ph.D. thesis, Universität Hamburg, 1986.

88. D. Kaufmann, O. Schallner, L.-U. Meyer, H.-H. Fick, and A. de Meijere, *Chem. Ber.* **1983**, *116*, 1377.

89. E. Vogel, H.-J. Altenbach, and C. D. Sommerfeld, *Angew. Chem.* **1972**, *84*, 986; *Angew. Chem. Int. Ed. Engl.* **1973**, *11*, 939; E. Vogel, H. J. Altenbach, and E. Schmidbaur, *Angew. Chem.* **1973**, *85*, 862; *Angew. Chem. Int. Ed. Engl.* **1973**, *12*, 838.

90. (a) R. Schwesinger and H. Prinzbach, *Angew. Chem.* **1972**, *84*, 990; *Angew. Chem. Int. Ed. Engl.* **1972**, *11*, 942. R. Schwesinger, H. Fritz, and H. Prinzbach, *Chem. Ber.*, **1980**, *112*, 3318. (b) R. Schwesinger and H. Prinzbach, *Angew. Chem.* **1972**, *85*, 1107; *Angew. Chem. Int. Ed. Engl.* **1973**, *12*, 989; R. Schwesinger, M. Breninger, B. Gallenkamp, K.-H. Müller, D. Hunkler, and H. Prinzbach, *Chem. Ber.* **1980**, *113*, 3127. (c) S. Kagabu and H. Prinzbach, *Angew. Chem.* **1975**, *87*, 248; *Angew. Chem. Int. Ed. Engl.* **1952**, *14*, 252. (d) S. Kagabu, C. Kaiser, R. Keller, P. G. Becher, K.-H. Müller, L. Knothe, G. Rihs, and H. Prinzbach, *Chem. Ber.* **1988**, *121*, 741. (e) B. Zipperer, K.-H. Müller, B. Gallenkamp, R. Hildebrand, M. Fletschinger, D. Burger, M. Pillat, D. Hunkler, L. Knothe, H. Fritz, and H. Prinzbach, *Chem. Ber.* **1988**, *121*, 757 and references cited therein.

91. D. L. Dalrymple and S. P. B. Taylor, *J. Am. Chem. Soc.* **1971**, *93*, 7098; H. Prinzbach and R. Schwesinger, *Angew. Chem.* **1972**, *84*, 988; *Angew. Chem. Int. Ed. Engl.* **1972**, *11*, 940; H. Prinzbach, D. Stusche, M. Breuninger, and J. Markert, *Chem. Ber.* **1976**, *109*, 2823 and references cited therein.

92. I. Erden, P. Gölitz R. Näder, and A. de Meijere, *Angew. Chem.* **1981**, *93*, 605; *Angew. Chem. Int. Ed. Engl.* **1981**, *20*, 583; M. Stöbbe, U. Beherns, G. Adiwidjaja, P. Gölitz, and A. de Meijere, *Angew. Chem.* **1983**, *95*, 904; *Angew. Chem. Int. Ed. Engl.* **1983**, *22*, 867; *Angew. Chem. Suppl.* **1983,** 1221; M. Stöbbe, P. Gölitz, I. Erden, S. Kirchmeyer, J. Höfer, and A. de Meijere, *Chem. Ber.*, submitted.

93. M. Engelhard and W. Lüttke, *Angew. Chem.* **1972**, *84*, 346; *Angew. Chem. Int. Ed. Engl.* **1972**, *11*, 310; M. Engelhard, Ph.D. thesis, Universität Göttingen, 1977.

94. R. T. Taylor and L. A. Paquette, *Angew. Chem.* **1975**, *87*, 488; *Angew. Chem. Int. Ed. Engl.* **1975**, *14*, 496.

95. W. Spielmann, D. Kaufmann, and A. de Meijere, *Angew. Chem.* **1978**, *90*, 470; *Angew. Chem. Int. Ed. Engl.* **1978**, *17*, 440.

96. A. de Meijere, D. Kaufmann, and O. Schallner, *Tetrahedron Lett.* **1974,** 3835.

97. D. Kaufmann and A. de Meijere, *Tetrahedron Lett.* **1974**, 3831.

98. D. Bosse and A. de Meijere, *Chem. Ber.* **1978**, *111*, 2223.

99. Compare J. M. Schulman, R. L. Disch, and M. L. Sabio, *J. Am. Chem. Soc.* **1986**, *108*, 3258 and references cited therein.

100. Compare H. Prinzbach, H.-P. Schal, and G. Fischer, *Tetrahedron Lett.* **1983**, *24*, 2147.

101. S. Kirchmeyer and A. de Meijere, *Helv. Chim. Acta* **1990**, *71*, in press.

9 The [n]Peristylane–Polyhedrane Connection*

LEO A. PAQUETTE

Evans Chemical Laboratories
The Ohio State University
Columbus, Ohio

Molecules can be pleasing to chemists for many reasons: structural intricacy as provided by nature; pivotal role in world culture; economic importance; or some complex function they perform. Aesthetic response is perhaps at its zenith when exquisite shape or structure is involved.[1] The fascination of the chemical community with dodecahedrane (**6**) is a case in point. This $C_{20}H_{20}$ polyhedrane has come to be regarded as the "Mount Everest of alicyclic chemistry."[2] Its unique geometry (all of its 20 identical methine carbon atoms possess perfect tetrahedral character), I_h symmetry (120 identity operations), encapsulation of a cavity incapable of solvation (approximate transannular distance of 4 Å), and the like have caused many individuals to attempt its total synthesis.[3–5] This feat, considered by many as a "challenge of substantial and significant proportions,"[4] was accomplished initially in 1982[6] and more recently in a very different way.[7]

At a somewhat earlier date, the relationship of the peristylanes **1**–**3** to their polyhedral counterparts **4**–**6** was formalized.[8] The first class of compounds is characterized by the interconnection of a smaller n-membered ring to one twice the original size at alternate carbon atoms of the latter. In the polycyclic $(CH)_{2n}$ systems represented by **3**–**6**, the central n-membered alicycle is linked to a pair of $n/2$-membered rings by alternate carbon atoms. Symmetry requires, of course, that n be an even integer.

*Dedicated to Paul von R. Schleyer for his pioneering work in adamantane chemistry for the last 30 years.

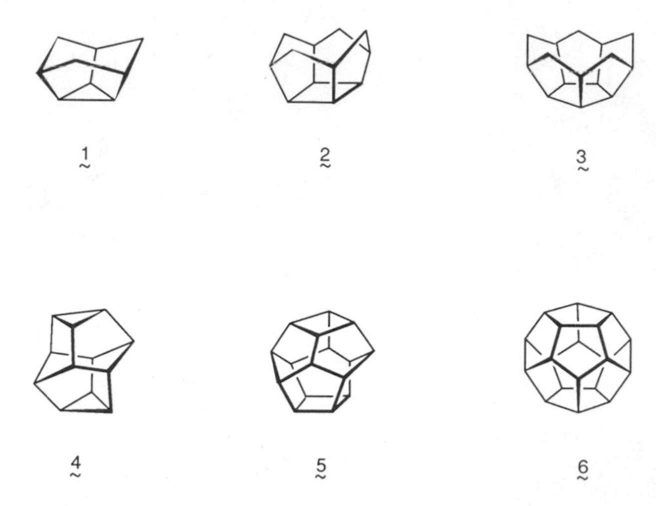

The promulgation of **1–3** as a subset of **4–6** served, on the one hand, to expand conceptually on the Eaton approach to dodecahedrane, which involved attempts to roof functionalized [5]peristylanes with a cyclopentane ring.[9] In addition, substantive interest was generated in the projected capping of **1**, **2**, and related molecules. These undertakings have proven not to be at all straightforward and indeed have not yet been carried to their logical conclusion in any instance.

In this chapter, discussion will focus first on those synthetic efforts devised to gain access successfully to each of the three peristylanes. Attention will then be given to the probable causative factors underlying the inability of these hemispherical molecules to be "roofed in." Finally, the successful alternative routes to the polyhedranes, realized by application of entirely different tactics, are summarized.

THE [3]PERISTYLANE AND *p*-[$3^2.5^6$]OCTAHEDRANE ASSEMBLIES

Synthesis of [3]Peristylanes

Three approaches to the parent [3]peristylane hydrocarbon (**1**) have been reported. The original Nickon–Pandit route took proper advantage of the appreciable kinetic tendency of carbene **8** for intramolecular C—H insertion.[10] Thus, pyrolysis of **7** gave **1**, a molecule initially called "triaxane" in an effort to capture its C_{3V} symmetry and to emphasize that the cyclopropane ring rests on three axial pillars fixed to a cyclohexane chair.[11] The hydrocarbon is characterized, as expected, by three types of carbon nuclei and four types of protons (ratio 3:3:3:3).

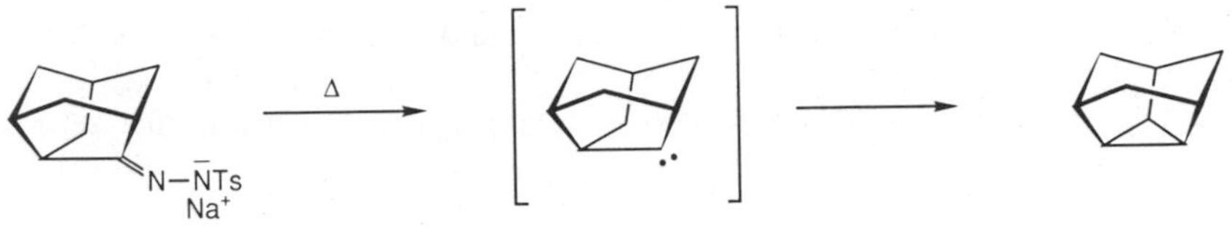

The other two routes to **1** begin from *endo*-bicyclo[3.2.1]oct-6-ene-3-carboxylic acid (**9**).[8] In this instance, heating of the derived aldehyde tosylhydrazone salt (**10a**) to 200–250°C at low pressure results in conversion to the cyclic azo compound (**11a**). Only at higher temperatures (500°C) is nitrogen loss effective and **1** produced. Recourse to the monodeuterated starting material **10b** made possible by analogous means the preparation of **12**.[8]

Taking advantage of the fact that reaction of **1** with thionyl chloride leads directly to chloro ketone **13**,[12] Garratt and White transformed this intermediate to dichloride **14**. Subsequent treatment of **14** with sodium naphthalenide gave **1**.[8]

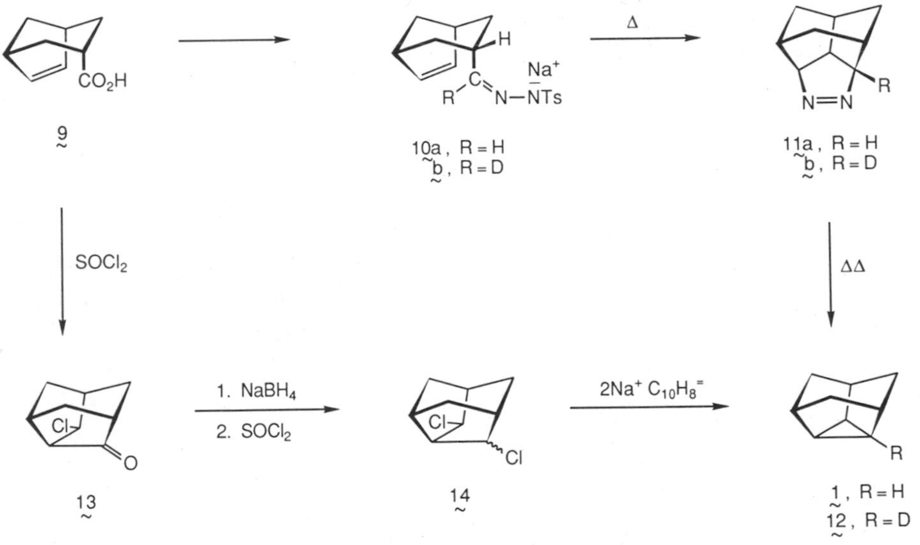

Recently, this work was followed by the design of a free radical route to the first chiral member of this compound class.[13] Following the stereocontrolled twofold alkylation of monoketal **15b**,[14] the carbonyl group in **16b** was reduced by the Wolff–Kishner method to give **17**. The relative ordering of the S_N2 reactions of **15b** is most important, since these two steps set the proper relative stereochemistry for ultimate radical cyclization. Oxa-di-π-methane rearrangement[15] of **18** furnished **19a** efficiently without need to block the hydroxyl group. Arrival at **20** materialized (75% yield) upon subjecting iodo ketone **19b** to the action of hexamethylditin under photochemical conditions. Clearly, cyclohexanol ring construction by intramolecular capture of a carbon-centered radical by a ketone carbonyl constitutes a useful means for setting a rim bond in place.

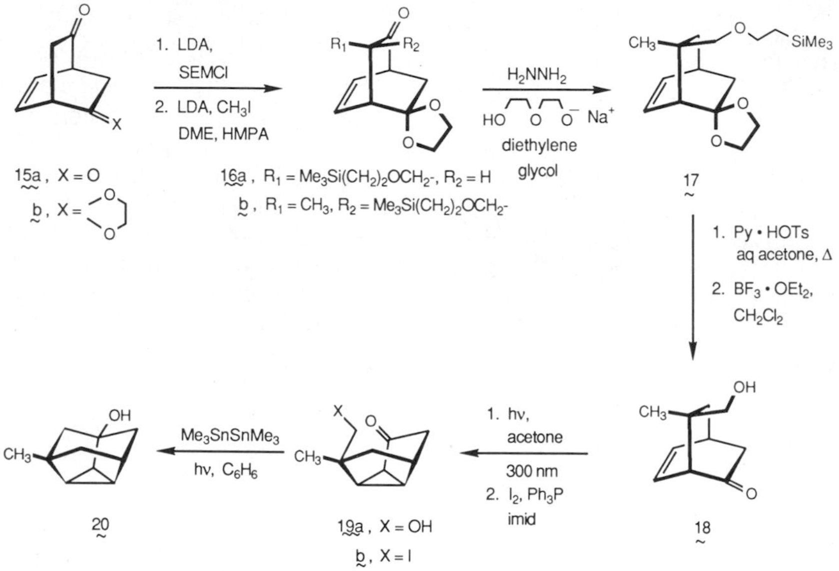

Alcohol **20** actually represents the second known oxygenated [3]peristylane. The first known example is acetate **21**, prepared by lead tetraacetate oxidation of **1**.[16] Whereas **20** is chiral and potentially resolvable, **21** is not.

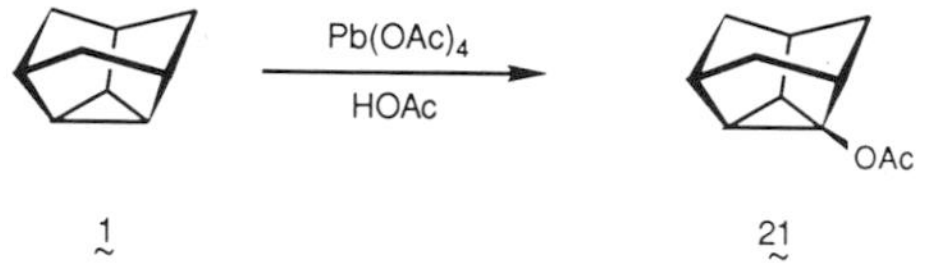

Attempts to Cap [3]Peristylane

Diketone **22**,[17] selected by Garratt et al. as their starting material,[18] was already recognized as a molecule in which the two carbonyl groups are so sterically compressed that they lack normal reactivity.[19] The situation is particularly exacerbated following nucleophilic attack at one of these sites. For this reason **22** was initially converted to **23**.[19] Subsequent condensation with the anion of diethyl (cyanomethyl)phosphonate then proceeded with its usual efficiency to give **24**. Stereospecific conversion to **25** was next achieved with magnesium in methanol. Once the remaining double bond had been epoxidized from its exo surface as in **26**, elaboration of the cyclopropane ring could be accomplished in three steps. In a remarkable reaction, heating **28** with potassium hydride in tetrahydrofuran (THF) resulted in decyanation and formation of **29**.

1. LiAlH(O*t* Bu)$_3$
2. 2N NaOH, MeOH
$(EtO)_2P(O)\overset{\ominus}{C}HCN$
22 23 24
Mg / CH$_3$OH
MCPBA
1. KH, THF
2. TsCl, py
25 26 27a, R = H
b, R = Ts
KH / THF
KH / THF / Δ
28 29
1. Dibal - H
2. air
TMSI
30 31

Unfortunately, the ether linkages in **28-30** proved less receptive to electrophile-induced cleavage than the constituent cyclopropane rings. The conversion of **28** to **31** is a typical example. Evidently, C—O bond scission is not kinetically attractive to these molecules because of the rather sizable nonbonded steric repulsions that would be brought into play in the event.

Further studies of **22** by Weber led to the discovery that the lithium salt of acetonitrile enters into efficient condensation (86%) with **22** to give **32**.[20] Although dehydration of this intermediate was never realized, the material could be channeled via **33** to **34**. As before, however, treatment of **34** with a variety of strong bases did not result in C—O bond scission to give **35**.

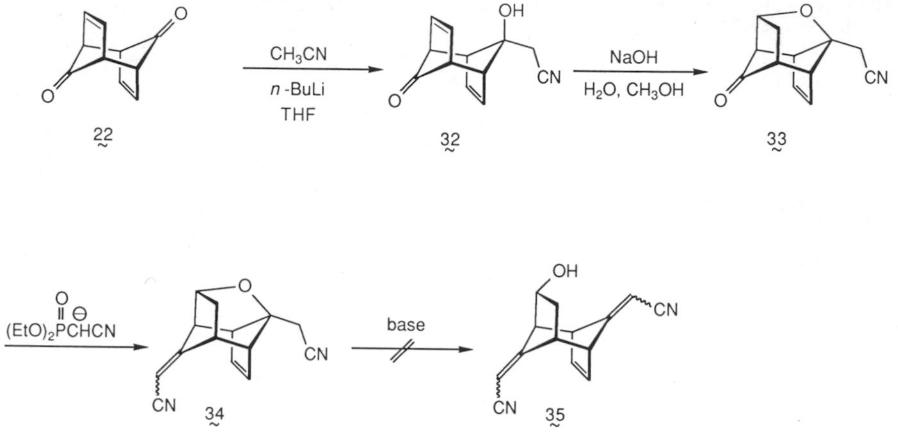

Synthetic Entry to 2,9-Dimethyl-*p*-[$3^2.5^6$]octahedrane

Hirao and co-workers nicely skirted the complications just described by taking proper advantage of a completely stereospecific acid catalyzed rearrangement noted earlier in their laboratories, namely, the quantitative conversion of **36** into **37**.[21] To this end, the known C_2-symmetric diketo diol

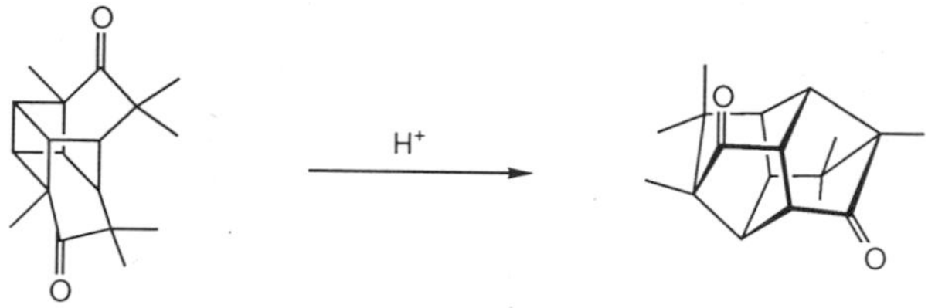

38a[22] was transformed into its dimesylate. When treated with boron trifluoride etherate and a small amount of 2,6-di-*t*-butylpyridine in benzene at room temperature, **38b** gave a mixture of diketones **39** (47%) and **40** (10%).[23] Sequential Wolff–Kishner reduction and ozonolysis of the major product made

41 available and set the stage for formation of the bis(tosylhydrazone). This penultimate intermediate was smoothly transformed into **42** under Bamford–Stevens conditions.

38a, R = H
b, R = $MeSO_2$

$BF_3 \cdot OEt_2$
C_6H_6, rt

39
40

1. $H_2NNHCONH_2$
2. KOH, 220°C
3. O_3; Me_2S

41

1. $TsNHNH_2$
2. KO*t*Bu, diglyme, 130°C

42

The C_2-symmetric nature of **42** was corroborated by its four-line proton and six-line carbon NMR spectra.

ROUTES TO [4]PERISTYLANE AND FUNCTIONALIZED DERIVATIVES

Preparation of [4]Peristylane and Its 2,6-Dione

Tricyclo[5.2.1.0^{2,6}]deca-2,5,8-triene (**43**), a molecule well known to capture dienophiles most often with complete below plane stereoselectivity,[24] reacts analogously with tosylacetylene to generate **44**.[25,26] Because of the strong tendency of *syn*-sesquinorbornatrienes such as **44** to experience rapid air oxidation,[27] this adduct was not isolated but epoxidized directly. The ensuing intramolecular [2 + 2] photocycloaddition of **45** not only confirmed the stereochemical course of the Diels–Alder reaction, but efficiently installed the necessary cyclobutane ring. Heating **46** with periodic acid in aqueous methanol led to **47**, the first known [4]peristylane derivative.

In order to arrive at the parent hydrocarbon, **47** was fully ketalized and subjected to reductive desulfonylation. Subsequent acid hydrolysis delivered the 1,4-dione **49**.[25,26] X-ray crystallographic analysis of this beautifully crystalline

Ts—≡—H
C_6H_6, Δ
MCPBA
43
44
45
Ts=CH_3—⟨benzene⟩—SO_2–
hν
350 nm
acetone
HIO_4
CH_3OH
H_2O
47
46

solid showed it to possess almost perfect C_{2v} symmetry in the solid state.[28] The density of **49** is substantive (1.42 g/cm^3). As a direct consequence of its pouchshaped ground-state conformation, dictated by the need of its inner methylene protons to move as far apart as possible, the two carbonyl groups are mutually compressed. Transannular ring closures such as that operating during conversion to **50** consequently operate with unusual ease.

1. Li,
$C_2H_5NH_2$
2. H_3O^+
Zn
sat. HCl
ether
48a, R = Ts
b, R = H
49
50
1. BF_3,
$(C_2H_5)_3SiH$
2. TsCl, py
$LiAlH_4$
THF
Δ
51a, R = H
b, R = Ts
2

Unwanted reactions of this type can, however, be avoided. The conversion of **49** to ditosylate **51b** and reductive displacement of the sulfonate ester residues, for example, leads uneventfully to **2**. [4]Peristylane (**2**) is a volatile hydrocarbon with a high melting point (>225°C). Its spectra clearly show the molecule to possess four-fold symmetry.

A somewhat modified, more expedient route to **49** has appeared more recently.[26] The adduct of **43** with (*Z*)-1,2-bis(phenylsulfonyl)ethylene (**52**),[29] namely, **53**, undergoes smooth epoxidation–desulfonylation to provide **55**. Diene epoxide prepared in this manner need not be purified prior to irradiation in acetone solution at 350 nm. The strained cage epoxide **56** undergoes oxidative cleavage in the presence of methanolic periodic acid to deliver the target compound.

SO_2Ph / SO_2Ph **52**, THF, 20°C; MCPBA; 1-2% Na/Hg, CH_3OH; hν 3500 Å, acetone; HIO_4, CH_3OH, H_2O

43 53 54 55 56 49

Functionalized Group Manipulations within [4]Peristylane-2,6-Dione

The replacement of the carbonyl oxygen atoms in **49** to various degrees by nitro groups was driven chiefly by a search for energetic compounds possessing elevated densities. As seen from the chemical reactivity of **58**, however, the molecular compactness of [4]peristylanes translates into an unusual facility for transannular cyclization.[30] The resultant dinitro bishomopentaprismane **59** is an interesting molecule. Single crystal X-ray analysis revealed its C—C single bonds to be highly strained. Moreover, the pair of nitro groups bring to **59** a density (1.63 g/cm) markedly enhanced relative to values exhibited by closely related structures.

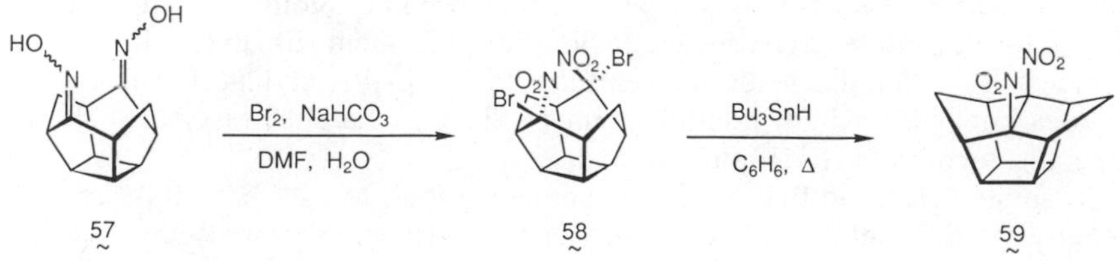

Although the formation of **59** proved to be an end result for many chemical reactions of these molecules, *endo,endo*-2,6-dinitro[4]peristylane (**62**) could be produced by reducing dioxime **57** first to the bis(hydroxylamine) **60**. Direct condensation of this intermediate with *N*-benzylidenebenzenesulfonamide and benzaldehyde made available bis(nitrone) **61**, ozonolysis of which did not induce closure.[31]

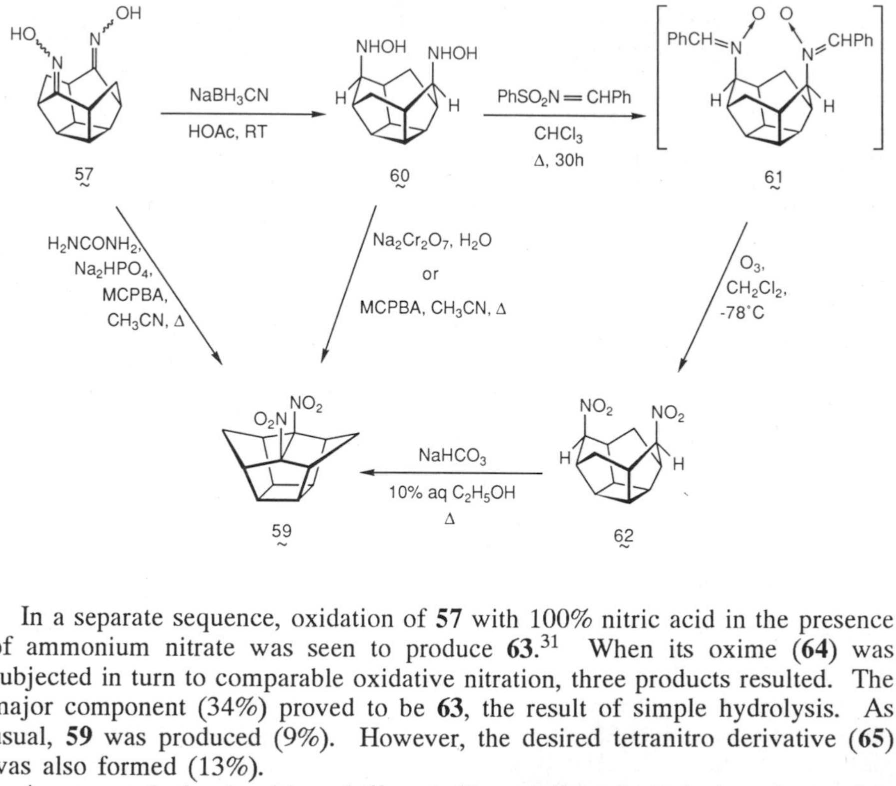

In a separate sequence, oxidation of **57** with 100% nitric acid in the presence of ammonium nitrate was seen to produce **63**.[31] When its oxime (**64**) was subjected in turn to comparable oxidative nitration, three products resulted. The major component (34%) proved to be **63**, the result of simple hydrolysis. As usual, **59** was produced (9%). However, the desired tetranitro derivative (**65**) was also formed (13%).

As expected, the densities of **62** and **65** are intimately linked to the number of nitro groups present: 1.54 and 1.70 g/cm^3, respectively.[31]

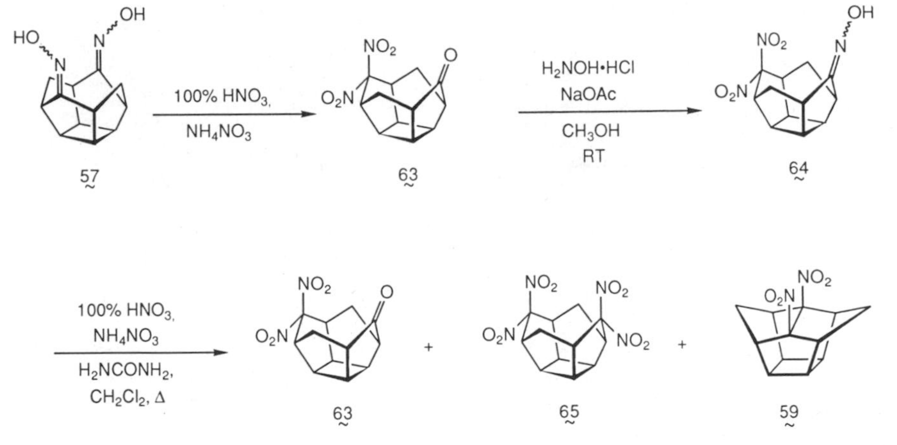

Extended Functionalization of the Fluted [4]Peristylane Perimeter

Very recently, [4]peristylanes functionalized at more than two rim positions have become available. To arrive at the 2,4,6-trione (**71**), Paquette and Shen first prepared epoxide **68** from two routes.[32] As seen with the parent triene

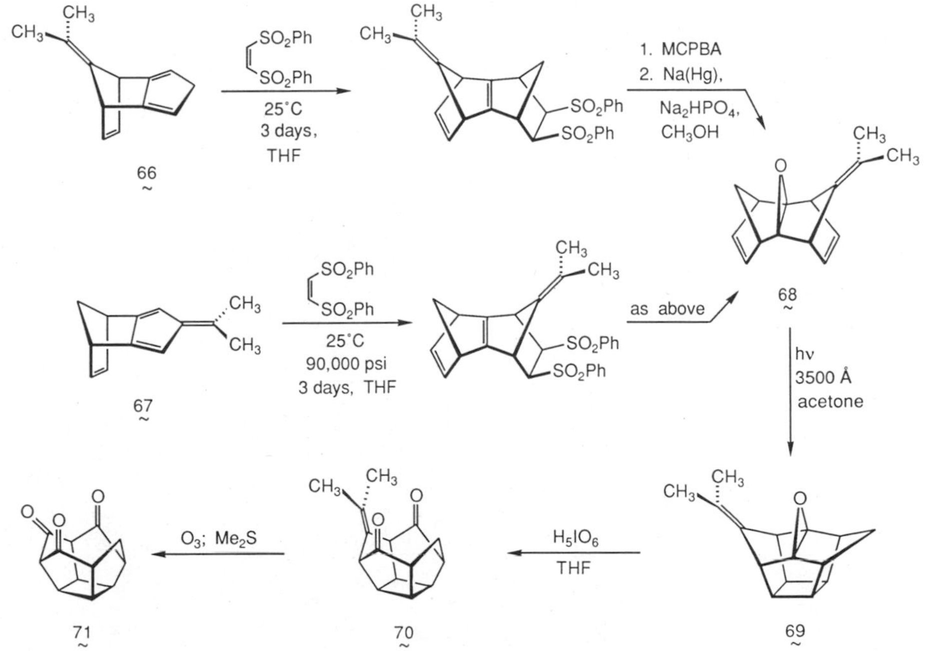

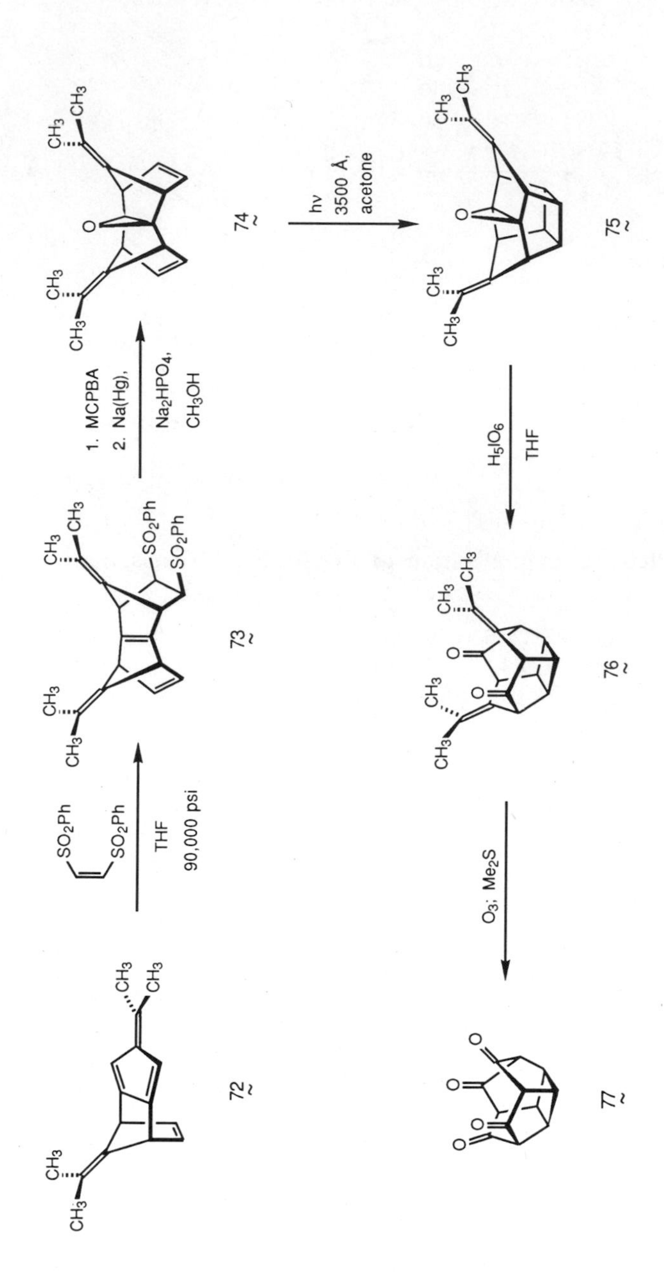

SO_2Ph
SO_2Ph
THF
90,000 psi
1. MCPBA
2. Na(Hg),
Na_2HPO_4,
CH_3OH
hν
3500 Å,
acetone
H_5IO_6
THF
O_3; Me_2S
72
73
74
75
76
77

(**43**), the isopropylidene derivatives **66** and **67** also cycloadd to **52** from below plane. While **66** enters into a room temperature Diels–Alder reaction at a rate closely comparable to that exhibited by **43**, high pressure conditions are warranted in the case of fulvene (**67**).

The pathway from **68** to **71** mirrors that developed earlier. It is important, however, that periodic acid cleavage *precede* ozonolysis for ultimate success. The colorless crystals of **71** do not melt up to 300°C.

With superfulvene (**72**)[33] as starting material, it proved an easy matter to arrive at epoxy diene **74** and its photochemically derived caged epoxide **75** with full preservation of C_{2v} character. The standard sequence of two consecutive oxidations on **75** concisely delivers the 2,4,6,8-tetraone **77**. This very insoluble colorless solid, mp >300°C, shows no spectroscopically detectable tendency to enolize.

Belting of the [4]Peristylane Nucleus

Diketone **49** has proven to be responsive to Wadsworth–Emmons condensation, providing the bis-homologated α,β-unsaturated diester **78** is a mixture of geometric isomers.[34] While catalytic hydrogenation of **78** afforded **79**, its carbomethoxy groups could not be engaged in Dieckmann or acyloin cyclization despite their apparent proximity. This unreactivity has been interpreted to

CH3OOC CH COOCH3 CH 78 — H2 / PtO2 → CH3OOC CH2 COOCH3 CH2 H H 79 —✗→ 80 (O, COOCH3, H, H); 79 —✗→ 81 (Me3SiO OSiMe3 H H)

79 ↓ 82a, R = H; b, R = THP (OR, OR, H, H) — Ph3P•Br2 → 83 (Br, Br, H, H)

83 —✗→ 84 (H, H); 83 — Na2S, C2H5OH → 85 (S, H, H) — MCPBA → 86 (O2 S, H, H)

COOCH$_3$ CH$_3$OOC CH CH — **87** — Zn, HCl, ether → CH$_3$OOC COOCH$_3$ — **88** — 1. Na-K, Me$_3$SiCl, ether; 2. FeCl$_3$, HCl, ether → H···O O — **89**

89 — Py•HBr$_3$, CHCl$_3$ → O O Br — **90** —✕→ O O — **91**

reflect the severe nonbonded steric interactions that arise when the carbon atoms to be linked again begin to align themselves in the proper fashion.

Alternatively, dibromide **83**, while showing no tendency for reductive coupling to **84**, does react with sodium sulfide under high dilution conditions to give **85**. Possible ring contraction of **85** or sulfone **86** via the Ramberg–Bäcklund reaction or Stevens rearrangement were to no avail, however.

In an alternate probe of the possibility of bridging [4]peristylane by a four-carbon chain, α-bromo diketone **90** was prepared by reinstallation of the transannular σ bond prior to peripheral ring formation.[34] Unfortunately, all attempts to achieve 1,4-dehydrobromination in **90** were thwarted.

For these reasons, the diester (**79**) was transformed into the crystalline diphenacyl derivative **92** with a view to accomplishing its twofold homo-Norrish cyclization.[20] However, many varied attempts to achieve the excited state **92** → **93** transformation proved equally unrewarding. Subsequent X-ray crystallographic analysis of **92** (Fig. 1)[35] showed both phenacyl side chains to

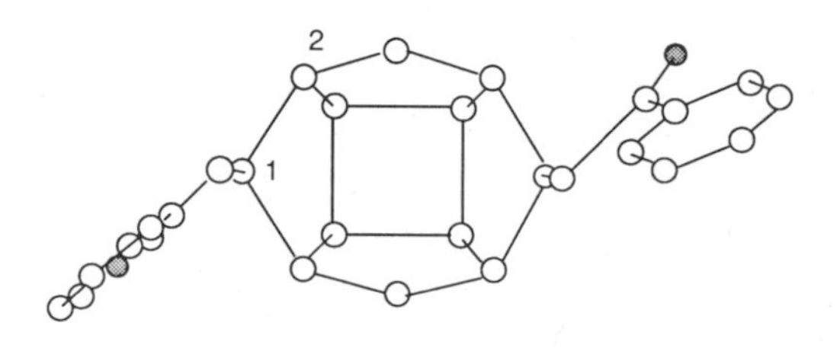

Fig. 1. Top view of **92** as determined crystallographically.

be planar 5 and pointing outwards from the center of the hemisphere. If the assumption is made that the long O—C nonbonded distances (4.14–4.31 Å) do not change a great deal from the crystalline to the solution state because of sterically impeded rotation, the inability to accomplish the excited state **92** → **93** transformation can be understood.

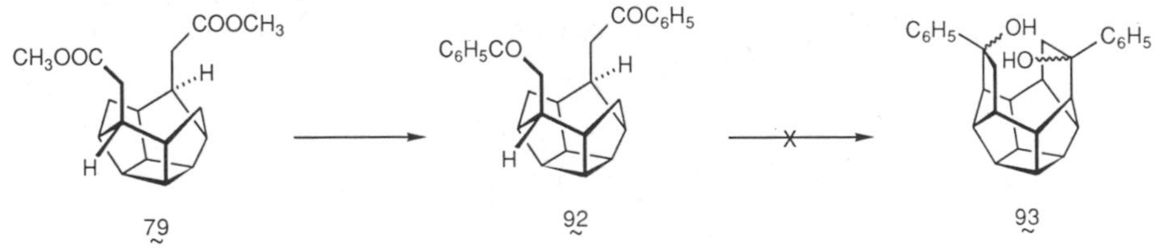

COOCH3 CH3OOC H H 79 → C6H5CO COC6H5 H H 92 —✕→ C6H5 OH HO C6H5 93

Weber also channeled effort in the direction of another photochemical pathway, one dealing with the possible double-barreled Norrish closure of diketone **97**.[20] Wittig olefination of **49** under salt-free conditions furnished **94**, hydroboration–oxidation of which led to diol **95**. Following arrival at diester

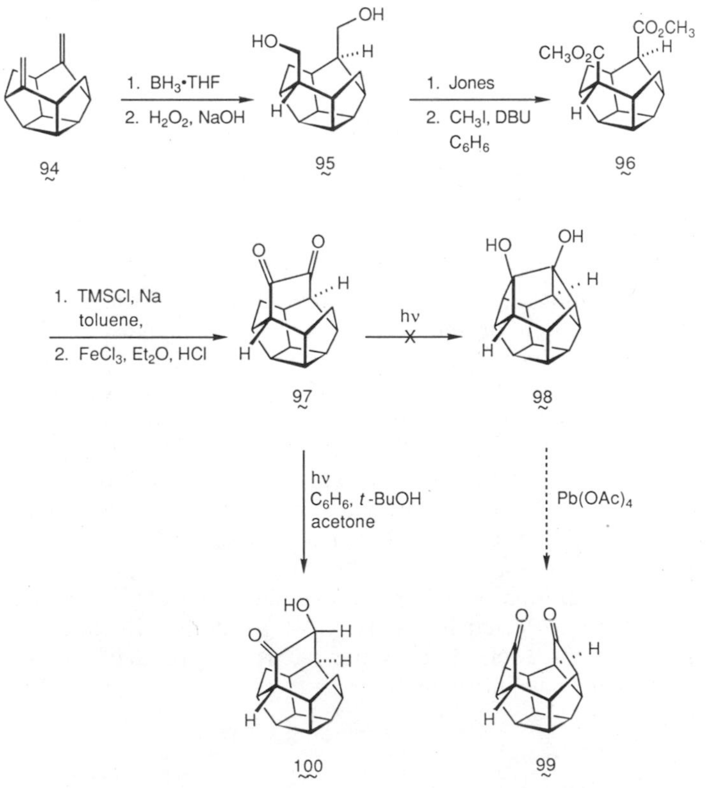

94 → 1. BH_3•THF 2. H_2O_2, NaOH → HO OH H H 95 → 1. Jones 2. CH_3I, DBU C_6H_6 → CH_3O_2C CO_2CH_3 H H 96

1. TMSCl, Na toluene, 2. $FeCl_3$, Et_2O, HCl → O O H H 97 —hν ✕→ HO OH H H 98

97 → hν C_6H_6, *t*-BuOH acetone → HO O H H H 100

98 ⇢ $Pb(OAc)_4$ ⇢ O O H H 99

96 whose carbomethoxyl groups have little or no ability to reside outside the cavity, acyloin cyclization proceeded smoothly to deliver ultimately the bright yellow crystalline **97**. Unfortunately, experimental conditions were not found for effecting its closure to diol **98**, a projected precursor of **99**. Photoreduction to **100** was encountered under certain circumstances, but formation of a new C—C bond was not seen.

As matters now stand, therefore, **97** represents the most advanced intermediate on the pathway to **5**, which remains a difficult synthetic challenge.

ELABORATION OF [5]PERISTYLANES

The Parent [5]Peristylane System

The foundation of the Eaton strategy for dodecahedrane construction was a convenient synthesis of a [5]peristylane larger fragment to which a cyclopentane "lid" would ultimately be bonded as illustrated in **101**.[36] In pioneering work,

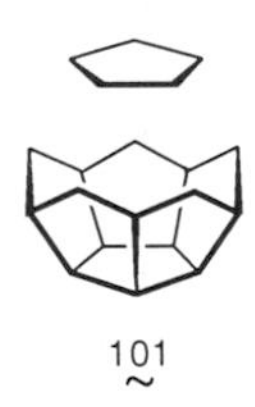

101

this group first demonstrated that the simple diketone **102** could be crafted into the parent hydrocarbon of the series (**110**).[37] Annulation to both carbonyl groups of the starting material[38,39] gave tetracycle **105**, whose catalytic hydrogenation led stereospecifically to **106**. Following introduction of a double bond in each of the two terminal rings, direct trans-skeletal reductive closure within **107** by zinc in acetic acid afforded the pentaquinane **108** and set the stage for installation of the 15th carbon atom. This was accomplished by treating **108** with ethyl formate under a strongly basic Claisen–Schmidt setting.

Standard ketalization conditions cause **109** to lose a molecule of water spontaneously. Saturation of this olefinic center, conversion to the bis(dithioketal), and desulfurization with Raney nickel delivered **110**. The high symmetry of this molecule is reflected in its ^{1}H NMR spectrum, which shows only four groups of absorptions, each of equal area. Furthermore, the ^{13}C NMR spectrum is characterized by only three lines.

1. $2LiCH_2CH_2CH_2OCH(OCH_2CH_3)CH_3$
2. aq $(NH_4)_2SO_4$
3. repeat step 1
4. $H_2O–H^+$

$CrO_3–H^+–H_2O$

HO OH
CH_2OH CH_2OH

102 103 104

PPA

$H_2–Pd/C$

105 106

1. ethylene glycol-H^+
2. 2 pyridinium bromide perbromide-THF, -10° to 0°
3. $NaOCH_3$-DMSO, 95°
4. H_2O-H^+

Zn, CH_3COOH
benzene

107 108

HCOOEt
KO*t*Bu
t-BuOH

1. ethylene glycol, H^+
2. H_2, Pd/C
3. H_3O^+
4. ethanedithiol, H^+
5. RaNi, EtOH

109 110

Advanced Functionalization of the [5]Peristylane Framework

Although **110** possesses useful functionality at three of its methylene groups, access to more highly oxygenated analogues was next sought.[40] When acetate **111** was discovered to undergo β-elimination of acetic acid readily, it proved an easy matter to introduce the malonate ester in Michael fashion from the exo direction as in **113**. Epimerization at this center could be realized by means of selenium chemistry. Catalytic hydrogenation is forced for obvious reasons to proceed from outside the hemisphere to deliver **115**.

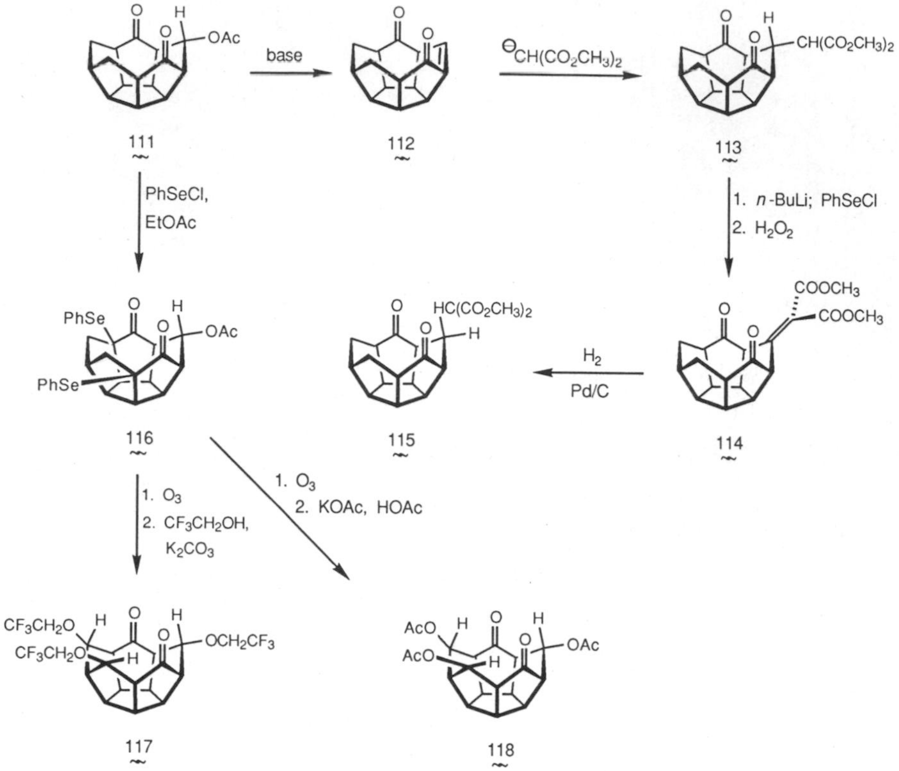

Acetate **111** offers many useful synthetic options of interest. For example, rapid reaction occurs when **111** is admixed with two equivalents of phenylselenenyl chloride in ethyl acetate solution.[40] The end product, isolated in 69% yield, is **116**. Oxidation of this bis(selenide) with ozone at low temperature, followed by dissolution in trifluoroethanol containing potassium carbonate and ethyl vinyl ether prior to warming, leads by elimination–readdition at three sites to **117**. Performance of the last step in potassium acetate and acetic acid near 20°C gives **118**. Consequently, access to [5]peristylanes functionalized at all five methylene carbon atoms can be realized. The open question is whether these intermediates are serviceable as precursors to dodecahedrane.

Efforts to Roof [5]Peristylanes

With the development of a reliable method for the synthesis of 2-alkylidene-1,3-cyclopentanediones,[41] the new protocol was applied to **111**.[42] Concurrently, a simple route to **119** was uncovered, and its condensation with *N*-(phenylthio)-succinimide and triethylamine was also found to lead to **120**. Oxidative elimination of phenylsulfenic acid from **120** proceeded smoothly to give the bright yellow ene-tetraone **121** in 73% yield. Diels–Alder addition of 2,3-dimethylbutadiene to its electron-deficient double bond then delivered **122**. Although the five-membered ring in this adduct finds itself correctly oriented vis-a-vis the [5]peristylane base, additional bond formation within **122** or more densely functionalized derivatives has not been reported.

1. MCPBA
2. 40°C

Et$_3$N

111 120 121 119 122

THE TRIQUINACENE DIMERIZATION PROBLEM

As with all three of the encapsulation schemes described above, the concept of triquinacene dimerization (see **123**), developed independently by Woodward[43] and by Jacobson[44] more than two decades ago, remains to be realized. Despite a projected exothermicity of 97 kcal/mol for sixfold C—C σ-bond construction, neither heat, light, nor transition metal catalysis has resulted in a chemical change. The complication appears to stem from statistical, entropic, and steric factors, which combine to disallow the proper mutual orientation of the two halves as illustrated.

123

A favored means for reducing to some degree these untoward influences is to install one or more interconnective bonds from the more congested surface of each structural segment. Compounds **86**, **90**, **97**, and **122** represent examples of this phenomenon. Since related strategy elements have been exercised in the triquinacene area, discussion of these findings is merited at this point.

Should one be dealing with a monofunctionalized triquinacene, it is necessary to recognize that the dimerization of such molecules will generate both meso and dl products if the monomer is itself racemic. The pinacolic coupling of (±)-**124** illustrates this point. Diols **125** and **126** are produced in approximately equal amounts.[45] On the other hand, submission of optically pure (+)-**124** to the same reaction conditions furnishes only **125** because of enforced enantiomer recognition. Also worthy of note in both processes is the strong preference for mutual exo bonding as usual.

Independent conversion of **125** and **126** to their thionocarbonates and subsequent heating with triethylphosphite gave **127** and **128**, respectively. Their subsequent total saturation made available *dl*- and *meso*-bivalvane (**129** and **130**, respectively). Both crystalline hydrocarbons have been examined crystallographically.[46] In the relevant dodecahedrane precursor, conformation **129′** is seen to be strongly favored. Not surprisingly then, all attempts to effect its proper dehydrogenative closure have come to naught.

1. Mg(Hg)x, Me_3SiCl; 2. EtOH, H_2O

124 → 125 (HO OH) + 126 (HO OH)

1. KH, CS_2; CH_3I
2. $(C_2H_5O)_3P$, Δ

127, 128

H_2, Pd/C

129′ ⇌ 129 (H, H); 130 (H, H)

125 / 126 → ($POCl_3$, py) 131 + 132

131 ⇌ 133 (N-methyltriazolinedione; 1. OH^- 2. MnO_2)

133

135 ← (CH_2N_2) 134 → (1. $CH_2{=}SO_2$ 2. Al_2O_3, CH_2Cl_2 3. OH^- 4. $(COCl)_2$) 136

MeOOC, OH, H — 135; HOOC, OH, H — 134; ClC(=O) — 136

135 + 136 → 137

137 → (hν, CH_3CN, sens) 138

137: H, COOMe, O=C–O

138: MeOOC, O

Diols **125** and **126** were purified chromatographically. A procedure not requiring this technique for separating the *meso* and *dl* series has also been devised.[47] The protocol involves direct dehydration of the **125/126** mixture to give **131/132** and exposure of these unpurified hexaenes to 0.5 equivalent of *N*-methyltriazolinedione. Only **131** enters into reaction since it uniquely can bond to the dienophile from the exo direction at both of its 1,3-diene termini. Adduct **133** crystallizes from solution leaving pure **132**. The reconversion of **133** to **131** is accomplished by a standard hydrolysis–oxidation procedure.

In an attempt to gain stereochemical control, Woodward and Repic prepared **134** in optically active condition and subsequently transformed it into **135** and into **136**.[48] Combination of these two intermediates (necessarily of identical configuration) as in **137** provided the opportunity for photochemical [2 + 2] cycloaddition. Unfortunately, further progress was stymied when **138** was found to be very resistant to hydrolysis with hydroxide ion, as well as other reagents normally capable of effecting ester hydrolysis.

With a similar theme in mind, Deslongchamps and Soucy transformed optically active acid **139** into (+)-2-formamidotriquinacene.[49] Condensation of **140** with the acid chloride of (+)-139 gave **141** from which the cyclic imidate salt (**142**) was produced. Again in this instance. cyclization to a dodecahedrane derivative could not be implemented.

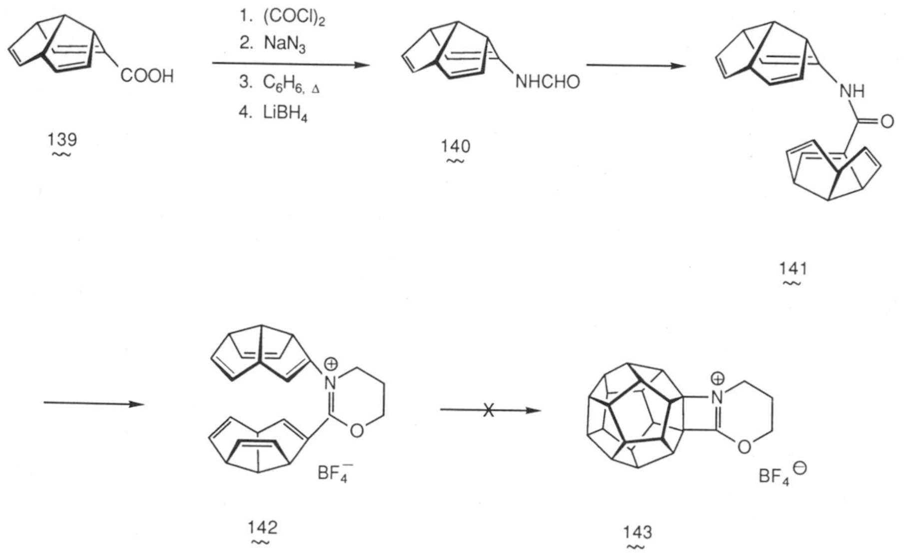

More recently, Serratosa and co-workers found it possible to prepare the C_3-triketone **145** in an enantioselective manner from optically pure **144**.[50] Despite this notable chemical intercorrelation, multiple aldol condensation of **145** under equilibrating conditions has not successfully led to the polyhydroxylated dodecahedrane stabilomer **146**.

In yet a different assault on the problem, dibromide (**147**) was coupled with dimercaptan **148** to generate a 3.5:1 mixture of the *anti*- and *syn*-triquinacenophanes **149** and **150**.[51] The identities of the two products were achieved by crystal structure analysis. Numerous efforts to induce **149** to cyclize were to no avail, despite a suggestion of feasibility from molecular orbital (MO) calculations.[52]

SYNTHESIS OF DODECAHEDRANE

The disadvantages associated with convergent schemes for arrival at **4–6** have now been rather well delineated. When a complex structural target such as dodecahedrane is involved, the inherent topological complexity needs to be examined alongside synthetic efficiency when a decision relative to a convergent or serial (stepwise) route is made.[53] We have seen above that the need to deal with the proper conjoining of two molecular segments precisely in the manner that will permit installation of the remaining interconnective bonds is not addressed well at all at the present time. Despite the application of ingenious tactics to counter the necessarily contrasteric alignment of the structural components, proper transannular bonding across the gap has not been achieved in any setting.

The Secododecahedrane Approach

These serious complications prompted consideration of a serial approach to **6** where the tactical disadvantages of nonconvergency would hopefully be overcome in large part by capitalizing on the symmetry of the target throughout the sequence. Following oxidative coupling of the inexpensive cyclopentadienide anion (**151**) to 9,10-dihydrofulvalene, a domino Diels–Alder reaction with dimethyl acetylenedicarboxylate was carried out to give **152**,[54] a molecule whose role it was to serve as the cornerstone of the spherical target molecule. Controlled reduction in the symmetry of **152** by conversion to diketo diester **153** provided a molecule possessing only a C_2 axis.[55]

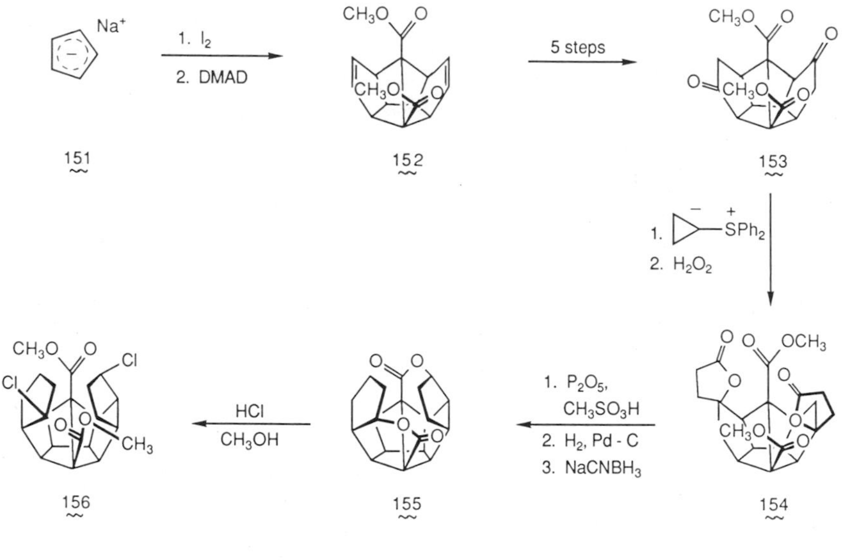

Bisspiroannulation of **153** with diphenylcyclopropylsulfonium ylide followed by hydrogen peroxide oxidation gave the axially symmetric dilactone **154**, rearrangement of which under strongly acidic conditions opened the way for arrival at **155**. Both **154** and **155** already contain all of the requisite 20 carbon atoms of dodecahedrane. Furthermore, dilactone **155** has 12 of the all-important cis-locked methine stereocenters properly installed.

The derived dichloro diester **156** undergoes a remarkable reductive alkylation when treated sequentially with lithium in liquid ammonia and then methyl iodide to provide keto ester **157**.[56] The intended role of the peripheral methyl groups in this early work was to guarantee by their positioning outside the sphere that the pair of carbonyl groups would be properly situated for continued utilization in the molecular construction process. In fact, irradiation of **157**

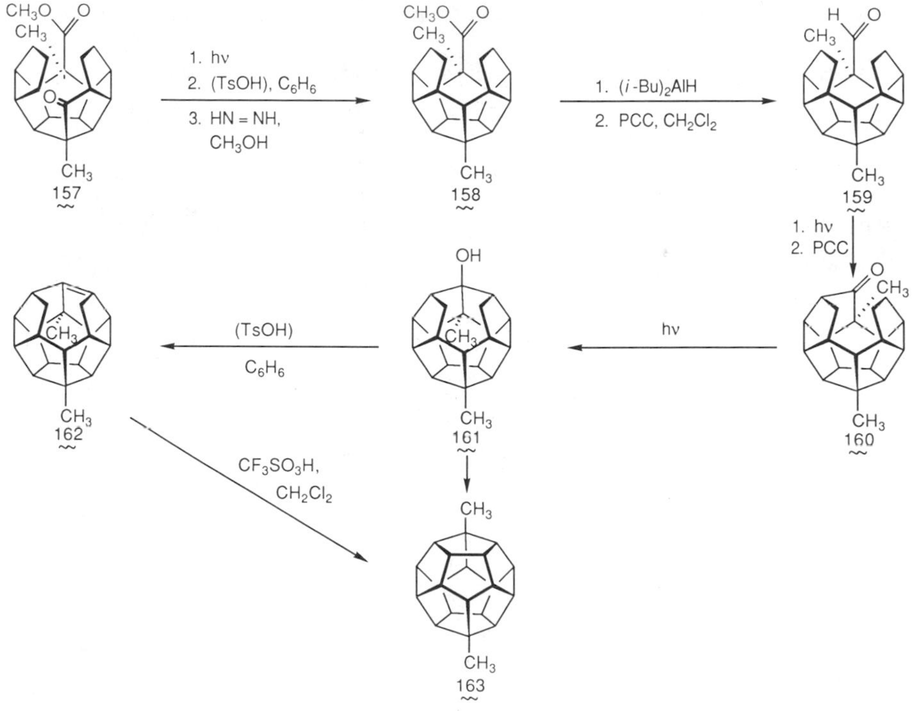

1. hν
2. (TsOH), C_6H_6
3. HN = NH, CH_3OH

1. $(i\text{-Bu})_2AlH$
2. PCC, CH_2Cl_2

1. hν
2. PCC

hν

(TsOH) C_6H_6

CF_3SO_3H, CH_2Cl_2

157 158 159 160 161 162 163

provided a tertiary alcohol, which was dehydrated and reduced with diimide to deliver the triseco ester **158**.[57] Since the carbomethoxy group in **158** is nonphotoactivable, reduction to the aldehyde level as in **159** was necessary prior to attaining more advanced sphere construction as in **160** and subsequently in **161**. The use of three discrete light-induced transformations while progressing from **157** to **161** is noteworthy. These maneuvers are a direct consequence of the heightened level of nonbonded steric interaction present in these molecules, which increases steadily in the progression from triseco to monoseco status.[58] The sizable amount of energy available in photochemical reactions enables these otherwise formidable energy barriers to be crossed.

Both **161** and its dehydration product **162** could be cyclized under strongly acidic conditions to the D_{3d}-symmetric dimethyldodecahedrane **163**.[57] The unexpected occurrence of a 1,2-methyl migration during installation of the final framework bond was substantiated by ^1H and ^{13}C NMR, as well as by X-ray crystallography.[57a,59]

When the dissolving metal reduction of **156** is performed with strict control of the amount of methyl iodide, the monoalkylated keto ester **164** results. This intermediate can be crafted as before into the secododecahedrene **165**.[60] Unlike **162**, however, **165** does not respond to trifluoromethanesulfonic acid in dichloromethane by isomerizing to **168**. Instead, a complex mixture is produced, the major constituent of which has been identified as the "isododecahedrane" **166** by X-ray crystal structure analysis. This remarkable transformation is the result of a transannular electrophilic substitution with inversion of configuration. This detour was circumvented by dehydrogenating **167** over palladium on charcoal. One sees, therefore, that the exceptional proximity of the two methylene carbon atoms in **165** is not sufficient by itself to allow facile ring closure!

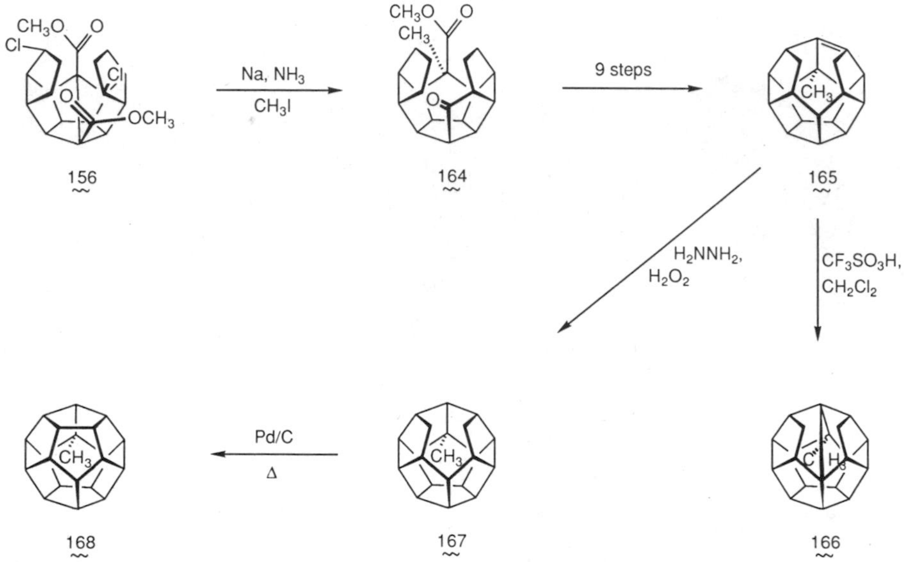

Although aldehyde **169** could be arrived at by a similar route, this intermediate proved too sensitive to deal with,[61] presumably as a direct consequence of its high enol content. Suitable resolution of this difficulty has mandated the introduction of a blocking group, and phenoxymethyl has served admirably in this capacity. This substituent is inert toward the reagents necessary to advance elaboration of dodecahedrane, causes no interference during the requisite bond construction (especially the photochemical steps), and is readily removed.

Once **171** was acquired by suitable reductive alkylation of **156** as before, it proved an easy matter to arrive at **173**. Sequential Birch reduction, acid hydrolysis, and chromate oxidation delivered β-keto aldehyde **174**. This intermediate responds inefficiently to retroaldol cleavage in alkaline solution.[61] A recent improvement bypasses this step as well as isolation of the bisseco ketone.[62] Direct irradiation of **174** results in photodecarbonylation and homo-Norrish cyclization to deliver **175** in 65% yield. Since the monoseco olefin derived from **175** is, like **165**, prone to uncontrolled carbocationic rearrangement, its diimide reduction was effected as a prelude to catalytic dehydrogenation.[61,63]

CH_3O O
$PhOCH_2$
171

1. hν
2. (TsOH), C_6H_6
3. H_2NNH_2, H_2O_2

CH_3O O
$PhOCH_2$
172

1. $(i\text{-Bu})_2AlH$
2. PCC, CH_2Cl_2
3. hν

HO CH_2OPh
173

1. Li, NH_3
C_2H_5OH
2. HCl, H_2O, THF
3. PCC, CH_2Cl_2

O
O CH
174

hν

OH
175

1. (TsOH), C_6H_6
2. H_2NNH_2, H_2O_2
3. Pt/Al_2O_3, Δ

6

The expectation that the ^1H (δ 3.35 in $CDCl_3$) and ^{13}C NMR (66.93 ppm in $CDCl_3$) spectra of **6** should be characterized by lone singlets has been borne out. Furthermore, the vibrational frequencies exhibited by dodecahedrane, namely, three IR active bands and eight Raman frequencies, agree fully with high rigidity for its interlinked methine units.[64] Crystals of the hydrocarbon are well formed (though twinned), dense (1.448 g/cm^3), and high melting (430 ± 10°C).[61b] Nonetheless, X-ray crystallographic analysis has proved possible.[65]

Isomerization of Pagodane

The beautiful earlier work of Schleyer in the adamantane field[66] led several groups to investigate Lewis acid promoted isomerization as a possible means of elaborating dodecahedranes. This thrust was supported by graph theoretical and molecular mechanics calculations that show **6** to be the $C_{20}H_{20}$ stabilomer by a wide margin.[58,67] Nonetheless, the early work was fraught with a great deal of disappointment. LeGoff and co-workers were unable to effect the conversion of basketene dimer **176** to dodecahedrane despite the potential release of 178

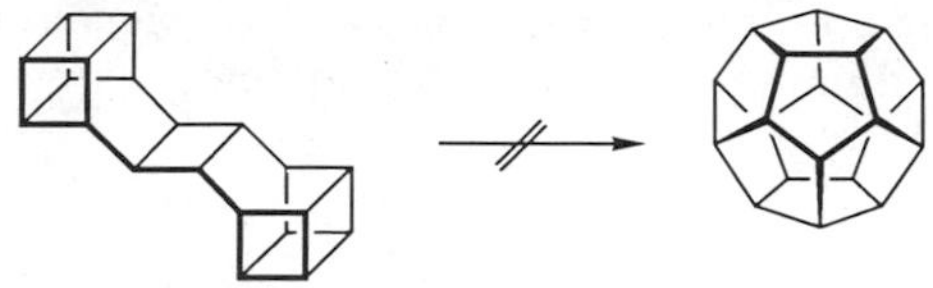

176 6

kcal/mol.[68] More recent efforts by Grubmüller and Schleyer have similarly been unfruitful.[2] In the belief that latent methyl groups might provide greater useful driving force to the desired isomerization, **177–181** were subjected to a wide range of catalysts. In no case was evidence garnered for conversion to one or more dimethyldodecahedrane isomers (**182**).

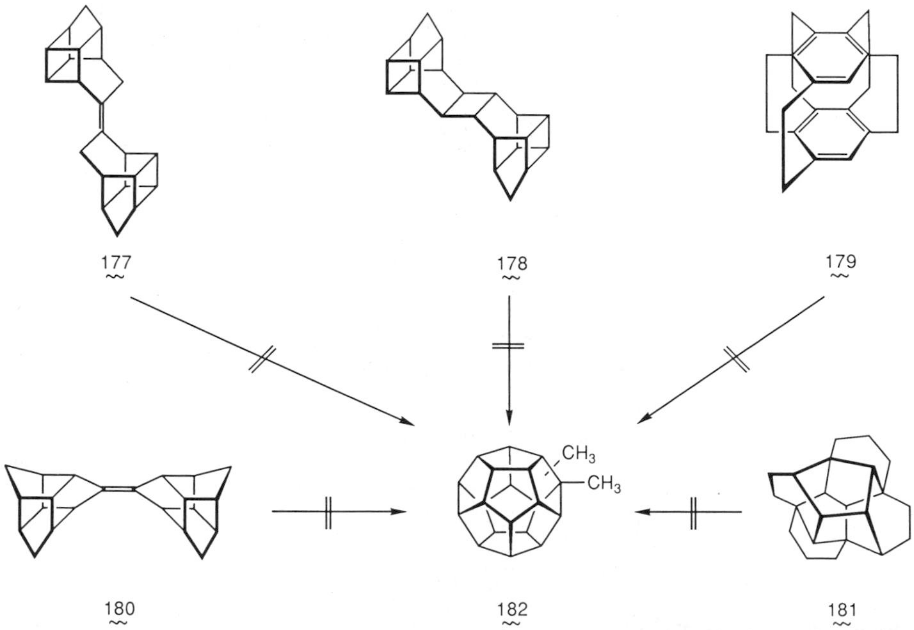

177 178 179

CH_3 CH_3

180 182 181

Prinzbach's attempts to take advantage of the availability of [1.1.1.1]pagodane (**190**) for gaining entry to dodecahedrane were more successful in this area. This undecacyclic $C_{20}H_{20}$ starting material was obtained from **183**, an

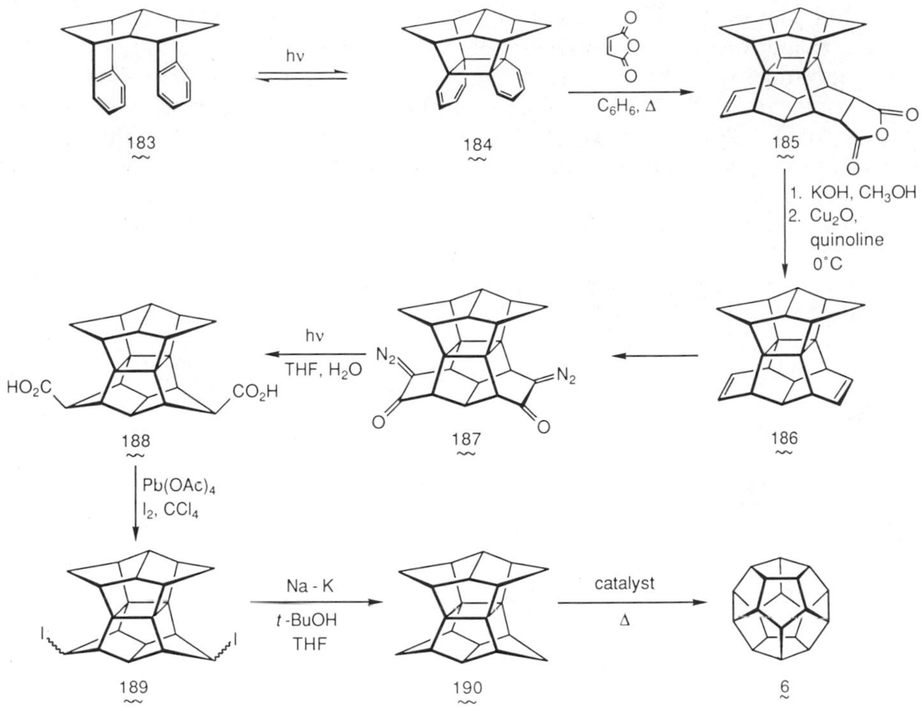

hν

C_6H_6, Δ

183 184 185

1. KOH, CH_3OH
2. Cu_2O, quinoline 0°C

hν

THF, H_2O

HO_2C CO_2H N_2 N_2 O O

188 187 186

$Pb(OAc)_4$
I_2, CCl_4

I I

Na - K
t-BuOH
THF

catalyst
Δ

189 190 6

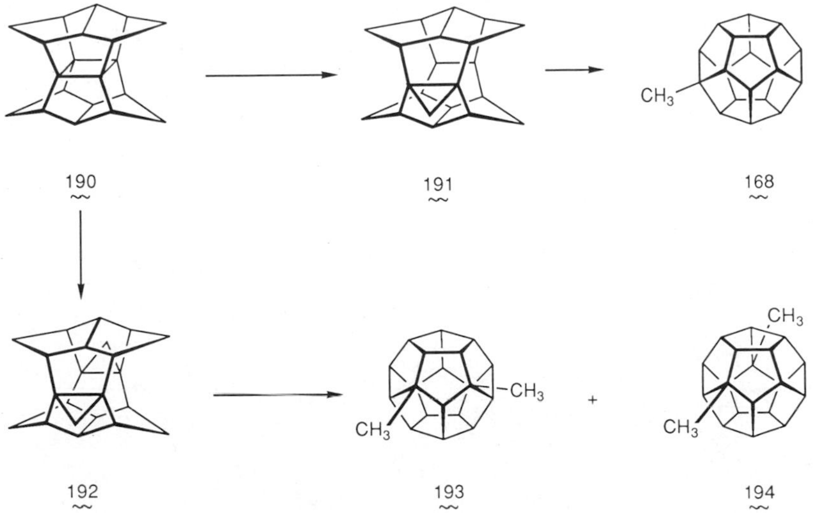

CH_3

190 191 168

CH_3 CH_3 + CH_3 CH_3

192 193 194

intermediate prepared in turn from isodrin.[69] Following excited-state [6 + 6] cycloaddition within **183**, tetraene (**184**) was treated with one equivalent of maleic anhydride. A domino Diels–Alder cycloaddition occurred to deliver **185** stereospecifically.[70] Following conventional conversion of **185** to diene **186**, ring contraction of the two olefinic bridges was accomplished via bis(α-diazoketone) **187**. Removal of the carboxyl groups in **188** constituted the remaining key steps in the sequence. In gas-phase isomerization experiments conducted in a flow apparatus over various catalysts at 250–450°C, pagodane is converted into complex, multicomponent mixtures from which dodecahedrane can be isolated in yields ranging from 0.1 to an optimized 8%.[71]

In other experiments, pagodane (**190**) was transformed into the cyclopropane derivatives **191** and **192**. When these are heated with 10% palladium on charcoal in a hydrogen atmosphere, hydrocarbon mixtures are produced in which the known[63] methylated dodecahedranes **168**, **193**, and **194** reside.[71]

CHEMICAL TRANSFORMATIONS OF DODECAHEDRANE

Electrophilic Monofunctionalization

The outcome of stirring **6** in liquid bromine solution overnight at room temperature is to produce monobromide **195** quantitatively.[62,72] This derivative

Cl
$FeCl_3$
CH_2Cl_2
Br
$AgBF_4$
CH_2Cl_2
Et_2O
F
196
195
197
$FeCl_3$,
C_6H_6
1. FSO_3H, SbF_5
CH_2Cl_2, -78°C;
CO
2. CH_2N_2
O
C—X
R
1. Dibal-H
2. PCC
198
199a, X = OH
b, X = OCH_3
c, X = NH_2
200a, R = CH_2OH
b, R = CHO

plays a pivotal role as intermediary to many other functionalized dodecahedranes including the chloro (**196**), fluoro (**197**), and phenyl (**198**) substituted compounds. Carbonylation of a solution of **195** in superacid provides an entry to the monocarboxylic acid (**199a**) and subsequently to **199b–200b**. Still more varied functionalization was realized by other avenues that allow for formation of oxygenated (**201, 202**) and nitrogen-containing systems (e.g., **203**).

CH_3 — 168 ← $(CH_3)_3Al$, hexane — Br — 195 → AgOTf, CH_3OH, CH_2Cl_2 → OCH_3 — 201

195 → $AgOCOCF_3$, CF_3COOH → OR — 202a, R = $COCF_3$; b, R = H

195 → 1. AgOTf, CH_3CN, Δ; 2. H_2O → $NHCOCH_3$ — 203

Methyl ester **199b** reacts near quantitatively with dimethylaluminum amide. Hoffmann rearrangement of the resulting carboxamide (**204**) and dehydration

CO_2CH_3 — 199b → Me_2AlNH_2, CH_2Cl_2 → CNH_2 (C=O) — 204 → $PhI(OCOCF_3)_2$, *tert*-BuOH, 50°C → $NHCO_2t$ Bu — 205

204 → $SOCl_2$, py → CN — 206

205 → HCl, EtOH, Et_2O → $NH_2 \cdot HCl$ — 207

206 → H_2, Pt, $CHCl_3$, C_2H_5OH → $CH_2NH_2 \cdot HCl$ — 208

gave rise to **205** and **206**, respectively.[73] Whereas acidic hydrolysis of **205** proceeded smoothly to provide **207**, hydrogenation of **206** over platinum in ethanol containing a small quantity of chloroform provided for direct in situ formation of amine hydrochloride **208**.

An unusual ring expansion was noted during reaction of bromide **195** with azidotrimethylsilane in the presence of stannic chloride.[73] Evidently, the intermediate azide undergoes Lewis acid induced migration of a framework C—C bond to nitrogen with loss of N_2 and ultimate 1,2 addition of Me_3SiN_3 to the strained imine so formed. Hydrogenation of **209** as before results in formation of the stable 1,1-diamine salt **210**.

Br
Me_3SiN_3
$SnCl_4$, CH_2Cl_2
0°C
N_3
NH
H_2, Pt
$CHCl_3$
EtOH
NH_2
NH
•HCl

195 209 210

Regiospecific 1,16-Disubstitution via the Dication

Many of the preceding reactions can be interpreted on the basis of transient intervention of the dodecahedryl cation (**211**) and, indeed, the species can be conveniently prepared from **196** in superacid solution at –78°C. The spectral properties of **211** show the species to be a static ion incapable of rapid 1,2-hydride shifting on the NMR time scale.[74] Because the cationic center in **211** is structurally precluded from attaining planarity and is consequently incapable of bending toward a neighboring C—H bond, bridging so as to transfer a hydride is precluded.

On the other hand, warming solutions of **211** promotes irreversible conversion to the unique, highly electron deficient 1,16-dication **213**. Its formation has been rationalized in terms of the loss of molecular hydrogen from the hypercarbon intermediate **212**.[74] This doubly charged species exhibits only two ^1H and three ^{13}C absorptions, in full agreement with its D_{3d} symmetry. Interestingly, **213** is stable at 0°C for several days and is formed as the exclusive product of ionization of the dibromide mixture **214**. This finding indicates that 1,2-hydride shifts materialize within the dication; however, these still proceed slowly on the NMR time sale.

Since the positive charges in **213** want to be as far as possible for electrostatic reasons, the way is paved for effective 1,16-functionalization for dodecahedrane. For example, quenching in methanol gives **215** from which **216** and **217** can be easily crafted.

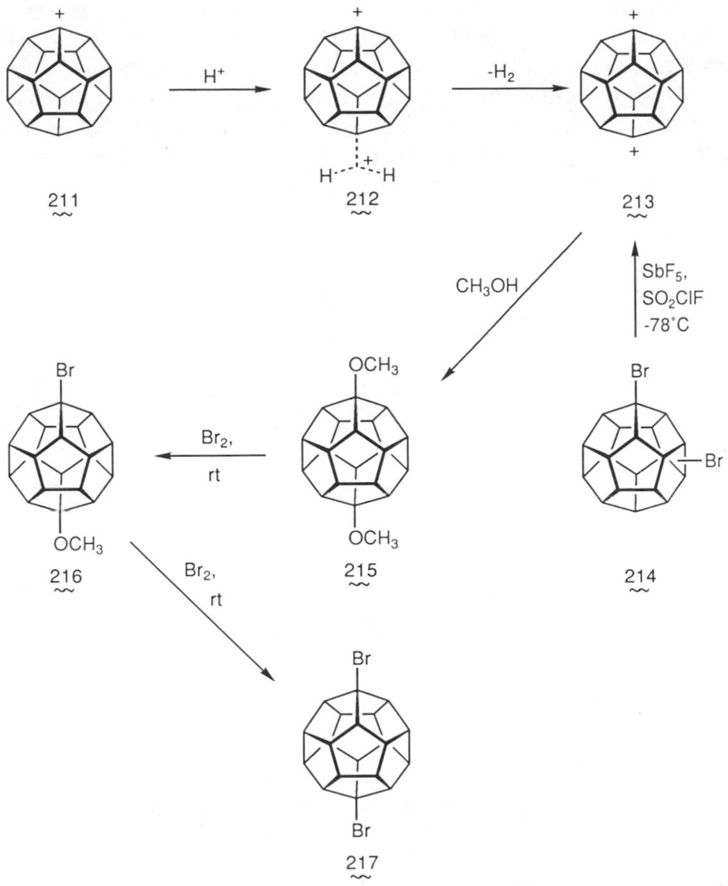

Synthesis of Cyclopropadodecahedranes

The preparation of cyclopropadodecahedranes could not be accomplished via either of the traditional means for fusing a three-membered ring across a C—C bond of a larger cyclic array since neither dodecahedrene nor a 1,20-bridgehead difunctionalized homododecahedrane were available. The new cyclopropanation procedure was developed around the finding that **6** reacts with dichlorocarbene under phase-transfer conditions to deliver the dichloromethyl derivative **218**.[75]

This substrate responds in two distinctly different ways to organolithium derivatives, the reactivity depending on the presence or absence of β-hydrogen atoms in the organometallic reagent. Thus, rapid reaction occurs in the presence of ethereal *t*-butyllithium at –100°C to give **219**. On the other hand, excess methyl- and phenyllithium act on **218** in ether solution at 0°C to furnish **220a** and **220b**, respectively. These divergent pathways have been rationalized in terms of cationic chloro carbenoid intermediates.[76]

219 ← *tert*-BuLi, Et_2O, -100°C — 218 ($CHCl_2$) — RLi, 0-20°C → 220a, R = CH_3; b, R = C_6H_5

When examined by X-ray crystallography,[75,76] **220b** was found to possess a distortion-free cyclopropane ring. Instead, the five-membered rings of the dodecahedrane core located in its immediate vicinity are nonplanar and clearly take up most all of the strain.

Homododecahedranone and the Degenerate 21-Homododecahedryl Cation

Expansion of the dodecahedrane core with incorporation of a functionalized carbon atom as in homododecahedranone (**221**) was accomplished from **218** by silver ion-promoted rearrangement-hydrolysis.[77] The precise structural features of **221**, elucidated by X-ray crystallographic analysis, denoted that solvolysis of the derived mesylate (**222b**) should occur at a rate comparable to 2-adamantyl mesylate. This was found to be the case.

The 21-homododecahedryl cation is of interest because it is a $(CH)_n^+$ species having the capacity of 21!/2 or 2.56 x 10^{19} degenerate arrangements simply by repetitive Wagner–Meerwein 1,2-carbon migration. To assess the level of scrambling, the d_1-mesylate **223** was prepared and subjected to acetolysis. Analysis of the d_1 acetate product (**224**) by ^{2}H NMR spectroscopy showed that the deuterium does indeed migrate around the sphere. However, these short-lived conditions do not allow for complete degeneracy to be attained. The data do rule out, however, the alternative possibility of ionization to localized carbenium–mesylate ion pairs for which backside stereospecificity would be mandated. Instead, ionization occurs with predominant, if not exclusive, adoption of a nonstereospecific process involving fully solvated cations.

Unfortunately, attempts to induce chloride **225** to ionize in Magic acid®

$CHCl_2$

218

$AgNO_3$
C_2H_5OH, H_2O
C_6H_6, Δ

O

H OR

H Cl

221

222a, R = H
b, R = SO_2CH_3
c, R = $COCH_3$

225

D OH

D OSO_2CH_3

HOAc
Δ

H OAc

D

223

224

solution at low temperature did not prove successful. Since two different cations were formed, it appears that the 21-homododecahedryl cation is irreversibly converted to more stable ions under long lived conditions.

Dodecahedrene

Positioning of a double bond within the dodecahedrane structure is certain to be accompanied by a significant increase in strain energy since the olefinic carbon atoms will likely exhibit a high degree of pyramidalization and near-tetrahedral geometry. To date, **226** has not yielded to preparative scale synthesis. However, its existence has been inferred as the product of an

O
O CF3
OR−
gas phase
202a
226

ion–molecule reaction in the gas phase.[78] Thus, Kiplinger has succeeded by working in a trapped ion cell under ion cyclotron conditions to achieve β-elimination of trifluoroacetic acid in **202a**. Since the reaction proceeds successfully with bases such as hydroxide and methoxide ion, but not others (e.g., ethoxide ion), it has proven possible to bracket the strain energy of **226**. The methoxide-promoted reaction is calculated to be exothermic by 41.5 kcal/mol. Comparable elimination with ethoxide would require an exothermicity of 38.5 kcal/mol. Accordingly, the amount of extra energy necessary to produce **226** appears to have boundary limits of 40 ± 3 kcal/mol. When this approximated heat of hydrogenation value is compared to that of other distorted olefins, there is every indication that the sp^2-hybridized centers in dodecahedrene have nearly tetrahedral character.[78]

CONCLUDING REMARKS

We have sought to provide an overview of the current status of polyhedrane chemistry, with particular emphasis on the difficulties associated with producing these structurally attractive molecules by capping the appropriate peristylane precursor. Although more broadly based in its development, the polyhedrane field today holds many similarities to the emergence of adamantane chemistry three decades ago. The phenomenal growth in our knowledge of adamantanes since that time was amply documented.[79] Can one expect new results in the polyhedron area to materialize as quickly? Of course, the development of any field is entirely dependent on the level of interest generated in the minds and laboratories of a reasonable number of senior investigators. Certainly, the pool of active practitioners is sizable at the present time. It is our hope that this chapter will entice still others to entertain the thought of making contributions to topographical organic chemistry.

REFERENCES

1. R. Hoffmann, *Am. Sci.* **1988**, *76*, 389.
2. P. Grubmüller, Ph.D. thesis, Friedrich-Alexander-Universität, Erlangen-Nürenberg, Federal Republic of Germany, 1979.
3. (a) G. Mehta, *J. Sci. Ind. Res.* **1978**, *37*, 256. (b) W. Grahn, *Chem. Unserer Zeit* **1981**, *15*, 52. (c) P. Laszlo, *Nouv. J. Chem.* **1983**, *7*, 69.
4. P. E. Eaton, *Tetrahedron* **1979**, *35*, 2189.
5. (a) L. A. Paquette, *Pure Appl. Chem.* **1978**, *50*, 2189. (b) L. A. Paquette, in *Organic Synthesis—Today and Tomorrow*, B. M. Trost and C. R. Hutchinson, Eds., Pergamon Press, New York, 1981, p. 335. (c) L. A. Paquette, *Proc. Natl. Acad. Sci. USA* **1982**, *79*, 4495. (d) L. A. Paquette, *Chem. Australia* **1983**, *50*, 138. (e) L. A. Paquette, in *Strategies and Tactics in Organic Synthesis*, T. Lindberg, Academic Press, New York, 1984, pp. 175 ff. (f) L. A. Paquette and A. M. Doherty, *Polyquinane Chemistry. Synthesis and Reactions*, Springer-Verlag, Heidelberg, 1987. (g) L. A. Paquette, *Chem. Rev.*, **1989**, *89*, 1051.
6. (a) R. J. Ternansky, D. W. Balogh, and L. A. Paquette, *J. Am. Chem. Soc.* **1982**, *104,* 4503. (b) L. A. Paquette, R. J. Ternansky, D. W. Balogh, and G. Kentgen, *J. Am. Chem. Soc.* **1983**, *105*, 5446.
7. W.-D. Fessner, B. A. R. C. Murty, J. Wörth, D. Hunkler, H. Fritz, H. Prinzbach, W. D. Roth, P. von R. Schleyer, A. B. McEwen, and W. F. Maier, *Angew. Chem. Int. Ed. Engl.* **1987**, *26*, 452.
8. P. J. Garratt and J. F. White, *J. Org. Chem.* **1977**, *42*, 1733.
9. (a) P. E. Eaton, R. H. Mueller, G. R. Carlson, D. A. Cullison, G. F. Cooper, T.-C. Chou, and E.-P. Krebs, *J. Am. Chem. Soc.* **1977**, *99*, 2751. (b) P. E. Eaton, A. Srikrishna, F. Uggeri, *J. Org. Chem.* **1984**, *49*, 1728. (c) P. E. Eaton, G. D. Andrews, E.-P. Krebs, and A. Kunai, *J. Org. Chem.* **1979**, *44*, 2824. (d) P. E. Eaton, W. H. Bunnelle, and P. Engel, *Can. J. Chem.* **1984,** *63*, 2612.
10. A. Nickon and G. D. Pandit, *Tetrahedron Lett.* **1968**, 3663.
11. A. Nickon and E. F. Silversmith, *Organic Chemistry: The Name Game*, Pergamon Press, New York, 1987.
12. J. E. Baldwin and W. D. Fogelsong, *J. Am. Chem. Soc.* **1968,** *90*, 4303.
13. L. A. Paquette, C. S. Ra, and T. W. Silvestri, *Tetrahedron,* **1989**, *45*, 3099.
14. R. K. Hill, G. H. Morton, J. R. Peterson, J. A. Walsh, and L. A. Paquette *J. Org. Chem.* **1985**, *50,* 5528.
15. M. Demuth and K. Schaffner, *Angew. Chem. Int. Ed. Engl.* **1982**, *21*, 820.
16. D. F. Covey and A. Nickon, *J. Org. Chem. 1977, 42,* 794.
17. (a) E. Baggiolini, E. G. Herzog, S. Iwasaki, S., R. Sehorta, and K. Schaffner, *Helv. Chim. Acta* **1967**, *50,* 297. (b) U. Klinsman, J. Gauthier, K. Schaffner, M. Pasternak, and B. Fuchs, *Helv. Chim. Acta* **1972**, *55*, 2643.
18. P. J. Garratt, C. W. Doecke, J. C. Weber, and L. A. Paquette *J. Org. Chem.* **1986**, *51*, 449.
19. (a) W. Ammann, F. J. Jaggi, and C. Ganter, *Helv. Chim. Acta* **1980**, *63*, 2019. (b) G. M. Ramos Tombo, S. Chakrabarti, and C. Ganter, *Helv. Chim. Acta* **1983**,

66, 914. (c) R. A. Pfund and C. Ganter, *Helv. Chim. Acta* **1979**, *62*, 228. (d) W. Ammann and C. Ganter, *Helv. Chim. Acta* **1977**, *60*, 1924.

20. J. C. Weber, Ph.D. thesis, The Ohio State University, 1987.
21. K. Hirao, M. Taniguchi, O. Yonemitsu, J. Flippen, and B. Witkop, *J. Am. Chem. Soc.* **1979**, *101*, 408.
22. (a) H.-D. Becker and A. Konar, *Tetrahedron Lett.* **1972**, 5177. (b) H.-D. Becker, *Liebigs Ann. Chem.* **1973**, 1675. (c) H. Iwakuma, H. Nakai, O. Yonemitsu, D. S. Jones, I. L. Karle, and B. Witkop, *J. Am. Chem. Soc.* **1972**, *94*, 5136. (d) T. Iwakuma, K. Hirao, and O. Yonemitsu, *J. Am. Chem. Soc.* **1974**, *96*, 2570.
23. K. Hirao, Y. Ohuchi, and O. Yonemitsu, *J. Chem. Soc. Chem. Commun.* **1982**, 99.
24. (a) L. A. Paquette, R. V. C. Carr, M. C. Bohm, and R. Gleiter, *J. Am. Chem. Soc.* **1980**, *102,* 1186. (b) M. C. Bohm, R. V. C. Carr, R. Gleiter, and L. A. Paquette, *J. Am. Chem. Soc.* **1980**, *102*, 7218. (c) L. A. Paquette, in *Stereochemistry and Reactivity of Pi Systems*, W. H. Watson, Ed., Verlag Chemie, Deerfield Beach, FL, 1983, pp. 41–73.
25. L. A. Paquette, A. R. Browne, C. W. Doecke, and R. V. Williams, *J. Am. Chem. Soc.* **1983**, *105,* 4113.
26. L. A. Paquette, H. Kunzer, K. E. Green, O. DeLucchi, G. Licini, L. Pasquato, and G. Valle, *J. Am. Chem. Soc.* **1986**, *108,* 3453.
27. L. A. Paquette and R. V. C. Carr, *J. Am. Chem. Soc.* **1980**, *102,* 7553.
28. P. Engel, J. W. Fischer, and L. A. Paquette, *Z. Kristallogr.* **1984**, *166*, 225.
29. O. DeLucchi, V. Luccini, L. Pasquato, and G. Modena, *J. Org. Chem.* **1984**, *49*, 596.
30. L. A. Paquette, J. W. Fischer, and P. Engel, *J. Org. Chem.* **1985**, *50*, 2524.
31. L. M. Waykole, C.-C. Shen, and L. A. Paquette, *J. Org. Chem.* **1988**, *53*, 4969.
32. L. A. Paquette, and C. C. Shen, *Tetrahedron Lett.,* **1988**, *29*, 4069.
33. L. A. Paquette, L. M. Waykole, C.-C. Shen, U. S. Racherla, R. Gleiter, and K. Litterst, *Tetrahedron Lett.*, **1988**, *29*, 4213.
34. L. A. Paquette, J. W. Fischer, A. R. Browne, and C. W. Doecke, *J. Am. Chem. Soc.* 1985, 107, 686.
35. P. Engel, J. C. Weber, and L. A. Paquette, *Z. Kristallogr.* **1986**, *177,* 229.
36. P. E. Eaton and R. H. Mueller, *J. Am. Chem. Soc.* **1972**, *94*, 1014.
37. P. E. Eaton, R. H. Mueller, G. R. Carlson, D. A. Cullison, G. F. Cooper, T.-C. Chou, and E.-P. Krebs, *J. Am. Chem. Soc.* **1977**, *99*, 2751.
38. P. E. Eaton, G. F. Cooper, R. C. Johnson, and R. H. Mueller, *J. Org. Chem.* **1972**, *37*, 1947.
39. P. E. Eaton, A. Srikrishna, and F. Uggeri, *J. Org. Chem.* **1984**, *49*, 1728.
40. P. E. Eaton, G. D. Andrews, E.-P. Krebs, and A. Kunai, *J. Org. Chem.* **1979**, *44*, 2824.
41. P. E. Eaton and W. H. Bunnelle, *Tetrahedron Lett.* **1984**, *25,* 23.
42. P. E. Eaton, W. H. Bunnelle, and P. Engel *Can. J. Chem.* **1984**, *62,* 2612.

43. R. B. Woodward, T. Fukunaga, and R. C. Kelly, *J. Am. Chem. Soc.* **1964**, *86*, 3162.

44. I. T. Jacobson, *Acta Chem. Scand.* **1967**, *21*, 2235.

45. (a) L. A. Paquette, I. Itoh, and W. B. Farnham, *J. Am. Chem. Soc.* **1975**, *97*, 7280. (b) L. A. Paquette, W. B. Farnham, and S. V. Ley, *J. Am. Chem. Soc.* **1975**, *97*, 7273.

46. J. Clardy, B. A. Solheim, J. P. Springer, I. Itoh, and L. A. Paquette, *J. Chem. Soc. Perkin Trans 2,* **1979**, 276.

47. L. A. Paquette, I. Itoh, and K. B. Lipkowitz, *J. Org. Chem.* **1976**, *41*, 3524.

48. O. Repic, Ph.D. thesis, Harvard University, 1976.

49. P. Deslongchamps and P. Soucy, P. *Tetrahedron* **1981**, *37*, 4385.

50. C. Almansa, A. Moyano, and F. Serratosa, *Tetrahedron* **1988**, *44*, 2657.

51. W. P. Roberts and G. Shohan, *Tetrahedron Lett.* **1981**, *22*, 4895.

52. W. P. Roberts, Ph.D. thesis, Harvard University, 1982.

53. (a) S. H. Bertz, *J. Am. Chem. Soc.* **1974**, *96*, 4671. (b) D. McNeil, B. R. Vogt, J. J. Sudol, S. Theodoropulos, and E. Hedaya, *J. Am. Chem. Soc.* **1974**, *96*, 4673. (c) L. A. Paquette, M. J. Wyvratt, H. C. Berk, and R. E. Moerck, *J. Am. Chem. Soc.* **1978**, *100*, 5845.

55. (a) L. A. Paquette, M. J. Wyvratt, O. Schallner, D. F. Schneider, W. J. Begley, and R. M. Blankenship, *J. Am. Chem. Soc.* **1976,** *98*, 6744. (b) L. A. Paquette, M. J. Wyvratt, O. Schallner, J. L. Muthard, W. J. Begley, R. M. Blankenship, and D. W. Balogh, *J. Org. Chem.* **1979**, *44*, 3616.

56. L. A. Paquette, D. W. Balogh, and J. F. Blount, *J. Am. Chem. Soc.* **1981**, *103,* 228.

57. (a) L. A. Paquette, D. W. Balogh, R. Usha, D. Kountz, and G. G. Christoph, *Science* **1981**, *211,* 575. (b) L. A. Paquette and D. W. Balogh *J. Am. Chem. Soc.* **1982**, *104,* 774.

58. P. R. Spurr, B. A. R. C. Murty, W.-D. Fessner, H. Fritz, and H. Prinzbach, *Angew. Chem. Int. Ed. Engl.* **1987**, *26,* 455.

59. G. G. Christoph, P. Engel, R. Usha, D. W. Balogh, and L. A. Paquette, *J. Am. Chem. Soc.* **1982**, *104,* 784.

60. L. A. Paquette, R. J. Ternansky, D. W. Balogh, and W. J. Taylor, *J. Am. Chem. Soc.* **1983**, *105,* 5441.

61. (a) R. J. Ternansky, D. W. Balogh, and L. A. Paquette, *J. Am. Chem. Soc.* **1982**, *104*, 4503. (b) L. A. Paquette, R. J. Ternansky, D. W. Balogh, and G. Kentgen, J. Am. Chem. Soc. **1983**, *105*, 5446.

62. L. A. Paquette, J. C. Weber, T. Kobayashi, and Y. Miyahara, *J. Am. Chem. Soc.* **1988**, *110*, 8591.

63. (a) L. A. Paquette, Y. Miyahara, and C. W. Doecke, *J. Am. Chem. Soc.* **1986**, *108*, 1716. (b) L. A. Paquette and Y. Miyahara, *J. Org. Chem.* **1987**, *52*, 1265.

64. O. Ermer, *Angew. Chem. Int. Ed. Engl.* **1977**, *6*, 411.

65. J. C. Gallucci, C. W. Doecke, and L. A. Paquette, *J. Am. Chem. Soc.* **1986**, *108*, 1343.

66. (a) P. von R. Schleyer, *J. Am. Chem. Soc.* **1957**, *79*, 3292. (b) E. Osawa, K.

Aigauri, N. Takaishi, Y. Inamoto, Y. Fujikura, Z. Majerski, P. von R. Schleyer, E. M. Engler, and M. Farcasiu, *J. Am. Chem. Soc.* **1977**, *99*, 5361.

67. T. Iizuka, M. Imai, N. Tanaka, T. Kan, and E. Osawa, *Gunmu Daigaku Kyoikugakubu Kiyo, Shizen Ragaku Hen* **1981**, *30*, 5, *Chem. Abstr.* **1982**, *97*, 126567m.

68. N. J. Jones, W. J. Deadman, and E. LeGoff, *Tetrahedron Lett.* **1973**, 2087.

69. (a) H. Prinzbach, G. Sedelmeier, C. Krüger, R. Goddard, H.-D. Martin, and R. Gleiter, *Angew. Chem. Int. Ed. Engl.* **1978**, *17*, 271. (b) G. Sedelmeier, Ph.D. thesis, Albert-Ludwigs-Universität, Freiburg, Federal Republic of Germany 1979.

70. (a) W.-D. Fessner, H. Prinzbach, and G. Rihs, *Tetrahedron Lett.* **1983**, *24*, 5857. (b) W.-D. Fessner, G. Sedelmeier, P. R. Spurr, G. Rihs, and H. Prinzbach, *J. Am. Chem. Soc.* **1987**, *109*, 4626.

71. W.-D. Fessner, B. A. R. C. Murty, J. Worth, D. Hunkler, H. Fritz, H. Prinzbach, W. D. Roth, P. von R. Schleyer, A. B. McEwen, and W. F. Maier, *Angew. Chem., Int. Ed. Engl.* **1987**, *26*, 452.

72. L. A. Paquette, J. C. Weber, and T. Kobayashi, *J. Am. Chem. Soc.* **1988**, *110*, 1303.

73. J. C. Weber and L. A. Paquette, *J. Org. Chem.*, **1988**, *53*, 5315.

74. (a) G. A. Olah, G. K. Surya Prakash, T. Kobayashi, and L. A. Paquette, *J. Am. Chem. Soc.* **1988**, *110*, 1304. (b) G. A. Olah, G. K. Surya Prakash, W.-D. Fessner, T. Kobayashi, and L. A. Paquette, *J. Am. Chem. Soc.*, **1988**, *110*, 8599.

75. L. A. Paquette, T. Kobayashi, and J. C. Gallucci, *J. Am. Chem. Soc.* **1988**, *110*, 1305.

76. L. A. Paquette, T. Kobayashi, M. A. Kesselmayer, and J. C. Gallucci, *J. Org. Chem.* **1989**, *54*, 2921.

77. L. A. Paquette, T. Kobayashi, and M. A. Kesselmayer, *J. Am. Chem. Soc.* **1988**, *110*, 6568.

78. J. P. Kiplinger, F. R. Tollens, A. G. Marshall, T. Kobayashi, D. Lagerwall, L. A. Paquette, and J. E. Bartmess, *J. Am. Chem. Soc.* **1989**, *111*, 6914.

79. Reviews: (a) R. C. Fort, *Adamantane—The Chemistry of Diamond Molecules*, Marcel-Dekker, New York, 1976. (b) E. M. Engler and P. von R. Schleyer, in *MTP International Review of Science, Vol. 5, Alicyclic Compounds*, W. Parker, Ed., Butterworths, London, 1973. (c) R. C. Bingham and P. von R. Schleyer, *Fortschr. Chem. Forsch.* **1971**, *18*, 1. (d) V. V. Sevast'yanova, K. M. Krayuskin, and A. G. Yurchenko, *Russ. Chem. Rev.* **1970**, *39*, 817. (e) R. C. Fort, Jr., and P. von R. Schleyer, *Chem. Rev.* **1964**, *64*, 277.

10 The Pagodane Route to Dodecahedrane

WOLF-DIETER FESSNER and HORST PRINZBACH

Department of Organic Chemistry and Biochemistry
University of Freiburg i. Br.
Federal Republic of Germany

INTRODUCTION

The task of the transliteration of the mathematically defined convex regular polyhedra—tetrahedron, cube, and pentagonal dodecahedron, the so-called Platonic—solids, into real (CH) molecular equivalents is a long standing challenge for organic chemists.[1,2] The relatively simpler but highly strained cubane[3] and a derivative of tetrahedrane[4] were the first goals to be achieved. Dodecahedrane (**1**) is the most complex of the Platonic series and, with the highest conceivable point group symmetry of I_h, stands on the border to perfect spherical symmetry. Numerous ingenious attempts to devise synthetic stratagems to arrive at **1** had failed for decades and as a result the elusive molecule was esteemed as the "Mount Everest of alicyclic chemistry."[5]

A breakthrough materialized when in 1982 the Paquette group announced a protocol that delivered the $C_{20}H_{20}$ hydrocarbon for the first time.[6] While this subject was extensively reviewed in past years,[7] in the following account we wish to summarize the efforts made in our laboratories from the time we entered this field in 1983[8] by taking advantage of a photoreaction, which was

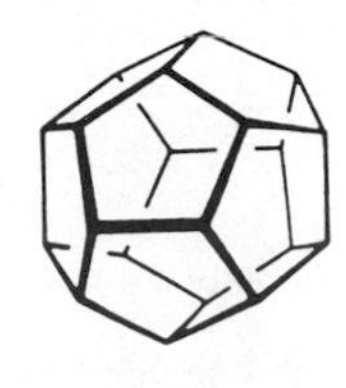

1

discovered in 1978.[9] The project, devised and pioneered by Fessner,[1,8] developed rapidly and was recently successful, due in part to our joining forces with our present Laureate Paul v. R. Schleyer.

SUMMARY OF CURRENT CONCEPTS

As in the total syntheses of complex natural products, the construction of the topologically spherical network of the dodecahedrane nucleus requires a high degree of stereochemical control, since conceptually, each of the methine units has to be installed in a contrathermodynamic fashion in which the newly formed bonds confine an unsolvated interior of the sphere and are incorporated within a perfectly eclipsed cis,syn environment. Although the target molecule is characterized as relatively strain-free, intermediates en route will suffer from the erection of an unparalleled level of torsional interactions and strong intramolecular nonbonded interactions around nonmade bonds or among the interior sites of the developing cage, which will increasingly exclude solvation.

Since a decade has passed since the publication of the last comprehensive review on this subject,[2] it seems appropriate to update this situation with a brief examination of both the old and the newly developed concepts, in order to give us a more concise assessment of our current knowledge so as to better design and evaluate alternative strategies.*

Thermodynamically Controlled Isomerization

Extensive analyses by graph theory and force-field calculations predicated the dodecahedrane framework as the $C_{20}H_{20}$ "stabilomer,"[10] that is, **1** is the most

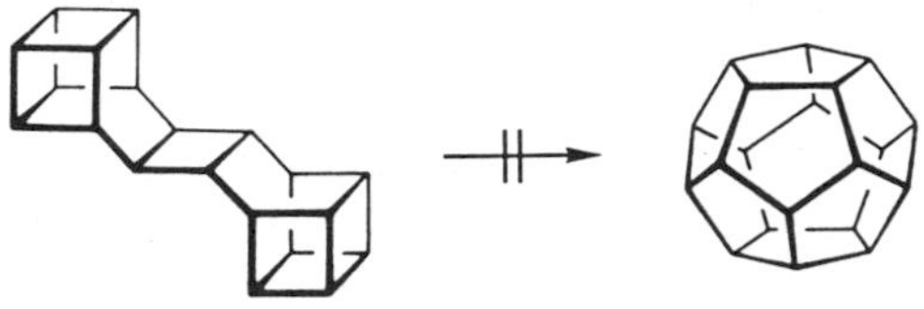

2 **1**

*The following summary and analysis sections are essentially a condensed edition of the earlier detailed review presented in Ref. 1.
Note added in proof: Some aspects of this review[1] have also been adapted in a more recent review: L. A. Paquette, *Chem. Rev.* **1989**, *89*, 1051.

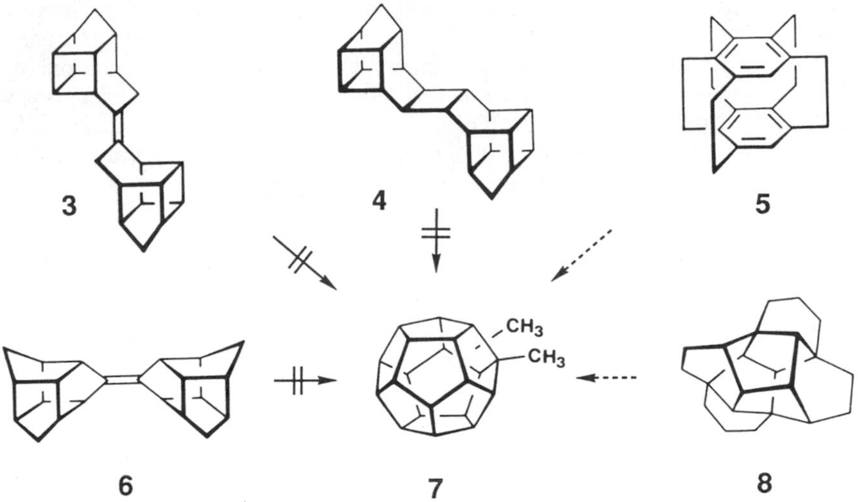

Scheme 1.

stable isomer among its class.[11] Adapting the early achievements of Schleyer in the synthesis of adamantane via thermodynamically controlled rearrangement of various $C_{10}H_{16}$ isomers,[12] the analogous isomerization of undecacycles on the $C_{20}H_{20}$ hypersurface into dodecahedrane as the final product seems feasible. Initial independent experiments with the *anti*-basketene dimer (**2**) in the LeGoff and Schleyer groups failed,[13,14] perceivably on account of the enormous concentration of strain energy in the polyfused cyclobutane rings (ΔE_{MM2} = 178 kcal/mol).[1]

Despite the introduction of auxiliary carbon atoms in order to gain kinetic assistance with the obscure rearrangement pathways to be traveled, the catalyzed treatment of various $C_{22}H_{24}$ hydrocarbons (**3**, **4**, and **6**) by Grubmüller[5] was similarly ineffective for their conversion into dimethyldodecahedranes (**7**) (Scheme 1).* More recently, the Erlangen group investigated some other globular precursors such as **5** or **8**, to no avail.[14]

C_{10} Dimerization

The early scheme of triquinacene dimerization (**9**; cf. Scheme 5, path IX), originally proposed by Müller,[15] Woodward and Fukunaga,[16] and Jacobson,[17] flourished in many ways because of its appealing simplicity. Although the net

*This is not surprising in view of the entropic disadvantage of rodlike precursors and since each molecular subunit of quinane **6** itself constitutes the pentacycloundecane stabilomer.[10]

process is predicted to be highly exothermic (ΔE_{MM2} = 97 kcal/mol),[1] its possible zipper-type progression requires an extraordinarily high-order transition state for all 12 trigonal reaction centers, which allows for the formation of the 6 interconnective C—C bonds from the more congested endo faces to occur. Consideration of statistical, entropic, and steric factors helps to explain why in fact all efforts involving heat, high pressure,[18] irradiation,[19] zeolite catalysis,[20] or transition metal complexation[21] were so far thwarted.

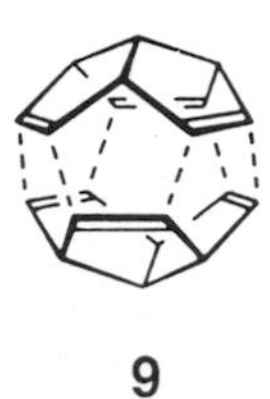

9

Similarly, Serratosa's hopes of achieving a convergent, reflexive "narcissistic" endo coupling between enantiomeric C_3-triquinanetriones (**10**) have not advanced beyond that of a Sisyphus' torment.[22] Force-field calculations reveal that the overall favorable thermodynamic situation for such an equilibrated multiple aldol process to **11** becomes quickly imbalanced by the developing severe steric demands.[1]

10 **11**

The need for at least one preformed endo,endo σ bond was realized at an early stage in order to facilitate triquinacene dimerization by proper

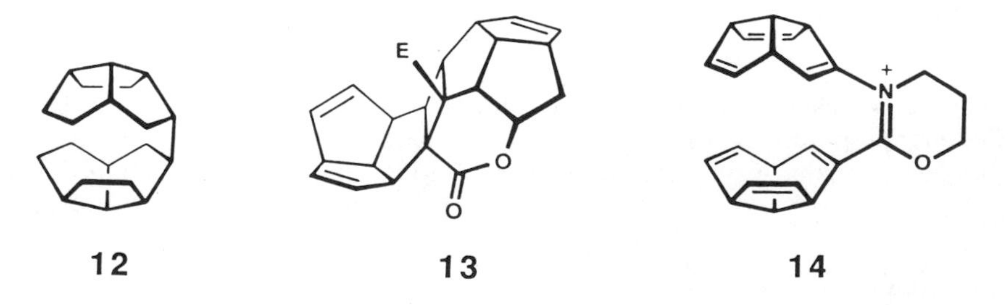

12 **13** **14**

preorientation. The requirement for coupling of enantiomerically homogeneous subunits was met with Paquette's bivalvanes (**12**),[23] Repic's diester (**13**),[24] and Deslongchamps' cyclic imidate salt (**14**).[25] Strong anti conformational preference,[26] however, precluded these substrates from further conversion towards dodecahedranes.

One step further towards complete orientational control over the triquinacene subunits was reported by Roberts and co-workers, who succeeded in the preparation of [3.3]triquinacenophanes (**15**).[27] Notwithstanding the proper anti juxtapositioning which was verified by X-ray analysis, an intramolecular installation of interconnective bonds is still lacking. Another triquinacenophane strategy based on diester **16** seems to be the objective of Camps et al.[28] who have not yet given details on their progress.

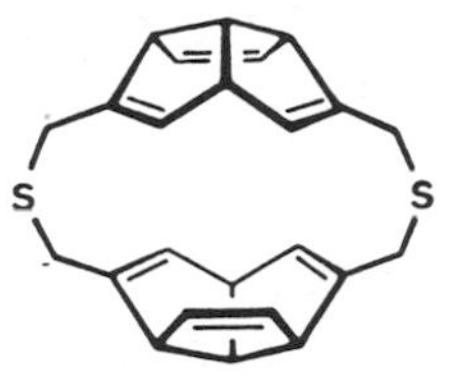

15 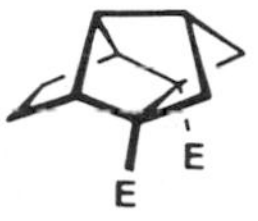 **16**

A somewhat related approach by Farnum et al. relies on the phane-type linkage of (homochiral) C_2-bicyclo[3.3.0]octadienes with *anti*-ethylene bridges (Scheme 5, path XII),[29] a strategy that was proposed by Pyrek following biomimetic considerations.[30] Once the $C_{20}H_{20}$ hexaene (**17**) is crafted, it will be interesting to see whether the parallel conformer, mandatory but disfavored by force-field calculations against an orthogonal conformation **17**′ ($\Delta E_{MM2} = -11$ kcal/mol),[1] can be at an advantage to reorganize into dodecahedrane.

The most simplistic dimerization scheme has been proposed by the Tolbert group (Scheme 5, path VIII). Yet, the all-*cis*-cyclopentanepentamethanol (**18**) did not serve as a suitable building block on account of the inoperability of nucleophilic displacements in a sterically crowded pentamesylate derivative.[31]

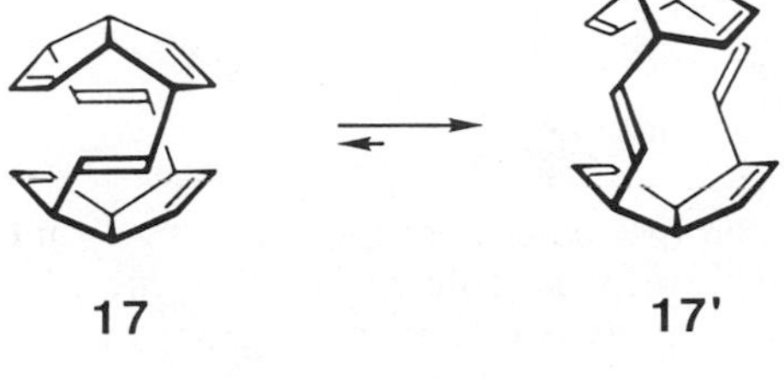

17 **17'** 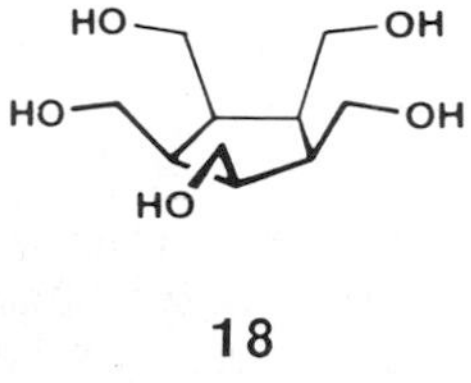 **18**

Convergent Elaboration from Polyquinane Fragments

The main direction of the Eaton work was founded on the concept of capping peristylane with a cyclopentane ring (**19**; Scheme 5, path VII). To this end, a peristylanedione was crafted following classic annelation methodology.[32] Once it was suitably functionalized along the fluted rim,[33] a cyclopentanedione unit could be correctly affixed (**20** → **21**).[34] All attempts to "nail down the roof" within the highly functionalized entity have proved unviable so far because of untamed effects from trans-skeletal steric congestion.

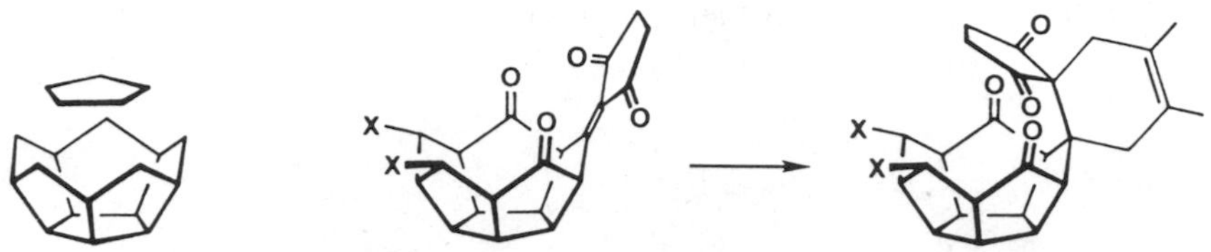

19 20 21

An elegantly designed synthesis of the bowl-shaped C_{16}-hexaquinacenedione (**22**)[35] was tested by Paquette for its quality as an entry to even higher polyquinanes (Scheme 5, path IV). The attachment of four more carbon atoms to **23** proved to be an easy matter, but acute steric complications prevented its transformation into dodecahedrane.[36]

An entirely different approach to the C_{16}-hexaquinane system (**24**) was ingeniously developed by the Eaton group,[37] but its usefulness as an eventual dodecahedrane precursor has yet to be demonstrated.

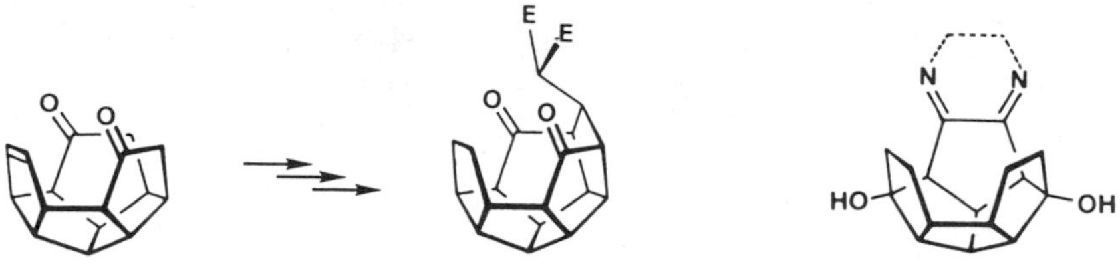

22 23 24

The Paquette group was led to abandon two other concepts. In a cyclization project employing tetraquinanebisepoxide **26** for anionic substitution chemistry, a change in relative alignment and reactivity of functional groups, which followed the first nucleophilic attack, caused an irreversible deviation of the product structure **27** from the expected *acs*-polyquinane nature **25** (Scheme 2).[38]

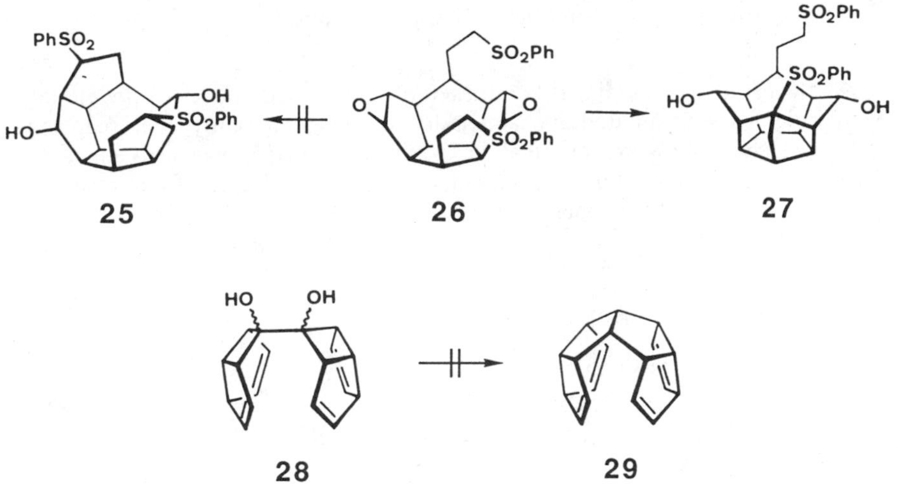

PhSO2 OH SO2Ph HO SO2Ph O O SO2Ph SO2Ph SO2Ph HO OH

25 26 27

HO OH

28 29

Scheme 2.

Skeletal rearrangement of the ketone coupling product (**28**) could not be directed towards C_{18}-hexaquinacene **29** (Scheme 5, path XI).[39]

Substantial enhancement of intramolecular transannular Michael reactivity was noted by Baldwin during investigations of cyclopentanone annelations to the conformationally flexible triketone **30** (Scheme 5, path III).[40] Due to the proclivity of the hexaquinane derivative **31** for rearrangements towards a thermodynamic equilibrium with isomers **32–34** (Scheme 3), this route has not advanced to the dodecahedrane level.

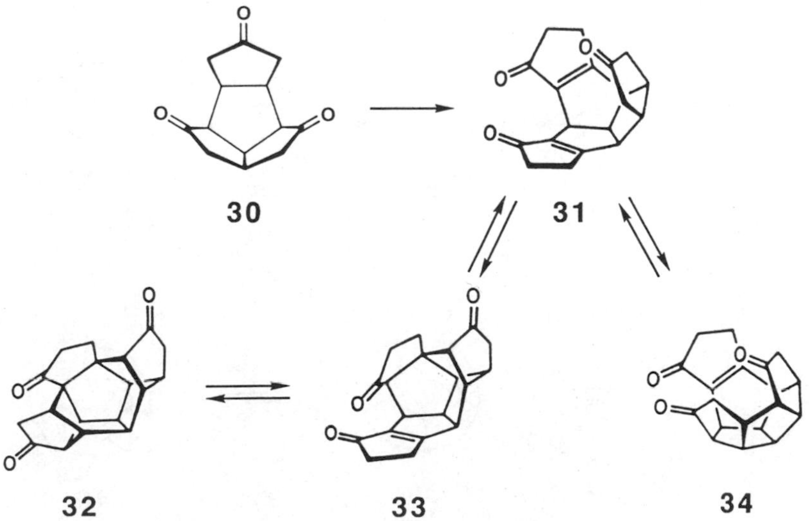

30 31

32 33 34

Scheme 3.

Serial Synthesis and Adjustment of C_{20} Precursors

A clever tactic for linear synthesis of dodecahedrane was developed by McKervey. Using an iterative annelation scheme with only four reagents, an applepeel shaped C_2 symmetrical hexaquinanedione **35** was arrived at.[41] As demonstrated by the close agreement of X-ray[41] and force-field results,[1] the molecule adopts a fully opened conformation because of intramolecular steric repulsions, a situation that so far has prevented its evolution into dodecahedrane.

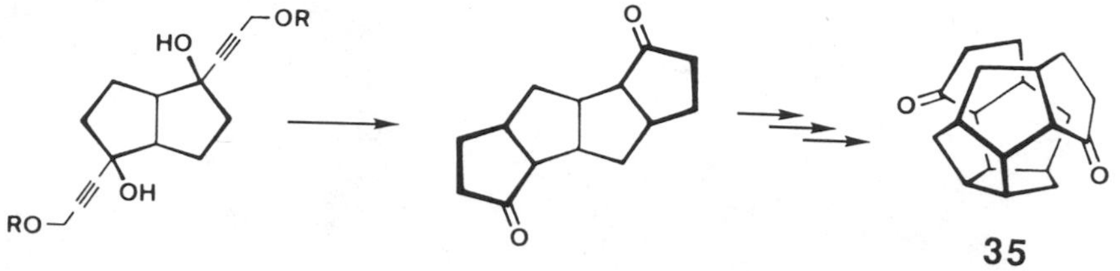

HO OR OH RO O O O O

35

The pincer Diels–Alder product of 9,10-dihydrofulvalene and dimethyl acetylenedicarboxylate (**36**) is the cornerstone of the successful Paquette synthesis (Scheme 5, paths V and VI).[42] The details of implementation of C_2 symmetry, followed by twofold cyclopentanone annelation[43] and successive further elaboration of the C_{20} framework (**37**) into dodecahedrane have been extensively documented.[6,7] Most notable among other features are notes of the extreme susceptibility of intermediates towards unwanted transannular cyclizations that occur at early spheroidal stages allowing conformational flexibility.[44] Also remarkable was the multiple advantageous use of photochemical homo-Norrish

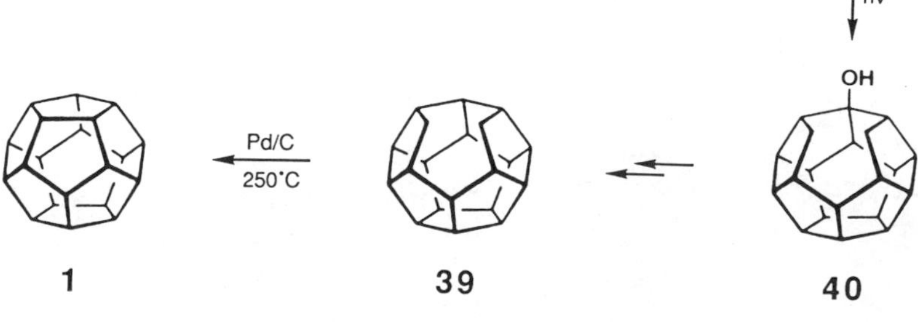

E E Cl E Cl E O

36 **37** **38**

hv

OH

Pd/C 250°C

1 **39** **40**

Scheme 4.

methodology for intramolecular ring cyclizations (e.g., **38** → **40**) and the installation of the final bond by controlled palladium catalyzed dehydrogenation (**39** → **1**) (Scheme 4).[45] Meanwhile, the original protocol was already sufficiently streamlined to allow the preparation of enough material for physical measurements[46] and exploration of fundamental dodecahedrane chemistry.[47]

A closely related approach by Mehta and Nair takes advantage of the spheroidal tetraquinadienedione **42**,[48] which is made available by flash vacuum pyrolysis of the hexacyclic diketone **41**.[48,49] The major difference from the Paquette synthesis is the C_{2v} design of twofold cyclopentanone annelation,[50] which is aimed at final aldol methodology for subsequent ring closures. Force-field calculational predictions[1] for an unfavorably twisted ground-state topography of diester **43**, due to strong repulsive interactions, and its potential usefulness for conversion into dodecahedrane will have to await experimental confirmation.

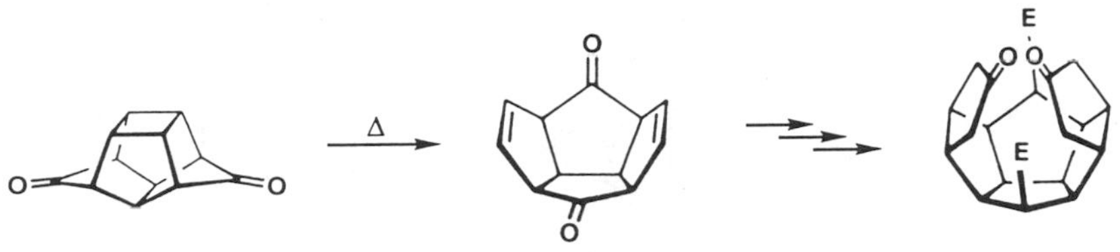

Δ

41 **42** **43**

RETROSYNTHETIC CONSIDERATIONS AND DEVELOPMENT OF THE PAGODANE STRATEGY

A closer inspection of typical structural examples from the plethora of diverse alternative approaches for dodecahedrane synthesis has convincingly demonstrated that deviations from assumed critical conformations and unexpected transannular reactions are the main causes for failure. Serious steric effects are best documented by X-ray structures of advanced synthetic intermediates from all the diverse routes, notably in those cases where structural information for unexpected reaction products is likewise available for comparison.

Thus, at the outset, a serial synthesis scheme appeared to be more promising than a convergent assembly. It would ideally involve a rapid buildup of a highly rigid framework that would eliminate the danger of deleterious transannular reactivity and would incorporate as many of the required 20 carbon atoms as soon as possible. Maximum exploitation of the exquisitely high symmetry of the target structure needed our utmost attention in order to

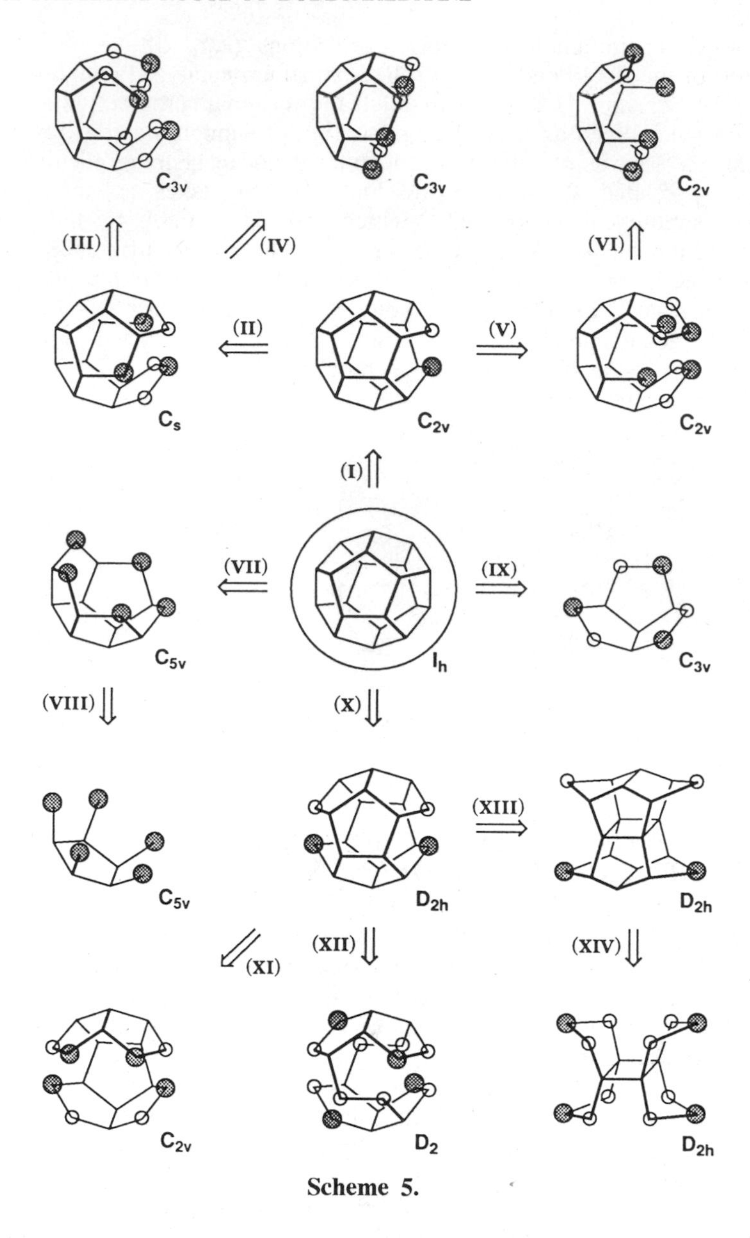

Scheme 5.

minimize the disadvantage of a lengthy serial approach. Final reorganization towards the ultimate polyquinane stabilomer should then most likely be assisted by thermodynamics.

A brief look into the retrosynthetic hierarchy (Scheme 5)[1] of the different approaches for dodecahedrane reveals that the more linear syntheses originate almost entirely from a monoseco precursor (path I), where only one of the bonds of dodecahedrane is dissected and where the construction therefore advances from a half-sphere. When one alternatively considers an ultimate bisseco precursor having a higher eightfold D_{2h} symmetry for superior synthetic efficiency (path X), it appears prima facie to be a dead end because there seems to be no immediate logical way for any further, chemically meaningful, dissections that simultaneously meet the symmetry requirements.

The picture improves significantly with a little mental trick: Inverting and interconnecting the central carbon atom pairs with formation of a cyclobutane ring (path XIII) constricts the molecule into a twofold bird cage system and thereby generates a novel polyquinane framework (**45**) that does indeed offer further opportunities for systematic disconnections (path XIV). The diversity of tactical methods that in principle can be applied for its construction (e.g., **44** and **46**) highlight the system as a promising key precursor.

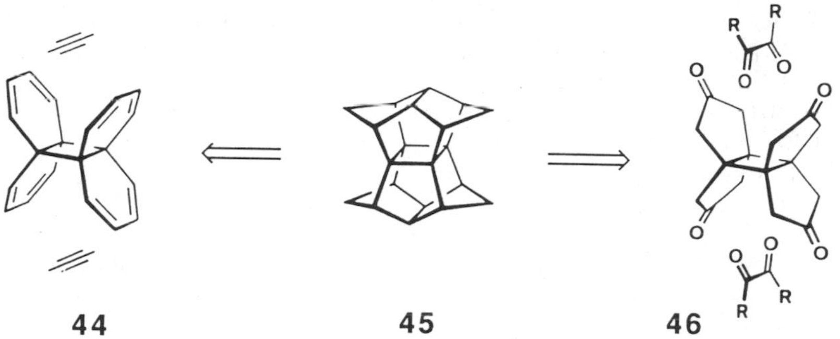

44 **45** **46**

The high complexity of this aesthetically appealing polycycle is apparent from the systematic IUPAC nomenclature* and from the total number of subsidiary ring systems as derived by graph analysis.[51] The intricate structure demands the creation of a more colloquial name. Because of the resemblance of **45** to the shape of an oriental temple and its reflexion in water, we have coined it *Pagodane*.[8,51] The nomenclature also strengthens its close relationship to dodecahedrane. Each molecule has the formula $C_{20}H_{20}$ and contains 12 multifused cyclopentane rings. As is evident from retrosynthetic analysis, structural alterations (**47**) required for the interconversion of **45** to **1** are limited only to the hydrogenolytic fission of two cyclobutane bonds with inversion at the reaction centers and concomitant opening of the molecular sphere, followed by two oxidative C—C bond formations between opposing methylene bridges. Energetically, this is a very favorable undertaking as force-field calculations place **45** by as much as 40 kcal/mol above the stabilomer **1**. Most of the strain energy within the pagodane system is taken up by the cyclobutane ring, making this structural unit an obvious target for chemical attack.

*Compound **45** is Undecacyclo[9.9.0.$0^{1,5}$.$0^{2,12}$.$0^{2,18}$.$0^{3,7}$.$0^{6,10}$.$0^{8,12}$.$0^{11,15}$.$0^{13,17}$.$0^{16,20}$]eicosane.
Compound **1** is Undecacyclo[9.9.0.$0^{2,9}$.$0^{3,7}$.$0^{4,20}$.$0^{5,18}$.$0^{6,16}$.$0^{8,15}$.$0^{10,14}$.$0^{12,19}$.$0^{13,17}$]eicosane.

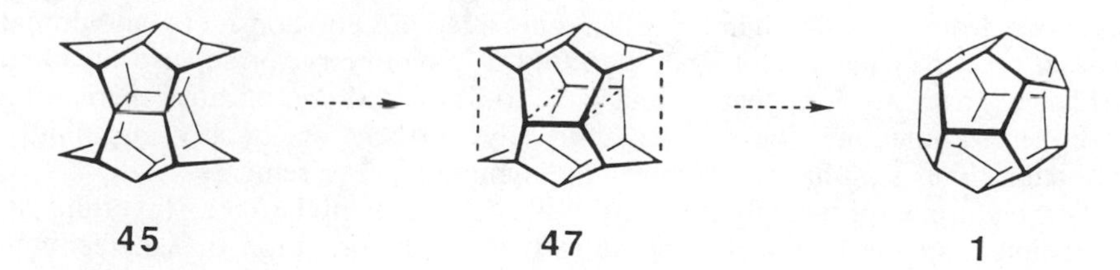

Our approach to the *Mount Everest of alicyclic chemistry* by way of the pagodane strategy is summed up in the tactical Scheme VI. A cornerstone of this synthesis is the insecticide isodrin (**48**),[52] which is still commercially available.* The molecule already contains 12 carbon atoms properly arranged into four cyclopentane rings, making up practically one half of the pagodane skeleton. In terms of an ascent to Mount Everest, **48** has to be regarded as constituting a solid base camp. Consecutive twofold benzoannelation (**49**), a [6π + 6π] photocycloaddition within the face-to-face oriented benzene rings (to **50**; a reaction that in fact triggered the whole pagodane project) and a domino Diels–Alder based elaboration into pagodane were published in detail[1,51] and shown to be adequate techniques for climbing the initial steep walls (Scheme 6). Therefore, after a short summary we will concentrate on further trails towards "the top."

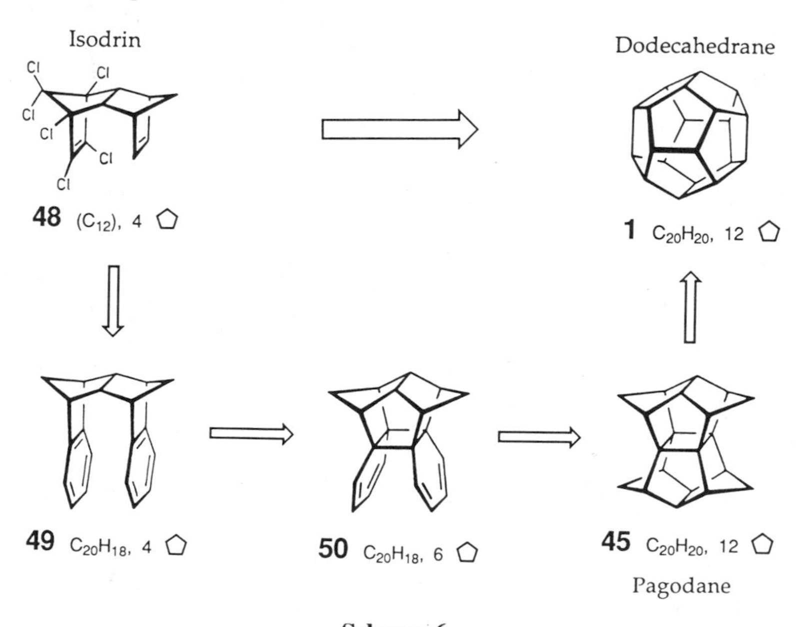

Scheme 6.

*By coincidence, isodrin also served as the starting material for the classic Woodward triquinacene synthesis.[16] At that time and at the beginning of our studies, isodrin was a cheap technical product that nowadays is marketed as a fine chemical.

CONSTRUCTION OF THE BRIDGED [3.3]ORTHOCYCLOPHANE

The first task is the dibenzoannelation of **48** which had to be performed in a stepwise manner since one of the norbornene-type bonds was protected by chlorine substituents. Monoannelation was described earlier as a five-step sequence with an overall yield of 38% **53** using the dimethyl acetal of tetrachlorocyclopentadienone as a benzene ring synthon.[53,54] An improvement was possible by substituting the previously used acetal with a more reactive sulfone derivative as the initial cycloadduct **51** spontaneously releases sulfur dioxide to an intermediate **52**, which aromatizes in situ to produce the benzo compound **54** (Scheme 7). After dehalogenation, this two-step alternative yielded up to 95% of the benzoolefin.

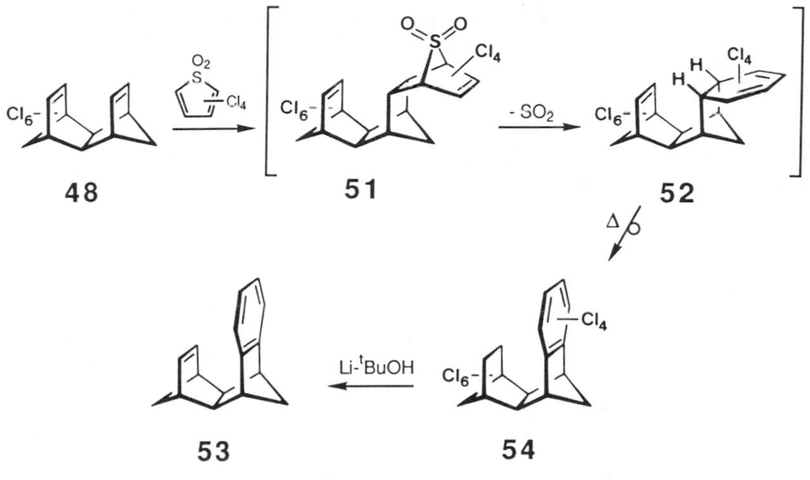

Scheme 7.

The second benzoannelation to **49** was executed analogously.[54] Again, the use of the sulfone building block raised the overall yields of the immediate cyclohexadiene precursor **55** by a factor of 2. The critical step was the final dehydrogenative aromatization as the first benzene ring exerts a very efficient shielding of the two inner hydrogen atoms that have to be eliminated (Fig. 3). After extensive variations of methods and conditions, an acceptable 70% yield could be reproducibly realized for this step when recourse was made to brute-

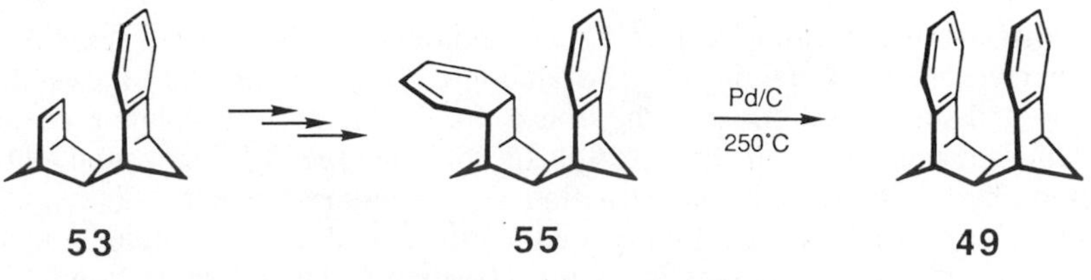

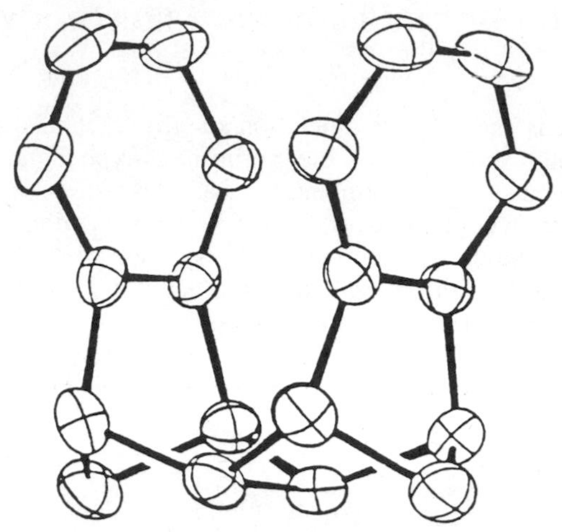

Fig. 1. The ORTEP view of one of the two crystallographically independent molecules of [3.3]orthocyclophane (**49**).

force palladium catalyzed dehydrogenation. Nevertheless, concomitant saturation to nonrecyclable components could not be fully suppressed.

The single crystal X-ray analysis of **49** (Figs. 1 and 3) revealed significant proximity effects.[9] The closest π–π distance of ~ 3 Å is much shorter than the van der Waals distance of 3.4 Å and the molecule alleviates strain by angle widening in the skeleton and by a pyramidalization at the benzoannelation positions. After all, the electronic situation between the almost parallel oriented orbitals approximates that of the transition state for the thermally symmetry-forbidden [6 + 6] cycloaddition.

As a consequence, characteristic spectroscopic properties arise. The NMR signals for the aromatic protons at 6.5 ppm appear at relatively high field due to the mutual shielding by the aromatic rings. The UV spectrum shows a broad long wavelength absorption starting from 310 nm, which is absent in the spectrum of indane and which is interpreted as a charge-transfer band. From the photoelectron spectroscopic analysis, a benzo–benzo resonance has been determined to be of equal magnitude as that operative in [2.2]paracyclophane.[9]

INTRAMOLECULAR [6π + 6π] PHOTOCYCLOADDITION

The next stage required a [6π + 6π] photocycloaddition between the face-to-face oriented benzene rings. The analogous reaction in the simple dienes, isodrin and the dechlorinated hydrocarbon, are very efficient processes.[55] Problems associated with the photolysis of analogous mono- and dibenzo compounds, especially with construction of a *syn-o,o'*-dibenzene moiety, arise because these reactions are highly endothermic and the σ-coupled butadiene units within the photoproducts are quite effective light absorbing competitors.

Consequently, we were confronted with the problem of selective excitation of the benzenoid chromophores of the starting material so as to limit possible competitive photochemical side reactions of the product. Furthermore, the kinetic stability of the product as a bridged benzene dimer was uncertain, as can be witnessed from the propensity of the syn and anti isomeric benzene dimers **56** and **57** for facile thermal fragmentation.[56]

56 **57**

The dibenzo compound was found to be resistant towards sensitized photodimerization. Direct excitation of the long wavelength transition was also unproductive. Success was finally achieved by irradiation with monochromatic 254-nm light.[54] Despite potential photochemical and thermal competition, a clean 7:3 photoequilibrium of **49:50** was established. A qualitative explanation for the position of this equilibrium refers to the relative absorption coefficients for both components at this wavelength. In contrast to the facile fragmentation of the unbridged *o,o'*-benzene dimers, to our good fortune, **50** proved to be extraordinarily stable, a property that was essential for the impending chemistry. In a thorough kinetic study, activation parameters for the clean thermal isomerization to **49** were determined as ~ 38 kcal/mol (Scheme 8). This activation barrier is 8 kcal/mol higher than that calculated for a diradical

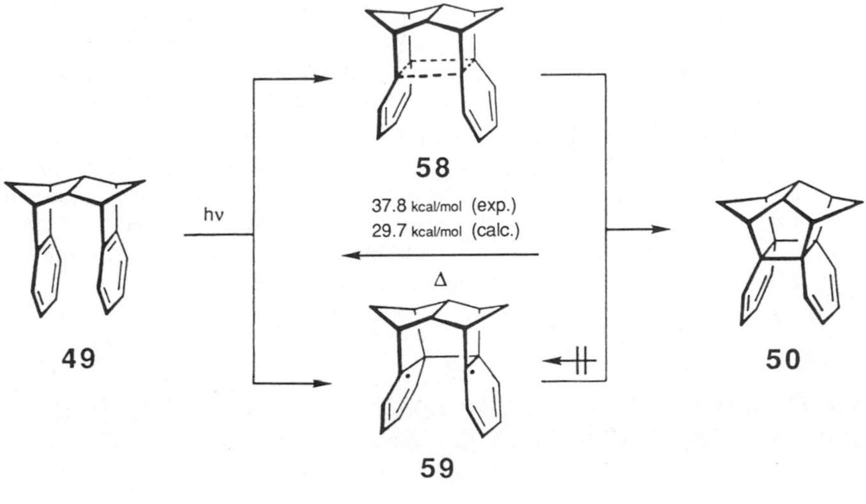

49 hv **58** 37.8 kcal/mol (exp.) 29.7 kcal/mol (calc.) Δ **59** **50**

Scheme 8.

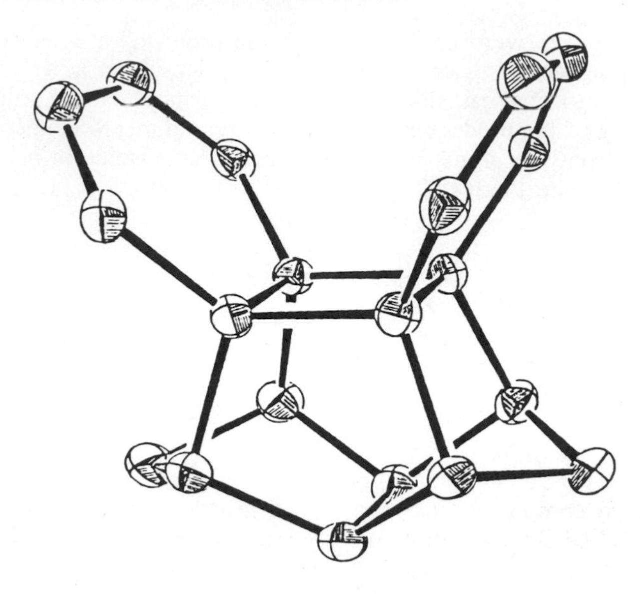

Fig. 2. The ORTEP view of one of the two crystallographically independent molecules of *syn-o,o'*-dibenzene compound **50**.

pathway. The discrepancy is interpreted as an indication that, due to the rigidity of the cage framework, a stepwise radical opening via **59** is prohibited and instead, the symmetry-forbidden, concerted mechanism (viz., **58**) is operative.[57]

From a consideration of the specific steric relationships within the unique *syn-o,o'*-dibenzene compound, there was some uncertainty as to the steric inhibition for a subsequent cycloaddition reaction exerted by the mutual shielding of the cyclohexadiene rings and the influence of the opposing methylene bridges. An X-ray crystal analysis of **50** (Figs. 2 and 3) allowed a more precise evaluation of the projected chemistry.[51] The structure determination assessed that the planar diene rings face each other at an angle of 67° with the transannular distance between the inner sp^2 carbon atoms amounting to only 3.06 Å. At 1.575 Å, the photochemically formed cyclobutane bonds are somewhat elongated with respect to the typical value of cyclobutanes (1.55–1.56 Å), an effect that is primarily ascribed to homoconjugative destabilization and is deducible from the PE analysis.[58]

DOMINO DIELS–ALDER BASED CONSTRUCTION OF THE SECOND CAGE: PAGODANES

At the inception of our project, it was recognized that the *syn*-tetraene arrangement in **50** was ideally suited for a domino Diels–Alder[42] reaction.

Conceptually, the addition of an acetylene equivalent to one of the equivalent diene units would meet our requirement in crafting the second cage of pagodane. By this multiple cycloaddition sequence (**60** → **61**), four new C—C bonds and two more cyclopentane rings would be created. Moreover, all six remaining methine groups would concurrently become adjusted stereospecifically in the desired all-cis,syn fashion. The last four cyclopentane rings could then be introduced by ring contractions at the olefinic bridges to develop the full D_{2h} symmetry of the [1.1.1.1]pagodane architecture.

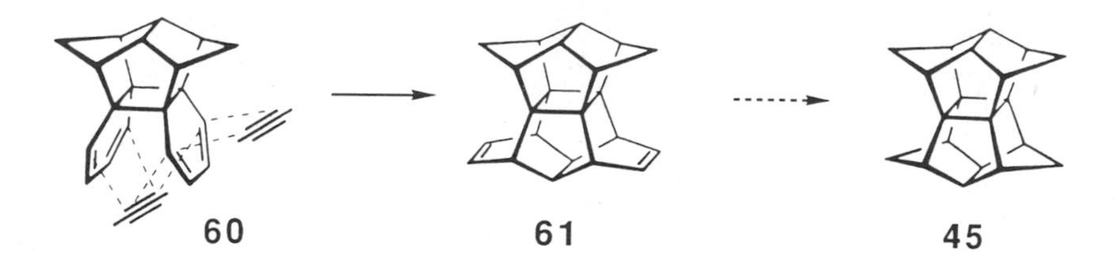

The space-filling van der Waals model of **50** (Fig. 3) illustrates that the *syn*-methylene hydrogen, which is situated directly over the exo face of the cyclohexadiene ring, will constitute a specific obstacle for external dienophile attack. On the other hand, the projection also leaves no doubt that an inside, pincer-type addition with ethylenic dienophiles is occluded and would only be possible, if at all, with sterically undemanding *sp*-hybridized dienophiles. Consequently, a competition between a more probable external cycloaddition and the thermal isomerization of **50** back to **49** had to be anticipated.

With maleic anhydride as a classic dienophile, the cycloaddition proceeded smoothly in refluxing benzene solution. A single product was formed stereospecifically and was isolated in quantitative yield. Its spectroscopic properties characterized it as the expected polycondensed monoadduct **63** resulting from an external, domino-type attack. With similar stereoselectivity, a series of other highly reactive dienophiles could be engaged in π-facially stereospecific domino cycloaddition to **50**.[1,59] Under a variety of conditions, no

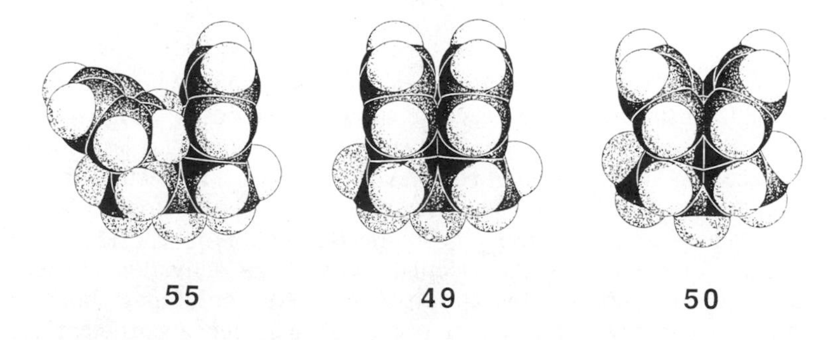

Fig. 3. Space-filling models of benzodiene **55**, dibenzo compound **49**, and photoadduct **50**.

primary adduct (e.g., **62**) or indications for its fleeting existence could be detected. The inner double bond in such intermediates obviously is perfectly oriented for a subsequent intramolecular addition. The intricate timing of the reaction course provokes the question for some degree of concertedness between the individual steps.

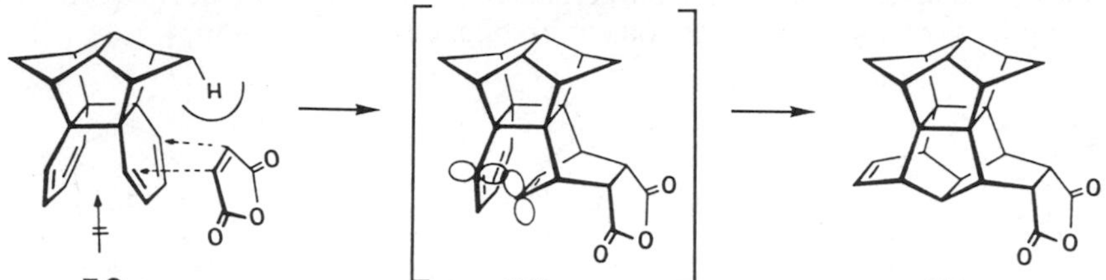

50 62 63

In the case of acetylenic dienophiles, competing formation of the pincer product was indeed observed, for example, **64** with dimethyl acetylene dicarboxylate in the order of 3%. The amount of 74% internal attack with the highly activated dicyanoacetylene to **65** can again be regarded as evidence for a higher concerted, simultaneous two-sided addition process.[59] This possibility remained unsupported, however, by a series of high-pressure studies, which rather favored the preceding formation of a relatively strong π complex to the inner diene faces as introductory step for a sequential cycloaddition.[60]

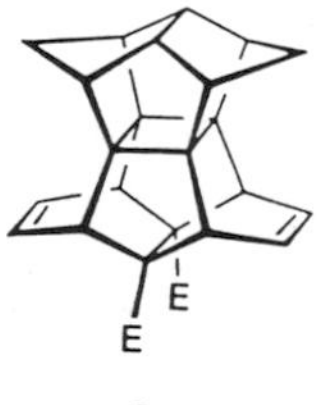

64 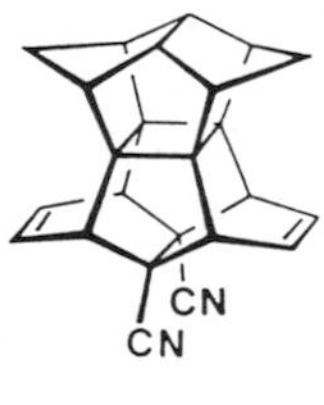 65

This view is further supported by a computational investigation with SCF–MO (self-consistent field–molecular orbital) methods[1,60] (Fig. 4), which clearly shows that the intramolecular cycloaddition **B** requires far less activation than any preceding intermolecular step (**A** or **C**). The respective π centers within a defined intermediate or at close transient stages are too far apart for a reasonable degree of cooperativity. A concerted double-pincer addition **D** is likewise clearly disfavored by the calculations with an activation barrier that exceeds the stepwise process by 40 kcal/mol. Not only must both diene moieties be considerably deformed in order to guarantee a sufficient orbital overlap, but the acetylene C—H bonds also have to be bent to a much higher degree, which would account for almost one-half of the excess energy.

Degradation of **63** was best performed by cuprous oxide decarboxylation.

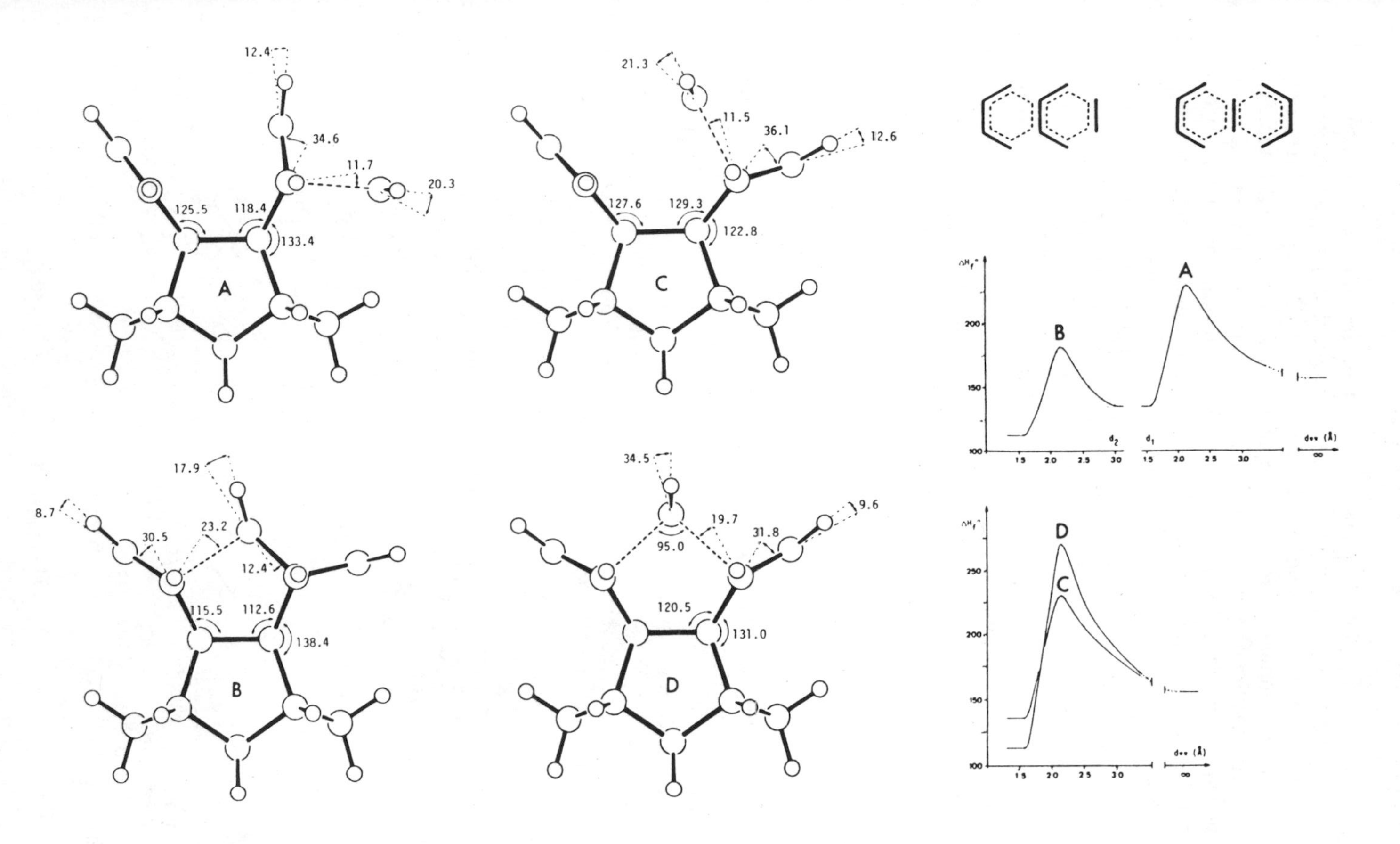

Fig. 4. Calculated reaction paths and transition states for constraint-synchronous cycloadditions of acetylene to tetraene **50** (modified neglect of differential overlap, MNDO). **A**: domino attack; **B**: intramolecular addition; **C**: single-sided pincer attack; **D**: double-sided pincer attack.

With acquisition of the [2.2.1.1]pagodadiene **61**, the economically important C_{2v} symmetry was reinstated. Its presence again allowed all subsequent chemical changes to be implemented concurrently at an independent pair of reaction centers and hence amounted to the manipulation of only a single type of functional group. Although high stereoselectivity in additions to the diene from the unhindered bottom face could be established by uniform epoxidation to **66**, the functional changes required within the contraction scheme must break the molecular symmetry into two subclasses of C_s and C_2 point groups. A specifically directed introduction of one of these symmetry elements or a separation was not called for, however, because both regiochemistries converge again with arrival at the [1.1.1.1]pagodane skeleton.

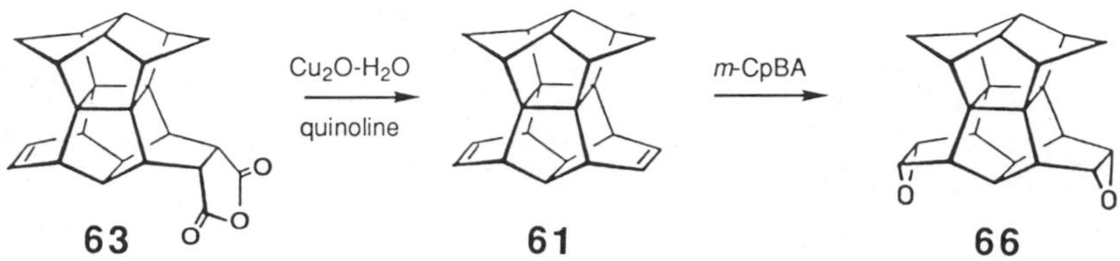

Cu_2O-H_2O, quinoline; *m*-CpBA

63 → **61** → **66**

Twofold ring contraction followed established methodologies.[51] To this end, diene **61** was submitted to exhaustive hydroboration, followed by Jones oxidation to regioisomeric diketones (**67**) (Scheme 9). Formylation for subsequent diazo group transfer could be adapted to a one-pot protocol that directly delivered bright yellow bis(diazoketones) (**68**). When methanolic solutions of **68** were subjected to photochemical Wolff ring contraction, practically uniform conversion to a highly crystalline C_{2v} diester took place.

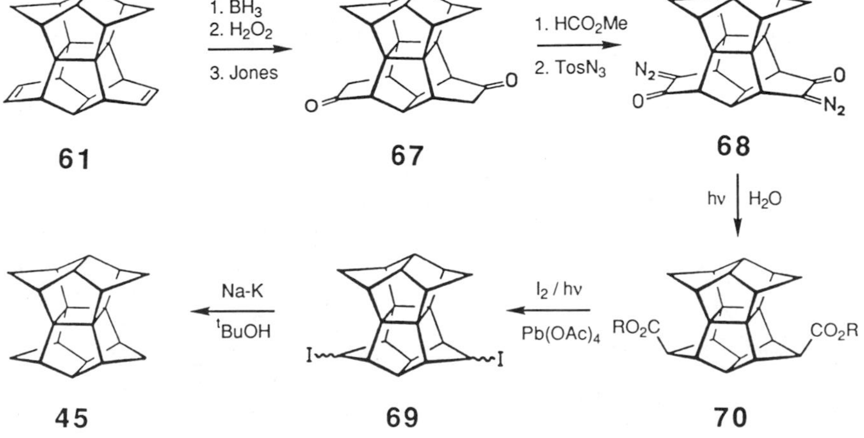

1. BH_3 2. H_2O_2 3. Jones; 1. HCO_2Me 2. $TosN_3$; hv, H_2O; I_2 / hv, $Pb(OAc)_4$; Na-K, ${}^tBuOH$

61, **67**, **68**, **70**, **69**, **45**

Scheme 9.

The thermodynamically unfavorable syn,syn stereochemistry of the ester groups in **70**, which must be interpreted as a result of the kinetically controlled anti capture of methanol to the relatively less hindered bottom side of the intermediate ketene stages, and the molecular dimensions of this first [1.1.1.1]pagodane derivative were acquired by X-ray analysis. Bis(iododecarboxylation) of the corresponding diacid, accessible by saponification of diester **70** or simply by photolysis of bis(diazoketones) (**68**) in aqueous tetrahydrofuran solution, removed the two extraneous carbon atoms to provide diiodides **69**. Alkali metal reduction ultimately delivered the title hydrocarbon **45**.

That access had been attained to the D_{2h} class of point group symmetry, a level rather uncommon at such a state of molecular intricacy, was immediately evident from the simple carbon and proton NMR spectra. Both spectra are characterized by only four groups of signals each, in response to the influence of three mirror planes that act upon the four interwoven norbornane units. Pagodane is a relatively high-melting compound that is only sparingly soluble in most organic solvents, but readily sublimes upon heating. The high stability of the molecular skeleton is attested to by the presence of a dominant molecular ion peak in the mass spectrum, while the few smaller fragments hardly rise above the noise level. As we discovered later, this intrinsic quality is shared fully only with the dodecahedrane nucleus.

Although the synthesis of pagodane from isodrin requires a total of roughly 45 functional changes, these could be concentrated into 14 one-pot manipulations, which were demonstrated to be amenable to routine large scale synthesis, making our protocol highly efficient. An overall yield of 24% was hereby obtained which implies remarkable averaged yields of 90% per step or 97% per functional change. Again, in terms of a Mount Everest expedition, this accomplishment considerably increases our prospect for further progression on our trail.

EVALUATION OF "ISOMERIZATION" PATHWAYS

As defined in the retrosynthetic introduction for a controlled transformation of pagodane into its isomer dodecahedrane, the ensuing task was the regiocontrolled scission of the two inner cyclobutane bonds with mandatory inversion at the participating carbon centers. At the outset, our hopes of selective transformations of pagodane were justified by force-field and MO calculations, which predicted an appreciable concentration of the molecular strain into the lateral cyclobutane bonds (b). A comparison of the results of the X-ray crystal analyses of the diester[8,51] and particularly of the parent [2.2.1.1]- and [1.1.1.1]pagodanes[61] indeed confirmed this effect (Fig. 5). Increasing with reduction of the bridge size or the introduction of endo substituents, the cyclobutane ring becomes unsymmetrically distorted due to the lengthening of the lateral bonds (b), while the frontal bonds (a) remain constant. For the last stage of our route, we planned to take advantage of this inherent bond strain because it is exactly these bonds (b) that have to be broken on the way.

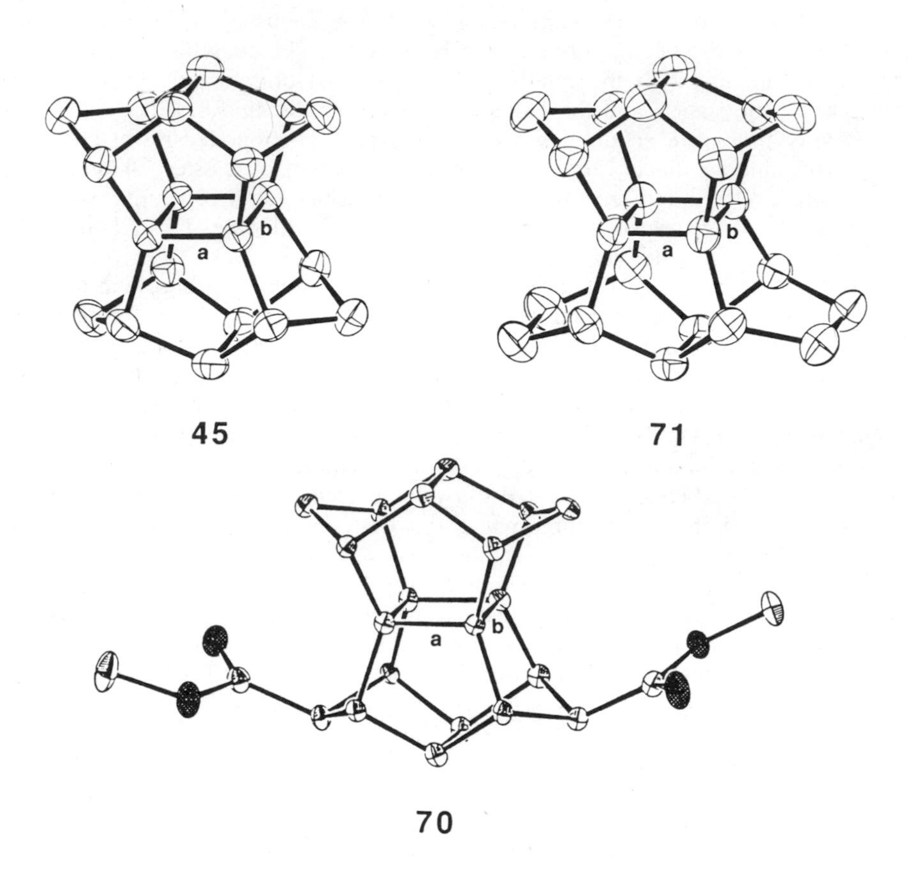

Fig. 5. The ORTEP plots of [1.1.1.1]pagodane (**45**), [2.2.1.1]pagodane (**71**), and dimethyl[1.1.1.1.]pagodane-4-*syn*,9-*syn*-dicarboxylate (**70**). Hydrogen atoms are omitted for clarity. Experimental cyclobutane bond lengths (a/b) are 1.549/1.573, 1.549/1.560, and 1.552/1.589 Å.

Principally, we envisioned three different approaches for this undertaking (Scheme 10).[1,62] Pathway A consists of a catalytic isomerization of pagodane to the by far energetically favored stabilomer. Alternative selective stepwise conversions were the thermal $2\sigma \rightarrow 2\pi$ cycloreversion of pathway B and the 1,2-addition pathway C. At one or the other later stages, the different products of these three pathways could eventually be interrelated chemically by similar procedures. The different possible pathways were investigated upfront by calculating energies and geometries of the respective products with the MM2 force-field method. Being fully aware of its deficiencies, the fast and sufficiently reliable estimate of energy and geometrical data received with this computational tool served as a kind of quantified molecular model and as a stimulating guide for devising experiments and interpreting their results.

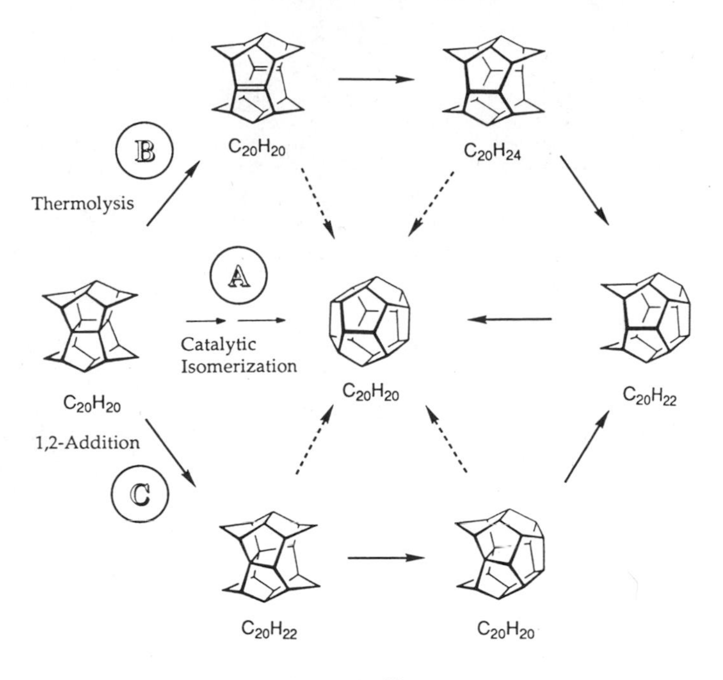

Scheme 10.

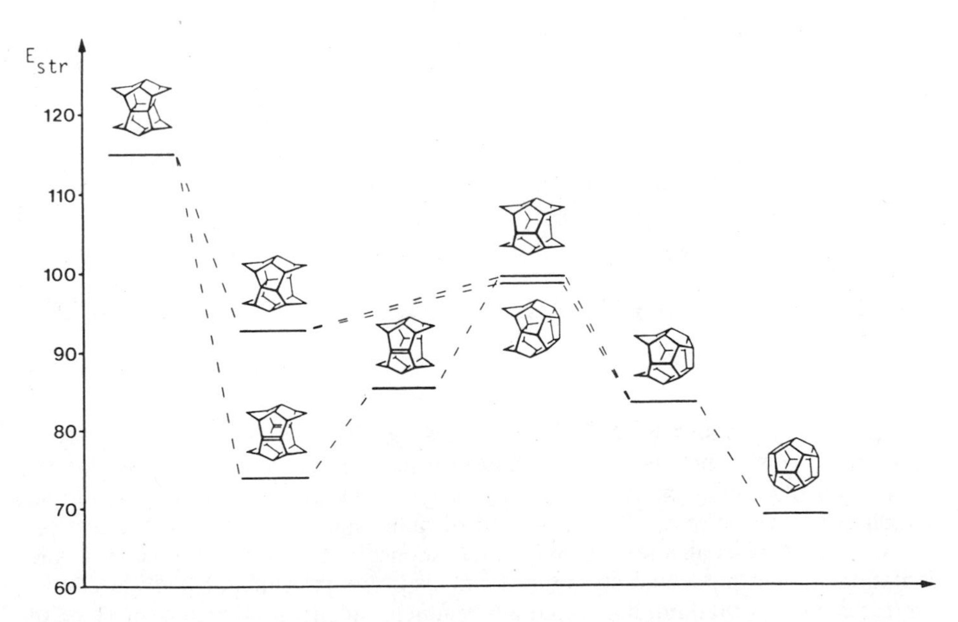

Fig. 6. Development of molecular strain for pathways B (lower entry) and C (MM2; kcal/mol).

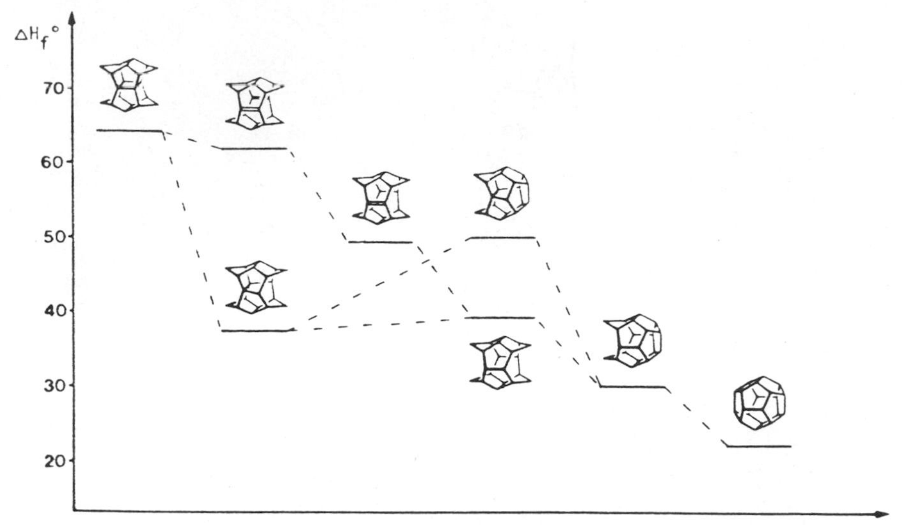

Fig. 7. Enthalpic trends for pathways B (upper entry) and C (MM2; kcal/mol).

The energy profiles for the two serial approaches show distinct differences. The kinetic preference, displayed in Figure 6 as the development of the inherent strain energies, is in favor of the 1,2-addition route C. In contrast, during a projected hydrogenation of the diene that would arise from a $2\sigma \rightarrow 2\pi$ cleavage of the alternative pathway B, a large amount of molecular strain is built-up. Thermodynamically, though, the latter sequence has the advantage of leading *continuously* downhill from pagodane to the stabilomer dodecahedrane (Fig. 7), while the thermodynamic relationship for a 1,2-addition procedure is discontinuous. We will see in the following sections how these numerical values are reflected in the actual chemical behavior of the targeted compounds.

PATHWAY A: CATALYTIC, THERMODYNAMICALLY CONTROLLED ISOMERIZATION

With Schleyer's adamantane synthesis as the archetype, a catalytic isomerization was the first choice because of its tempting simplicity. No matter what kind of product mixtures one would get from such a one-step operation, which could well be described as a kind of molecular roller coaster, the distinct properties of dodecahedrane should unquestionably facilitate its detection and isolation. Lewis acids and strong protic acids in the presence of hydride donors were tested first. Aluminum bromide treatment slowly produced a mixture of several new components, whose identity is still uncertain. The proton NMR spectra showed neither the characteristic singlet of dodecahedrane nor other low-field signals typical for geometries of enhanced spherical nature. Protonation of **45** with sulfuric acid/hexane or with triflic acid in dichloromethane solution only promoted unspecific decomposition.[1]

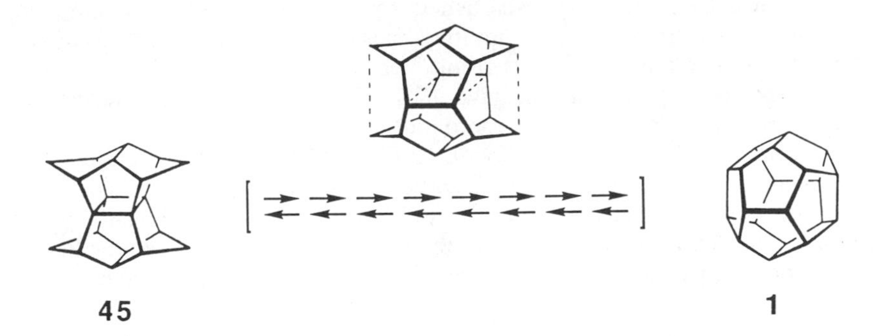

With noble metal catalysts that were specifically developed for the isomerization of hydrocarbons in the oil refinery industry and in the synthesis of higher adamantoids,[63] however, among the multitude of compounds produced, trace amounts of **1** were detected for the first time. While collaborating with the Schleyer group, comparable results were obtained from our initial experiments, which were conducted with intimate mixtures in the condensed phase,[1] and by similar experiments in the gas phase.[64] After enrichment by crystallization, five major products were isolated by vapor phase chromatography and analyzed for molecular composition and by NMR spectroscopy. While the occurrence of $C_{20}H_{22}$ and $C_{20}H_{24}$ compounds was not at all surprising, their structural formulation as the expected symmetrical [4]peristylane **76**, which might have resulted from hydrogenolysis of the longest C—C bond of pagodane, or as secododecahedranes **39** or **75** was excluded by the NMR data. Besides firm identification of the main component of the reaction mixture as the C_{2v} symmetrical dihydropagodane **72**, the tentative assignment of another constituent

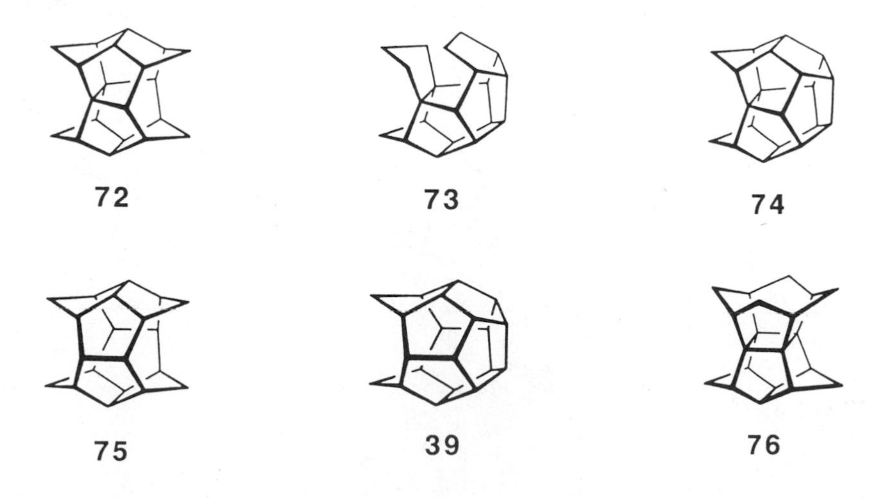

as the C_s polycycle **73** suggests that the latter might arise as a product of the former along the proposed scheme C (via **74**). After enrichment by simple fractional crystallization, the identity of an average of 2–3% dodecahedrane produced was unequivocally established by its characteristically ultrasimple NMR spectra and by gas chromatography–mass spectroscopy (GC/MS) analysis. Later improvements by the Maier group[64] enabled us to increase the amount of **1** to 8%, but an experiment that gave 17% of the desired compound could, to our greatest frustration, never be reproduced.

As seen from our discussion on p. 387, the mechanistic alternative of a rearrangement cascade involving carbocationic transitions is highly questionable. Supported by the occurrence of a multitude of eicosanes with higher hydrogen content, a sequential hydrogenation–dehydrogenation scheme,[65] equivalent to a combination of pathways B and C, rather seems to be in agreement with the experimental facts.

PATHWAY B: THERMAL 2σ → 2π CYCLOREVERSION

Clearly not satisfied with these results, as disappointing as they were on one hand but encouraging on the other, we turned to the more laborious serial approaches. A report of Fukunaga in 1977 especially aroused our interest, since he described a thermal isomerization of the bird cage **78** in the gas phase, although only modest yields of 35% of **79** were achieved.[66] According to a recent thorough reinvestigation,[57] which was in good agreement with pertinent force-field calculations (Scheme 11),[1] the thermolysis results in a selective

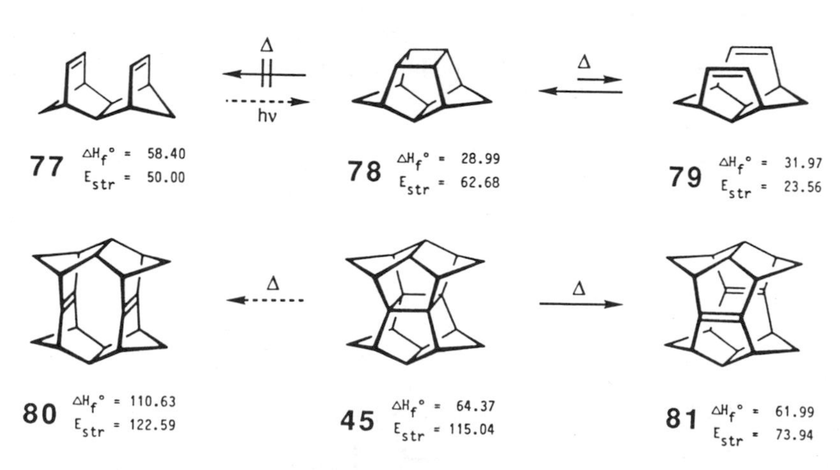

Scheme 11. Heat of formation and strain energy for [2 + 2] cycloreversion products of **78** and **45** (MM2; kcal/mol).

cleavage of the lateral bonds with reversible formation of the less strained **79** (no **77**) in a narrow temperature range above 400°C. By comparison of the energy data for the respective processes of the simple cage and the "twofold" pagodane system, the desired ring opening of the latter to **81** should have better prospects.

Not surprisingly, the extensively bridged pagodane network proved to be even more stable thermally than **78**. Above 600°C the only monomeric product isolated was identified as naphthalene![1,62] The yield of naphthalene, isolated from experiments with total conversion at 750°C, reached a remarkable 60%. What was the reason for the occurrence of such an extensive fragmentation? The answer comes from inspection of potential stabilization pathways open to the dienes. Indeed, because of the high rigidity of the molecular skeleton, a reverse reaction via an assumed diradical (**82**) for the kinetically and thermodynamically favored pathway could become more efficient than product formation (Scheme 12). This would enhance the likelihood that the by far less favorable reaction channel via diradical **83** becomes productive, from which the highly strained isomeric diene **80** escapes by fragmentation. From this diene, two symmetry equivalent, consecutive [4 + 2] cycloreversions can deliver two identical $C_{10}H_{10}$ triene fragments (**84**), which are probable

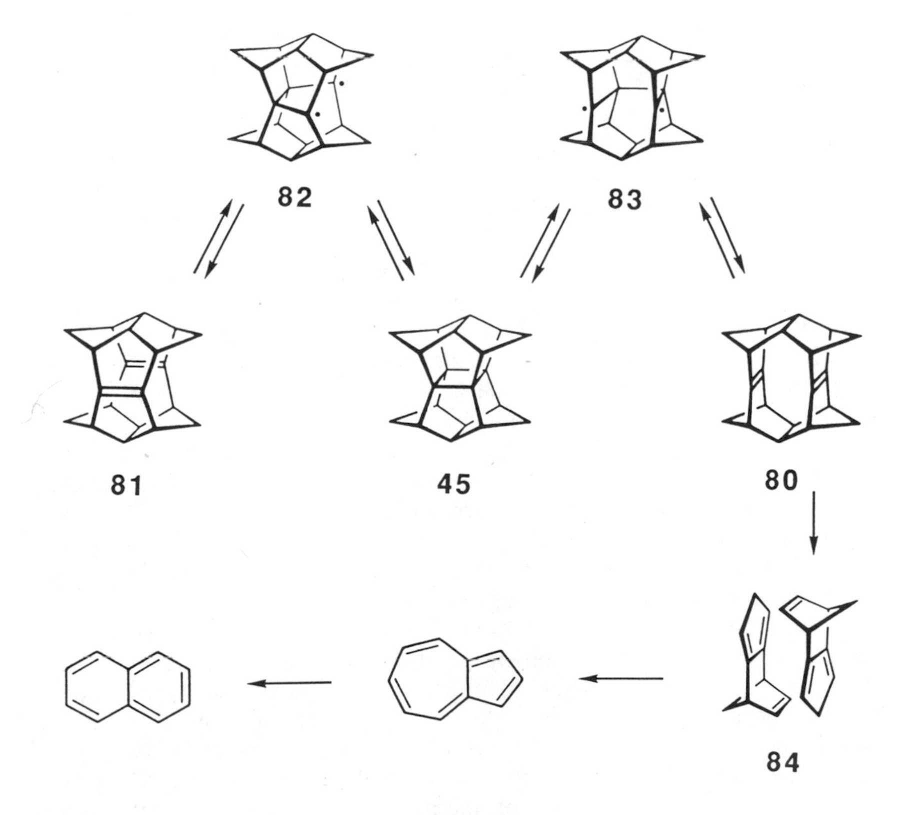

Scheme 12.

naphthalene precursors by rearrangement via azulene. This explanation was further substantiated by control experiments. At 650°C, under conditions where conversion of **45** is limited to 60%, a nearly quantitative formation of naphthalene was observed for both the C_{10} triene **84**, as well as for azulene. Still, the outcome seems useful in one respect. What a beautiful cume problem: How to get naphthalene from pagodane?

PATHWAY C: OXIDATIVE 1,2-ADDITION ACROSS THE CYCLOBUTANE RING

Our next option was the scission of the cyclobutane ring by regioselective addition to **45** under inversion at opposite centers. The reactivity of the central cyclobutane ring is accentuated by it being part of several small ring propellane units. For the more strained propellanes, it is known that they are not only capable of a facile [2 + 2] cycloreversion of the type already encountered in the previous section, but also that they rapidly add electrophiles or radicals across their central C—C bond.[67] As shown earlier from the energy profiles, from this point on, a further development towards dodecahedrane would require a secondary additive (**72** → **75**) or dehydrogenative conversion (**72** → **74**) leading energetically to some kilocalories per mol *uphill* for both the thermodynamic as well as the strain values (Scheme 13). Still, the geometrical prerequisites for either a secondary C—C bond formation or further bond breaking to a bissecododecahedrane seemed reasonable. Alternatively, 1,4-elimination methodology (**72** → **81**) would constitute a tool equivalent to the directly unachievable $2\sigma \rightarrow 2\pi$ cleavage.

The previously proposed 1,4-diradical **82**, which is formed thermally upon breaking one of the weakened cyclobutane bonds, could indeed be efficiently captured by controlled hydrogenolysis.[1,62] By heating a mixture of **45** with palladium on carbon in an autoclave under hydrogen pressure, a 95% yield of the high-melting C_{2v} symmetrical decacyclic polyquinane **72** was obtained. None of the possible isomeric dihydro compounds was detected, as was expected from the relatively higher enthalpies of formation calculated. Specifically this carbon framework originates because of its unique quality as a thermodynamic sink, as will be recurrently found in subsequent conversions at the $C_{20}H_{22}$ polyquinane level. The twofold element of a trans fused bicyclo[3.3.0]octane unit appears to be particularly well accommodated within the spherical network of **72**, although an isolated trans ring junction is ~ 6 kcal/mol less stable than the isomeric cis fusion.[68] Obviously, the reduction of overall torsional strain easily compensates for these minor increments.

An X-ray structure determination of **72**, carried out also as a check on the calculated geometry data, was unrewarding on account of random crystallographic disorder relative to the lost plane of symmetry (Fig. 8).[1] A structure simulating that of a fully ring opened diene (namely, **81**) resulted instead with oversized ellipsoids of thermal vibration for the "unsaturated" carbon atoms.

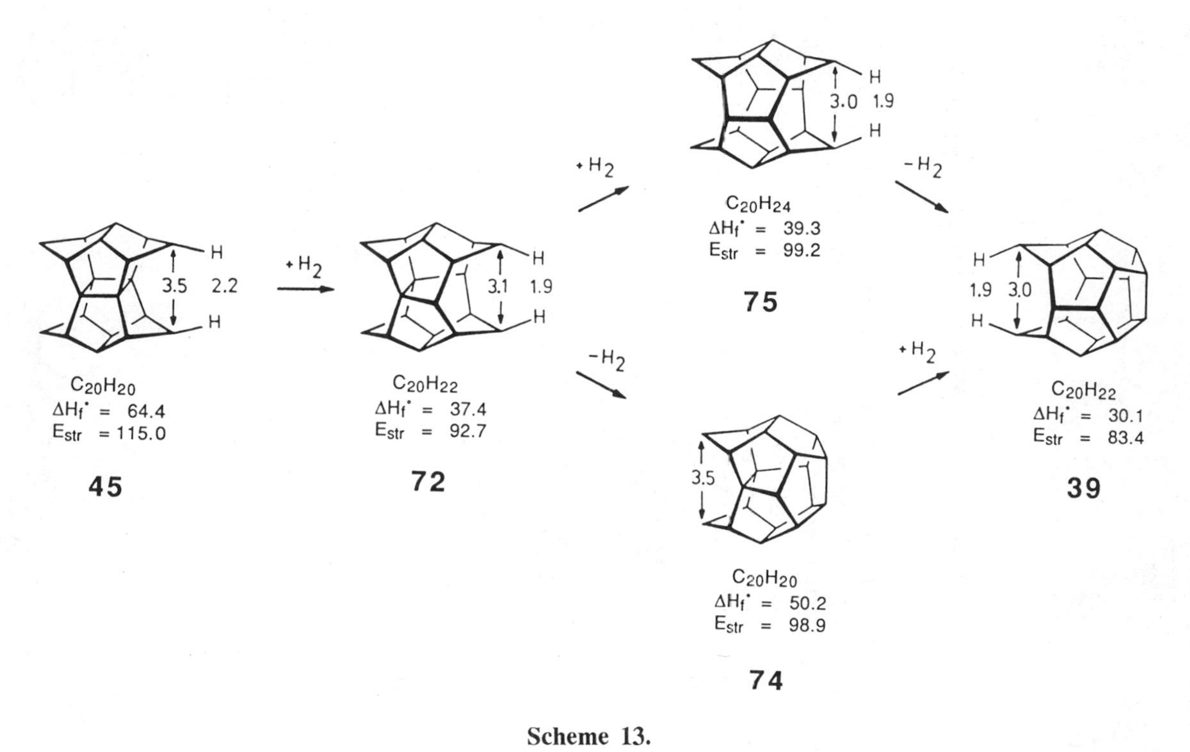
H
3.5 2.2
H
$C_{20}H_{20}$
ΔH_f° = 64.4
E_{str} = 115.0
45
+H2
H
3.1 1.9
H
$C_{20}H_{22}$
ΔH_f° = 37.4
E_{str} = 92.7
72
+H2
-H2
H
3.0 1.9
H
$C_{20}H_{24}$
ΔH_f° = 39.3
E_{str} = 99.2
75
3.5
$C_{20}H_{20}$
ΔH_f° = 50.2
E_{str} = 98.9
74
-H2
+H2
H
1.9 3.0
H
$C_{20}H_{22}$
ΔH_f° = 30.1
E_{str} = 83.4
39

Scheme 13.

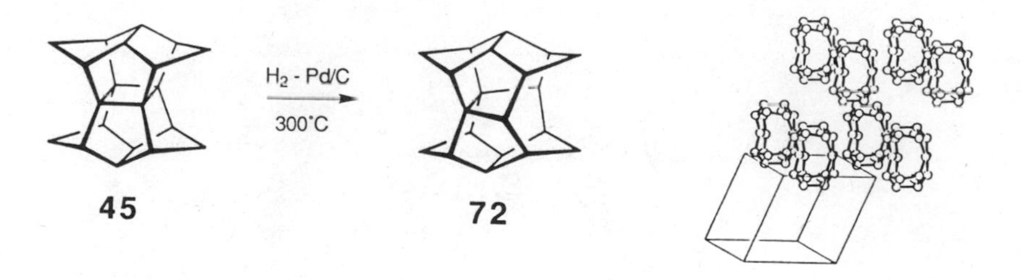

Fig. 8. Packing diagram with averaged structure obtained for secopagodane (**72**).

This pivotal diene **81** could indeed be synthesized shortly after the failure of the X-ray experiments. Pagodane reacted only very sluggishly when treated with excess of elemental bromine under thermal conditions. With the aid of concurrent illumination, according to a procedure developed previously for the **78** → **79** conversion,[49a] **45** was converted rapidly into a single dibromide in quantitative yield (Scheme 14). Its nature as C_{2v} symmetrical **85**, resulting from the anticipated regioselective 1,2 addition across one of the prestrained cyclobutane bonds, was evident from spectral data. Noteworthy is the propensity of this halide to undergo solvolysis, by necessity via an S_N1 mechanism. Interestingly, **85** readily expels bromine upon warming or irradiating with the recovery of **45**, a finding that explains why a vast excess of bromine is required for its synthesis. In contrast, heating of **85** in the presence of zinc in dimethylformamide results in smooth 1,4-bromine elimination and the formation of the ring opened nonacyclic diene (**81**) with reconstitution of the D_{2h} symmetry status.[69]

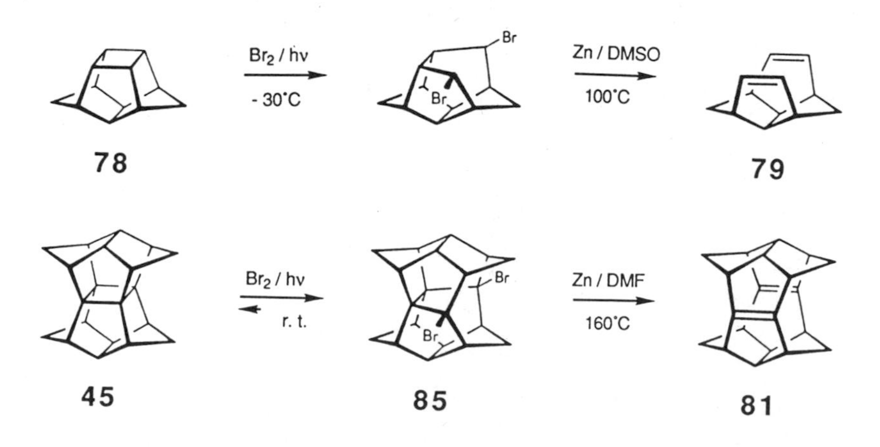

Scheme 14.

While culmination at the D_{2h} bisseco level of elaboration gave us the opportunity to install the final framework bonds, several observations made in conjunction with the surrounding chemistry stipulated a brief recourse to more detailed investigations of conceivable electron deficient intermediates.

ONE-ELECTRON OXIDATION OF PAGODANES: UNUSUALLY PERSISTENT RADICAL CATIONS

The ready addition of bromine, and the ease with which electrons are expelled from the highest occupied MO as determined by photoelectron studies,[70] led us to investigate the response of pagodane towards oxidizing media. Added fascination arose from the fact that the thermally symmetry forbidden $2\sigma \rightarrow 2\pi$ isomerization becomes symmetry allowed and rapid on the corresponding radical cation level, a phenomenon that is well established, for example, for the catalytic quadricyclane–norbornadiene rearrangement.[71] Hence, this possibility was considered as a promising alternative to the stepwise ring opening via dibromides.

Cyclovoltammetric oxidation of **45** by Heinze and co-workers[72] resulted in the observation of an irreversible wave at 1.2 V (vs. Ag/AgCl) suggestive of an ECE (electrochemical, chemical, electrochemical) mechanism, where the initial oxidation step (→ **86**) is accompanied by a rapid chemical process immediately following formation of the radical cation (→ **88**), and consecutively by a secondary further oxidation step to a dication (**89**) (Scheme 15). Convincing evidence for the postulated intermediacy of the $2\sigma \rightarrow 2\pi$ rearranged opened radical cation **88** was gained from parallel experiments with the corresponding

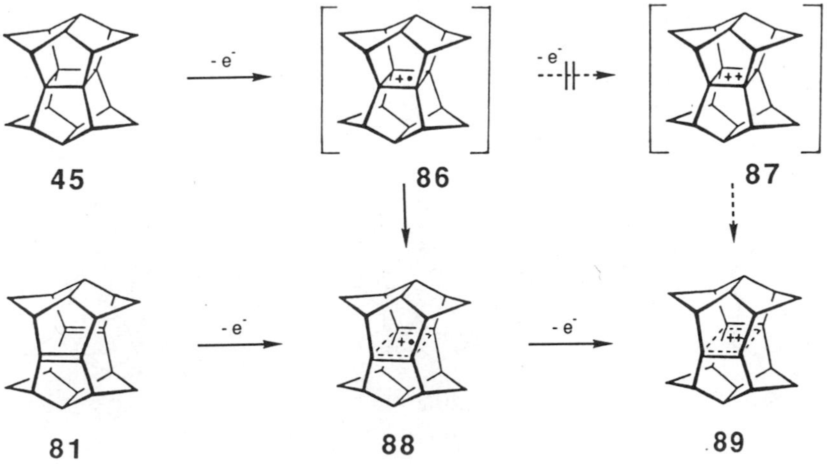

Scheme 15.

diene **81**. The unsaturated valence isomer was shown to be reversibly oxidized at much lower appearance voltage (0.66 V) to the same radical cation, which subsequently underwent further oxidation to a dication at 1.2 V. The dication, however, which is formed by the second oxidation, reacts so rapidly with solvent impurities that its reduction could no longer be observed by cyclic voltammetry. A mass spectroscopic investigation using charge stripping techniques reassured the facile formation of a dicationic species, but could not distinguish between the postulated initial radical cation isomers **86** and **88**.[73]

When generated under carefully controlled conditions in dichloromethane solution in the Gerson laboratories and inserted into an ESR probe,[72] the deep blue radical cation **88**, which gives rise to a strong temperature independent spectrum, proved to be unusually persistent with a lifetime of 2 days at ambient temperature. Nine groups of lines are indicative of a hyperfine interaction with

- e⁻

71 **90** **91**

eight statistically equivalent protons expected from a D_{2h}-symmetrical species. The crucial influence of the specific geometrical constitution became evident, when the bishomologous [2.2.1.1]pagodane radical cation **91** was found to decay considerably faster than the parent [1.1.1.1] species.

Our original anticipation that the valence isomerization of **45** into **81** might be facilitated on a preparative level using radical cation methodology could not be reduced to practice. Pagodane reacted rather sluggishly with ammoniumyl cation salts in chloroform with equimolar amounts of the reagents.[62] From these experiments as well as from an electrolysis experiment conducted in dichloromethane, only the 1,4-dichloride (**92**) (and hydrolysis products, e.g., **93**) was isolated, which presumably is formed by halide abstraction from the solvent at intermediate cationic stages.

$Ar_3N\ SbCl_6$ / $CHCl_3$; H_2O

Cl Cl OH HO

45 **92** **93**

RESPONSE TO SUPERACIDS: THE PAGODANE DICATIONS

Information about the nature of the dicationic species (**89**), suspected from the cyclovoltammetric studies, was gained by experiments conducted in collaboration with the Olah group. The experiments had originally been aimed at the catalytic, thermodynamically controlled conversion of a pagodyl cation to the dodecahedryl cation. Instead, they furnished an unexpected dicationic species with unusual properties.[61] When a sample of [1.1.1.1]pagodane (**45**) was treated with SbF_5 or Magic acid® in SO_2ClF solution, four line proton and carbon NMR spectra evolved, characteristic of a species that has retained the overall D_{2h} symmetry. The individual shielding effects can only be interpreted by the assumption of a dication with delocalization of the charge within the central cyclobutane plane of the molecule. The dication was found to be stable even at ambient temperature for several hours. The same dicationic species was obtained by either oxidation of **45** or diene **81**, as well as by ionization of dibromo compound **85** (Scheme 16). The structural information was supplemented by an X-ray structure determination (Fig. 9) of the single monomeric product (**94**) which arose through quenching with methanol. In addition, the bishomologous dication **95** could be generated and quenched under the same conditions. Although this is quite reminiscent of the situation for the respective radical cations, **95** proved to be far less stable than its more constrained counterpart.

As the NMR spectra of the former dication (**89**) were unaffected by temperature in a range from –130 to +20°C, the D_{2h} symmetry of the ionic species is unlikely that of an averaged structure resulting from a rapid equilibration process between classical dications **96–101**, but is rather due to a static minimum geometry of **89**. Considering structural requirements from the different

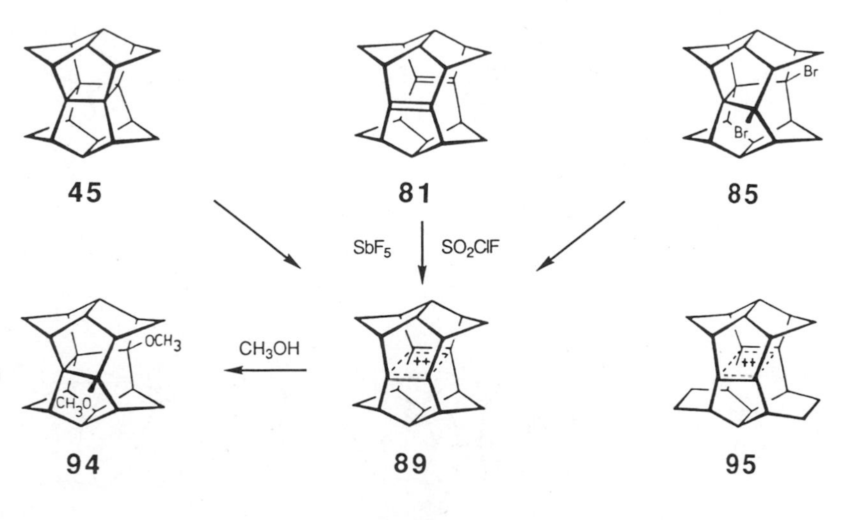

Scheme 16.

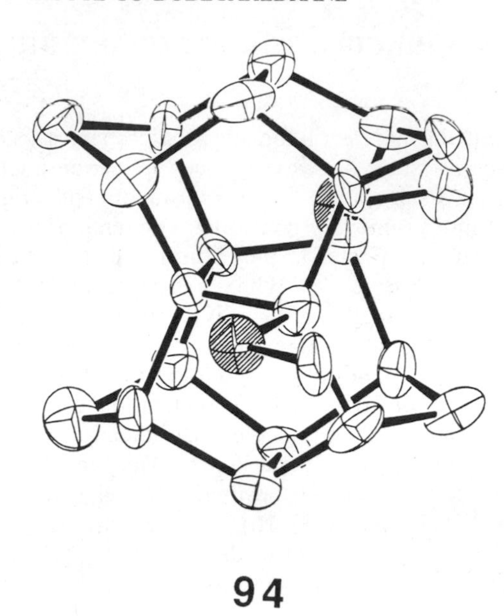

Fig. 9. ORTEP view of the dication quenching product 2,12-dimethoxysecopagodane (**94**). Oxygen atoms are shaded and hydrogen atoms omitted for clarity.

precursors and the quenching product, and a detailed carbon chemical shift analysis, the dications **89** and **95** have to be interpreted as novel $4c$ - $2\pi e$ (bis)homoaromatic pericyclic species, which embody a topological equivalent to a two-electron Woodward–Hoffmann transition state for cycloaddition between ethylene and ethylene dication.

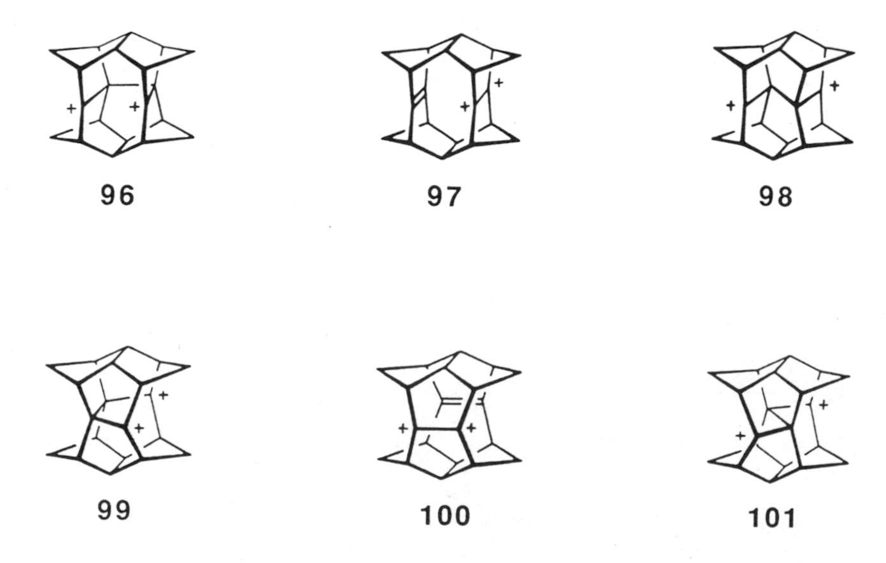

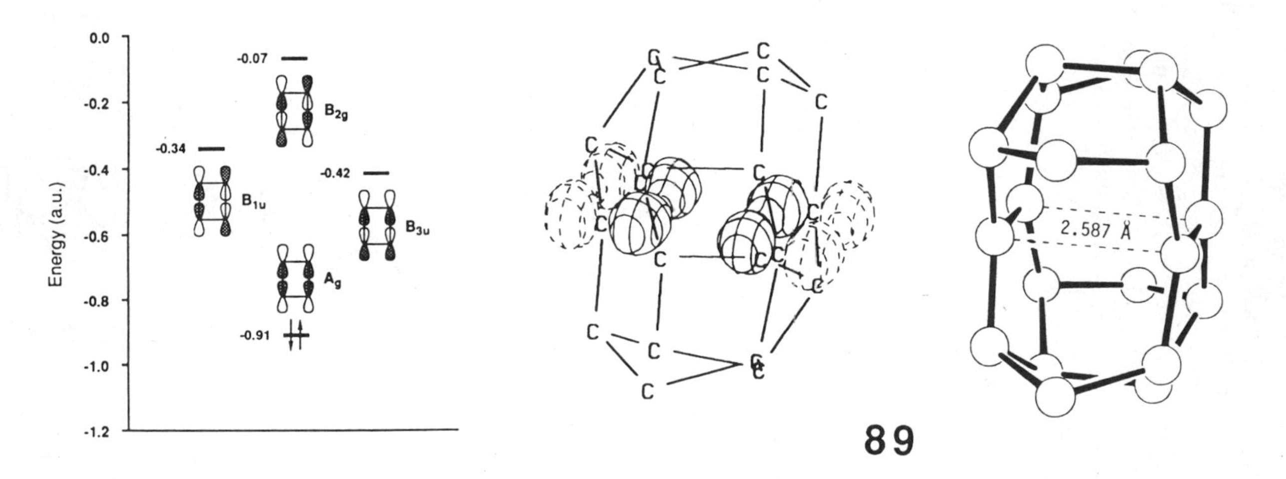

Fig. 10. Molecular orbital diagram of rectangular cyclobutane dication, $(CH_2)_4^{2+}$ (STO-3G); Jorgensen plot of the HOMO (scaled down for clarity) and MNDO optimized structure of dication **89**.

A detailed exploration of the energy potential surface for the parent [1.1.1.1]pagodane dication with several semiempirical SCF–MO methods[1,61] leaves little doubt that the nonclassical D_{2h} symmetrical system of **89** is the true energy minimum surpassing in this respect any classical dication structure. The rectangular geometry for the centers engaged in this unique bonding situation suggests a considerable π character for the shorter framework bonds and a bond order of roughly one half for the other type of transannular C—C contacts. In the π-MO diagram shown in Figure 10, the lowest in-phase bonding orbital [equivalent to the highest occupied molecular orbital (HOMO) of pagodane dication] is occupied by two electrons, which rationalizes a profitable bonding interaction.

As an additional candidate for entry into a thermodynamically controlled rearrangement and as a supplement to the dication investigations, the secopagodyl monocation **104** was studied under superacid conditions.[74] Whereas direct protonation of pagodane with different complex superacid systems was unrewarding (likely for kinetic reasons), cation **104** smoothly emerged upon protonation of the diene valence isomer (**81**), ionization of chloride (**102**), and protolytic ionization of secopagodane (**72**) (Scheme 17). The high stability

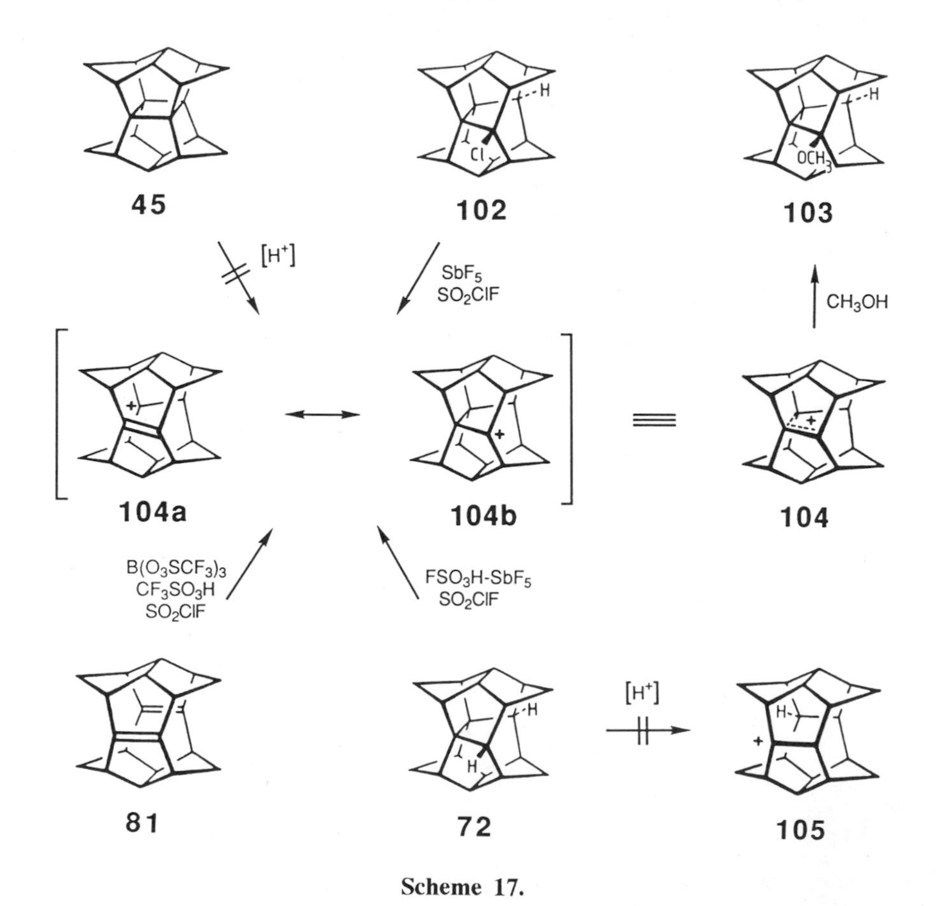

Scheme 17.

associated with cation **104** is due to strong hyperconjugative stabilization of the electron deficient center (**104a** ↔ **104b**), as documented by the unusual carbon chemical shift values, owing to a perfectly coplanar alignment of the empty *p* orbital with the strained transannular C—C single bond.

In conjunction with increasing experimental and theoretical knowledge regarding the stability of related spherical polyquinane cations like **105** or **106**[75] and a pronounced retardation of 1,2-bond migrations among a dodecahedrane nucleus (**107** and **108**),[76] the above results qualify the chances for an access via thermodynamically driven carbocation rearrangement cascades as highly implausible.

The detailed studies on skeletal isomerizations in "adamantaneland"[12] have illustrated that the complex situation is characterized by bond shifts among a rather flexible network of methylene and methine units on a pathway towards a framework containing only fully staggered interactions. Translocation of a positive charge among secondary and tertiary carbocations is facilitated by the small globular size of the tricyclic systems containing favorable alignment of migrating bonds. In contrast, the involvement of polyquinanes is dictated by the ultimate requirement for exclusively methine constituents in the target, tolerating only a balanced situation between quaternary and secondary centers in precursor structures. Cationic skeletal isomerizations are hampered by the relatively rigid interconnections of cyclopentane rings around a larger sphere, the requirement for a pathway against increasing torsional strain and by the prevalence of tertiary carbocations devoid of suitably oriented migratory bonds.[76] Carbon–carbon bond shifts involving such a tertiary carbenium ion moreover would imply the creation of a quaternary center and concurrently a deviation from the sphere.

Cationic rearrangements of *hemi*spherical polyquinanes, when freed from the rigorous constraints set upon them by sphericality and when bearing extraneous methylene units, promptly deviate from polyquinane structures: For example, the C_{3v}-triquinane **109** (the fully saturated molecular half of dodecahedrane) delivers *adamantane*.[77]

$AlBr_3$

109

REACTIVITY OF UNSATURATED BISSECODODECAHDRANES: CONSEQUENCES OF HYPERSTABILITY

Utilization of the pivotal D_{2h} $C_{20}H_{20}$ diene **81**, accessible in 16 high-yielding steps from isodrin, in a hydrogenation–dehydrogenation scheme for arrival at the parent dodecahedrane, or after functionalization at the unsaturated centers for symmetrically 1,2-, 1,6-di-, or 1,2,16,17-tetrasubstituted dodecahedranes, appeared to be an easy matter that would make pentagonal dodecahedranes available in a directed, linear synthesis with a minimum of 18 steps from isodrin, fewer than the number of constituent carbon atoms.

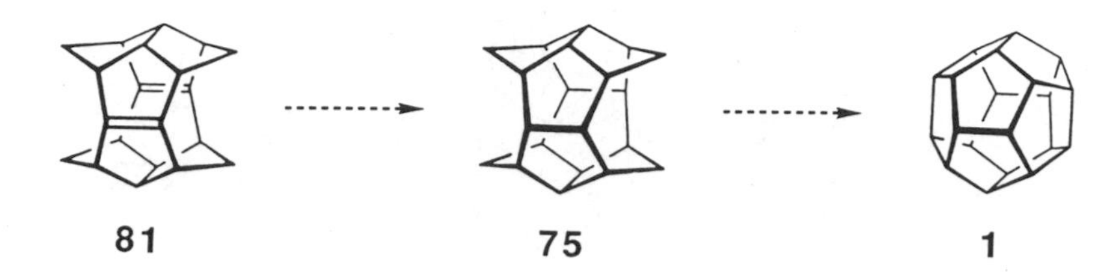

The consequence of perfectly colinear π orbitals within the bissecopagodane ≡ bissecododecahedrane environment profoundly influences the chemical behavior of diene **81**. The molecular geometries calculated with classical force field versus SCF–MO methods differ considerably in the extent of out-of-plane bending predicted for the double bonds (MM2: 8.9°, MNDO: 16.2°),[1] due to effects of strong transannular electron repulsion, which are not adequately accounted for in the simple mechanical approach. The strong enforced electronic interaction is also attested to by a large splitting of 2 eV in the photoelectron spectrum.[70] Owing to this proximity effect, the chemical behavior of **81** is characterized by rapid transannular capture of a diversity of electrophilic reagents (cf. cation **105**),[78] and by an efficient sensitized or direct photochemical [2 + 2] cycloaddition to **45** (Scheme 18).[79]

The marked propensity of **81** towards transannular bonding causes adverse results in attempts of catalytic olefin saturation.[69] Depending on the reaction conditions, varying amounts of **45** or **72** are formed concomitantly with sluggish 1,2-hydrogenation to **110** (Scheme 19). Under no conditions, though, could the formation of the fully saturated D_{2h} bissecododecahedrane (**75**) be detected. Clean saturation of one double bond can be achieved by diimide reduction. Although diimide has a strong hydrogenation potential because of its thermodynamic relationship with the molecular nitrogen released, **75** could not be formed even under forcing conditions.

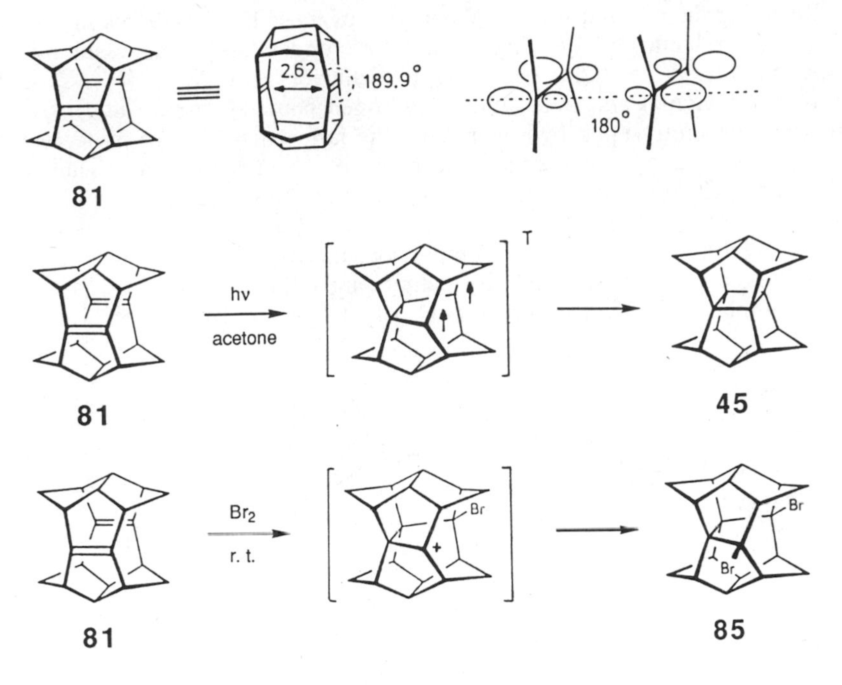

Scheme 18.

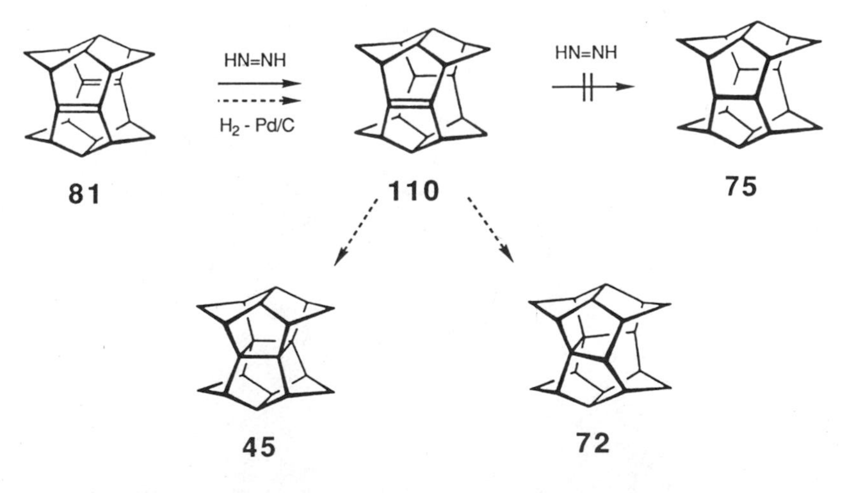

Scheme 19.

Recalling the calculated relative energies of these hydrocarbons, the undesired chemical reluctance of **110** can be understood as resulting from hyperstability,[1,69] a term defined by Schleyer and Maier.[80] Caused by progressive vicinal and transannular hydrogen contacts around the increasingly spherical geometry upon hydrogenation, the large increment of molecular strain energy leads to a strong kinetic decrease of the reaction rate, which is not adequately offset by the thermodynamic relationships (Fig. 11). While the gain in enthalpy of hydrogenation is calculated to be only ~ 11 kcal/mol for both bisseco olefins **81** and **110**, for comparison, the experimentally determined heat of hydrogenation for the sterically demanding tetramethylethylene is still some 26 kcal/mol. The greater reactivity of the diene **81**, which still can be

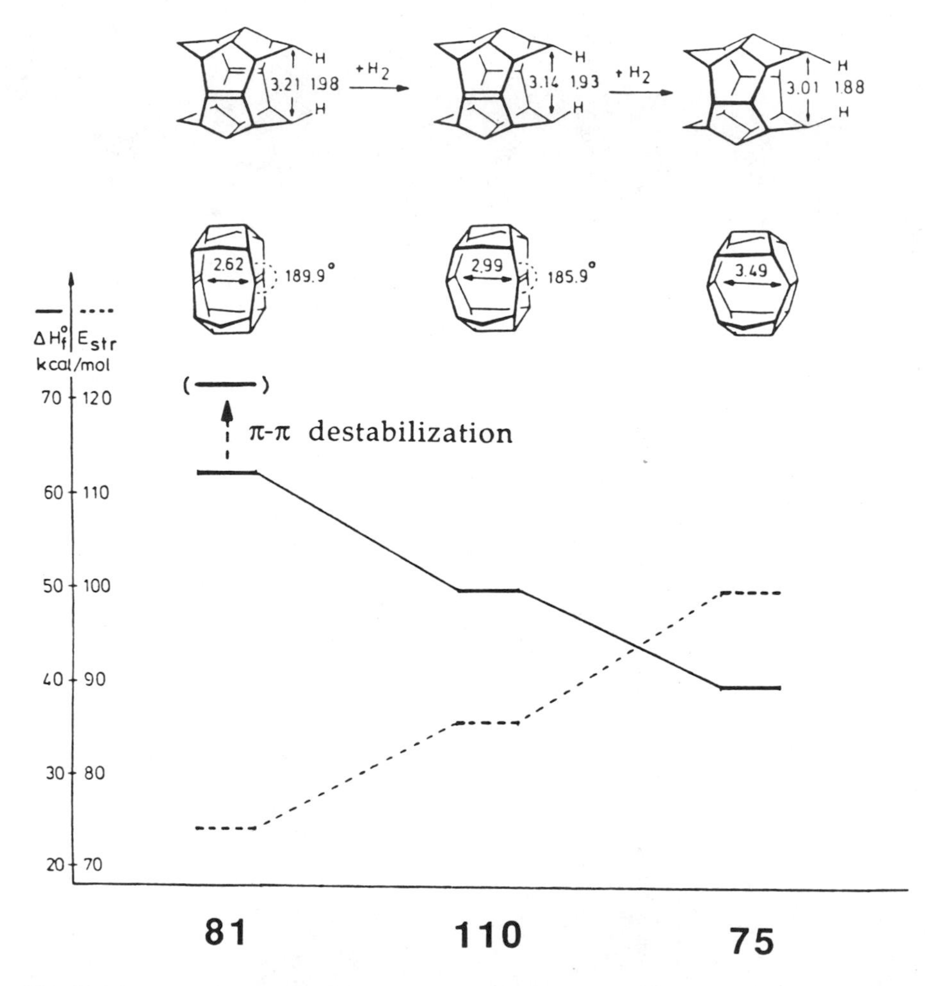

Fig. 11. Hyperstability resulting from counterbalanced trends for enthalpy of formation and strain energy during hydrogenation of diene **81**. π–π Destabilization is neglected in MM2 scheme.

hydrogenated despite similar force-field data, allows a rough estimate to be made as to the energetic contribution of the inherent π–π destabilization.

Reagents with more driving force can surmount this hyperstability barrier. Epoxidation to **111**, **112**, and **113** is effected in high yield (Scheme 20).[69] A difference in relative reaction rates can be exploited into a selective preparation of the ene-epoxide **111**. Surprisingly, even the saturated monoepoxide **113** yields the D_{2h} bisepoxide (**112**) upon exhaustive peracid oxidation,[81] conceivably via a regiospecific oxidative insertion into the strain activated opposite methine units.

Whereas cis hydroxylation of **81** with permanganate does not stop before oxidative cleavage to ene-diketone **117**, the hydroxylated bissecododecahedranes **115** and **116** can be acquired upon treatment of olefins **81** and **110** with osmium tetroxide in good yields (Scheme 21).[82] Unexpectedly, the saturated diol **116** is also prone to transannular rearrangement even under conditions of mild acid catalysis—presumably due to minimum overlap of the vacant *p* with the sp^3 backlobe of the vis-a-vis C—H σ bond in an intermediate cation of type **105**, and driven by the relief of strong nonbonded and torsional forces. Anionic chemistry does not have to deal with such competition: Ethers **118** and **119** are formed in high yields.

Crystals of the tetrabenzyloxy derivative (**119**) proved suitable for an X-ray structure analysis, which provided quantitative geometrical data for the D_{2h} bissecododecahedrane skeleton (Fig. 12). Unusually close C···C and H···H contacts of 2.95 and 1.83 Å at the cramped intramolecular gaps are convincing reason for severe bond angle distortions and a deformation of the transcavity diameter with values of 3.69 and 4.24 Å for depth and height, respectively.[82]

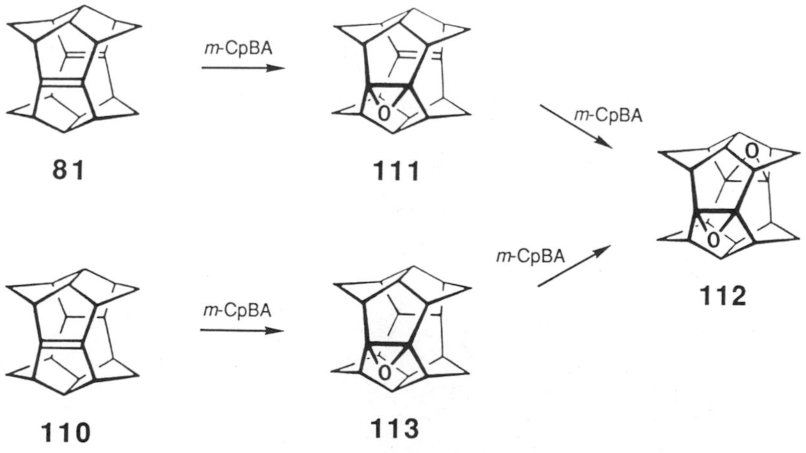

81 —*m*-CpBA→ **111** —*m*-CpBA→ **112**

110 —*m*-CpBA→ **113** —*m*-CpBA→ **112**

Scheme 20.

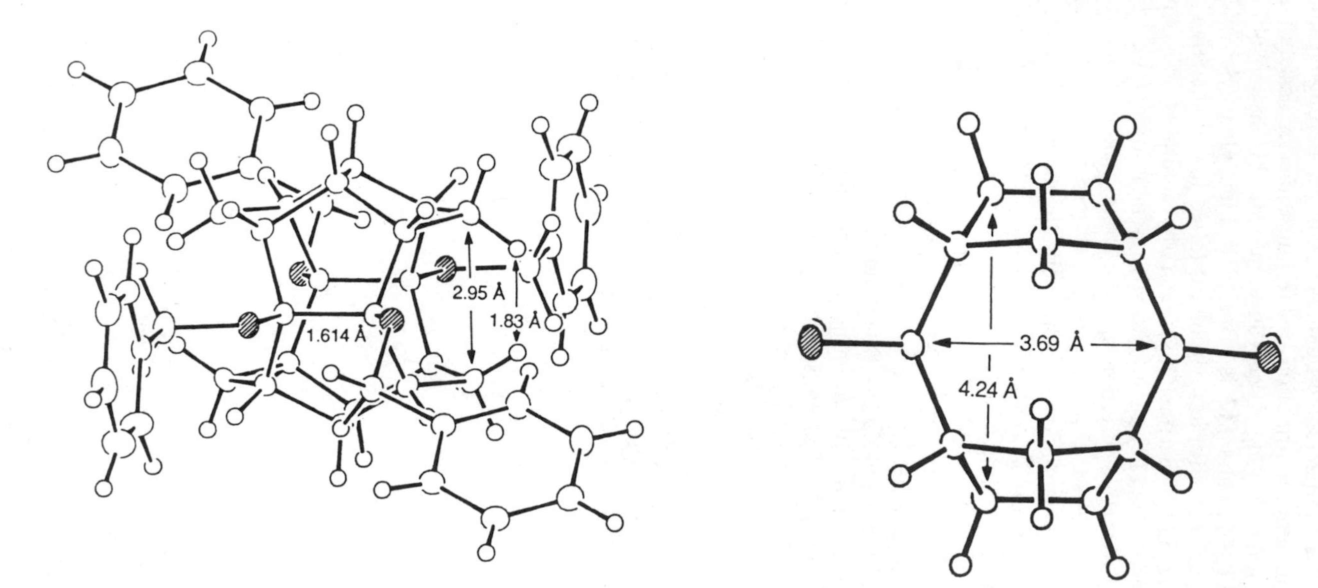

Fig. 12. The ORTEP views and selected molecular dimensions of tetra(benzyloxy)-bissecododechedrane (**119**) (oxygen atoms are shaded and benzyl groups are omitted in the drawing on the right.

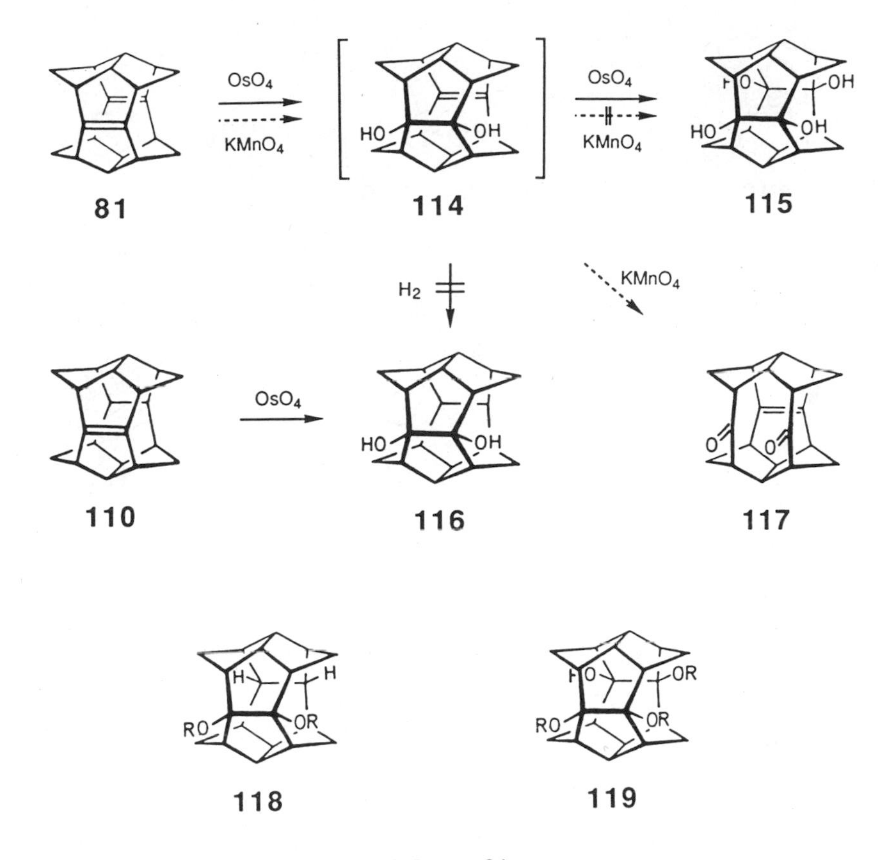

Scheme 21.

With similar success, olefins **81** and **115** can be selectively mono- and bis-cyclopropanated by carbene additions (Scheme 22).[69] Excessive dichloro-carbene regiospecifically functionalizes the hydrocarbon portion of monoadduct **120** with a carbon substituent to **122** by an insertion process[8b] reminiscent to that observed with perepoxidation. Following dehalogenation, the cyclopropane rings of **124** can be opened selectively by careful treatment with hydrogen over a palladium catalyst, leading to regioisomeric dimethylbissecododecahedranes **125** and **126**.[64] The analogous conversion of the mono(cyclopropa) derivative **121** to **123** is somewhat erratic because of competing opposite dehydrogenation, which opens the way for subsequent transannular rearrangements (cf. Scheme 25).

The bis(cyclopropa) derivative **124** is also prone to cationic isomerizations similar to its unsaturated counterparts (Scheme 23).[83] Interestingly, during the course of transannular capture of protic acids (to **127** and **128**) a rarely encountered contrathermodynamic charge migration is formally required from a (hyperconjugatively stabilized) tertiary (**129**) to a primary carbocation (**130**), which is ultimately captured by the nucleophile.

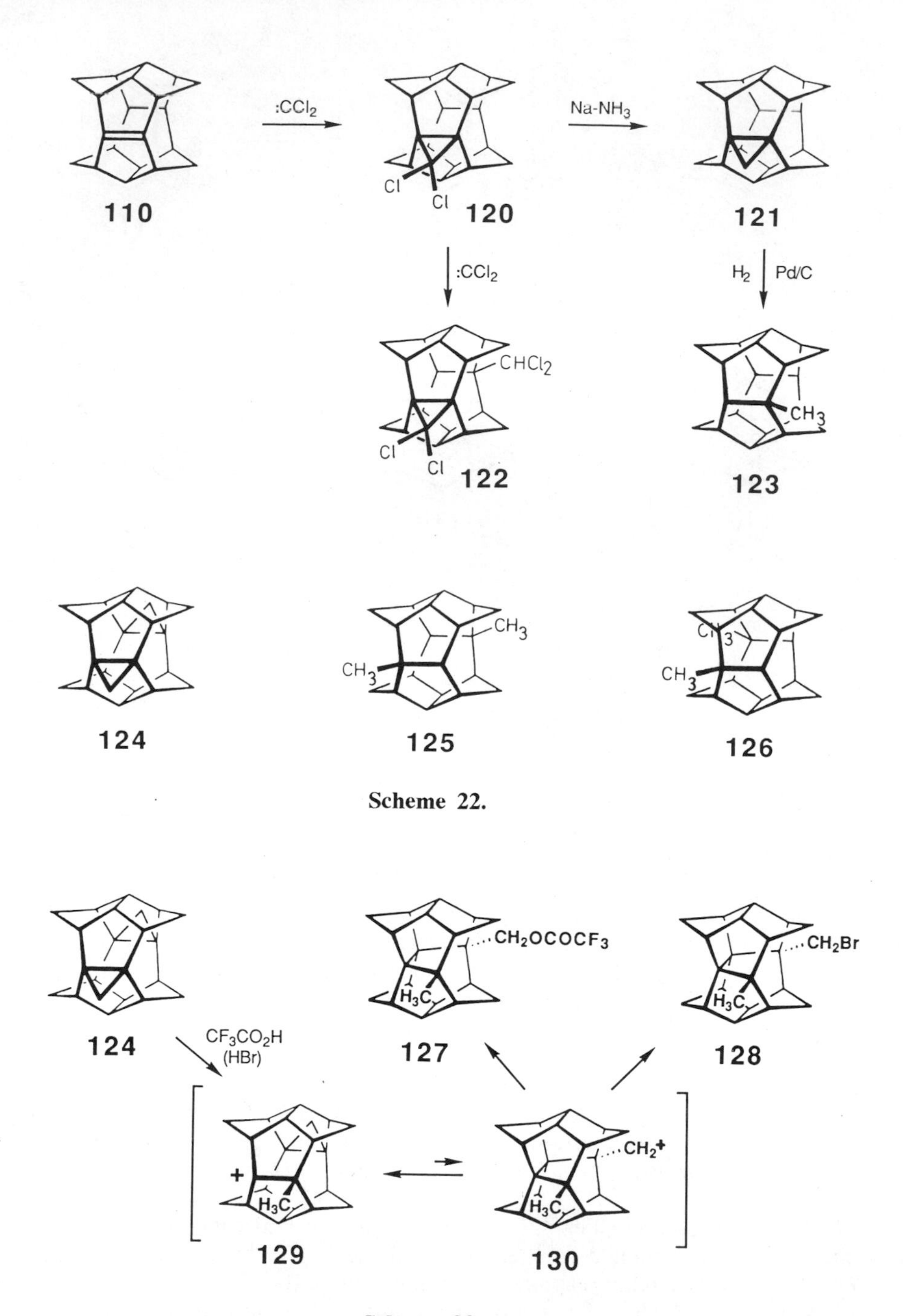

Scheme 22.

Scheme 23.

Due to their restricted steric accessibility, the olefinic bonds of **81** so far proved unreactive as 2π components in Diels–Alder reactions but could be engaged in dipolar cycloaddition, for example, with benzonitriloxide to give the monoadduct **131**. Whereas the parent ketene does not seem to be sufficiently reactive towards **81**, there is evidence that addition of dichloroketene occurs (**132**).[83]

Ph N O PhC≡N→O $Cl_2C{=}C{=}O$ Cl Cl O

131 **81** **132**

DEHYDROGENATIVE BOND FORMATION: FAILURE AND SUCCESS OF DODECAHEDRANE SYNTHESIS

During the attempted dehydrogenation of oxygen substituted bisseco precursors **115** (**119**) or **116** (**118**), taking recourse to the pioneering development of the Paquette synthesis, it became quickly apparent that despite highly favorable thermodynamic relationships and close spatial proximity of the hydrogen atoms to be extruded, the desired process is kinetically so heavily inhibited that elimination/hydrogenolysis of the substituents becomes by far the dominant reaction pathway.[82] The formation of secopagodane (**72**) as the major end result attests to the probable intermediacy of hyperstable, unsaturated species of the type **104**.

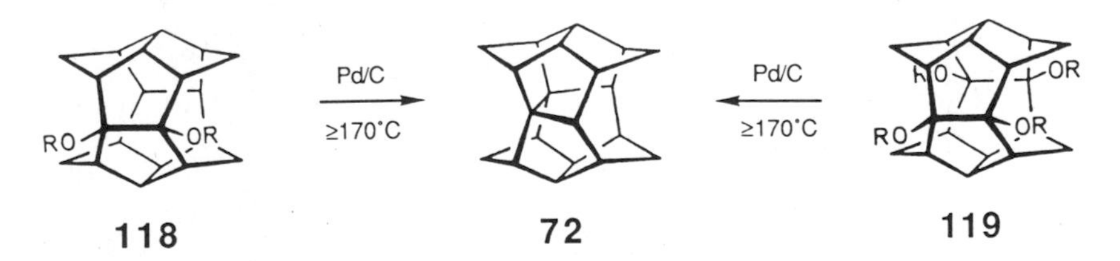

Not unexpectedly, the dehydrogenation of methyl substituted bissecododecahedranes **123** or **125**/**126** equally led to a complex mixture of hydrocarbons, containing only small amounts of the known monomethyl- (**133**) (Scheme 24) or D_{3d} and C_{2v} dimethyldodecahedranes (**136** and **137**).[64] Competition can be restrained when the cyclopropa derivatives **121** or **124** are submitted directly to carefully adjusted reaction conditions. A remarkable 30% of D_{3d} dimethyldodecahedrane (**136**) could thus be achieved or, allowing for extensive demethylation, 50% of the parent dodecahedrane (**1**) (Scheme 25). The wide margin of prevalence of the D_{3d} (**136**) over the C_{2v} product (**137**) clearly

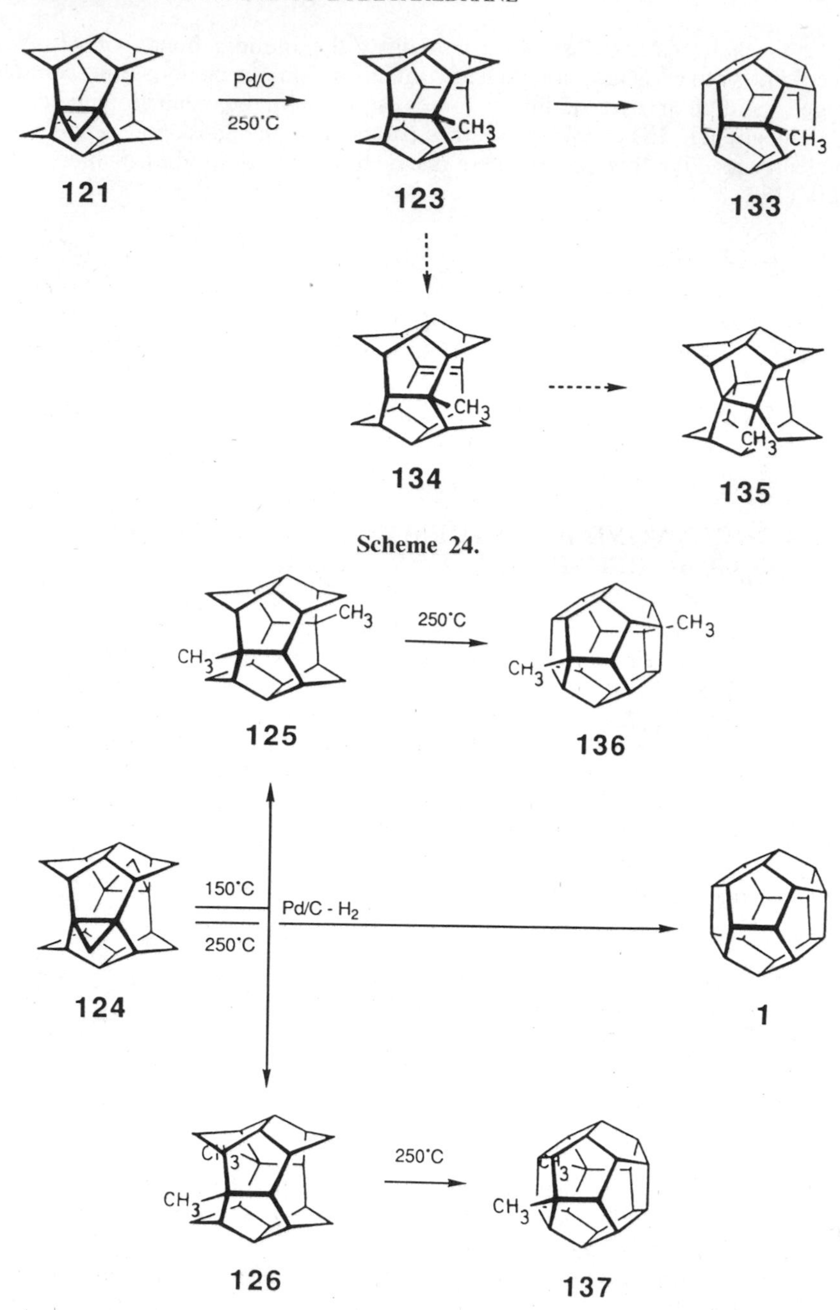

Scheme 24.

Scheme 25.

demonstrates the necessity for a 1,2- or 1,4-blocking scheme against transannular bonding for a pagodane strategy that ultimately relies on a dehydrogenative installation of the final two bonds.

OUTLOOK: FURTHER SHAPING OF THE PAGODANE STRATAGEM

So far, the overall efficiency of our linear dodecahedrane approach was rather impressive. Still, the actual situation proved to be unsatisfactory due to problems with reproducing the exact timing of demethylation in the final heterogenous dehydrogenations and to the separation difficulties, when carried out on a larger scale, regarding the sparingly soluble isomeric compounds produced. Obviously, the ultimate preparative application of bisseco-dodecahedranes, featuring diverse substitution patterns for the synthesis of variously substituted dodecahedranes, depends decisively on the design and application of appropriate alternative techniques for creation of the last two C—C-bonds.

It was planned in the early stages of this project[1] to circumvent the problems inherent in dehydrogenative procedures by making use of ester or derived ketone functionalities,[51] which can be conducted along the established route. The photochemical *homo*-Norrish cyclopentane formation, successfully applied in the Paquette synthesis, was especially appealing since a broad variety of substitution patterns are tolerated and two easily manipulable hydroxyl functions become available. To our dismay, we learned that the placement of the C—H bonds, which are to be inserted directly above the carbonyl π plane (**138**), constitutes a rather unfavorable situation, since the partially emptied oxygen *n*-orbital is sterically inaccessible for hydrogen abstraction to occur (Scheme 26).[84] Also, a *homo*-enolization scheme (**140**), well known for certain other cage compounds,

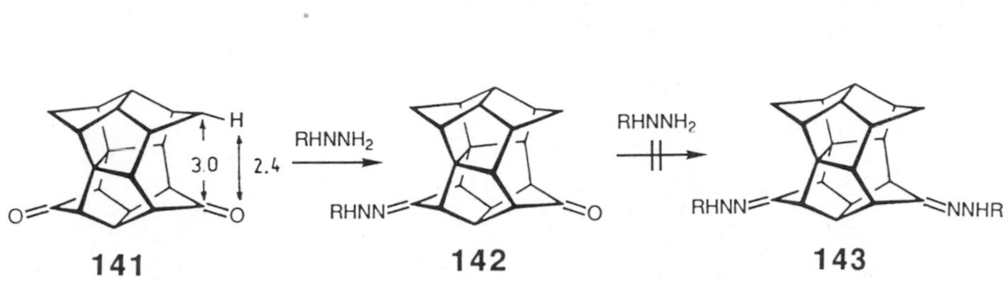

H H X X X X O O hv HO X X X OH base X X X X O O

138 **139** **140**

H 3.0 2.4 O O RHNNH$_2$ RHNN O RHNNH$_2$ RHNN NNHR

141 **142** **143**

Scheme 26.

has not yet furnished the desired dihydroxydodecahedranes (**139**).[85] Alternative transannular carbene insertion was wrecked by a kind of steric constraint typical for the chemistry of half-cage compounds: The diketone (**141**) readily forms a tosylhydrazone (**142**) at the pagodane-type carbonyl group but, even under forcing conditions, proved absolutely reluctant towards further condensation at the sterically encumbered seco side. A derivatization preceding the cyclobutane ring opening was no solution since the conditions necessary for the opening could not be reconciliated with the reactivity of the individual carbene precursors.

Consequently, subsequent efforts were focused towards further activation at all methylene bridges in order to facilitate the two ultimate bond formations.[86] These endeavors have just recently led to a real breakthrough: By extensive experimentation, efficient protocols for making multifunctionalized pagodanes evolved in which the mode of substitution can be deliberately adjusted to the needs of individual C—C bond forming methods, for example, for photochemical or reductive cyclizations or for nucleophilic additions or S_N2 displacements. From this reservoir, a series of dodecahedranes (**144**) emerged via an aldol variant,[87] dodecahedranes with their outer sphere regiospecifically modified at up to eight positions and even unsaturated dodecahedranes, which are isolable in spite of the extensive pyramidalization expected for the olefinic carbon atoms.[1,88] Thus, a new chapter of dodecahedrane chemistry[47] was opened.

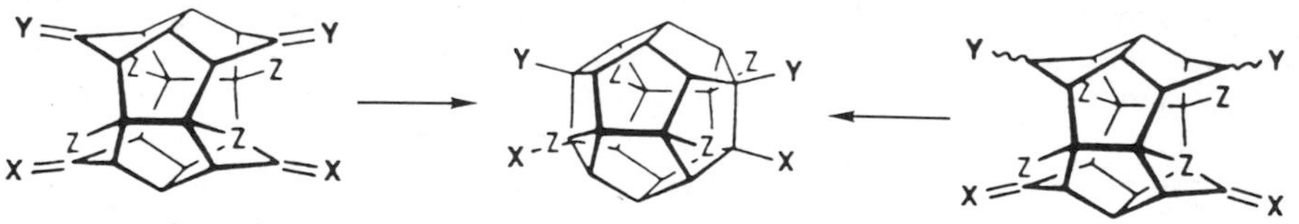

144

As we have tried to demonstrate in this chapter, the high overall efficiency and the broad structural and functional variability qualifies our original tactical scheme as a comprehensive approach towards making classes of compounds that combine novel molecular architecture with novel chemistry, moving us toward goals as high or even higher than the *Mount Everest of alicyclic chemistry.* It is as well tempting as justified to close with a quotation of our earlier statement:[8b] The reward for our efforts has been multifaceted, in which the expected as well as the unexpected, the successes as well as the failures, have added up to an adventure of lasting fascination.

ACKNOWLEDGMENTS

Special appreciation goes to the dedicated and talented graduate (G. Lutz, J.-P. Melder, and R. Pinkos) and postdoctoral co-workers (B.A.R.C. Murty and P.R. Spurr) who have shared with us the efforts and the excitement. We acknowledge with gratitude the efforts of our many colleagues, quoted in joint works, who contributed to individual facets of this work. This research program was made possible by the continuing financial support from the Deutsche Forschungsgemeinschaft, the Fonds der Chemischen Industrie and the BASF AG, as well as the generous cooperativity of the Ciba–Geigy AG.

REFERENCES

1. W.-D. Fessner, Ph.D. thesis, Universität Freiburg, Federal Republic of Germany, 1986.
2. (a) G. Mehta, *J. Sci. Ind. Res.* **1978**, *37*, 256. (b) P. E. Eaton, *Tetrahedron* **1979**, *35*, 2189. (c) W. Grahn, *Chem. Unserer Z.* **1981**, *15*, 52.
3. P. E. Eaton and T. W. Cole, *J. Am. Chem. Soc.* **1964**, *86*, 962.
4. G. Maier, S. Pfriem, U. Schäfer, and R. Matusch, *Angew. Chem. Int. Ed. Engl.* **1978**, *17*, 520.
5. P. Grubmüller, Ph.D. thesis, Universität Erlangen-Nürnberg, Federal Republic of Germany, 1979.
6. (a) R. J. Ternansky, D. W. Balogh, and L. A. Paquette, *J. Am. Chem. Soc.* **1982**, *104*, 4502. (b) L. A. Paquette, R. J. Ternansky, D. W. Balogh, and G. Kentgen, *J. Am. Chem. Soc.* **1983**, *105*, 5446.
7. (a) L. A. Paquette, *Pure Appl. Chem.* **1978**, *50*, 1291. (b) L. A. Paquette in *Organic Synthesis—Today and Tomorrow*, B. M. Trost and C. R. Hutchinson, Eds., Pergamon Press, New York, 1981, p. 335. (c) L. A. Paquette, *Proc. Natl. Acad. Sci. USA* **1982**, *79*, 4495. (d) L. A. Paquette, *Chem. Aust.* **1983**, *50*, 138. (e) L. A. Paquette, in *Strategies and Tactics in Organic Synthesis,* T. Lindberg, Ed., Academic Press, New York, 1984, p. 175.
8. (a) W.-D. Fessner, H. Prinzbach, and G. Rihs, *Tetrahedron Lett.* **1983**, *24*, 5857. (b) H. Prinzbach and W.-D. Fessner, in *Organic Synthesis: Modern Trends,* O. Chizhov, Ed., Blackwell Scientific, Oxford, 1987, p. 23.
9. H. Prinzbach, G. Sedelmeier, C. Krüger, R. Goddard, H.-D. Martin, and R. Gleiter, *Angew. Chem. Int. Ed. Engl.* **1978**, *17*, 271.
10. S. A. Godleski, P. v. R. Schleyer, E. Osawa, Y. Inamoto, and Y. Fujikura, *J. Org. Chem.* **1976**, *41*, 2596.
11. T. Iizuka, M. Imai, N. Tanaka, T. Kan, and E. Osawa, *Science Reports of the Faculty of Education, Gunma University* **1981**, *30*, 5; *Chem. Abstr.* **1982**, *97*, 126567m.
12. (a) P. v. R. Schleyer, *J. Am. Chem. Soc.* **1957**, *79*, 3292. (b) E. Osawa, K.

Aigauri, N. Takaishi, Y. Inamoto, Y. Fujikura, Z. Majerski, P. v. R. Schleyer, E. M. Engler, and M. Farcasiu, *J. Am. Chem. Soc.* **1977**, *99*, 5361.

13. N. J. Jones, W. D. Deadman, and E. LeGoff, *Tetrahedron Lett.* **1973**, 2087.
14. P. v. R. Schleyer, personal communication.
15. D. M. Müller, *Chem. Weekblad* **1963**, *59*, 334.
16. R. B. Woodward, T. Fukunaga, and R. C. Kelly, *J. Am. Chem. Soc.* **1964**, *86*, 3162.
17. I. T. Jacobson, *Acta Chem. Scand.* **1967**, *21,* 2235.
18. See Ref. 2b.
19. (a) D. Bosse and A. de Meijere, *Angew. Chem. Int. Ed. Engl.* **1974**, *86*, 706. (b) L. A. Paquette, J. D. Kramer, P. B. Lavrik, and M. J. Wyvratt, *J. Org. Chem.* **1977**, *42,* 503.
20. W. F. Maier, personal communication.
21. P. W. Codding, K. A. Kerr, A. Oudeman, and T. S. Sorensen, *J. Organomet. Chem.* **1982**, *232*, 193.
22. E. Carceller, M. L. Garcia, A. Moyano, M. A. Pericas, and F. Serratosa, *Tetrahedron* **1986**, *42,* 1831.
23. (a) L. A. Paquette, I. Itoh, and W. B. Farnham, *J. Am. Chem. Soc.* **1975**, *97*, 7280. (b) L. A. Paquette, I. Itoh, and K. B. Lipkowitz, *J. Org. Chem.* **1976**, *41,* 3524.
24. O. Repic, Ph.D. thesis, Harvard University, 1976.
25. P. Deslongchamps and P. Soucy, *Tetrahedron* **1981**, *37*, 4385.
26. J. Clardy, B. A. Solheim, J. P. Springer, I. Itoh, and L. A. Paquette, *J. Chem. Soc. Perkin Trans. 2* **1979**, 296.
27. (a) W. P. Roberts and G. Shoham, *Tetrahedron Lett.* **1981**, *22*, 4895. (b) W. P. Roberts, Ph.D. thesis, Harvard University, 1982.
28. P. Camps, J. Castane, and M. T. Santos, *Chem. Lett.* **1984**, 1367.
29. (a) D. G. Farnum and T. A. Monego, *Tetrahedron Lett.* **1983**, *24*, 1361. (b) T. A. Monego, Ph.D. thesis, Michigan State University, 1982.
30. J. S. Pyrek, *Pol. J. Chem.* **1979**, *53*, 1557.
31. (a) L. M. Tolbert, J. C. Gregory, and C. P. Brock, *J. Org. Chem.* **1985**, *50, 548.* (b) C. P. Brock, J. C. Gregory, and L. M. Tolbert, *Acta Cryst.* **1986**, *C42*, 1063.
32. (a) P. E. Eaton and R. H. Mueller, *J. Am. Chem. Soc.* **1972**, *94*, 1014. (b) P. E. Eaton, R. H. Mueller, G. R. Carlson, D. A. Cullison, G. F. Cooper, T.-C. Chou, and E.-P. Krebs, *J. Am. Chem. Soc.* **1977**, *99*, 2751. (c) P. E. Eaton, A. Srikrishna, and F. Uggeri, *J. Org. Chem.* **1984**, *49*, 1728.
33. P. E. Eaton, G. D. Andrews, E.-P. Krebs, and A. Kunai, *J. Org. Chem.* **1979**, *44*, 2824.
34. (a) P. E. Eaton and W. H. Bunelle, *Tetrahedron Lett.* **1984**, *25*, 23. (b) P. E. Eaton, W. H. Bunelle, and P. Engel, *Can. J. Chem.* **1984**, *62*, 2612.
35. L. A. Paquette, R. A. Snow, J. L. Muthard, and T. Cynkowski, *J. Am. Chem. Soc.* **1979**, *101*, 6991.
36. (a) R. L. Sobczak, M. E. Osborn, and L. A. Paquette, *J. Org. Chem.* **1979**, *44*, 4886. (b) M. E. Osborn, S. Kuroda, J. L. Muthard, J. D. Kramer, P. Engel, and L. A. Paquette, *J. Org. Chem.* **1981**, *46*, 3379.

37. P. E. Eaton, R. S. Sidhu, G. E. Langford, D. A. Cullison, and C. L. Pietruszewski, *Tetrahedron* **1981**, *37*, 4479.

38. N. J. Hales and L. A. Paquette, *J. Org. Chem.* **1979**, *44*, 4603.

39. R. J. Ternansky, Ph.D. thesis, The Ohio State University, 1982.

40. (a) J. E. Baldwin and P. L. M. Beckwith, *J. Chem. Soc. Chem. Commun.* **1983**, 279. (b) J. E. Baldwin, P. L. M. Beckwith, J. D. Wallis, A. P. K. Orrell, and K. Prout, *J. Chem. Soc. Perkin Trans. 2* **1984**, 53.

41. M. A. McKervey, P. Vibuljan, G. Ferguson, and P. Y. Siew, *J. Chem. Soc. Chem. Commun.* **1981**, 912.

42. (a) D. McNeil, B. R. Vogt, J. J. Sudol, S. Theodoropulos, and E. Hedaya, *J. Am. Chem. Soc.* **1974**, *96*, 4673. (b) L. A. Paquette and M. J. Wyvratt, *J. Am. Chem. Soc.* **1974**, *96*, 4671. (c) L. A. Paquette, M. J. Wyvratt, H. C. Berk, and R. E. Moerck, *J. Am. Chem. Soc.* **1978**, *100*, 5845.

43. L. A. Paquette, M. J. Wyvratt, O. Schallner, J. L. Muthard, W. J. Begley, R. M. Blankenship, and D. Balogh, *J. Org. Chem.* **1979**, *44*, 3616.

44. (a) L. A. Paquette, W. J. Begley, D. Balogh, M. J. Wyvratt, and D. Bremner, *J. Org. Chem.* **1979**, *44*, 3630. (b) D. W. Balogh and L. A. Paquette, *J. Org. Chem.* **1980**, *45*, 3038. (c) D. W. Balogh, L. A. Paquette, P. Engel, and J. F. Blount, *J. Am. Chem. Soc.* **1981**, *103*, 226.

45. (a) L. A. Paquette, Y. Miyahara, and C. W. Doecke, *J. Am. Chem. Soc.* **1986**, *108*, 1716. (b) L. A. Paquette and Y. Miyahara, *J. Org. Chem.* **1987**, *52*, 1265.

46. (a) J. C. Gallucci, C. W. Doecke, and L. A. Paquette, *J. Am. Chem. Soc.* **1986**, *108*, 1343. (b) I. Santos, D. W. Balogh, C. W. Doecke, A. G. Marshall, and L. A. Paquette, *J. Am. Chem. Soc.* **1986**, *108*, 8183.

47. (a) L. A. Paquette, J. C. Weber, and T. Kobayashi, *J. Am. Chem. Soc.* **1988**, *110*, 1303. (b) G. A. Olah, G. K. S. Prakash, T. Kobayashi, and L. A. Paquette, *J. Am. Chem. Soc.* **1988**, *110*, 1304. (c) J. C. Weber and L. A. Paquette, *J. Org. Chem.* **1988**, *53*, 5315.

48. G. Mehta and M. S. Nair, *J. Am. Chem. Soc.* **1985**, *107*, 7519.

49. (a) G. Sedelmeier, W.-D. Fessner, R. Pinkos, C. Grund, B. A. R. C. Murty, D. Hunkler, G. Rihs, H. Fritz, C. Krüger, and H. Prinzbach, *Chem. Ber.* **1986**, *119*, 3442. (b) L. A. Paquette, K. Nakamura, and P. Engel, *Chem. Ber.* **1986**, *119*, 3782.

50. G. Mehta and K. R. Reddy, *Tetrahedron Lett.* **1988**, *29*, 3607.

51. W.-D. Fessner, G. Sedelmeier, P. R. Spurr, G. Rihs, and H. Prinzbach *J. Am. Chem. Soc.* **1987**, *109*, 4626.

52. S. B. Soloway, A. M. Damiana, J. W. Sims, H. Bluestone, and R. E. Lidov, *J. Am. Chem. Soc.* **1960**, *82*, 5377.

53. K. Mackenzie, *J. Chem. Soc.* **1965**, 4646.

54. G. Sedelmeier, Ph.D. thesis, Universität Freiburg, Federal Republic of Germany, 1979.

55. (a) R. C. Cookson and E. Crundwell, *Chem. Ind. (London)* **1958**, 1004. (b) G. Jones, W. G. Becker, and S.-H. Chiang, *J. Am. Chem. Soc.* **1983**, *105*, 1269.

56. (a) J. F. M. Oth, H. Röttele, and G. Schröder, *Tetrahedron Lett.* **1970**, *61*. (b) C. Cometta-Morini, Ph.D. thesis, ETH Zürich, 1986. (c) N. C. Yang, B. J. Hrnjez, and M. G. Horner, *J. Am. Chem. Soc.* **1987**, *109*, 3158. (d) N. C. Yang, T. Noh, H. Gan, S. Halfon, and B. J. Hrnjez, *J. Am. Chem. Soc.* **1988**, *110*, 5919.

57. W. v. E. Doering, W. R. Roth, R. Breuckmann, L. Figge, H.-W. Lennartz, W.-D. Fessner, and H. Prinzbach *Chem. Ber.* **1988**, *121*, 1.

58. R. Gleiter, H. Zimmermann, W.-D. Fessner, and H. Prinzbach *Chem. Ber.* **1985**, *118*, 3856.

59. W.-D. Fessner, C. Grund, and H. Prinzbach, *Tetrahedron Lett.*, **1989**, *30*, 3133.

60. (a) F.-G. Klärner, U. Artschwager-Perl, W.-D. Fessner, C. Grund, R. Pinkos, J.-P. Melder, and H. Prinzbach, *Tetrahedron Lett.*, **1989**, *30*, 3137. (b) W.-D. Fessner and H. Prinzbach, *Twelfth Austin Symposium on Molecular Structure*, Austin, TX 1988, P4.

61. G. K. S. Prakash, V. V. Krishnamurthy, R. Herges, R. Bau, H. Yuan, G. A. Olah, W.-D. Fessner, and H. Prinzbach, *J. Am. Chem. Soc.* **1986**, *108*, 836; *J. Am. Chem. Soc.* **1988**, *110*, 7764.

62. W.-D. Fessner, B. A. R. C. Murty, and H. Prinzbach, *Angew. Chem. Int. Ed. Engl.* **1987**, *26*, 451.

63. W. Burns, M. A. McKervey, T. R. B. Mitchell, and J. J. Rooney, *J. Am. Chem. Soc.* **1978**, *100*, 906.

64. W.-D. Fessner, B. A. R. C. Murty, J. Wörth, D. Hunkler, H. Fritz, H. Prinzbach, W. D. Roth, P. v. R. Schleyer, A. B. McEwen, and W. F. Maier, *Angew. Chem. Int. Ed. Engl.* **1987**, *26*, 452.

65. (a) M. A. McKervey, J. J. Rooney, and N. G. Samman, *J. Chem. Soc. Chem. Commun.* **1972**, 1185. (b) V. Amir-Ebrahimi and J. J. Rooney, *J. Chem. Soc. Chem. Commun.* **1988**, 260.

66. T. Fukunaga and R. A. Clement, *J. Am. Chem. Soc.* **1977**, *42*, 270.

67. (a) P. E. Eaton and G. H. Temme III., *J. Am. Chem. Soc.* **1973**, 95, 7508. (b) K. B. Wiberg, *Acc. Chem. Res.* **1984**, *17*, 379.

68. J. W. Barrett and R. P. Linstead, *J. Chem. Soc.* **1936**, 611.

69. P. R. Spurr, B. A. R. C. Murty, W.-D. Fessner, H. Fritz, and H. Prinzbach, *Angew. Chem. Int. Ed. Engl.* **1987**, *26*, 455.

70. E. Heilbronner and H.-D. Martin, personal communications.

71. (a) R. W. Hoffmann and W. Barth, *J. Chem. Soc. Chem. Commun.* **1983**, 345. (b) H. D. Roth, *Acc. Chem. Res.* **1987**, *20*, 343.

72. H. Prinzbach, B. A. R. C. Murty, W.-D. Fessner, J. Mortensen, J. Heinze, G. Gescheidt, and F. Gerson, *Angew. Chem. Int. Ed. Engl.* **1987**, *26*, 457.

73. T. Drewello, W.-D. Fessner, A. J. Kos, C. B. Lebrilla, H. Prinzbach, P. v. R. Schleyer, and H. Schwarz, *Chem. Ber.* **1988**, *121*, 187.

74. G. K. S. Prakash, W.-D. Fessner, G. A. Olah, G. Lutz, and H. Prinzbach, *J. Am. Chem. Soc.* **1989**, *111*, 746.

75. G. K. S. Prakash, personal communication.

76. (a) G. A. Olah, G. K. S. Prakash, W.-D. Fessner, T. Kobayashi, and L. A. Paquette, *J. Am. Chem. Soc.* **1988**, *110*, 8599. (b) L. A. Paquette, T. Kobayashi, M. A. Kesselmayer, *J. Am. Chem. Soc.* **1988**, *110*, 6568.

77. L. A. Paquette, G. V. Meehan, and S. J. Marshall, *J. Am. Chem. Soc.* **1969**, *91*, 6779.

78. B. A. R. C. Murty, G. Lutz, and H. Prinzbach, unpublished results.

79. B. A. R. C. Murty, P. R. Spurr, R. Pinkos, C. Grund, W.-D. Fessner, W. R. Roth, and H. Prinzbach, *Chimia* **1987**, *41*, 32.

80. W. F. Maier and P. v. R. Schleyer *J. Am. Chem. Soc.* **1981**, *103*, 1891.

81. R. Pinkos, G. Lutz, B. A. R. C. Murty, and H. Prinzbach, unpublished results.

82. G. Lutz, D. Hunkler, G. Rihs, and H. Prinzbach, *Angew. Chem. Int. Ed. Engl.* **1989**, *28*, 298.

83. G. Lutz, W.-D. Fessner, and H. Prinzbach, unpublished results.

84. S. Ariel, S. Evans, N. Omkaram, J. R. Scheffer, and J. Trotter, *J. Chem. Soc. Chem. Commun.* **1986**, 375.

85. P. R. Spurr, R. Pinkos, and H. Prinzbach, unpublished results.

86. (a) J.-P. Melder, H. Fritz, and H. Prinzbach, *Angew. Chem. Int. Ed. Engl.* **1989**, *28*, 300. (b) R. Pinkos, G. Rihs, and H. Prinzbach, *Angew. Chem. Int. Ed. Engl.* **1989**, *28*, 303.

87. J.-P. Melder, R. Pinkos, H. Fritz, and H. Prinzbach, *Angew. Chem. Int. Ed. Engl.* **1989**, *28*, 305.

88. O. Ermer, *Aspekte von Kraftfeldrechnungen*, Wolfgang Baur Verlag, München, 1981, p. 454.

Author Index

Subject Index